D0161687

Modeling Differential Equations
in Biology

Modeling Differential Equations in Biology

Clifford Henry Taubes
Harvard University

CAMBRIDGE
UNIVERSITY PRESS

CAMBRIDGE UNIVERSITY PRESS
Cambridge, New York, Melbourne, Madrid, Cape Town, Singapore, São Paulo, Delhi

Cambridge University Press
The Edinburgh Building, Cambridge CB2 8RU, UK

Published in the United States of America by Cambridge University Press, New York

www.cambridge.org
Information on this title: www.cambridge.org/9780521708432

First edition© Prentice Hall Inc. 2001
Second edition© C. Taubes 2008

First published by Prentice Hall 2001

Printed in the United Kingdom at the University Press, Cambridge

A catalogue record for this publication is available from the British Library

ISBN 978-0-521-70843-2

Contents

Preface

This book is a compendium of chapters for a course on differential equations and their applications in the biological sciences that I developed at Harvard University. The book and the course roughly follow a wonderful book by Edward Beltrami called *Mathematics for Dynamic Modeling* published by Academic Press, which is a book written for students who already have a reasonably sophisticated mathematics background. This book covers many of the topics in Beltrami's book (and shamelessly borrows some of his examples), but it is designed for life science students who have had only the basics of calculus, which is to say that students should have a good intuitive feel for the meaning of differentiation and integration, and they should be at home integrating and differentiating sines, cosines, powers, and exponential functions. Note that the book is not really aimed at potential applied mathematicians; instead, my goal is to introduce to future experimental biologists some potentially useful tools and modes of thought.

The material here is organized into 28 chapters with accompanying articles from the current (circa late 1990s) biology research literature that illustrates the utility of the mathematics. (A few of the supporting articles come from the geology and earth science literature, geology being a "hobby" of mine.) I have supplied a paragraph or so of commentary about each of the illustrative articles. I don't require students to understand the biology in these articles (goodness knows how little I understand); rather I mean for the articles to make a convincing case that the mathematics from the course is relevant to specific areas of current biological research.

The chapters are grouped into three parts. Part 1, which consists of Chapters 1 through 12, covers the basics of ordinary differential equations with a single time derivative. The focus is on equations with one and two unknowns, and the material covered includes phase plane analysis, equilibria, and stability. Some of the chapters in Part 1 are closely modeled on material that Otto Bretscher and Robin Gottlieb developed for the second semester of a basic two-semester calculus sequence at Harvard. A few of these chapters also closely follow parts of the book *Calculus* by Hughes-Hallett, Gleason, et al. The reader needing more discussion of the basics of phase plane analysis will find it in the latter. Part 2, consisting of Chapters 13 through 22, introduces differential equations in space and time, which is to say partial differential equations. The diffusion and advection equations are the focus here. (My students found the material in Chapters 16 through 19 to be the most difficult of the 28 chapters; even so, they were fascinated by the possibilities for the applications.) Part 3 contains the remaining Chapters 22 through 28 and returns to the milieu of the ordinary differential equation, investigating some of the intriguing dynamical properties that can occur.

Extra exercises with solutions are provided at the end of the book. The grouping of the exercises corresponds to the three major parts of the book.

Here is a more detailed table of contents:

Chapter 1: *Introduction.*   Provides the author's view of the role of mathematics in biology.

Chapter 2: *Exponential Growth with Appendix on Taylor's Theorem.*   The simplest and simultaneously most ubiquitous differential equations are discussed. The Appendix introduces Taylor's expansions.

- *Reading 2.1: HIV-1 Dynamics in Vivo: Virion Clearance Rate, Infected Cell Life-Span, and Viral Generation Time* by A. S. Perelson, A. U. Neumann, M. Markowitz, J. M. Leonard, and D. D. Ho in *Science* **271** (1996) 1582–1586.

- *Reading 2.2: Left Snails and Right Minds* by S. J. Gould in *Natural History*, April 1995, 10–18.

- *Reading 2.3: Cloning of inv, a Gene that Controls Left/Right Asymmetry and Kidney Development* by T. Mochizuki, et al. in *Nature* **395** (1998) 177–181.

- *Reading 2.4: A New Spin on Handed Asymmetry* by K. J. Vogan and C. J. Tabin in *Nature* **397** (1999) 295–298.

Exercises for Chapter 2

Chapter 3: *Introduction to Differential Equations.*   The discussion concerns differential equations for one unknown function of a single variable.

- *Reading 3.1: Biologists Sort the Lessons of Fisheries Collapse* by R. Holmes in *Science* **264** (1994) 1252–1253.

- *Reading 3.2: New Study Provides Some Good News for Fisheries* by M. Barinaga in *Science* **269** (1995) 1043.

- *Reading 3.3: Population Dynamics of Exploited Fish Stocks at Low Population Levels* by R. A. Meyers, N. J. Barrowman, J. A. Hutchings, and A. A. Rosenberg in *Science* **269** (1995) 1106–1108.

Exercises for Chapter 3

Chapter 4: *Stability in a One-Component System.* The discussion concerns stability analysis of equilibrium solutions to differential equations for one unknown function of one variable.

- *Reading 4.1: Tempo and Mode of Speciation* by M. L. Rosenzweig in *Science* **277** (1997) 1622–1623.

  Exercises for Chapter 4

Chapter 5: *Systems of First-Order Differential Equations.* This chapter introduces the subject of differential equations for two or more unknown functions of a single variable. In the process, the notion of a function of two or more variables is introduced.

- *Reading 5.1: Left Snails and Right Minds* by S. J. Gould in *Natural History*, April 1995, 10–18.

  Exercises for Chapter 5

Chapter 6: *Phase Plane Analysis.* This chapter provides the details of phase plane analysis for differential equations for two unknown functions of a single variable.

- *Reading 6.1: Red Grouse and Their Predators* by S. Thirgood and S. Redpath in *Nature* **390** (1997) 547.

- *Reading 6.2: Wolves, Moose, and Tree Rings on Isle Royale* by B. E. McLaren and R. O. Peterson in *Science* **266** (1994) 1555–1558.

  Exercises for Chapter 6

Chapter 7: *Introduction to Vectors.* This discussion introduces vector notation and the basic rules for manipulating vectors.

- *Reading 7.1: Power Laws Governing Epidemics in Isolated Populations* by C. J. Rhodes and R. M. Anderson in *Nature* **381** (1996) 600–602.

- *Reading 7.2: Allometry and Simple Epidemic Models for Microparasites* by G. A. De Leo and A. P. Dobson in *Science* **379** (1996) 720–722.

- *Reading 7.3: Scope of the AIDS Epidemic in the United States* by P. S. Rosenberg in *Science* **270** (1995) 1372–1375.

  Exercises for Chapter 7

Chapter 8: *Equilibrium in Two-Component, Linear Systems.* This chapter starts the discussion of stability analysis for equilibrium points of differential equations for two unknown functions of a single variable. The linear case is considered here.

- *Reading 8.1: Better Protection of the Ozone Layer* by M. K. W. Ko, N-D. Sze, M. J. Prather in *Nature* **367** (1994) 505–508.

  Exercises for Chapter 8

Chapter 9: *Stability in Nonlinear Systems.* Stability analysis for equilibrium solutions to fully nonlinear differential equations for two unknown functions of a single variable is treated here. Along the way, the notions of 2×2 matrices, their traces, and determinants are introduced as well as the notion of a partial derivative.

  Exercises for Chapter 9

Chapter 10: *Non-Linear Stability Revisited.*   This chapter reviews the basics of stability analysis. It also continues the discussion of partial derivatives and introduces second- and higher-order partial derivatives. This chapter also contains a brief discussion of integration of functions of two variables.

• Reading 10.1: *Hopes for the Future: Restoration Ecology and Conservation Biology* by A. P Dobson, A. D. Bradshaw, and A. J. M. Baker in *Science* **277** (1997) 515–522.

• Reading 10.2: *Effects of Disturbance on River Food Webs* by J. T. Wootton, M. S. Parker, and M. E. Power in *Science* **273** (1996) 1558–1561.

Exercises for Chapter 10

Chapter 11: *Matrix Notation.*   This chapter serves as an aside to introduce the basics of addition, subtraction and multiplication of matrices. Eigenvalues and eigenvectors are touched on.

Exercises for Chapter 11

Chapter 12: *Remarks about Australian Predators.*   With the help of Reading 12.1, the discussion here illustrates an application of phase plane and stability analysis.

• Reading 12.1: *The Case of the Missing Meat Eaters* by T. Flannery in *Natural History*, June 1993, 22–24.

• Reading 12.2: *Uniting Two General Patterns in the Distribution of Species* by I. Hanski and M. Gyllenberg in *Science* **275** (1997) 397–400.

Exercises for Chapter 12

Chapter 13: *Introduction to Advection.*   This chapter begins the discussion of differential equations for an unknown function of both time and space. A tautological conservation equation is derived, with the advection equation providing the simplest example.

• Reading 13.1: *Malaria: Focus on Mosquito Genes* by P. Aldhous in *Science* **261** (1993) 546–548.

• Reading 13.2: *Research Community Swats Grasshopper Control Trial* by B. Goodman in *Science* **260** (1993) 887.

• Reading 13.3: *Hosts and Parasitoids in Space* by H. C. J. Godfray and M. P. Hassel in *Nature* **386** (1997) 660–661. Also: *Insect Parasitoid Species Respond to Forest Structure at Different Spatial Scales* by J. Roland and P. D. Taylor in *Nature* **386** (1997) 710–713.

• Reading 13.4: *Scope of the AIDS Epidemic in the United States* by P. S. Rosenberg in *Science* **270** (1995) 1372–1376.

Exercises for Chapter 13

Chapter 14: *Diffusion Equations.* The diffusion equation is introduced and discussed here as a second example of the previous chapter's tautological conservation equation.

- *Reading 14.1: Counting Polymers Moving through a Single Ion Channel* by S. M. Bezrukev, I. Vodyanoy, and V. A. Parsegian in *Nature* **370** (1994) 279–281.

Exercises for Chapter 14

Chapter 15: *Two Key Properties of the Advection and Diffusion Equations.* This chapter focuses on the predictive nature of the advection and diffusion equations and then on the fact that the linear versions of these equations obey the superposition principle.

- *Reading 15.1: Diffusional Mobility of Golgi Proteins in Membranes of Living Cells* by N. B. Cole, C. L. Smith, N. Sciaky, M. Terasaki, M. Edidin, and J. Lippincott-Schwartz in *Science* **273** (1996) 797–801.

- *Reading 15.2: Past Temperatures Directly from the Greenland Ice Sheet* by D. Dahl-Jensen, K. Mosegaard, N. Gundestrup, G. D. Clow, S. J. Johnsen, A. W. Hansen, and N. Balling in *Science* **282** (1998) 268–271.

Exercises for Chapter 15

Chapter 16: *The No Trawling Zone.* Boundary constraints for the diffusion equation are introduced here with an example from ocean ecology. This chapter also starts a more general discussion of the separation of variables technique.

- *Reading 16.1: Fishing for Answers: Deep Trawls Leave Destruction in Their Wake—But for How Long?* by J. Raloff in *Science News* **150** (1996) 268–271.

Exercises for Chapter 16

Chapter 17: *Separation of Variables.* The Laplace equation for a function of two space variables is introduced here as the equilibrium version of a diffusion equation. The technique of separation of variables is then presented in the Laplace equation context.

Exercises for Chapter 17

Chapter 18: *The Diffusion Equation and Pattern Formation.* The chapter starts by describing a potential use of the diffusion equation to model pattern formation in developing embryos. The chapter then centers on stability criteria for an equilibrium solution to the diffusion equation.

- *Reading 18.1: Generic Modelling of Cooperative Growth Patterns in Bacterial Colonies* by E. Ben-Jacob, O. Schochet, A. Tenenbaum, I. Cohen, A. Czirok, and T. Vicsek in *Nature* **368** (1994) 46–49.

- *Reading 18.2: Dynamics of Stripe Formation* by H. Meinhardt in *Nature* **376** (1995) 722–733. *A Reaction-Diffusion Wave on the Skin of the Marine Angelfish* Pomacanthus by S. Kondo and R. Asai in *Nature* **376** (1995) 765–768. Also *Letters to Nature* by T. Hofer and P. K. Maini in *Nature* **380** (1996) 678; *Letters to Nature* by S. Kondo and R. Asai in *Nature* **380** (1996) 678.

- *Reading 18.3: Complex Patterns in a Simple System* by J. E. Pearson in *Science* **261** (1993) 189–192.

- *Reading 18.4: Direct and Continuous Assessment by Cells of Their Position in a Morphogen Gradient* by J. B. Gurdon, A. Mitchell, and D. Mahony in *Nature* **376** (1995) 520–521. Also *Activin Signalling and Response to a Morphogen Gradient* by J. B. Gurdon, P. Harger, A. Mitchell, and P. Lemaire in *Nature* **371** (1994) 487–492.

Exercises for Chapter 18

Chapter 19: *Stability Criterion.* This chapter provides a more detailed discussion of the stability criteria for an equilibrium solution to the diffusion equation.

Exercises for Chapter 19

Chapter 20: *Summary of Advection and Diffusion.*   This chapter revisits the key issues of the previous chapters on the advection and diffusion equations. The maximum principle is also introduced here.

- *Reading 20.1: Helping Neurons Find Their Way* by J. Marx in *Science* **268** (1995) 971–973.

- *Reading 20.2: New Protein Appears to be Long-Sought Neural Inducer* by Marcia Baringa in *Science* **262** (1993) 653-654. Also *Neural Induction by the Secreted Polypeptide Noggin* by T. M. Lamb, A. K. Knecht, W. C. Smith, S. E. Stachel, A. N. Economides, N. Stahl, G. D. Yancopolous, and R. M. Harland in *Science* **262** (1993) 713–718.

- *Reading 20.3: Mobility of Photosynthetic Complexes in Thylakoid Membranes* by C. W. Mullineaux, M. J. Tobin, and G. R. Jones in *Nature* **390** (1997) 421–424.

Exercises for Chapter 20

Chapter 21: *Traveling Waves.* This chapter introduces the notion of a traveling wave solution to a nonlinear diffusion equation. The context of the discussion is the article that is discussed in Reading 21.1.

- *Reading 21.1: Hantavirus Outbreak Yields to PCR* by E. Marshall in *Science* **262** (1993) 832–836. Also *US Braces for Hantavirus Outbreak* by C. Holden in (Science Scope), *Science* **280** (1998) 993.

Exercises for Chapter 21

Chapter 22: *Traveling Wave Velocities.*   This chapter continues the discussion of traveling waves; the issue here is the prediction of the speed of the traveling wave.

- *Reading 22.1: Travelling Waves in Vole Population Dynamics* by E. Ranta and V. Kaitala in *Nature* **390** (1997) 456.

- *Reading 22.2: Control of Spiral-Wave Dynamics in Active Media by Periodic Modulation of Excitability* by O. Steinbock, V. Zykov, and S. Müller in *Nature* **366** (1993) 322–324.

Exercises for Chapter 22

Chapter 23: *Periodic Solutions.* The subject here is the Poincaré-Bendixson theorem, which gives sufficient conditions for certain differential equations for two unknown functions of time to admit a cyclical solution.

- *Reading 23.1: Wolves, Moose, and Tree Rings on Isle Royale* by B. E. McLaren and R. O. Peterson in *Science* **266** (1994) 1555–1558.

- *Reading 23.2: Snowshoe Hare Populations Squeezed from Below and Above* by N. C. Stenseth in *Science* **269** (1995) 1061–1062. Also *Impact of Food and Predation on the Snowshoe Hare Cycle* by C. J. Krebs, S. Boutin, R. Boonstra, A. R. E. Sinclair, J. N. M. Smith, M. R. T. Dale, K. Martin, and R. Turkington in *Science* **269** (1995) 1112–1115.

Exercises for Chapter 23

Chapter 24: *Fast and Slow.* This chapter discusses a simple differential equation that models "switching" behavior, in which a system swings rapidly between two very long lived, almost static configurations.

- *Reading 24.1: Disparate Rates of Molecular Evolution in Cospeciating Hosts and Parasites* by M. S. Hafner, P. D. Sudman, F. X. Villablanca, T. A. Spradling, J. W. Demastes, and S. A. Nadler in *Science* **265** (1994) 1087–1090.

Exercises for Chapter 24

Chapter 25: *Estimating Elapsed Time.* This chapter considers the following problem: Estimate the time required for a differential equation solution to attain some specified value.

Exercises for Chapter 25

Chapter 26: *Switches.* Here, the model from Chapter 24 is used to explain the switching behavior that genes exhibit in a developing embyo. The context of the discussion comes from the article in Reading 26.1.

- *Reading 26.1: Thresholds in Development* by J. Lewis, J. M. W. Slack, and L. Wolpert in *J. Theor. Biol.* **65** (1977) 579–590.

Exercises for Chapter 26

Chapter 27: *Testing for Periodicity.* This chapter considers the following problem: How are periodic components to time-dependent data recognized? The discussion introduces Fourier components and the power spectrum.

- *Reading 27.1: A Pervasive Millennial-Scale Cycle in North Atlantic Holocene and Glacial Climates* by G. Bond, W. Showers, M. Cheseby, R. Lotti, P. Almasi, P. deMenocal, P. Priore, H. Cullen, I. Hajdas, and G. Bonani in *Science* **278** (1997) 1257–1266.

Exercises for Chapter 27

Chapter 28: *Causes of Chaos.*   This chapter gives a heuristic motivation for the appearance of truly "chaotic" phenomena in certain differential equations for three or more functions of time.

- *Reading 28.1: Chaos, Persistence, and Evolution of Strain Structure in Antigenically Diverse Infectious Agents* by S. Gupta, N. Ferguson, and R. Anderson in *Science* **280** (1998) 912–915.

- *Reading 28.2: Controlling Chaos in the Brain* by S. J. Schiff, K. Jerger, D. H. Duong, T. Chang, M. L. Spano, and W. L. Ditto in *Nature* **370** (1994) 615–620.

- *Reading 28.3: Predicting and Producing Chaos* by P. Kareiva in *Nature* **375** (1995) 189–190. Also *Experimentally Induced Transitions in the Dynamic Behavior of Insect Populations* by R. F. Costantino, J. M. Cushing, B. Dennis, and R. Desharnais in *Nature* **375** (1995) 227–230.

Acknowledgments

It is an honor to thank the many individuals who helped me over the years develop the material for this book. First, I am very proud to acknowledge the generosity of the many scientists and science writers who allowed me to reprint their articles in this book. Their commitment to the open dissemination of knowledge is exemplary. Of course, in this regard, I also owe a great debt to the journals *Science, Nature, Science News, Natural History*, and the *Journal of Theoretical Biology* for allowing me to reprint their copyrighted articles. Some original articles included color figures, which we were unable to reproduce in color here.

A large debt is also owed to Robin Gottlieb and Otto Bretscher for allowing the use in this book of my modifications of sections from their calculus notes. On a related theme, I am proud to acknowledge the profound influence on this work of Edwin Beltrami's book, "Mathematics for Dynamical Modeling"; indeed, I first found many of the texts examples in Beltrami's book.

I thank the many course assistants that worked with me over the years in Harvard University's Math 19. Their input to this book's precurser, the original Math 19 course notes, was invaluable; especially their role in uncovering the many errors, typographical and otherwise. I also thank the many students who took Math 19, for they taught me biology and how one might teach math to biologists. Of course, they also found many errors in my presentation and in the Math 19 course notes.

I gratefully acknowledge Svetlana Alpert's help over the years putting together course material for Math 19. Many others from the Harvard University Mathematics Department also helped in this regard, and I thank them as well.

Thanks are more than due to those at Prentice Hall, most especially George Lobell and Betsy Williams, for their work to put my course notes into book form. Moreover, I owe an extra debt to George for agreeing to publish my book and for his support during its preparation.

Finally, I wish to acknowledge these scientists who reviewed and commented on the early drafts of this book: Daniel E. Bentil, University of Vermont; David Isaacson, Rensselaer Polytechnic Institute; Stephen Krone, University of Idaho; and Michael Li, Mississippi State University. Their input was invaluable.

Clifford Henry Taubes

1
Introduction

This chapter and the subsequent 27 chapters are about differential equations and how they are applied by biologists. As the branch of mathematics called *differential equations* is a direct application of ideas from calculus, and as this is a mathematics text, I should begin by telling you a little bit about what is meant by the term *differential equation*. However, I'll digress first to begin an argument for including mathematics in the tool kit of a working biologist.

1.1 Modeling in the Biological Sciences

First, I freely admit to not being a biologist. In fact, until I started teaching the course on which this book was based, I knew very little of recent work in biology. I took biology in high school, dissected a worm and a frog, and happily found other interests. Subsequently, I kept minimally abreast of the subject by reading articles from popular science journals such as *Science News* and *Scientific American*. However, since I started teaching this course, I have endeavored to educate myself about modern biology and have found it to be a glorious thing. In fact, I would be happy to argue the case that our understanding of biology now ranks as the (or at least one of the) crowning achievements of human knowledge.

My recent and ongoing education in biology has taught me the following lesson: With some notable exceptions, biology at the cusp of the twenty-first century is very much an experimentally driven science. Life is *extremely* complicated, and sorting out these complications is the task at hand. This is to say that the *laboratory* rules the field. At the risk of some exaggeration, one might say that it is somewhat premature in most fields of biology to spend too much energy with theory. In fact, I think that the following situation is common: You are trying to guess how a particular biological process works. You come up with a good proposal for the process. But, does nature use your proposal? You do some experiments and you find, lo and behold, that nature maybe uses your proposal, but probably not; and in any event, nature has found 20 completely different ways to work the process and is using all 20 *simultaneously*. (On the other hand, there are certain subfields of biology that could use, perhaps, more experimentation and less theory. Population biology is a particular example, for in this field, controlled experiments on macroscopic life forms are not easy to devise.)

With the preceding understood, where is the place for mathematics in biology? The answer to this question necessarily requires an understanding of what modern mathematics is. In this regard, I should say that term *mathematics* covers an extremely broad range of subjects. Even so, a unifying definition might be as follows:

Mathematics consists of the study and development of methods for prediction.

Meanwhile, a science such as biology has, roughly, the following objective:

To find useful and verifiable descriptions and explanations of phenomena in the natural world.

To be useful, a description need be nothing more than a catalog or index. But an explanation is rarely useful without leading to verifiable *predictions*. It is here where mathematics can be a great help. In practice, working biologists use mathematics as a tool to facilitate the development of predictive explanations for observed phenomena. This is how mathematics will be viewed in this text. (The use of mathematics as a tool to make predictions of natural phenomena is called **modeling** and the resulting predictive explanation is often called a **mathematical model**.)

At this point, it is important to realize that a vast range of mathematics has found biological applications. Two, in particular, are **differential equations** and **probability/statistics**. This text is concerned almost solely with differential equations; almost nothing will be said about probability and statistics. However, this is not to say that the latter are less important. In fact, they are extremely important, and you should take a course on probability and statistics (or experimental design) if you haven't done so already.

1.2   Equations

The preceding discussion about predictions is completely abstract, and so a digression may prove useful to bring the discussion a bit closer to earth. In particular, consider what is meant by a prediction: You measure in your lab certain quantities—numbers really. Give these measured quantities letter names such as a, b, c, etc. A prediction can take the form of a formula that allows you to determine the value for the quantity c

by measuring only the quantities a and b. Such a formula might involve simply an algebraic equation that relates a and b to c.

For example, if you lived in Greece some twenty-five hundred years ago, you might discover that the length, c, of the hypotenuse of a right triangle can be predicted from the measured lengths, a and b, of the other two sides. Indeed, if you were Pythagoras, your thoughts might be rendered in modern mathematics as

$$c = \sqrt{a^2 + b^2}. \tag{1}$$

Or, you might determine that the area, A, of a disk can be predicted with knowledge of its radius, r, using the equation

$$A = \pi r^2. \tag{2}$$

These are examples of **algebraic equations** in that they involve simple expressions between what is known [a and b in (1)] and what is to be predicted [c in (1)]. A famous and modern algebraic equation is Einstein's formula

$$E = mc^2, \tag{3}$$

which describes how the total energy (E) of a body at rest can be computed if you know its mass (m) and the speed of light ($c \approx 3$ hundred million meters per second). An algebraic equation with applications to biology describes how the weight of a body (say w) would change if it had weight w_0 and you hypothetically scaled its length, width, and height by the same factor, say s. This formula asserts that

$$w = s^3 w_0. \tag{4}$$

1.3   Differential Equations

Differential equations can arise when studying quantities that depend on some auxiliary variable. For example, it is typical in the biological sciences to study time-dependent phenomena. A doctor can be concerned with the amount of a certain medicinal drug in the body as a function of time. That is, there is a function that depends on the variable $t = $ time, and its value at time t, say $f(t)$, is the concentration of the medicine at time t in the blood.

Here is another example: An environmental scientist can be concerned with the concentration of mercury in clams along a certain stretch of river. Here, the concentration might depend on the distance downstream. Thus, the concern is with a function that depends on the variable $x = $ distance downstream and its value at distance x, say $f(x)$, is the concentration of mercury in clams that are found at distance x. By the way, this concentration might depend on both position *and* time—a more complicated situation that shall concern us later in the course.

Here is a third example: A developmental biologist studying fly embryos might be concerned with the level of a certain molecular growth factor as a function of distance from the embryo head. Here, the function in question is the level of the growth factor as a function of the variable that measures the distance from the head of the embryo. Of course, this function can also depend on time as well as position; and it most probably does since live embryos develop as time progresses.

For a fourth example, an epidemiologist might consider the number of deaths from a certain disease as a function of age at death. Here, the variable is the age, α, at death, and the number of deaths of people at age α from the disease gives a function. You could denote the latter by $N(\alpha)$. By the way, this example illustrates an important point: The variable in question need not be time nor a position, but some entirely different quantity. Indeed, the same epidemiologist might consider the average number of heart attack victims in a particular locale, as a function of the level of cholesterol in the victim. In this case, the variable, call it c, is the level of cholesterol, and the function in question assigns to each value of c the number $n(c)$, which is the yearly average of heart attack sufferers with cholesterol level c.

With these examples understood, you can say that a ***differential equation*** for a function (or for some collection of functions) of a variable (or collection of variables) is simply an algebraic equation that involves both the function and its derivatives. For example, the equation

$$\frac{dp}{dt} = p \tag{5}$$

is a differential equation. This example is discussed in great detail in the next chapter.

1.4   Continuity, Differentiation, and Derivatives

This subsection should be a review and can be skipped if you desire. The subject is the derivative. Consider a function, say p, of a variable, say t. This variable t can measure the passage of time, or position along some line from a starting point, or something entirely different. The function p assigns to each value of t a number, $p(t)$. For the most part, this book will deal with functions that are, first, continuous functions of the variable. This means that near any parameter value t, the values of p are close to the number $p(t)$, and as the parameter value moves closer to the given value t, then the values of p come closer and closer to $p(t)$ and, in fact, limits to $p(t)$ in the sense that the values of p don't make discrete jumps as the parameter is varied. (Another way to say this is that the graph of p can be drawn without lifting pen from paper.)

In any event, if p is a continuous function of t, then we can try to define its derivative. The latter, as you should recall, is a function, called $\frac{dp}{dt}$, whose value at parameter value t is *defined* by a limit process:

$$\frac{dp}{dt}(t) \equiv \lim_{\Delta \to 0} \frac{p(t + \Delta) - p(t)}{\Delta}. \tag{6}$$

(Here, the extra line in the equal sign signifies that this equality serves to define the expression on the left-hand side.) Another way to view the derivative of p is as the function whose value is the slope of the graph of p.

If you were to choose functions at random, then you would find some for which there is no limit on the right side as Δ tends to zero, or for which the limit is infinite. [For example, try the function $f(t) = \sqrt{t}$ at $t = 0$] Such functions won't concern us here, and all the functions that are discussed in this book will be implicitly taken to be differentiable—in fact, the derivatives will be assumed to be differentiable and those derivatives will be differentiable, and so on.

1.5   Continuity and Differentiability in Biology

It is important to realize that continuous functions may not be appropriate for describing certain biological processes. For example, the function that assigns to a time t the number $N(t)$ of live bacteria in a petri dish is not a continuous function. Indeed, this function can only take values that are whole numbers, that is, $N(t) = 0, 1, 2, \ldots$. Thus, $N(t)$ is either 3 or 4 but never the number $\pi = 3.1416\ldots$, let alone 3.5.

Nonetheless, it is sometimes a reasonable ***approximation*** to reality to pretend that $N(t)$ is continuous. For example, if the number of bacteria is measured in units of 1 million, then $N(t)$ can take on values that differ by 0.000001; in this case, the approximation that $N(t)$ is continuous might not look so bad. In particular, if the ***experimental error*** in counting bacteria is greater than 1, then little is lost by modeling $N(t)$ (here, measured in units of 1 bacterium) as a continuous function. Indeed, in this case, the discreteness of N is effectively invisible. In any event, be forewarned that the use of continuous functions in the biological sciences constitutes a modeling approximation that may or may not make sense in any given application.

In particular, here is a general rule of thumb that summarizes the preceding discussion:

- *If the true function under discussion jumps in value, then its replacement with a continuous function is reasonable when the experimental error is larger than any of the jumps.*

Of course, one can also consider whether the assumption of differentiability makes sense in any given situation. The rule of thumb here is usually the following:

- *Once the step to a continuous function is made, the step to differentiability rarely adds trauma.*

2

Exponential Growth

Suppose that we are interested in the values of some quantity, say p, as a function of time, t. A differential equation for p is an equation that equates the time derivative of p to some function of p.

2.1 The Simplest Equation

The simplest differential equation reads

$$\frac{d}{dt}p = 0, \tag{1}$$

which asserts that the quantity p is constant in time. A less trivial example would be

$$\frac{d}{dt}p = c, \tag{2}$$

where c is some constant, say 2 or -3.4 or 107. (When this equation arises in the real world, the constant c is usually determined by some experimental measurement.) This last equation asserts that the rate of change of p is constant with time. Equation (2) can be solved fairly easily by integrating both sides:

$$p(t) - p(0) = \int_0^t \frac{dp}{dt} = ct. \tag{3}$$

The first equality here is just the fundamental theorem of calculus. Thus, the solution to (2) has

$$p(t) = p(0) + ct. \tag{4}$$

2.2   The Exponential Growth Equation

The next simplest differential equation has

$$\frac{d}{dt}p = p, \tag{5}$$

or, more generally,

$$\frac{d}{dt}p = ap, \tag{6}$$

where a is some given number. For example, $a = 3$ or -5.56 etc. These equations assert that the rate of change of p is proportional to p itself.

For example, suppose $p(t)$ represents the number of bacteria in a petri dish at time t. Suppose that the dish is well sugared, so that the bacteria don't lack for food. One expects that each bacteria will fission into two bacteria at a regular rate, say once every 20 minutes. If no bacteria die, then the rate of change of p is equal to $0.05p$ in units of bacteria per minute. Put differently, if there are $p(t)$ bacteria at time t, then one expects that roughly one out of every 20 bacteria are undergoing fission at any given minute. Thus, the population of bacteria in the petri dish at time t is governed by Equation (6) with $a = 0.05$ bacteria per minute.

The general solution to (6) is

$$p(t) = p(0)e^{at}. \tag{7}$$

Here, $p(0)$ is the value of p at time 0. Alternately, you can write (7) as

$$p(t) = p(t_0)e^{a(t-t_0)}, \tag{8}$$

where t is any convenient time and $p(t_0)$ is the value of p at that time. Do you believe that (7) and (8) are the same? Don't let me con you. Check that they are the same by using (7) to solve for $p(t_0)$ and then plugging the result into (8).

By the way, notice that when $a > 0$, the quantity $p(t)$ is increasing with time, and when $a < 0$, then $p(t)$ decreases with time. These are important features of (6) that will play a key role in future discussions.

The validity of (7) [or (8)] can be established by plugging the right-hand side of (7) [or (8)] into (6) and differentiating. Here, you should remember that

$$\frac{d}{dt}e^{at} = ae^{at}.$$

2.3   An Important Remark

It is important for you to remember that (7) or (8) is the general form of the solution to (6). For example, when you see the equation $\frac{d}{dt}p = -5p$, you should immediately think "Aha!, so $p(t) = p(0)e^{-5t}$." It is also important for you to realize that the solution $p(t)$ to (6) is increasing with time if $a > 0$ and decreasing with time if $a < 0$.

2.4   A Generalization of the Exponential Growth Equation

There is a simple generalization of (6) that arises often: This is the equation

$$\frac{d}{dt}q = aq + c, \tag{9}$$

where a and c are constants and q is the function that we want to find. For example, you might find (9) with $a = 5$ and $c = -1.1$, or $a = -10.2$ and $c = -33$. You should view (9) as a combination of Equations (2) and (6). Indeed, when $a = 0$, this equation is the same as (2) except that here, our unknown function is named q while in (2), its name is p. Alternately, when $c = 0$, Equation (9) is the same as (6) except that here again, the unknown function has a different name.

By the way, you should realize that the name of the unknown function has no substantive bearing on the form of the solution to a differential equation. For example, changing p to q in the exponential growth equation doesn't change the nature of the solutions one iota. There is a lesson here: There is no universal notation for the unknown function in a differential equation; and none is employed in these chapters.

In any event, let us consider (9) in the case where $a \neq 0$. In this case, the general solution is

$$q(t) = \left[q(0) + \frac{c}{a}\right]e^{at} - \frac{c}{a}. \tag{10}$$

This last equation can be checked in two ways: First, the right side of (10) can be differentiated to see $\frac{d}{dt}q = [q(0) + \frac{c}{a}]ae^{at}$. This uses (8). Meanwhile, you can also evaluate $aq + c$ by substituting the right side of (10) for q. Or, you can change variables to $p(t) \equiv [q(t) + \frac{c}{a}]$ and, by substituting into (9), verify that $p(t)$ has to obey the exponential growth equation

$$\frac{d}{dt}p = ap \tag{11}$$

when and only when the function $q(t)$ obeys (9).

The fact that the substitution $p(t) = [q(t) + \frac{c}{a}]$ changes (9) to (11) shows that (6) and (9) are really the different manifestations of the same equation.

2.5   When Does the Exponential Growth Equation Arise?

The exponential growth equation $\frac{d}{dt} p = ap$ arises often in the sciences. There are two reasons for this. Here is the first reason: Suppose that $p(t)$ represents the population of identical particles or creatures at time t that *do not interact* with each other. Use a_b to denote the birth rate and a_d to denote the death rate. Then, the population $p(t)$ will obey an exponential equation where the coefficient a is equal to the difference, $a_b - a_d$. Note that if a_b is the birth rate (measured, say, in births per second), then $1/a_b$ is the average time between births. Likewise, $1/a_d$ is the average time between deaths. For example, if the birth rate is 4 per day, then, all things being equal, on average there will be a birth every quarter day.

Notice that these quantities a_b and a_d, and hence the quantity a, can, in principle, be determined by experiments on some small number of creatures (or particles) in isolation. This is what makes the exponential equation so useful. Experiments done with small numbers of creatures or particles can be used to *predict* the behavior of large numbers of particles. This is the great utility of (6) and (9). They allow you to *predict* the behavior of *large* numbers of creatures or particles in terms of quantities that have been measured from experiments with *small* numbers of particles. [However, the large numbers of particles or creatures must not interact with each other in a substantive way. If they do, then all bets are off *vis-à-vis* the applicability of (6) or (9).]

The second reason for the ubiquitous appearence of (6) [and also (9)] has to do with Taylor's theorem.

2.6   Taylor's Theorem

Generally speaking, functions can be arbitrarily complicated. For example, imagine the function

$$f(x) = \cos\left(\sin\left(\frac{2}{\sin(x^3) + 4 + x^2 \cos^2 x}\right)\right). \tag{12}$$

What would you do with the differential equation

$$\frac{d}{dt} x = f(x) \tag{13}$$

for a function x of t, where f is given by (12)?

(Note that here the unknown function is x and not p nor even q. Remember that there is no universal "name" for the unknown function and so you can name it what you like.)

Taylor's theorem gives a method for dealing with a horrible function such as f in (12). Suppose that you are interested in the solution $x(t)$ to (13) only for x near some point x_0 of interest. If this happens to be the case (and often it is), then you only need to know about the function f near the point x_0. For example, x_0 might be zero, or it might be 12.33, or it might be -21.677.

In any event, the strategy is to sacrifice some *accuracy* for *solvability*. You replace the horrible function $f(x)$ with an approximate function, a function $g(x)$ that is very close to $f(x)$ as long as x is close to x_0. The function $g(x)$ should be such that you *can* solve the equation

$$\frac{d}{dt}x = g(x). \tag{14}$$

Here is the key point:

> **The solution, $x(t)$, to (14) will behave much like the solution to (13) for those times t where $x(t)$ from (14) is near to x_0.**

That is, you are interested in the solution to (13), but (13) is way too hard to solve. So, instead, you solve an easier equation, (14), and observe that for times t where the solution to (14) is near to x_0, then the solution to (14) will provide a reasonable approximation to the desired solution to (13). Thus, you will gain some knowledge at the expense of some accuracy. [You lose accuracy because you have replaced (13) by the approximation (14).]

With the preceding understood, Taylor's theorem tells you how to approximate any function $f(x)$ in the vicinity of any given value x_0 by a simpler function. Indeed, the simplest Taylor's approximation replaces $f(x)$ with the *constant* function

$$g(x) = f(x_0). \tag{15}$$

This is called the **zeroth-order Taylor's approximation**. (In the mathematics literature, you might see this called the "zeroth-order Taylor's expansion," or even the "zeroth-order Taylor's series.") Then, instead of solving (13), you would find $x(t)$ that solved

$$\frac{d}{dt}x = f(x_0) \tag{16}$$

with the starting value $x(0) = x_0$. The solution to (15) is

$$x(t) = x_0 + f(x_0)t, \tag{17}$$

which is a reasonable approximation to the solution to (13) as long as $|t|$ is small.

In certain cases, the approximation in (17) is not good enough. Typically, this occurs when $f(x_0) = 0$, for in this case (17) gives no hint as to the behavior of the solution to the true equation (13). When $f(x_0) = 0$, then you should replace the function f by the **first order Taylor's approximation**,

$$g(x) = \left.\frac{df}{dx}\right|_{x=x_0} (x - x_0). \tag{18}$$

This approximation beats the approximation in (15) because not only does $g(x_0)$ agree with $f(x_0)$ (which is zero), but also their first derivatives agree at x_0. This means that g in (18) approximates f over a larger range of x values than does g in (15).

Anyway, with g given by (18), then the approximate equation (14) reads

$$\frac{d}{dt}x = \left.\frac{df}{dx}\right|_{x=x_0}(x-x_0),\tag{19}$$

which is the same form as (9) if you set

$$a = \left.\frac{df}{dx}\right|_{x=x_0}\tag{20a}$$

$$c = \left.\frac{df}{dx}\right|_{x=x_0}x_0.\tag{20b}$$

The solution $x(t)$ to (19) with $x(0) = x_0$ is a good approximation to the solution to (13) as long as t is such that (19)'s solution stays close to x_0. Thus, our generalized exponential in Equation (9) arises as a sort of *universal* approximation to a more complicated equation [such as in (13)].

By the way, there is an nth-order Taylor's theorem that approximates $f(x)$ near x, by

$$g_n(x) = f(x_0) + \left.\frac{df}{dx}\right|_{x=x_0}(x-x_0) + \frac{1}{2}\left.\frac{d^2 f}{dx^2}\right|_{x=x_0}(x-x_0)^2 + \cdots$$

$$+ \frac{1}{n(n-1)\cdots 1}\left.\frac{d^n f}{dx^n}\right|_{x=x_0}(x-x_0)^n,$$

This $g_n(x)$ is the simplest function whose first n deriviatives at x_0 agree with the first n derivatives of f at x_0. (For $n > 1$, these nth order expansions are not used in later chapters.)

2.7   Doubling Times and Half-Lives

One of the main points of this chapter is that the linear equation, (6), has solutions of the form $p(t) = p(0)e^{at}$, where $p(0)$ gives the value of p at time $t = 0$. If a is positive, then p grows with time and in this case, the constant a here determines the "doubling time" of the population, this being the time, t_2, where the population is twice its original value. Thus t_2 is determined by the requirement that $p(t_2) = 2p(0)$. In particular, if you take the expression $p(t) = p(0)e^{at}$, replace t on the right-hand side by t_2 while replacing $p(t)$ on the left-hand side by $2p(0)$, and then take the natural logarithm of both sides of the resulting equality, you should find that

$$\ln(2) + \ln p(0) = at_2 + \ln p(0).\tag{21}$$

Cancel the $\ln p(0)$ from both sides of this last equation and also divide both sides by a to find

$$t_2 = \frac{\ln(2)}{a}.\tag{22}$$

For example, if $a = 3$, then the doubling time is $\ln(2)/3$.

On the other hand, if a is negative, then $p(t)$ will decrease with time, and you can ask for the time, $t_{1/2}$, where the population is half of its original size. This $t_{1/2}$ is called the half-life. As with the doubling time, the half-life can be computed in terms of the constant a that appears in (6). Since $t_{1/2}$ is determined by the condition that $p(t_{1/2}) = p(0)/2$, the same sort of manipulation that derived (22) finds

$$t_{1/2} = \frac{\ln(2)}{-a}. \tag{23}$$

[You should try to derive (23) by yourself to check whether you really understand things.]

2.8   Sums of Exponential Functions

Suppose a quantity that you wish to measure, say the level of AIDS virus in a patient's blood, has been predicted, using a mathematical model, to be given by the function

$$f(t) = e^{-t} + e^{-4t} \quad \text{for } t \geq 1. \tag{24}$$

Here, t is measured in days after starting a particular drug therapy. If you want to test this model, you might try to measure the actual viral load each day as function of time t (starting at $t = 1$) and compare the resulting function [call it $f_{\exp}(t)$] with the prediction in (24). Of course, there is some experimental error in any measurement, and so $f_{\exp}$ can never be determined exactly. Question: How accurate must $f_{\exp}$ be measured in order for you to be reasonably sure that the correct answer is the function in (24) as opposed to some other function, say $g(t) = e^{-t}$?

Exponentials fall off quite fast, and the larger the coefficient before t in the exponent, the faster the fall-off. Would you believe that at $t = 1$, the second term in (24) is approximately 5% the size of the first term? Moreover, at $t = 2$, the second term in (24) is approximately 1/4 of 1% of the first term. Thus, the second term in (24) will be very difficult to measure.

The lesson: Beware of sums of exponentials. In a sum of exponentials where all of the exponents are negative multiples of t, the term with the smallest (in absolute value) multiple of t will dominate the sum at large t. If there are any exponentials with positive multiples of t, then the term with the largest positive multiple of t in the exponent will dominate at large times.

2.9   Lessons

Here is what you should remember from this first lesson:

- The exponential growth equation $\frac{d}{dt}x = ax$ is solved by $x(t) = x(t_0)e^{a(t-t_0)}$, where t_0 is any convenient time.

- Its generalization, $\frac{d}{dt}x = ax + c$, is solved by

$$\dot{x}(t) = -\frac{c}{a} + \left(x(t_0) + \frac{c}{a}\right)e^{a(t-t_0)},$$

where t_0 is any convenient time.

- Remember that any function $f(x)$ is approximated for x near some point x_0 of interest by its zeroth-order Taylor's approximation, $g_0(x) = f(x_0)$. And, if $f(x_0) = 0$, you should consider approximating for x near x_0 by the first-order Taylor's approximation, $g_1(x) = \frac{df}{dx}\Big|_{x=x_0}(x-x_0)$. (This is the simplest function that has the same value as f at $x = x_0$ and that has the same derivative of f at $x = x_0$.)

- Beware sums of exponentials functions of time. The term in the sum with the least negative or most positive exponential will dominate the sum at large times.

2.10   Appendix on Taylor's Approximations

Here is another view about Taylor's series: Suppose that $f(x)$ is a function of the variable x. In general, as x varies, the function f can be quite complicated. But suppose that for some reason we are only interested in f for values of x near to some particular point x_0. For example, near $x_0 = 5$ or near $x_0 = -1$ or $x_0 = 0$.

Zeroth-order Taylor's series: Find the constant function $g(x) = c$ (here, c is a constant) that best approximates f near $x = x_0$. *Answer:* Take c to equal $f(x_0)$. For example, if $f(x) = \sin(x)$ and $x_0 = 5$, then take $g(x) = \sin(5)$. For a second example, take $f(x) = \sin(x)$ and $x_0 = 0$. Then $g(x) = \sin(0) = 0$.

First-order Taylor's series: A somewhat better approximation to $f(x)$ for x near x_0 can be had by replacing $f(x)$ by the function $g(x) = c + bx$, where c and b are both appropriately chosen constants. The strategy for choosing c and b is to require that $g(x_0)$ and $f(x_0)$ agree and also require that their derivatives agree at x_0 too. Thus the graphs of f and g also have the same slope at x_0. That is, you require

$$f(x_0) = g(x_0) = c + bx_0 \quad \text{and} \quad \frac{df}{dx}\Big|_{x=x_0} = \frac{dg}{dx}\Big|_{x=x_0} = b. \tag{25}$$

Thus,

$$g(x) = f(x_0) + \frac{df}{dx}\Big|_{x=x_0}(x - x_0). \tag{26}$$

For example, if $f(x) = \sin(x)$ and if $x_0 = 5$, then $\frac{df}{dx}\Big|_{x=x_0} = \cos(5)$ and so

$$g(x) = \sin(5) + \cos(5)(x - 5) \tag{27}$$

is the first-order Taylor's series for the function $\sin(x)$ near the point $x_0 = 5$.

Higher-order Taylor's series: These don't play much of a role in this course, but just for your information, the nth-order Taylor's expansion of a function $f(x)$ near a point x_0 is an approximation of f by a function $g(x)$ of the form $g(x) = c + b_1 x + b_2 x^2 + \cdots + b_n x^n$, where the numbers b_k are determined from the kth derivative of f at the point x_0. To be precise, $b_k = (k(k-1)(k-2) \cdots 1) f^k(x_0)$, where $f^k(x)$ is the kth derivative of f at the point x.

Precision: Typically, the first-order Taylor's expansion for a function f is a better approximation than the zeroth-order, and the second better than the first, etc. However, when trying to use the Taylor's approximation for f to get an approximate solution to a differential equation of the form $\frac{dx}{dt} = f(x)$, we shall stick to first-order expansions, since higher-order expansions typically yield approximate equations that are themselves unsolvable in closed form. The first-order Taylor's expansion for f yields the approximate equation $\frac{dx}{dt} = c + bx$, which has the closed form solution $x(t) = -\frac{c}{b} + \left(x(0) + \frac{c}{b}\right) e^{bt}$.

READINGS FOR CHAPTER 2

READING 2.1

HIV-1 Dynamics in Vivo: Virion Clearance Rate, Infected Cell Life-Span, and Viral Generation Time

Commentary: This article contains a fairly simple model of the interaction between HIV virus and the human immune system. The model is used to obtain (from patient data) an estimate on the total number of viral particles in the patient. The results are interesting and got a great deal of press when they came out. In any event, for the purpose of this chapter, you should focus attention on the equation numbered (4). This equation is for the variable V_1, which stands for the plasma concentration of virions in the infected pool. Since c in this equation is supposed to be a constant, this is our standard exponential equation.

    This article also illustrates an important issue about sums of exponentials: Suppose that you have a sum of exponential functions of time, with different exponents such as $ae^{\alpha t} + be^{\beta t}$, where α and β and also a and b are constants. Suppose, for the sake of argument, that $\alpha > \beta$. Then, very quickly, the term with $e^{\alpha t}$ will be much greater than $e^{\beta t}$, and so this sum will look very much like $ae^{\alpha t}$. For example, if the constants a and b are nearly the same, while α is positive and five times bigger than $|\beta|$, then after one unit of time, the $e^{\beta t}$ term will be only 2% of the $e^{\alpha t}$ term. Likewise, if α and β are negative, and $|\beta|$ is five times bigger than $|\alpha|$, then the term with $e^{\beta t}$ will be 2% of the $e^{\alpha t}$ term after one unit of time.

    In this article, Equation (6) is just such a sum of exponentials, where you can see from Table 1 that the constant c is predicted to be on the order of six to ten times the constant δ. Thus, after one day, the e^{-ct} terms will be roughly 2% of the $e^{-\delta t}$ terms. This raises the following question: In the article, the constants c and δ are determined from experimental data by finding that pair of c and δ that makes the exponential sum in Equation (6) look most like the experimental data that is graphed in Figure 1. Are the values for c believable? For, if c is as large as claimed, the terms with e^{-ct} in Equation (6) are indistinguishable from zero after one day if the experimental error is more than a few percent.

HIV-1 Dynamics in Vivo: Virion Clearance Rate, Infected Cell Life-Span, and Viral Generation Time

Science, **271** (1996) 1582–1586.
Alan S. Perelson, Avidan U. Neumann,
Martin Markowitz, John M. Leonard, David D. Ho

A new mathematical model was used to analyze a detailed set of human immunodeficiency virus-type 1 (HIV-1) viral load data collected from five infected individuals after the administration of a potent inhibitor of HIV-1 protease. Productively infected cells were estimated to have, on average, a life-span of 2.2 days (half-life $t_{1/2}$ = 1.6 days), and plasma virions were estimated to have a mean life-span of 0.3 days ($t_{1/2}$ = 0.24 days). The estimated average total HIV-1 production was 10.3×10^9 virions per day, which is substantially greater than previous minimum estimates. The results also suggest that the minimum duration of the HIV-1 life cycle in vivo is 1.2 days on average, and that the average HIV-1 generation time—defined as the time from release of a virion until it infects another cell and causes the release of a new generation of viral particles—is 2.6 days. These findings on viral dynamics provide not only a kinetic picture of HIV-1 pathogenesis, but also theoretical principles to guide the development of treatment strategies.

HIV-1 replication in vivo occurs continuously at high rates [1, 2]. Ho *et al.* [1] found that when a protease inhibitor was administered to infected individuals, plasma concentrations of HIV-1 decreased exponentially, with a mean $t_{1/2}$ of 2.1 ± 0.4 days. Wei *et al.* [2] and Nowak *et al.* [3] found essentially identical kinetics of viral decay after the use of inhibitors of HIV-1 protease or reverse transcriptase. The viral decay observed in these studies was a composite of two separate effects: the clearance of free virions from plasma and the loss of virus-producing cells. To understand the ki-

netics of these two viral compartments more precisely, we closely monitored the viral load in five HIV-1-infected patients after the administration of a potent protease inhibitor. Using a mathematical model for viral dynamics and nonlinear least squares fitting of the data, we obtained separate estimates of the virion clearance rate, the infected cell life-span, and the average viral generation time in vivo.

Ritonavir [4, 5] was administered orally (600 mg twice daily) to five infected patients, whose baseline characteristics are shown in Table 1. After treatment, we measured HIV-1 RNA concentrations in plasma at frequent intervals (every 2 hours until the sixth hour, every 6 hours until day 2, and every day until day 7) by means of an ultrasensitive modification [1, 5] of the branched DNA assay [6]. Each patient responded with a similar pattern of viral decay: an initial lag followed by an approximately exponential decline in plasma viral RNA (see Figure 1 for examples).

After ritonavir was administered, a delay in its antiviral effect was expected because of the time required for drug absorption, distribution, and penetration into the target cells. This pharmacokinetic delay could be estimated by the time elapsed before the first drop in the titer of infectious HIV-1 in plasma (Table 1 and Figure 1B). However, even after the pharmacokinetic delay was accounted for, a lag of ~1.25 days was observed before the plasma viral RNA concentration fell (Figure 1). This additional delay is consistent with the mechanism of action of protease inhibitors, which render newly produced virions noninfectious but do not inhibit either the

---

Author affiliations: Alan S. Perelson, Avidan U. Neumann, Martin Markowitz: Theoretical Division, Los Alamos National Laboratory, Los Alamos, NM 87545, USA. Martin Markowitz, David D. Ho: Aaron Diamond AIDS Research Center, 455 First Avenue, New York, NY 10016, USA. John M. Leonard: Pharmaceutical Products Division, Abbott Laboratories, Abbott Park, IL 60064, USA.

Table 1.   Summary of HIV-1 clearance rate, infected cell loss rate, and virion production rate for the five patients. Base-line values are average of measurements taken at days −7, −4, −1, and 0; each virion contains two RNA copies. Pharmacologic delay was estimated from the first drop in plasma infectivity for patients 102, 105, and 107. Delay was estimated by best fit of viral load to Equation 6 for patients 103 and 104. Lower and upper 68% confidence intervals were calculated by a bootstrap method [22] in which each experiment was simulated 100 times. Total virion production was calculated from plasma and extracellular fluid volumes estimated from body weights, assuming that plasma and extracellular fluid are in equilibrium.

				Estimates from fit of V(t) to plasma viral load								
				Virion clearance				Infected cell loss				Total virion production $(10^9$ per day)
Patient	Base-line values		Pharm. delay (hours)	c (day^{-1})	Confidence interval		$t_{1/2}$ (days)	δ (day^{-1})	Confidence interval		$t_{1/2}$ (days)	
	CD4 cells (per mm³)	Plasma virions $(10^3$ per ml)			Lower	Upper			Lower	Upper		
102	16	294	2	3.81	1.93	7.03	0.18	0.26	0.24	0.30	2.67	12.9
103	408	12	6	2.73	2.04	3.70	0.25	0.68	0.63	0.73	1.02	0.4
104	2	52	2	3.68	2.53	6.19	0.19	0.50	0.47	0.54	1.39	2.9
105	11	643	6	2.06	1.42	3.76	0.34	0.53	0.48	0.60	1.31	32.1
107	412	77	2	3.09	2.56	4.55	0.22	0.50	0.48	0.52	1.39	3.1
Mean	170	216	3.6	3.07	2.10	5.05	0.24	0.49	0.46	0.54	1.55	10.3
±SD	196	235	2.0	0.64	0.42	1.34	0.06	0.13	0.13	0.14	0.57	11.7

production of virions from already infected cells or the infection of new cells by previously produced infectious virions [7]. In our previous study [1], this additional delay was missed because measurements were less frequent (every 3 days), and the results were fitted to a single exponential, which was sufficient to provide minimum estimates of HIV-1 kinetics. In contrast, in the present study, we obtained 15 data points during the first 7 days, which allowed a careful analysis of the results by means of a new mathematical model of viral kinetics.

We assumed that HIV-1 infects target cells (T) with a rate constant k and causes them to become productively infected cells (T^*). Before drug treatment, the dynamics of cell infection and virion production are represented by

$$\frac{dT^*}{dt} = kVT - \delta T^* \tag{1}$$

$$\frac{dV}{dt} = N\delta T^* - cV \tag{2}$$

where V is the concentration of viral particles in plasma, δ is the rate of loss of virus producing cells, N is the number of new virions produced per infected cell during its lifetime, and c is the rate constant for virion clearance [8]. The loss of infected cells could be the result of viral cytopathicity, immune elimination, or other processes such as apoptosis. Virion clearance could be the result of binding and entry into cells, immune elimination, or nonspecific removal by the reticuloendothelial

system.

We assumed that ritonavir does not affect the survival or rate of virion production of infected cells, and that after the pharmacological delay, all newly produced virions are noninfectious. However, infectious virions produced before the drug effect are still present until they are cleared. Therefore, after treatment with ritonavir,

$$\frac{dT^*}{dt} = kV_1 T - \delta T^* \tag{3}$$

$$\frac{dV_1}{dt} = -cV_1 \tag{4}$$

$$\frac{dV_{NI}}{dt} = N\delta T^* - cV_{NI} \tag{5}$$

where V_1 is the plasma concentration of virions in the infectious pool [produced before the drug effect; $V_1(t = 0) = V_0$], V_{NI} is the concentration of virions in the noninfectious pool [produced after the drug effect; $V_{NI}(t = 0) = 0$], and $t = 0$ is the time of onset of the drug effect. In our analyses, we assumed that viral inhibition by ritonavir is 100%, although the model can be generalized for nonperfect drugs [9].

Assuming that the system is at quasi steady state before drug treatment [10] and that the uninfected cell concentration T remains at approximately its steadystate value, T_0, for 1 week after drug administration [1, 5], we find from Equations 3 through 5 that the total concentration of plasma virions, $V = V_1 + V_{NI}$, varies as

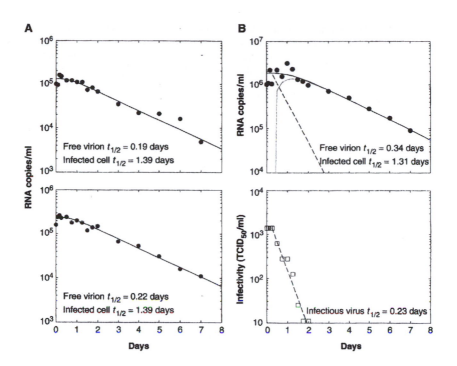

Fig. 1 (A) Plasma concentrations (copies per milliliter) of HIV-1 RNA (circles) for two representative patients (upper panel, patient 104; lower panel, patient 107) after ritonavir treatment was begun on day 0. The theoretical curve (solid line) was obtained by nonlinear least squares fitting of Equation 6 to the data. The parameters c (virion clearance rate), δ (rate of loss of infected cells), and V_0 (initial viral load) were simultaneously estimated. To account for the pharmacokinetic delay, we assumed $t = 0$ in Equation 6 to correspond to the time of the pharmacokinetic delay (if measured) or selected 2, 4, or 6 hours as the best-fit value (see Table 1). The logarithm of the experimental data was fitted to the logarithm of Equation 6 by a nonlinear least squares method with the use of the subroutine DNLS1 from the Common Los Alamos Software Library, which is based on a finite difference Levenberg-Marquardt algorithm. The best fit, with the smallest sum of squares per data point, was chosen after eliminating the worst outlying data point for each patient with the use of the jackknife method. (B) Plasma concentrations of HIV-1 RNA (upper panel; circles) and the plasma infectivity titer (lower panel; squares) for patient 105. (Top panel) The solid curve is the best fit of Equation 6 to the RNA data; the dotted line is the curve of the noninfectious pool of virions, $V_{NI}(t)$; and the dashed line is the curve of the infectious pool of virions, $V_1(t)$. (Bottom panel) The dashed line is the best fit of the Equation for $V_1(t)$ to the plasma infectivity data. TCID$_{50}$, 50% tissue culture infectious dose.

$$V(t) = V_0 \exp(-ct) + \frac{cV_0}{c - \delta} \times$$

$$\left\{ \frac{c}{c - \delta} [\exp(-\delta t) - \exp(-ct)] - \delta t \exp(-ct) \right\} \tag{6}$$

which differs from the Equation derived by Wei *et al.* ([2]; see [11]). Allowing T to increase necessitates the use of numerical methods to predict $V(t)$ but does not substantially alter the outcomes of the analyses given below [12].

Using nonlinear regression analysis Figure 1, we estimated c and δ for each of the patients by fitting Equation 6 to the plasma HIV-1 RNA measurements (Table 1) [12]. The theoretical curves generated from Equation 6, using the best-fit values of c and δ, gave an excellent fit to the data for all patients (see Figure 1 for examples). Clearance of free virions is the more rapid process, occurring on a time scale of hours. Values of c ranged from 2.06 to 3.81 day^{-1}, with a mean of 3.07 plus minus 0.64

day^{-1} (Table 1). The corresponding $t_{1/2}$ values for free virions ($t_{1/2} = \ln 2/c$) ranged from 0.18 to 0.34 days, with a mean of 0.24 ± 0.06 days (~6 hours). Confirmation of the virion clearance rate was obtained from an independent experiment that measured by quantitative cultures [13] the rate of loss of viral infectivity in plasma for patient 105 (Figure 1B). The loss of infectious virions occurred by first-order decay, with a rate constant of 3.0 day^{-1}, which is within the 68% confidence interval of the estimated c value for that patient (Table 1).

At steady state, the production rate of virus must equal its clearance rate, cV. Using the estimate of c and the pretreatment viral concentration V_0, we obtained an estimate for the rate of virion production before ritonavir administration. Each patient's plasma and extracellular fluid volumes were estimated on the basis of body weight. Total daily virion production and clearance rates ranged from 0.4×10^9 to 32.1×10^9 virions per day, with a mean of 10.3×10^9 virions per day released into the extracellular fluid (Table 1) [14]. The rate of loss of virus-producing cells, as estimated from the fit of Equation 6 to the HIV-1 RNA data, was slower than that of free virions. Values of delta ranged from 0.26 to 0.68 day^{-1}, with a mean of 0.49 ± 0.13 day^{-1}; the corresponding $t_{1/2}$ values were 1.02 to 2.67 days, with a mean of 1.55 ± 0.57 days (Table 1). A prediction of the kinetics of virus-producing cells can be obtained by solving Equation 3 [15].

Several features of the replication cycle of HIV-1 in vivo could be discerned from our analysis. Given that c and δ represent the decay rate constants for plasma virions and productively infected cells, respectively, then $1/c$ and $1/\delta$ are the corresponding average life-spans of these two compartments. Thus, the average life-span of a virion in the extracellular phase is 0.3 ± 0.1 days, whereas the average life-span of a productively infected cell is 2.2 ± 0.8 days (Table 2). The average viral generation time τ is defined as the time from the release of a virion until it infects another cell and causes the release of a new generation of viral particles; hence, τ should equal the sum of the average life-span of a free virion and the average life-span of a productively infected cell. This relation, $\tau = 1/c + 1/\delta$, can be shown formally (Table 2). The average value of τ for the patients was 2.6 ± 0.8 days (Table 2).

By a heuristic procedure, we found minimal estimates for the average duration of the HIV-1 life

cycle and of its intracellular or eclipse phase (from virion binding to the release of the first progeny). The duration of the HIV-1 life cycle, S, is defined as the time from the release of a virion until the release of its first progeny virus; we estimated S by the lag in the decay of HIV-1 RNA in plasma (Figure 1) after the pharmacologic delay (Table 1) is subtracted. The shoulder in the RNA decay curve is explained by the fact that virions produced before the pharmacologic effect of ritonavir are still infectious and capable of producing, for a single cycle, viral particles that would be detected by the RNA assay. Thus, the drop in RNA concentration should begin when target cells interact with drug-affected virions and do not produce new virions. These "missing virions" would first have been produced at a time equal to the minimum time for infection plus the minimum time for production of new progeny. The estimated values for S were quite consistent for the five patients, with a mean duration of 1.2 ± 0.1 days (Table 2). In steady state, $1/c = 1/NkT_0$ is the average time for infection (Table 2, legend); if this average time is assumed to be greater than the minimal time for infection, then a minimal estimate of the average duration of the intracellular phase of the HIV-1 life cycle is given by $S - (1/c) = 0.9$ days [16].

Previous studies that used potent antiretroviral agents to perturb the quasi steady state in vivo provided a crude estimate of the $t_{1/2}$ of viral decay in which the life-span of productively infected cells could not be separated from that of plasma virions [1, 2]. Our results show that the average life-span of a productively infected cell (presumably an activated CD4 lymphocyte) is 2.2 days; thus, such cells are lost with an average $t_{1/2}$ of ~1.6 days (Figure 2). The life-spans of productively infected cells were not markedly different among the five patients (Table 2), even though individuals with low CD4 lymphocyte counts generally have decreased numbers of virus-specific, major histocompatibility complex class I-restricted cytotoxic T lymphocytes [17].

The average life-span of a virion in blood was calculated to be 0.3 days. Therefore, a population of plasma virions is cleared with a $t_{1/2}$ of 0.24 days; that is, on average, half of the population of plasma virions turns over approximately every 6 hours (Figure 2). Because our analysis assumed that the antiviral effect of ritonavir was complete and that target cells did not recover during treatment, our esti-

Table 2.   Summary of virion life-span $(1/c)$ and infected cell life-span $(1/\delta)$, duration of the viral life cycle (S) and of the intracellular phase $[S - (1/c)]$, and average viral generation time (τ) for the five patients. The values for S and for the minimal estimate of $S - (1/c)$ were obtained by a heuristic procedure and should be viewed as rough estimates (S was estimated by the length of the shoulder on graphs of RNA copies versus time, as in Figure 1). Confidence intervals for τ were obtained by a bootstrap method on a model with τ and $1/\delta$ as parameters. SDs reflect the differences between patients and not the accuracy of the estimates. The fate of a large population of virions was followed to estimate the in vivo value of τ. For a system in quasi steady state, the average generation time can be defined as the time required for V_0 particles to produce the same number of virions in the next generation. After a protease inhibitor is administered, all newly produced virions are assumed to be noninfectious. To keep track of the number of noninfectious particles, we assumed for the purposes of this calculation that noninfectious particles are not cleared and act as a perfect marker, recording the production of virions after one round of infection. Thus, from Equation 5, dV_{NI}/dt equals $N\delta T^*$. We also assumed that before the drug is given, there are no infected cells [that is, $T^*(0) = 0$], so that only new infections are tracked. Under these circumstances, τ is the average time needed for V_0 virions to produce V_0 noninfectious particles after ritonavir administration. After treatment, no further infectious particles are produced and hence the number of infectious particles V_1 declines exponentially [that is, $V_1(t) = V_0 \exp(-ct)$, where $t = 0$ is the time at which the drug takes effect after pharmacokinetic delays]. The existing infectious particles infect cells, and the number of infected cells T^* varies as given by the solution of Equation 3, with the initial condition $T^*(0) = 0$. At any given time t, the mean number of virions produced from the initial V_0 virions is $V_{NI}(t) = P(t)V_0$, where $P(t)$ is the (cumulative) probability that a virion is produced by time t. The probability density of a virus being produced at time t is $p(t) = dP/dt$, and thus the average virion production time τ equals $\int_0^\infty tp(t)\, dt$. Hence, $\tau = (1/V_0) \int_0^\infty t(dV_{NI}/dt)\, dt = (1/V_0) \int_0^\infty tN\delta T^*\, dt$. Substituting the solution of Equation 3 for T^* and integrating yields $\tau = [(1/\delta) + (1/c)]$. Because the system is at steady state [10], $c = NkT(0)$, thus the clearance rate and the rate of new cell infection are coupled. Thus, the viral generation time can also be viewed as the time for an infected cell to produce N new virions—that is, its life-span $(1/\delta)$ plus the time for this cohort of N virions to infect any of the T_0 uninfected target cells $[1/(NkT_0)]$.

Patient	1/c (days)	1/δ (days)	S (days)	S – (1/c) (days)	τ (days)
102	0.3	3.8	1.2	0.9	4.1
103	0.4	1.5	1.0	0.6	1.8
104	0.3	2.0	1.2	0.9	2.3
105	0.5	1.9	1.3	0.8	2.4
107	0.3	2.0	1.2	0.8	2.3
Mean	0.3	2.2	1.2	0.9	2.6
±SD	0.1	0.8	0.1	0.1	0.8

mates of the virion clearance rate and infected cell loss rate are minimal estimates [12, 16]. Consequently, the true virion $t_{1/2}$ may be shorter than 6 hours. For example, Nathanson and Harrington [18] found that monkeys clear the Langat virus from their circulation on a time scale of ~30 min. Thus, the total number of virions produced and released into the extracellular fluid is at least 10.3×10^9 particles per day [14]; this rate is about 15 times our previous minimum estimate [1]. At least 99% of this large pool of virus is produced by recently infected cells [1, 2] (Figure 2). At quasi steady state, the virion clearance rate cV equals the virion production rate $N\delta T^*$. Because c has similar values for all patients studied (Table 1), the degree of plasma

viremia is a reflection of the total virion production, which in turn is proportional to the number of productively infected cells T^* and their viral burst size N. The average generation time of HIV-1 was determined to be 2.6 days, which suggests that ~140 viral replication cycles occur each year, about half the number estimated by Coffin [19].

It is now apparent that the repetitive replication of HIV-1 (left side of Figure 2) accounts for ≥99% of the plasma viruses in infected individuals [1, 2, 19], as well as for the high destruction rate of CD4 lymphocytes. The demonstration of the highly dynamic nature of this cyclic process provides several theoretical principles to guide the development of treatment strategies:

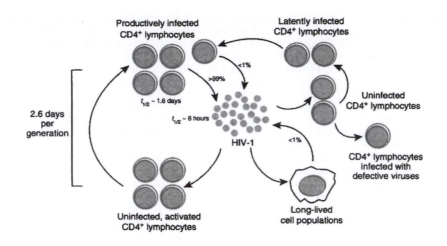

Fig. 2 Schematic summary of the dynamics of HIV-1 infection in vivo. Shown in the center is the cell-free virion population that is sampled when the viral load in plasma is measured.

1) An effective antiviral agent should detectably lower the viral load in plasma after only a few days of treatment.

2) On the basis of previous estimates of the viral dynamics [1, 2] and data on the mutation rate of HIV-1 (3.4×10^{-5} per base pair per replication cycle) [20] and the genome size (104 base pairs), Coffin has cogently argued that, on average, every mutation at every position in the genome would occur numerous times each day [19]. The larger turnover rate of HIV-1 described in our study makes this type of consideration even more applicable. Therefore, the failure of the current generation of antiviral agents, when used as monotherapy, is the inevitable consequence of the dynamics of HIV-1 replication. Effective treatment must, instead, force the virus to mutate simultaneously at multiple positions in one viral genome by means of a combination of multiple, potent antiretroviral agents. Moreover, because the process of producing mutant viruses is repeated for ~140 generations each year, early and aggressive therapeutic intervention is necessary if a marked clinical impact is to be achieved [21].

3) From our study and previous reports [1, 2, 5], it is now clear that the "raging fire" of active HIV-1 replication (left side of Figure 2) could be put out by potent antiretroviral agents in 2 to 3 weeks. However, the dynamics of other viral compartments must also be understood. Although they contribute ≤1% of the plasma virus, each viral compartment (right side of Figure 2) could serve as the "ember" to reignite a high rate of viral replication when the therapeutic regimen is withdrawn. In particular, we must determine the decay rate of long-lived, virus-producing populations of cells such as tissue macrophages, as well as the activation rate of cells latently carrying infectious proviruses. This information, someday, will enable the design of a treatment regimen to block de novo HIV-1 replication for a time sufficient to permit each viral compartment to "burn out."

REFERENCES

1. D. D. Ho *et al.*, *Nature* **373**, 123 (1995).
2. X. Wei *et al.*, *ibid.*, p. 117.
3. M. A. Nowak *et al.*, *ibid.* **375**, 193 (1995).
4. D. J. Kempf *et al.*, *Proc. Natl. Acad. Sci. U.S.A.* **92**, 2484 (1995).
5. M. Markowitz *et al.*, *N. Engl. J. Med.* **333**, 1534 (1995).
6. C. Pachl *et al.*, *J. Acquired Immune Defic. Syndr.* **8**, 446 (1995); Y. Cao *et al.*, *AIDS Res. Hum. Retroviruses* **11**, 353 (1995).
7. D. L. Winslow and M. J. Otto, *AIDS* **9** (suppl. A), S183 (1995).
8. The rate of virion production is expressed as the product $N\delta$ to convey either of two possibilities: (i) HIV-1 is produced continuously

at an average rate given by the total production of virus particles, N, divided by the cell life-span, $1/\delta$; or (ii) N virions are produced in lytic bursts occurring at the rate of cell death, delta.

9. The effect of a nonperfect drug can be modeled by simply adding the term $(1 - \eta)N\delta T^*$ to Equation 4 and multiplying the first term in Equation 5 by the factor η, where η represents the drug's inhibitory activity [for example, $\eta = D/(D + EC_{50})$, where D is the plasma concentration of drug and EC_{50} is the concentration required for 50% effectiveness].

10. In quasi steady state, $dT^*/dt = 0$ and $dV/dt = 0$. Thus, $kV_0T_0 = \delta T_0^*$ and $N\delta T_0^* = cV_0$, where the subscript 0 indicates a steady-state value. Combining these equations yields $NkT_0 = c$. Each virion infects cells at rate kT_0, with each infection leading on average to the production of N new virions. At steady state, the production of new virions at rate NkT_0 must balance the virion clearance at rate c.

11. Equation 6 differs from $V(t) = [V_0/(c - \delta)][c \exp(-\delta t) - \delta \exp(-ct)]$, the Equation introduced by Wei *et al.* [2] for analysis of the effects of drug treatment on viral load. Their analysis was based on Equation 1 and Equation 2 and the assumption that no new infections occur after drug treatment ($k = 0$ after treatment). Equation 6 is a new model appropriate for protease inhibitors, which do not prevent infections arising from preexisting mature infectious virions. Because of the symmetry between c and δ in the Wei *et al.* equation, the virion clearance rate cannot be distinguished from the infected cell death rate by data fitting.

12. Because our parameter estimates are based on the assumption of complete inhibition of the production of new infectious virions and no increase in target cells, we expect our parameter estimates to be minimal estimates. Generalizing our model to relax these two assumptions, we can show that δ is always a minimal estimate and that, with target cell growth, c is typically a minimal estimate. We tested how the estimates of c and δ depend on the assumption that ritonavir is 100% effective as follows: We generated viral load data assuming different drug effectivenesses with $c = 3$ and $\delta = 0.5$. With these "data," we used our fitting procedure to estimate c and δ

under the assumption that the drug is 100% effective. For data generated with $\eta = 1.0$, 0.99, 0.95, and 0.90, we estimated $c = 3.000$, 3.003, 3.015, and 3.028, respectively, and $\delta = 0.500$, 0.494, 0.470, and 0.441, respectively. Thus, our estimate of c remains essentially unchanged, whereas that of δ is a slight underestimate (for example, for $\eta = 0.95$, $\delta = 0.47$ rather than the true 0.5). Consequently, if a drug is not completely effective, cell life-spans may be somewhat less than we estimate. If the target cells are allowed to increase by the maximum factor observed in the five patients (that is, fivefold), we find that the derived values of c and δ are minimal estimates. Thus, for example, for data generated with $\eta = 1$, with $c = 3.00$ and $\delta = 0.500$, we find that our fitting procedure yields estimates of $c = 2.76$ and $\delta = 0.499$.

13. D. D. Ho, T. Moudgil, M. Alam, *N. Engl. J. Med.* **321**, 1621 (1989).

14. Virions that are not released into the extracellular fluid are not included in this estimate. Thus, the total production in the body is even larger.

15. The solution to Equation 3 is

$$T^*(t) = \frac{T_0^*}{c - \delta}[c \exp(-\delta t) - \delta \exp(-ct)]$$
$$(7)$$

If cellular RNA data were obtained, this Equation could be fitted to those data, and the parameter estimates for c and delta could be verified for consistency with the viral kinetics.

16. In principle, more accurate estimates of the duration of the intracellular or eclipse phase of the viral life cycle can be obtained with a model that explicitly includes a delay from the time of infection until the time of viral release. For example, Equation 2 can be replaced by

$$\frac{dV}{dt} = N\delta \int_o^\infty T^*(t - t')w(t')\,dt' - cV \quad (8)$$

where $w(t')$ is the probability that a cell infected at time $t - t'$ produces virus at time t. Explicit solutions to our model, with $w(t')$ given by a gamma distribution, will be published elsewhere (A. S. Perelson *et al.*, in preparation). Alternatively, if virally producing cells T_p rather than infected cells T^* are to be tracked, Equation 1 can be replaced by

$$\frac{dT_p}{dt} = kT \int_o^\infty V(t - t')\omega(t')\, dt' - \delta T_p \quad (9)$$

Models of this type can also be solved explicitly when $\omega(t')$ is given by a gamma function. M. Nowak and A. Herz (personal communication) have solved this model for the case where $\omega(t')$ is a delta function, in which case the delay simply adds to the pharmacologic delay and Equation 6 is regained after this combined delay. Analysis of current data by nonlinear least squares estimation has so far not allowed accurate simultaneous estimation of c, δ, and the intracellular delay. However, the qualitative effect of including the delay in the model is to increase the estimate of c, which is consistent with our claim that the values of c in Table 1 are minimal estimates. Higher values of c (hence lower values of $1/c$) will lead to increased estimates of the intracellular delay, $S - (1/c)$. Thus our estimate of the duration of the intracellular phase, as derived above and given in Table 2, is still a minimal estimate.

17. A. Carmichael, X. Jin, P. Sissons, L. Borysiewicz, *J. Exp. Med.* **177**, 249 (1993).
18. N. Nathanson and B. Harrington, *Am. J. Epidemiol.* 85, 494 (1966).
19. J. M. Coffin, *Science* **267**, 483 (1995).
20. L. M. Mansky and H. M. Temin, *J. Virol.* **69**, 5087 (1995).
21. D. D. Ho, *N. Engl. J. Med.* **333**, 450 (1995).
22. B. Efron and R. Tibshirani, *Stat. Sci.* **1**, 54 (1986).
23. We thank the patients for participation; A. Hurley, Y. Cao, and scientists at Chiron for assistance; B. Goldstein for the use of his nonlinear least squares fitting package; and G. Bell, T. Kepler, C. Macken, E. Schwartz, and B. Sulzer for helpful discussions and calculations. Portions of this work were performed under the auspices of the U.S. Department of Energy. Supported by Abbott Laboratories, grants from NIH (NO1 AI45218 and RR06555) and from the New York University Center for AIDS Research (AI27742) and General Clinical Research Center (MO1 RR00096), the Aaron Diamond Foundation, the Joseph P. Sullivan and Jeanne M. Sullivan Foundation, and the Los Alamos National Laboratory Directed Research and Development program.

READING 2.2

Left Snails and Right Minds

Commentary: For the purposes of this class, the article makes the observation that most species of snails consist of animals whose shells curl almost exclusively in one direction. For example, Gould points out that in India, the conch shell Turbinella pyrum is venerated as a symbol of Vishnu, and the exceedingly rare left-handed shells are worth their weight in gold. In this article, Gould points out that most snail species are almost exclusively right handed.

What is the simplest mathematical model that predicts the appearance of such a bias in population handedness? Here is one such model: This model (like all models) is built on some reasonable biological assumptions that are then translated into mathematics.

- Assume that the probability of a snail of type A (= left or right) breeding with one of type B is proportional to the product of the number of type A times the number of type B.

- Assume that two lefts always produce a left offspring, two rights always produce a right offspring, and a left-right wedding produces left and right offspring with equal probability.

All of these assumptions are open to question. You should question them! In particular, the first assumption has the following implication: Given a mate choice between a snail of the same handedness, or of the opposite handedness, a snail will choose its own kind to mate with twice as often as the opposite kind. (Do you see why?) Can this be a reasonable assumption? In any event, the verification of these assumptions requires observations of real snail populations.

Whether you believe the assumptions or not, note that they have the following effect: When there are more left than right-handed snails, there will be more than the proportionate number of left offspring, and vice versa. For example, if the ratio of left to right is, say 2 : 1, the ratio of left babies to right babies will be roughly 5 : 2.

Here, (Left × Left) + 1/2(Left × Right) = (2 × 2) + (2 × 1)/2 = 5, while 1/2(Left × Right) + (Right × Right) = (2 × 1)/2 + (1 × 1) = 2. Thus, there is a sort of positive feedback that promotes left-handed snails over right when there are more left than right, and vice versa.

Here is how to model this situation: Let $p(t)$ denote the percentage of left-handed conch shells at time t. Here, we can assume that $t = 0$ is a very long time ago (hundreds of millions of years, perhaps). An equation that has behavior that is roughly consistent with the preceding assumptions is

$$\frac{dp}{dt} = \alpha p(1 - p)(p - 1/2),$$

where α is some positive constant.

Note that the rate of change of p is negative when $p < 1/2$ and positive when $p > 1/2$. Also, the rate of change of p is zero when $p = 0$ and when $p = 1$, and also when $p = 1/2$. (You should ask yourself why this is reasonable.) The preceding equation is, perhaps, the simplest equation you can write for $p(t)$ that has zero when $p = 0$ or 1 and $\frac{dp}{dt}$ negative or positive depending on whether p is less than or greater than 1/2. Such was my motivation for writing this equation.

Left Snails and Right Minds

Natural History, April 1995, pages 10–18.
Stephen Jay Gould

> What immortal hand or eye
> Could frame thy fearful symmetry?

William Blake's familiar inquiry about the creation of tigers raises a vital question that we may pose literally, although the poet's intention may have been more metaphorical. Why does symmetry, particularly our own bilateral style of mirror images around a central axis, predominate among animals of complex anatomical design? Why do we come in equivalent right and left halves? And why do we get so fascinated by the minor departures, usually more of function than overt form, that loom so large in our culture: the predominance of right-handedness and the difference between "right" and "left" brains?

A few major groups of organisms do not present a basically bilateral symmetry, including my own favorite subject for research, the gastropods, or snails. The soft body of a snail is tolerably bilateral when pulled from the shell and stretched out, but the animal houses this body in a shell built by winding a tube in one direction around an axis of coiling. The snail shell may therefore be the most familiar nonbilateral form among "higher" animals.

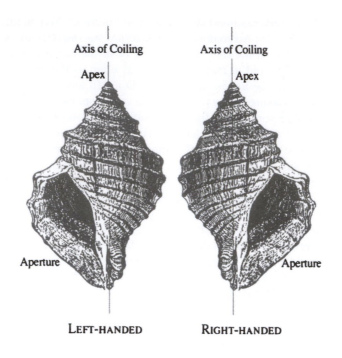

Axis of Coiling

Apex

Aperture

LEFT-HANDED

Axis of Coiling

Apex

Aperture

RIGHT-HANDED

Fig. 1

A tube can be wound around a vertical axis in either of two directions, designated as right- and left-handed. If we hold a snail in our conventional position, with the apex at the top and the aperture (or opening for the body) at the bottom, then we call the direction of coiling right-handed if the aperture lies to the right of the axis of coiling when we view the specimen face to face, and left-handed if the aperture lies to the left of the axis of coiling. (All this should be much clearer in the illustration above than in any words I can supply.)

But this naming is truly arbitrary, for snails know nothing about apex up and aperture down (in life, most snails carry their shells more or less horizontal to the ground). If we draw the specimen apex down (as French scientific illustrators have always done), then the apertures of right-handed specimens open to the left of the axis of coiling.

In India, for example, the conch shell *Turbinella pyrum* is venerated as a symbol of Vishnu. (In the *Bhagavad-Gita*, Vishnu, in the form of his most celebrated avatar, Krishna, blows his sacred conch shell to call the army of Arjuna into battle.) The exceedingly rare left-handed specimens of this shell are particularly treasured and used to sell for their weight in gold. But Hindus interpret the apex as the bottom of the shell and therefore call this rare form right-handed. Perhaps they treasure these rare shells because only these specimens, in the Indian version of an arbitrary decision, match the style of dominant handedness in human beings (and, I suppose, anthropomorphic deities).

A purist might forgive snails for departure from the bilateral paradigm if only they honored an even more inclusive symmetry by growing right- and left-handed spirals in equal numbers. But snails remain twisted and awry on this criterion as well–for right-handed shells vastly outnumber lefties not only in the sacred conch of India but in virtually all species and groups. Right-handed shells are called dextral, from the Latin *dexter*, meaning "right," and memorialized in our language by a host of prejudicial terms invented by the right-handed majority to honor their predominance. Right is dexterous, not to mention "correct" in many languages—awright buddy. The law, by the way, is *droit* in French and *Recht* in German, both meaning "right." (The language police will never regulate these essays, but it remains fair and historically interesting to point out that the "rights of man," noble as the sentiment may be, embody two linguistic prejudices of unfairly dominant groups.) Left-handed shells are called sinistral, from the Latin *sinister*, meaning "left"—also denigrated in our languages as sinister or *gauche*, a French lefty. I shall, for the rest of the essay, use this terminology by calling right-handed shells dextral, and lefties sinistral. I also can't help wondering if we didn't make our initial arbitrary decision to call a snail's apex "up" because this orientation would then allow us to designate this overwhelmingly more common direction of coiling as "right."

The vast majority of forms grow a dextral shell, although a few sinistral specimens have been found in most species. For example, in *Cerion*, the West Indian land snail that forms the subject of my own technical research, only six sinistral specimens have ever been found, out of millions examined (while, as stated above, a lefty *Turbinella* in India was literally worth its weight in gold). A few species grow exclusively or predominantly sinistral shells, but related species of the same group are usually dextral. We often exact a price from these rare sinistrals by giving them names to match their apostasy—as in *Busycon contrarium* or *B. perversum*, the technical monikers variously awarded to the most common sinistral species of northern Atlantic waters. A few groups of species (notably the family Clausiliidae) are predominantly sinistral but, again, all closely related lineages are dextral. In short, dextral snails greatly predominate (at a far higher frequency than human righties versus lefties) at all levels: individuals within a species, species within a lineage, and lineages within larger groups.

At this point, astute and inquisitive readers will be asking the obvious questions, "Why? What conceivable advantage does dextrality hold over coiling in the other direction?" I can only report that these inquiries are both appropriate and fascinating—and that we don't have a clue about the answers. (I would not even assume that the questions should be posed in terms of putative advantages. The two modes might be entirely equivalent in functional terms, with dominant dextrality only a historical legacy of what happened to arise first.) I'm sorry to wimp out on such interesting questions, but I can at least quote, on the same subject and to the same point, that greatest of all prose stylists in nat-

ural history, D'Arcy Wentworth Thompson (from his book *On Growth and Form*, first published in 1917 and still vigorously in print): "But why, in the general run of shells, all the world over, in the past and in the present, one direction of twist is so overwhelmingly commoner than the other, no man knows."

This essay, instead, shall take another turning on the subject of directionality in coiling—namely, the history of illustrations for snail shells in zoological treatises. Let me begin with a figure that I first considered both anomalous and amusingly in error. The plate reproduced below (Fig. 2) is from a famous work in natural history, published in 1681 by one of Britain's finest physicians and zoologists, Nehemiah Grew: *Musaeum Regalis Societatis, or a description of the natural and artificial rarities belonging to the Royal Society, whereunto is subjoyned the comparative anatomy of stomachs and guts.* (They did love long titles back then, and we will ignore the appendix, with its remarkable illustrations of vertebrate intestines, all stretched out and circling the pages.)

Note that all but one of these shells are sinistral in Grew's engraving. The exception, shown at the bottom, is conventionally dextral. Has the world turned? Those shells are labeled "wilk," or "whelk" in our modern spelling (a common name for conchlike shells)–and nearly all whelks are dextral, including the species shown here. The exception, drawn dextrally, gives the story away by the name imprinted above: "Inverted Wilk Snail." In other words, the shell labeled "inverted" is, in life, a rare sinistral named according to an old tradition for derogatory designation of the unusual. (At least "inverted" seems milder than "perverted," as in *Busycon perversum*.)

Obviously, Grew printed his snails in mirror image from their actual constitution. I initially assumed that Grew had committed a simple error and laughed at his fellowship with snail men throughout the history of illustration, for we are still making the same mistake today. In the current version, an offspring of modern technology, a snail may appear with reversed coiling because the photograph has been made from a negative inadvertently turned over before printing. Any expert paying explicit attention will notice the error, but we fallible mortals often let something this global slip by—for a reversed snail doesn't look grievously wrong if you don't have your eye and mind directly attuned

Square Wilk

Long Square Wilk

Thick Lipp'd Wilk

Triangular Wilk

Inverted Wilk Snail

Fig. 2 In an illustration from Nehimiah Grew's 1681 volume, only the bottom snail is shown coiling to the right.

to the issue of symmetry.

Any professional snail man can give you his list of embarrassments in this category. A dear late colleague, one of the world's leading experts on snails, published a beautiful wraparound dust jacket photo of reversed shells for the cover of his popular book. I must also admit (and how wonderfully unburdening after all these years of hiding such a shameful secret) that my own first publication on snails included several photographs of a newly discovered protoconch (embryonic shell) of an important genus—all published from reversed negatives. (I received the sweetest and most diplomatic letter from a colleague asking me if these dextral shells really had sinistral protoconchs and urging me to publish separately on such an important

Fig. 3 Dextral-coiling species are consistently shown as sinistral in a 1719 edition of Michele Mercati's *Metallothea.*

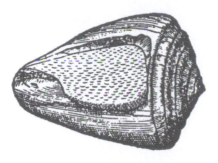

Fig. 4 A dextral snail, shown with sinistral coiling in *Musaeum Metallicum,* by sixteenth-century Italian naturalist Ulisse Aldrovandi.

finding, or suggesting that maybe, just maybe, I had made the old error of reversed printing.) Baseball players make a proper distinction between physical errors, which can happen to anyone anytime and should engender no shame, and mental errors—bonehead judgments, forgetting the rules—which should never occur. Ordinary and honorable errors of fact are unavoidable in science, a field that thrives on self-correction, and properly defines its own progress by such improvement. I have never written an essay, and never will, without this analogue of a physical error. But printing a snail backward is a mental error. No excuses possible.

So much for my first thoughts about Grew's mistake. But as soon as I remembered a scholar's first obligation—to drag oneself from judgment within a smug present, considered better, and to place oneself, so far as possible, into the life and of a person under consideration—I immediately realized that the issue could not be so simple. All media for printed illustration in pre-nineteenth-century treatises of natural history—woodblock printing, engraving on metal plates, lithography—require the initial production of an inverted image. That is, the engraver must carve a mirror image figure into his metal plate, so that the paper, placed atop the inked plate before pressing down to print, will receive the figure in proper orientation. Needless to say, all printers know this rule perfectly well; nothing could be more fundamental to their work.

Therefore, a printer who wants to engrave an ordinary dextral shell must carve a sinistral image upon his metal plate. Clearly, Grew's printer drew the snails onto his plate as he saw them, rather than

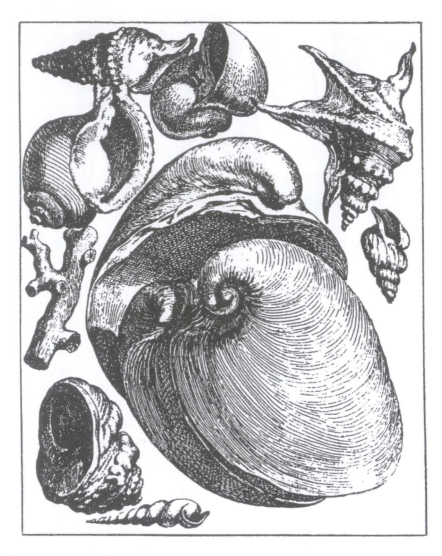

Fig. 5 Dextral snail species are depicted with left-handed coiling in a volume on fossil snails by seventeenth-century paleontologist Augustino Scilla.

reversed—and the result is an inverted image in the printed book: ordinary dextral snails look sinistral, and the lone sinistral looks like a conventional dextral.

But how did this happen? and why? This oddity cannot be the result of a simple fool's error of the baldest kind, for the engraver surely knew his rules and must have etched his letters and numbers in proper reversed order onto his plate, for all writing and numeration is correct in the printed version. Many scenarios suggest themselves, and we do not have enough evidence to decide: perhaps Grew supplied his printer with sketches already reversed but forgot to pass this information along; perhaps Grew provided all the sketches on a single sheet (without the words), and the printer then erred in pasting the sketch upon his plate recto rather than the proper verso. (I am assuming that engravers worked by affixing a sketch, drawn on transparent paper, directly onto their plate and then carving through.)

But we should also consider a hypothesis of a fundamentally different kind. Perhaps we should not be so quick to assume, from our arrogant present, that these "primitives" of the seventeenth century must have been making an error at par with boo-boos still occasionally committed by modern photography. Perhaps the reversed shells of Grew's illustrations are not errors at all, but representations of a convention then followed and now abandoned.

I shall defend this more generous alternative in concluding my essay, but I had not considered this solution when I first saw Grew's plate about ten years ago. I simply stored this little "fact" away in my mental file of oddities in natural history. I must have labeled this item "Grew's funny mistake," for I never considered the possibility that reversed snails could be anything but an error, however committed.

As their primary virtue and utility, such mental files can lurk in the brain (wherever and however this remarkable organ stores such information) without disturbing one's thinking and planning in any manner. The files just hang around, waiting for some trigger to transport them into consciousness. (I would, for this reason, defend such ancient practices as rote learning for the basic chronology of human history and for reading the classics, particularly Shakespeare and the Bible, with a view to memorizing key passages.) I love antiquarian books in natural history, and my eyes do inevitably wander, for professional reasons, to pictures of snails. Thus, my "Grew mistake" file has been accessed quite a few times during the past decade. But I never had any project in mind, and I had devised the wrong preliminary conclusion about Grew's reversals. In fact, it took three or four random repetitions to make the subject explicit as a worthy topic, to force a revision of my own initial error, and to perceive the larger theme about science and human perception that could convert such a trivium (the depiction of snail coiling) into a decent subject for an essay.

A few years after my reading of Grew's book, I purchased a copy of my personal favorite among beautiful and important works in natural history, Michele Mercati's *Metallotheca*. Mercati (1541–

1593), director of the Vatican botanical garden, also became curator of the papal collection of minerals and fossils organized under the aegis of the imperial pope Sixtus V, whose taxes impoverished the papal lands while building Rome in splendor. (I also love the man's name—the fifth instar of a guy named "six"; Sixtus I, a second-century figure, was the sixth bishop of Rome after Peter and took his name accordingly.) Mercati prepared a series of gorgeous engravings for a catalog of the Vatican collection, but this work never appeared in his lifetime (perhaps because Sixtus V died unexpectedly in 1590). But the plates hung around in the Vatican's vast storehouses for nearly a century and a half, until J. M. Lancisi finally published them, along with Mercati's text and many new engravings, in 1719 as the *Metallotheca*. (If a *bibliothèque* is a library, then a *metallothèque* is a collection of metals and other objects of the mineral kingdom.)

The *Metallotheca* contains numerous plates of fossil snails in a chapter called *Lapides Idiomorphoi* (or stones that look like living things—Mercati, along with many sixteenth-century scholars, did not interpret fossils as remains of organisms but as manifestations of "plastic forces" inherent in rocks). In all plates—so we are in the presence of a conscious generality, not an individual error—dextral snails appear as sinistral engravings.

But assumptions die hard, even if never founded on anything sensible. I couldn't call reversed printing a simple error any more, so I opted for the next line of defense within the bias of progress: I assumed that such indifference to nature's factuality must represent a curious archaism of the bad old days (for Mercati goes way back to the sixteenth century)—and thus not worthy of much intellectual attention. Again, I stored the observation on the back shelves, in the stacks of my mentatheca.

More random encounters since then have finally shaken up my false assumption, for I have noted sinistral illustrations of dextral shells again and again in works published before 1700. In fact, almost all snail illustrations from this period are reversed, so we must be observing a conscious convention, not an occasional error. By contrast, I have almost never seen a reversed illustration in works, say, from Linnaeus's time (early to mid-eighteenth century) onward, except as real and infrequent errors. Therefore, and interestingly, the obvious hypothesis that photography ushered in the change must be false. I simply do not know (but would dearly love to have the answer) why a convention of drawing snails in reversed coiling yielded to the conviction that we should depict them as we see them.

To shorten my chronicles of personal discovery, two further examples finally convinced me that older illustrations had drawn snails with reversed coiling on purpose. I first consulted as close to an "official" source as the sixteenth century can provide—the *Musaeum Metallicum* (another account of a major fossil and rock collection) by the Italian naturalist Ulisse Aldrovandi (1522–1605), who, in competition with his Swiss colleague Konrad Gesner (1516–1565), wrote the great compendia that pulled together all available knowledge about animals—ancient and modern, story and observation, myth and reality, human use and natural occurrences. My edition of Aldrovandi's posthumous work on fossils dates from 1648 and illustrates all snails as sinistrally coiled, although the figures depict dextral species.

If the standard source still doesn't completely convince, then seek an author with special expertise. I therefore consulted my copy of one of the great works in late-seventeenth-century paleontology, *De Corporibus Marinis Lapidescentibus* (On Petrified Marine Bodies) by Augustino Scilla (my Latin edition dates from 1747, but Scilla first published his work in Italian in the 1690s). I decided on Scilla as a final test case because he was a painter by trade, a leading figure on the *seicento* in Sicily, and he engraved his own plates. All his snails are dextral species, and all are engraved with sinistral coiling. Clearly, if standard sources and noted artists all drew snails in mirror image from their natural occurrence, they must have been following a well-accepted convention of the time, not making an error.

But why would earlier centuries have adopted a convention so foreign to our own practices? Why would these older illustrators have chosen to depict specimens in mirror image when they surely knew the natural appearance of these shells? Did they devise this convention in order to make life easier for a profession founded on the principle that one carves in reverse in order to print in the desired orientation? But if so, what aid could be provided by the convention of printing snails in reverse? I suppose that an engraver could then paste a picture directly on his plate and cut through with maximal visibility (whereas the usual technique forced him

to invert the drawing before affixing it to the plate, thus making him view the sketch through the backside of the paper; but papers of adequate transparency must have been available, and I wonder if the usual technique really imposed any great hardship. Or did engravers mechanically copy an original figure in reverse orientation and then paste this copy onto the plate? If so, a convention permitting reversed printing would allow engravers to omit a time-consuming step).

Whatever the reason, the very existence of the convention does, I think, teach us something important: the conceptual world of pre-eighteenth-century zoology must have accorded little importance to the orientation of a shell. These men were not stupid, and they were not primitive. If they were willing to sacrifice what we would call "accuracy" for some gain in ease of production (or for some other reason not now apparent to us), then they must have held a notion of accuracy quite different from ours. The recovery of "fossil" thought patterns from such intriguing hints as this small, but previously unnoted, change in a practice of illustration provides the kind of intellectual lift that keeps scholars going.

The greatest impediment to such recovery—one that infested my own first thoughts on this issue and precluded any movement toward a proper solution after I had made my initial and accurate observations—lies in lamentable habits imposed by the twinned biases of progress and objectivity. We assume that we now do things better than at any time in the past, and that our improvements record increasing objectivity in shedding old prejudices and learning to view the world more accurately. We therefore interpret our predecessors, especially when their views differ from ours, as weighted down by biases and lacking in data—in short, as pretty darned incompetent compared with us. We therefore do not take them seriously, and we view their differences from us as crudity and error. Thus, we cannot understand the interesting reasons for historical changes in practice, and we cannot recover the older systems, coherent in their own terms (and often based on a fascinatingly different philosophy of nature), that made the earlier procedures so reasonable.

The key, in this case, lies in realizing that an apparent error in past practice represents a convention, now foreign to our concepts but evidently pursued for conscious reasons by our predecessors. We must still overcome one obstacle in striving to view the past more sympathetically (thereby gaining insight into present styles of thinking). We might understand that printing snails in reverse represented a convention, not an error, but still hold (via the bias of progress) that the history of changing conventions must record a pathway to greater accuracy in representation. We might, for example, hold (in utmost naivete) that our predecessors once drew what they wanted to see, whereas we now photograph what actually is.

Two arguments should convince us that history marks no path from stilted convention to raw accuracy. First, I have talked with many professional photographers, and all recognize as a canard the old claim that their technology gave us objective precision, where only subjective drawing reigned before. Technological improvements in photography do make older styles of prevarication less possible. (In my book *The Mismeasure of Man*, I showed how one pioneer eugenicist doctored his pictures of supposedly retarded people to make them look more benighted. His retouchings are so crude that no one today, with a lifetime of experience in looking at good photographs, would be fooled. But he got away with his ruse in 1912, for few people then had enough experience to recognize a doctored photo, and retouching represented an accepted art for repairing crude shots in any case.) But other technological improvements make all manner of fooling around with photos ever more possible and elaborate (just think of Woody Allen as Zelig, or Tom Hanks as Forrest Gump, artificially incorporated into the great events of twentieth century history by trick photography). Who can balance the gains and losses? Why speak of these changes as gains and losses at all? We have not dispensed with conventions for accuracy; we have only adopted different conventions.

Second—and the clinching argument that made me decide to write this essay—we have not, even today, abandoned all conventions for reversed illustration. In fact, one highly prestigious and technologically "cutting edge" field continues to present upside-down photographs, just as our forebears drew their snails right to left. How many readers realize that conventional photos of moons and planets are upside-down? (If you doubt my claim, compare the full moon on a clear night with the photograph in your old astronomy text.) Modern astronomers, of course, are no more fools than the old snail illustrators. They present photographs upside down to match what one sees in conventional refracting telescopes. (Or, rather, they print the photos as they are taken through such telescopes. Is this convention any different from carving a snail as one sees it onto an engraving plate and then producing the paper image in reverse?)

Clearly, astronomers feel that the trouble taken to print photographs from refracting telescopes upside down (thus rendering the object as it exists in the sky) would not be worth the gain. In fact, one might argue that reversing the photo would sow confusion rather than provide benefit, for with the exception of our moon, we cannot see features of other moons and planets with our naked eye, and therefore know these bodies primarily as seen through refracting telescopes—that is, upside down. I must suppose that the old snail illustrators also regarded the direction of coil as unimportant for illustration—and I would like to know why. I would also like to know what triggered the change from an accepted convention to a no-no.

I shall not, either in this forum or anywhere, resolve the age-old riddle of epistemology: how can we "know" the "realities" of nature? I will, rather, simply end by restating a point well recognized by philosophers and self-critical scientists, but all too often disregarded at our peril. Science does progress toward more adequate understanding of the empirical world, but no pristine, objective reality lies "out there" for us to capture as our technologies improve and our concepts mature. The human mind is both an amazing instrument and a fierce impediment—and the mind must be interposed between observation and understanding. Thus, we will always "see" with the aid (or detriment) of conventions. All observation is a partnership between mind and nature, and all good partnerships require compromise. The mind, we trust, will be constrained by a genuine external reality; this reality, in turn, must be conveyed to the brain by our equally imperfect senses, all jury-rigged and cobbled together by that maddeningly complex process known as evolution.

READING 2.3

*Cloning of inv, a Gene that Controls Left/Right Asymmetry
and Kidney Development*

Commentary: The explanation for left/right handedness asymmetry may require
more than mathematical input. It turns out that the underlying biology is not well
understood and so the assumptions behind mathematical explanations need to be care-
fully questioned. In particular, I include this article to demonstrate a case where a
simple mathematical model (see the previous commentary) explains the phenomena
while the story nature writes is most probably much more complicated.

In any event, this article concerns the observation that the arrangement of a mammal's internal organs is distinctly asymmetric with respect to the left and the right. For
example, almost all of us have our hearts on the left side. However, there are a few
rare individuals whose internal organs are organized with the reverse of the standard
chirality. This article discusses the genetic basis for this asymmetry in mice. Here is
a very rough outline of the story: In development, the body plan starts out left/right
symmetric, but the symmetry is eventually broken. Before the asymmetry in the embryo is observed, certain genes are expressed asymmetrically; among these are *nodal*
and *lefty*. A certain mutation in mice, called *inv*, inverts the pattern of expression for
nodal and *lefty*. This article identifies the precise gene that is disrupted in these *inv*
mutants. This is done by inserting a "minigene," which codes for the protein coded by
the suspect gene; the reappearance of this protein restores the normal asymmetry.

Cloning of inv, a Gene that Controls Left/Right Asymmetry and Kidney Development

Nature, **395** (1998) 177–181.

Toshio Mochizuki, Yukio Saijoh, Ken Tsuchiya, Yasuaki Shirayoshi, Setsuo Takai, Choji Taya,
Hiromichi Yonekawa, Kiyomi Yamada, Hiroshi Nihei, Norio Nakatsuji, Paul A. Overbeek,
Hiroshi Hamada, Takahiko Yokoyama

Most vertebrate internal organs show a distinctive left/right asymmetry. The *inv* (*inversion of
embryonic turning*) mutation in mice was created previously by random insertional mutagenesis [1]; it produces both a constant reversal of
left/right polarity (situs inversus) and cyst formation in the kidneys [2]. Asymmetric expression
patterns of the genes *nodal* and *lefty* are reversed
in the *inv* mutant [3–6], indicating that *inv* may
act early in left/right determination. Here we

identify a new gene located at the *inv* locus. The
encoded protein contains 15 consecutive repeats
of an Ank/Swi6 motif [7, 8] at its amino terminus.
Expression of the gene is the highest in the kidneys and liver among adult tissues, and is seen in
presomite-stage embryos. Analysis of the transgenic genome and the structure of the candidate
gene indicate that the candidate gene is the only
gene that is disrupted in *inv* mutants. Transgenic
introduction of a minigene encoding the candi-

Author affiliations: Toshio Mochizuki, Ken Tsuchiya, Hiroshi Nihei: Department of Medicine, Kidney Center.
Yukio Saijoh, Hiroshi Hamada: Institute for Molecular and Cellular Biology, Osaka University, and Yukio Saijoh,
Hiroshi Hamada, Takahiko Yokoyama: Crest, Japan Science and Technology Corporation. Yasuaki Shirayoshi,
Norio Nakatsuji: Mammalian Development Laboratory, National Institute of Genetics. Setsuo Takai, Kiyomi Ya-
mada: Department of Genetics, Research Institute, International Medical Center of Japan. Choji Taya, Hiromichi
Yonekawa: Department of Laboratory Animal Science, The Tokyo Metropolitan Institute of Medical Science.

date protein restores normal left/right asymmetry and kidney development in the *inv* mutant, confirming the identity of the candidate gene.

Organ primordia of vertebrae are symmetrical at first, and then acquire distinctive left/right asymmetry. In mice, embryonic turning and heart looping are the earliest manifestations of left/right asymmetry. Before morphological asymmetry develops, several genes, including *nodal* and *lefty* in mice, are expressed asymmetrically [3–6, 9, 10]. Asymmetric expression of *nodal* and *lefty* is reversed [3–6] in the *inv* mutant. The *inv* mutation produces a constant reversal of the left/right asymmetry, contrasting with several experimental and genetic mutations that randomize alterations in left/right asymmetry [11–14]. In addition, the mutation accompanies cyst formation in the kidney and jaundice [2].

The *inv* mutation was found in a family of transgenic mice into which the tyrosinase minigene had been introduced [1, 2], and is thought to be created by insertional mutagenesis [2]. We previously identified single-copy probes (named p3.2H and p2.3H) that flank the integrated tyrosinase minigene [2, 15]. Analysis of these probes showed that a region next to p3.2H had been deleted whereas a region containing p2.3H had become duplicated (Figure 1A), and also indicated that intrachromosomal inversion might have occurred [15].

To characterize the *inv* locus further, we screened the ICRF yeast artificial chromosome (YAC) library [16] by hybridization with the flanking probe p3.2H [2]. We identified three positive clones, IVCRFFy902E0682, ICRFFy902G0572 and ICRFFy902G022 (abbreviated as E0682, G0571 and G022) (Figure 1B). Cosmid libraries were constructed from these YACs. We identified two cosmids, E121 and G16, that mapped to opposite sides of the transgenic insert (Figure 1B). The distance between the two cosmid is about 400 kilobases (kb) in the wild-type genome (Figure 1B). An inversion, if present, should make one of these clones map onto different regions of the *inv* chromosome [15]. Fluorescent *in situ* chromosome hybridization (FISH) using E121 or G16 showed that both probes are located on chromosome 4 at band B of the wild-type and of the *inv* chromosome (Figure 1C). The results do not support the presence of an intrachromosomal inversion in the *inv* genome, which had been proposed [17] as a possible explanation for the *inv* phenotype.

We used genomic walking to determine the exact size of the deletion in the *inv* locus (Figure 2a). A single-copy probe (G160 E1) is present in the *inv* genome, whereas probe H3-5 is missing, defining the extent of the genomic deletion between these two probes.

To identify transcribed regions in the genomic deletion, we screened a complementary DNA library derived from mouse embryos at embryonic day (E) 17.5 (Clontech) with cosmids E120 and EG5 (Figure 2a) and identified one cDNA clone (clone 28). We screened a mouse cDNA library derived from E7.0 embryos (Clontech) several times to obtain a full-length cDNA (Figure 2b). The total size of the cDNA, based on the sequence, is ~5.6 kb. The size corresponds well to the 6.0-kb transcript size estimated from northern blot analysis (Figure 3a, Figure 3b).

We next analysed the genomic structure of this candidate gene. We screened a bacterial artificial chromosome (BAC) library (Research Genetics) with probe HS3.5 [15] and obtained six clones (B15, C18, C19, D5, G15, I17 and O19) (Figure 1C). We also screened a C57Bl/6 genomic library with the 5' region of the cDNA and obtained three positive clones (F1, F3 and F21). These BACs and phages as well as cosmid clones were characterized by hybridization with genomic and cDNA probes, polymerase chain reaction (PCR) and pulsed-field gel analysis, and a physical map of the region surrounding the *inv* locus was made (Figure 2a).

We used Southern blotting to hybridize parts of the cDNA to these BACs, cosmids and phages, and identified fragments in which exons of the candidate gene were located (Figure 2a). Exon-intron boundaries were confirmed by sequencing.

To assay for missing exons in the *inv* mutant, we performed Southern hybridizations using four cDNA probes (Figure 2), F1-R3, H2, H3 and 3E-2. Southern hybridizations with F1-R3 and 3E-2 showed that the *inv* homozygote retains the probed regions (Figure 2c). Hybridizations using H2 and H3 showed that several bands are missing in the *inv* homozygote. Combined with the Southern hybridization pattern produced by using the flanking probe p3.2H (a wild-type band at 9.5 kb and the *inv* band at 20 kb, [2]), these hybridization results confirm that exons 4-12 of the candidate gene are missing in the *inv* genome (Figure 2a).

Northern hybridization showed that the gene is expressed as early as E7 (Figure 3a). Whole-mount

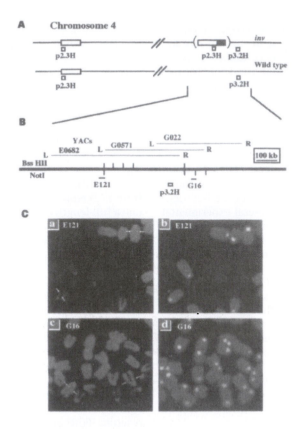

Fig. 1 The *inv* transgenic integration site and FISH. A, Wild-type and *inv* mutant chromosomes 4. The open rectangle and black rectangle within the parentheses indicate the duplicated genomic region and the integrated tyrosinase minigene [15], respectively. Probes p3.2H and p2.3H are shown as open squares under the lines. The duplicated regions on the *inv* genome separate 5–10% of the length of chromosomes 4. B, YACs surrounding the transgenic integration site. The top three lines indicate three YACs (G022, G0571 and E0682). L, left arm of the YAC; R, right arm of the YAC. The bottom line shows the wild-type genome. Short vertical bars above and below the line indicate BssHII sites and NotI restriction sites, respectively. Short horizontal bars indicate cosmid clones E121 and G16. C, FISH. a, c, Mouse metaphase chromosomes from cultured fibroblasts of a heterozygous *inv* female mouse after FISH using cosmid E121 (a) or G16 (c) as a probe. b, d, The same chromosomes were stained with PI to identify chromosomes and their banding (b, d). All signals are located on chromosome 4 at band B (arrows).

in situ hybridization also showed that the candidate gene is expressed in presomite-stage embryos (Figure 3c) before asymmetrical expression of *nodal* and *lefty*. In the adult, the highest level of expression is seen in kidney and liver (Figure 3b). Genomic analysis and expression patterns of the gene indicated that it is a good candidate for encoding the *inv* protein.

To determine whether the candidate gene can correct the phenotype seen in *inv* mutants, we gen-erated transgenic mice carrying a minigene (GGS-*inv*) (Figure 4a). As results of whole-mount in situ hybridization indicated that the gene may be expressed symmetrically and almost ubiquitously, we decided to use a CAG promoter, which consists of a cytomegalovirus immediate early (CMV-IE) enhancer and chick β-actin promoter [18].

One GGS-*inv*-positive mouse (V15) was heterozygous for the *inv* mutation. This founder mouse (*inv*/+, GGS-*inv*) was mated to *inv*/+ females to test

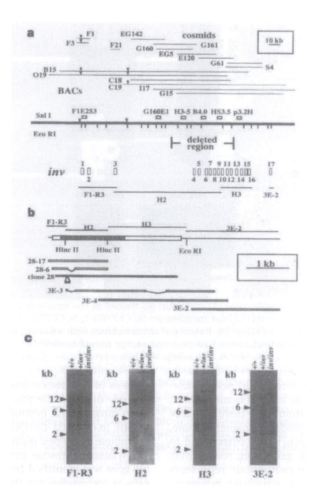

Fig. 2 Genomic organization of the *inv* gene, cDNA clones and Southern hybridization. a, Phage, cosmid and BAC clones at the *inv* locus. The bold line indicates the wild-type genome. Lines above the bold line show BACs, cosmids and lambda clones (F1, F3, F21). Open rectangles above the bold line indicate probes for genomic walking. EcoRI and SalI sites are shown as short vertical bars or arrows under or above the wild-type genome, respectively. The deleted region is also indicated below. Exons are shown as numbered boxes, and the lines beneath the exons indicate the regions of binding of the cDNA probes. b, Probes for Southern hybridizations, structure of the cDNA, and cDNA clones. Probes for Southern hybridizations are F1-R3, H2, H3 and 3E-2. In the cDNA structure, a boxed area indicates a coding region of the cDNA and a single line indicates an untranslated region. The shaded region indicates repeats of the Ank/Swi6 motif. The overlapping cDNA clones are 28-17, 28-6, clone 28, 3E-3, 3E-4 and 3E2. Clone 28 contains extra sequence and lambda 28-6 and lambda 3E-3 lack sequences. c, Southern blot analyses of wild-type (+/+), heterozygous (+/*inv*) and homozygous (*inv*/*inv*) DNA with probes F1-R3, H2, H3 and 3E-2. Genomic DNAs were digested with EcoRI.

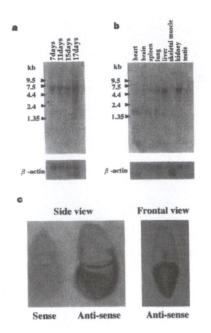

Fig. 3 Northern blot analysis of embryonic (a) and adult (b) tissues of the mouse using the candidate *inv* cDNA. Clontech multiple tissue northern (MTN) blot membranes (a, CL7763-1; b, CL7762-1) were hybridized with a32 P-labelled 3.0-kb EcoRI fragment of clone 28. Results of hybridization with a human beta-actin probe are shown below. c, Whole-mount in situ hybridization to a presomite-stage mouse embryo.

whether the GGS-*inv* minigene could restore normal left/right asymmetry in *inv*/*inv* homozygotes. In the offspring resulting from these matings, three types of coat colour were produced, namely, albino, light brown and dark brown, which correspond to the +/+, *inv*/+ and *inv*/*inv* genotypes, respectively (Figure 4b). Dark brown coat colour represents a gene-dosage effect of the tyrosinase minigene. These genotypes were confirmed by PCR (Figure 4c).

We obtained 10 *inv* homozygotes among 53 offspring from 7 litters. Three of the homozygotes lacked the transgene and died after birth. The remaining seven *inv* homozygotes contained the GGS-*inv* transgene, and they all grew normally. We studied the visceral organs of three *inv*/*inv*, GGS-*inv* animals. In all cases, the laterality defects seen in the *inv* mutants were corrected (Figure 4d, Figure 4e, Figure 4f). In the thorax, the right lung had more lobes than the left lung and the apex of the heart pointed to the left (Figure 4d). Stomach and spleen (Figure 4e) were located on the left side, and the right kidney was positioned anterior to the left one (Figure 4f). Within a couple of days after birth, the

kidneys of *inv* homozygous mice were filled with severely dilated collecting ducts (Figure 4h, Figure 4k) compared with normal kidneys (Figure 4g, Figure 4j). In contrast, the kidneys of the *inv*/*inv*, GGS-*inv* mouse developed normally (Figure 4i, Figure 4l). The transgenic rescue experiments show that the *inv* gene was identified correctly and that only one gene is responsible for the phenotype seen in the *inv* mutant.

The cDNA of the candidate gene has a long open reading frame which encodes 1,062 amino acids (Figure 5). The most distinctive feature of this candidate protein is the presence of 15 successive repeats of the Ank/Swi6 motif, which consists of about 33 amino acids (Figure 5). This motif has been found in vertebrate and invertebrate proteins, including ankyrins [7, 8, 19], developmental-regulatory proteins [20, 21] and cell-cycle-regulatory proteins [22-24]. A BLAST search showed that the candidate protein shares the highest homology with ankyrins [7, 8]. Like ankyrin, our candidate protein has a short amino-acid sequence before the Ank/Swi6 repeats. How-

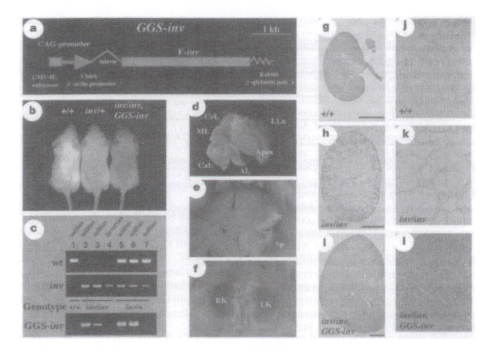

Fig. 4 Transgenic rescue of *inv/inv* by introducing the candidate *inv* gene. a, The GGS-*inv* construct. b, Coat colour of 2-week-old mice from three different genotypes. c, Situs types and genotypes of seven pups. Genotypes were determined by PCR analysis of wild-type and *inv* alleles and of the transgene, GGS-*inv*). There are three *inv/inv* pups (2-4). Two of them (2,3) contain a GGS-*inv* transgene and have the situs solitus phenotype, whereas the third (4) does not contain a GGS-*inv* transgene and has the situs inversus phenotype. d-f, Visceral organs of *inv/inv*, GGS-*inv* mice. All viscera in d, the thorax (lung, heart) and e, f, the abdomen (liver, stomach, spleen and intestine (e) and kidney (f)) showed normal morphologies and positions. AL, accessory lobe; CaL, caudal lobe; CrL, cranial lobe; ML, medial lobe; LLu, left lung; Sp, spleen; St, stomach; RK, right kidney; LK, left kidney. g-l, Histological analysis of kidney (5 days post-natal) in a wild-type embryo (g, j), an *inv* homozygote (h, k), and an *inv* homozygote containing the GGS-*inv* transgene (5 weeks post-natal i, l). Scale bar indicates 1.0 mm (in g, h, k).

ever, the candidate protein does not contain the spectrin-binding and regulatory domains found in ankyrins. We found no homology of the carboxy-terminal half of the candidate *inv* gene with other, previously described, genes. The candidate protein does not contain a signal peptide or transmembrane domain, indicating that it is probably an intracellular protein.

The genomic deletion in *inv* eliminates most of its coding sequences. Thus the *inv* mutant is probably a loss-of-function mutant. Another, less likely, possibility is that the remaining Ank/Swi6 repeat, either alone or fused with another protein, has a dominant-negative function.

Our transgenic rescue experiments indicate that localized expression of the *inv* protein may not be necessary, as the CAG promoter directs ubiquitous expression of the gene. As the *inv* protein is probably an intracellular protein, it may control the establishment of left/right asymmetry by affecting cell polarity rather than by its asymmetric distribution in embryos.

Injection of Vg1 into the R3 cell in Xenopus embryos leads to a constant reversal of left/right asymmetry, and broadening the expression pattern of Vg1 in aberrant cells on the right side of the embryos causes the *inv* phenotype [25, 26]. It is not known how this aberrant expression is induced in the absence of the *inv* protein.

Although the mechanisms by which the *inv* pro-

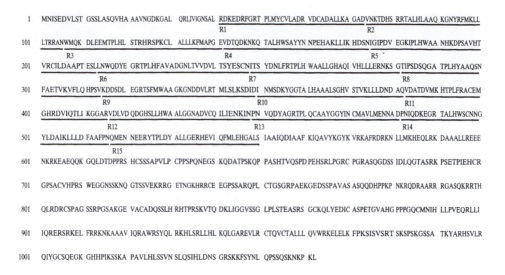

```
   1   MNISEDVLST GSSLASQVHA AAVNGDKGAL  QRLIVIGNSAL RDKEDRFGRT PLMYCVLADR VDCADALLKA GADVNKTDHS RRTALHLAAQ KGNYRFMKLL
                                                         R1                                 R2
 101   LTRRANWMQK DLEEMTPLHL STRHRSPKCL  ALLLKFMAPG EVDTQDKNKQ TALHWSAYYN NPEHAKLLIK HDSNIGIPDV EGKIPLHWAA NHKDPSAVHT
                   R3                                R4                               R5
 201   VRCILDAAPT ESLLNWQDYE GRTPLHFAVADGNLTVVDVL TSYESCNITS YDNLFRTPLH WAALLGHAQI VHLLLERNKS GTIPSDSQGA TPLHYAAQSN
                   R6                               R7                                R8
 301   FAETVKVFLQ HPSVKDDSDL EGRTSFMWAA GKGNDDVLRT MLSLKSDIDI NMSDKYGGTA LHAAALSGHV STVKLLLDND AQVDATDVMK HTPLFRACEM
                   R9                               R10                               R11
 401   GHRDVIQTLI KGGARVDLVD QDGHSLLHWA ALGGNADVCQ ILIENKINPN VQDYAGRTPL QCAAYGGYIN CMAVLMENNA DPNIQDKEGR TALHWSCNNG
                   R12                              R13                               R14
 501   YLDAIKLLLD FAAFPNQMEN NEERYTPLDY ALLGERHEVI QFMLEHGALS IAAIQDIAAF KIQAVYKGYK VRKAFRDRKN LLMKHEQLRK DAAALLREEE
                   R15
 601   NKRKEAEQQK GQLDTDPPRS HCSSSAPVLP CPPSPQNEGS KQDATPSKQP PASHTVQSPD PEHSRLPGRC PGRASQGDSS IDLQGTASRK PSETPIEHCR

 701   GPSACVHPRS WEGGNSSKNQ GTSSVEKRRG ETNGKHRRCE EGPSSARQPL CTGSGRPAEKGEDSSPAVAS ASQQDHPPKP NKRQDRAARR RGASQKRRTH

 801   QLRDRCSPAG SSRPGSAKGE VACADQSSLH RHTPRSKVTQ DKLIGGVSSG LPLSTEASRS GCKQLYEDIC ASPETGVAHG PPPGQCMNIH LLPVEQRLLI

 901   IQRERSRKEL FRRKNKAAAV IQRAWRSYQL RKHLSRLLHL KQLGAREVLR CTQVCTALLL QVWRKELELK FPKSISVSRT SKSPSKGSSA TKYARHSVLR

1001   QIYGCSQEGK GHHPIKSSKA PAVLHLSSVN SLQSIHLDNS GRSKKFSYNL QPSSQSKNKP KL
```

Fig. 5 Deduced amino-acid sequence of the *inv* candidate gene. Underlining indicates 15 repeats of the Ank/Swi6 motif.

tein specifies left/right asymmetry and by which loss of the *inv* protein reverses left/right asymmetry still remain to be established, our identification of the gene involved will be useful in elucidating these mechanisms.

Methods

Construction of cosmid libraries. Yeast DNA was partially digested with *Sau3*AI and was dephosphorylated and ligated with *BamHI-XbaI*-digested sCos 1 vector (Stratagene). Ligated products were packaged using Gigapack II (Stratagene) and infected using NM554 or XL1BlueMR.

Suppressive hybridization for cDNA screening. Cosmid EG5 or E120 was digested with NotI and released from the vector. Inserts were labelled with [^{32}P]dCTP using the random priming method (Strategene). ^{32}P-labelled probe was mixed with 50 μg of mouse *Cot1* DNA (Gibco) and denatured by boiling. The mixture was incubated at 65°C for 30 min and then hybridized to filters.

FISH analysis. Cosmid clones G16 and E121 were labelled by nick-translation with biotin-16-dUTP (Boehringer). Mouse chromosomes were prepared from early passages of fibroblasts from a heterozygous *inv* female mouse. Standard FISH methods [27] were used with some modifications. Hybridized DNA probes were detected by fluorescence

avidin DCS (Vector). We reduced background signals by competition with mouse Cot1 sequences (added in hybridization solution). The slides were finally stained with 4′,6′-diamino-2-phenylindole (DAPI) and propidium iodide (PI).

Whole-mount *in situ* hybridization. Embryos were collected from randomly bred ICR mice. To determine stages of development, we took the day on which a vaginal plug was detected to be E0.5. Digoxigenin-labelled (Boehringer) probe was generated with SP6 RNA polymerase (antisense) or T7 RNA polymerase (sense) from a 2.5-kb *EcoRI-HindIII* fragment of clone 28. Whole-mount *in situ* hybridizations were performed at high stringency [28].

Northern and Southern hybridization. Multiple-tissue northern blots (Clontech) were hybridized with clone 28, which was labelled by random-priming using the Prime it II kit (Stratagene) or an oligonucleotide labelling kit (Pharmacia) and [^{32}P]dCTP. Membranes were hybridized in express hybridization buffer (Clontech), washed with 2 × SSC (SSC is 0.15 M NaCl, 30 mM sodium citrate, pH 7.0), 0.1% SDS at 65°C, and exposed overnight. For northern hybridization, the final wash was 2 × SSC, 0.1% SDS at 50°C.

The GGS-*inv* construct and production and identification of transgenic mice. The GGS-*inv* transgene contains the full coding sequence of the *inv* candidate gene (*F-inv*) linked to the CAG promoter (Figure 4a). A 6.0-kb fragment of GGS-*inv* was microinjected into embryos at the one-cell stage collected from *inv* heterozygous matings to generate transgenic mice, using standard techniques [29].

Genotypes were determined by Southern hybridization or by PCR analysis using the following primers: 11220 (5'-TGAGTTGTGTGTGGAGGAATCTCTT-3'), 11291 (5'-GTGATGTGCTGGAAGATGGAAATTG-3'), 4633 (5'-GACTACCACAGCTTGCTGTATCAGAG-3'), B-28 (5'-TTTCTTTCTAGAAGGAGATGGAC-3') and E-28 (5'-GCCGCTGCAGTTAATGGAGATAAA-3'). Primer pairs 4633 and 11220 were used to detect the *inv* mutant locus, 11291 and 11220 to detect the wild-type locus, and B-28 and E-28 to detect the GGS-*inv* transgene.

Tissue processing and histological analysis. Kidneys were fixed in Bouin's fixative, dehydrated and embedded in Paraplast. Serial sections (6-μm thickness) were stained with haematoxylin and eosin using standard procedures.

REFERENCES

1. Yokoyama, T. et al. Conserved cysteine to serine mutation in tyrosinase is responsible for the classical albino mutation in laboratory mice. *Nucleic Acids Res.* **18**, 7293–7298 (1990).

2. Yokoyama, T. et al. Reversal of left-right asymmetry: a situs inversus mutation. *Science* **263**, 679–681 (1993).

3. Meno, C. et al. Left-right asymmetric expression of the TGF beta-family member in mouse embryos. *Nature* **381**, 151–155 (1996).

4. Meno, C. et al. Two closely-related left-right asymmetrically expressed genes, *lefty*-1 and *lefty*-2: their distinct expression domains, chromosomal linkage and direct neuralizing activity in Xenopus embryos. *Genes Cells* **2**, 513–524 (1997).

5. Collignon, J., Varlet, I. & Robertson, E. J. Relationship between asymmetric *nodal* expression and the direction of embryonic turning. *Nature* **381**, 155–158 (1996).

6. Lowe, L. A. et al. Conserved left-right asymmetry of *nodal* expression and alterations in murine situs inversus. *Nature* **381**, 158–161 (1996).

7. Lux, S. E., John, K. M. & Bennett, V. Analysis of cDNA for human erythrocyte ankyrin indicates a repeated structure with homology to tissue-differentiation and cell-cycle control proteins. *Nature* **344**, 36–42 (1990).

8. Bennett, V. Ankyrins, adapters between diverse plasma membrane proteins and the cytoplasm. *J. Biol. Chem.* **267**, 8703–8706 (1992).

9. Levin, M., Johnson, R. L., Stern, C. D., Kuehn, M. & Tabin, C. A molecular pathway determining left-right asymmetry in chick embryogenesis. *Cell* **82**, 803–814 (1995).

10. Isaac, A., Sargent, M. G. & Cooke, J. Control of vertebrate left-right asymmetry by a snail-related zinc finger gene. *Science* **275**, 1301–1304 (1997).

11. Yost, H. J. Regulation of vertebrate left-right asymmetries by extracellular matrix. *Nature* **357**, 158–161 (1992).

12. Hummel, K. P. & Chapman, D. B. Visceral inversion and associated anomalies in the mouse. *J. Hered.* **50**, 9–13 (1959).

13. Layton, W. M. Random determination of a developmental process. Reversal of normal visceral asymmetry in the mouse. *J. Hered.* **67**, 336–338 (1976).

14. Supp, D. M., Witte, D. P., Potter, S. S. & Brueckner, M. Mutation of an axonemal dynein affects left-right asymmetry in inversus viscerum mice. *Nature* **389**, 963–966 (1997).

15. Yokoyama, T., Harrison, W., Elder, F. F. B. & Overbeek, P. A. in *Developmental Mechanisms of Heart Disease* (eds Clark, E. B. & Takao, A.) 513–520 (Futura, New York, 1995).

16. Larin, Z., Monaco, A. P. & Lehrach, H. Yeast artificial chromosome libraries containing large inserts from mouse and human DNA. *Proc. Natl Acad. Sci. USA* **88**, 4123–4127 (1991).

17. Klar, A. J. S. A model for specification of the left/right axis in vertebrates. *Trends Genet.* **10**, 392–396 (1994).

18. Niwa, H., Yamamura, K. & Miyazaki, J. Efficient selection for high-expression transfectants with a novel eukaryotic vector. *Gene* **108**, 193–200 (1991).

19. Peter, L. L. & Lux, S. E. Ankyrins: structure and function in normal cells and hereditary spherocytes. *Semin. Hematol.* **30**, 85–118 (1993).

20. Yochem, J., Weston, K. & Greenwald, I. The Caenorhabditis elegans lin-12 gene encodes a transmembrane protein with overall similarity to Drosophila Notch. *Nature* **335**, 547–550 (1988).

21. Yochem, J. & Greenwald, I. glp-1 and lin-12,

genes implicated in distinct cell-cell interactions in C. elegans, encode similar transmembrane proteins. *Cell* **58**, 887–898 (1989).

22. Andrews, B. J. & Herskowitz, I. The yeast SWI4 protein contains a motif present in developmental regulators and is part of a complex involved in cell-cycle-dependent transcription. *Nature* **342**, 830–833 (1989).

23. Aves, S. J., Durkacz, B. W., Carr, A. & Nurse, P. Cloning, sequencing and transcriptional control of the Schizosaccharomyces pombe cdc10 'start' gene. *EMBO J.* **4**, 457–463 (1985).

24. Breeden, L. & Nasmyth, K. Similarity between cell-cycle control genes of budding yeast and fission yeast and the Notch gene of Drosophila. *Nature* **329**, 651–654 (1987).

25. Hyatt, B. A., Lohr, J. A. & Yost, H. J. Initiation of vertebrate left-right axis formation by ma-

ternal Vgl. *Nature* **385**, 62–65 (1996).

26. Hyatt, B. A. & Yost, H. J. The left-right coordinator; the role óf Vgl in organizing left-right axis formation. *Cell* **93**, 37–46 (1998).

27. Lawrence, J. B., Villnave, C. A. & Singer, R. H. Sensitive, high resolution chromatin and chromosome mapping in situ: presence and orientation of two closely integrated copies of EBV in a lymphoma cell line. *Cell* **52**, 51-61 (1988).

28. Wilkinson, D. G. in *In Situ Hybridization: A Practical Approach* (ed. Wilkinson, D. G.) 75–84 (IRL, Oxford, 1992).

29. Hogan, B., Costantini, F. & Lacy, L. in *Manipulating the Mouse Embryo: A Laboratory Manual* 153–203 (Cold Spring Harb. Lab., Cold Spring Harbor, 1986).

READING 2.4

A New Spin on Handed Asymmetry

Commentary: Here is another article which describes some recent discoveries about the underlying biological basis for left/right asymmetry. As with the previous commentary, I have enclosed this one to impress upon you that the underlying biology is subtle. In particular, be sure to question the underlying assumptions of any give mathematical explanation for the asymmetry.

The article here is a "News and Views" article from *Nature* which describes some ideas of S. Nonaka and collaborators (in *Cell* **95** (1998) 829–837) about the origin of left/right asymmetry in mice. Roughly said, the idea is that certain cells in a key place in an as yet undifferentiated mouse embryo have cilia which sweep in a preferential direction. (They observe this sweeping motion.) The sweeping motion of the cilia (tiny hair-like structures which protrude out from the cell wall) causes a current in the extracellular fluid which, in turn, causes a left/right asymmetry in the concentrations of various signaling molecules. The resulting asymmetry is the proximate cause of the later asymmetric development of the embryo.

A New Spin on Handed Asymmetry

Nature, **397** (1999) 295–298.

Kyle J. Vogan, Clifford J. Tabin

In vertebrates, most internal organs develop asymmetrically with respect to the body's midline, with the liver on the right and the stomach on the left, for example. This phenomenon, known

as left-right (L-R) asymmetry, raises an intriguing question—how is the early bilateral symmetry of the embryo first broken such that the L-R axis is always orientated the same way? A report by Nonaka

---

Author affiliations: Kyle J. Vogan and Clifford J. Tabin: Department of Genetics, Harvard Medical School.

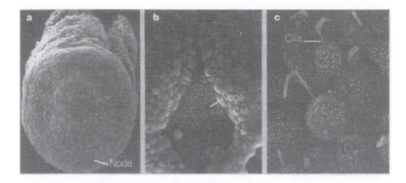

Fig. 1 Monociliated cells on the ventral surface of the mouse node. a, The node is at the apex of the egg cylinder, and the head process, which will give rise to the notochord, extends anteriorly. b, Close-up showing the cup-like shape of the node. The anterior is orientated towards the top. c, Higher magnification showing the monocilia on each cell in the head process and on the ventral surface of the node. (Micrographs courtesy of K. Sulik and T. Poe, Univ. North Carolina.)

et al. [1] in *Cell* now provides a remarkable insight into this problem. They show that the mouse node, a cup-shaped cavity in the embryo's midline, uses anticlockwise rotation of cilia to create a directional flow of extraembryonic fluid. The authors propose that this 'nodal flow' concentrates critical L-R determinants to one side of the node, activating distinct downstream signalling pathways on the left and right sides of the embryo.

People with defective cilia in their airways and immotile sperm cells often have mirror-image reversals of the L-R axis (Kartagener's syndrome) [2]. Moreover, the *inversus viscerum* (*iv*) mouse has L-R patterning defects that have been linked to a mutation in an axonemal dynein protein (dyneins are critical force-generating components of ciliary motors) [3]. Such findings have led to the suggestion that cilia are critical for establishing L-R pattern, and two embryonic structures have become particularly attractive candidates for the site at which they might act. The first is the node, which is an important source of patterning signals in the early embryo, and is also the site of the earliest known molecular asymmetries [4, 5]. The second is the notochord, a structure at the midline that has been proposed to act as a barrier to prevent asymmetric signals from crossing over to the opposite side [6, 7]. Both the ventral side of the node and the notochord consist of monociliated cells [8, 9] (Figure 1), and the *iv* gene product itself is highly expressed at the node during early development [3].

Despite these suggestive findings, the function of nodal cilia remained unclear. For one, nodal cilia do not share the '9 + 2' organization of microtubule doublets typical of most motile cilia [8, 9]. Moreover, attempts to assess the motion of these cilia directly yielded equivocal results [8, 9]. Although one group [8] reported motility, they saw no directionality to suggest that these cilia could bias L-R pattern.

To address these issues, Nonaka *et al.* [1] generated a mouse lacking the kinesin superfamily member KIF3B, a force-generating motor protein that was suspected to be critical for the assembly of cilia. Strikingly, mutant embryos showed randomized L-R asymmetry and were completely devoid of cilia. Inspired by this result, the authors examined the motility of nodal cilia in normal embryos, and made a dramatic discovery—nodal cilia rotate anticlockwise, a type of motion that is never seen with conventional cilia. Furthermore, by tracking the motion of fluorescently labelled beads, Nonaka *et al.* found a net leftward flow of extraembryonic fluid in the node region. This flow, the authors note, can arise without any input of L-R positional information, so could provide the initial bias that is needed to generate the L-R axis.

This work offers an elegant solution to a vexing problem—namely, how does the embryo orientate the L-R axis relative to the other axes of the body? It was proposed that morphological L-R asymmetry might be linked to the chirality of a

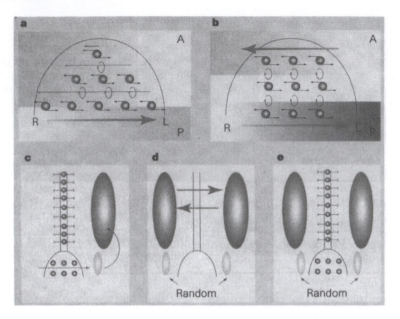

Fig. 2 Proposed roles of cilia during left-right (L-R) patterning. a, On the ventral surface of the node, cilia (blue circles) rotate anticlockwise. Because of the shape of the node, there are more cilia towards the posterior (P). Nonaka *et al.* [1] propose that interference (red circles) between the currents set up by successive rows of cilia limits flow to a weak rightward current across the anterior (A) and central regions of the node, but leaves a strong leftward flow in the posterior (bold red arrow), which concentrates a uniformly produced factor on the left side (green shading). b, Alternatively, the currents generated by individual cilia could result in interference throughout much of the node, leaving a net rightward flow at the anterior and a net leftward flow at the posterior. c, Wild-type cilia may both initiate L-R asymmetry at the node (red arrow), and act as a barrier (red lines) to prevent the diffusion of L-R factors (green and purple) across the midline. d, In *kif3b* mutants there are no cilia, so L-R initiation is random and signals can cross the midline. e, In *iv* mutants the cilia are thought to be immotile, so L-R initiation is random but the midline barrier is still intact.

hypothetical 'F' molecule, which is aligned in cells with respect to the anteroposterior and dorsoventral axes [10]. But this leaves unanswered the question of how the F molecules might be orientated in this way. The ciliary model circumvents this problem because the cilia themselves carry information about the dorsoventral axis (they all project ventrally) and, unlike the hypothetical F molecule, there is no need to align them further. The orthogonal flow of extraembryonic fluid at the node is derived from anticlockwise rotation of the cilia, which is presumably determined by the chirality of the molecules that drive their motion.

How, then, is information about the anteroposterior axis integrated in the system? Anticlockwise motion of the cilia should produce flow with equal and opposite left and right vector compo-

nents. Paradoxically, however, Nonaka et al. find no evidence for rightward flow. To explain this, they propose that differences in the width of the node along the anteroposterior axis lead to more cilia towards the posterior end, which synergistically intensify the leftward flow at the posterior edge (Figure 2a). They propose that this strong, narrow, leftward flow is balanced by a much broader and weaker rightward flow, the effects of which could be negated by passive diffusion. Alternatively, if equal and opposite leftward and rightward flows do exist (Figure 2b), anteroposterior information could be imparted to the system by localizing L-R determinants either anterior or posterior to the node itself. In this model, molecules that diffused into the node from its anterior side would meet a rightward current and be concentrated to the right, whereas factors that originated posterior to the node would experience a leftward flow and be concentrated to the left.

Somewhat unexpectedly, the *kif3b* and *iv* mutations have distinct effects on the expression of downstream L-R patterning genes. These differences can be explained if the cilia at the midline also act as a barrier during L-R signalling (Figure 2c). In the *kif3b* mutant, where the cilia fail to form, both initiator and barrier functions would be lost (Figure 2d). By contrast, the iv mutation is predicted to cripple the motor that drives the cilia, rendering them immotile. These cilia could, nevertheless, still act as a midline barrier (Figure 2e).

How general is this mechanism of L-R determination? Because rotating the chick node by 180° at an early stage of development does not affect L-R patterning, it had been suggested that the initial L-R decision is made outside the node [11]. However, rotating the node through 180° should not perturb the uniform, anticlockwise motion of the cilia so, if the same biasing mechanism operates in the chick, normal L-R patterning should still occur. Assuming that rotation of nodal cilia is indeed a universal mechanism for determining L-R asymmetry in vertebrates, we now have a broad conceptual outline that describes the key steps in L-R patterning— from the initial breaking of symmetry at the node, through a cascade of signals that culminates in the direct induction of asymmetrically biased morphogenetic events [12]. As we continue to fill in the details of this outline, we will discover further how spatial orientation is controlled during the assembly of an embryo and, judging by the extraordinary pace at which this story has unfolded over the past few years, we should not have long to wait.

REFERENCES

1. Nonaka, S. *et al. Cell* **95**, 829–837 (1998).
2. Afzelius, B. A. *Science* **193**, 317–319 (1976).
3. Supp, D. M., Witte, D. P., Potter, S. S. & Brueckner, M. *Nature* **389**, 963–966 (1997).
4. Levin, M., Johnson, R. L., Stern, C. D., Kuehn, M. & Tabin, C. *Cell* **82**, 803–814 (1995).
5. Levin, M. *et al. Dev. Biol.* **189**, 57–67 (1997).
6. Danos, M. C. & Yost, H. J. *Development* **121**, 1467–1474 (1995).
7. Levin, M., Roberts, D. J., Holmes, L. B. & Tabin, C. *Nature* **384**, 321 (1996).
8. Sulik, K. *et al. Dev. Dyn.* **201**, 260–278 (1994).
9. Bellomo, D., Lander, A., Harragan, I. & Brown, N. A. *Dev. Dyn.* **205**, 471–485 (1996).
10. Brown, N. A. & Wolpert, L. *Development* **109**, 1–9 (1990).
11. Pagan-Westphal, S. M. & Tabin, C. J. *Cell* **93**, 25–35 (1998).
12. Harvey, R. P. *Cell* **94**la, 273–276 (1998).

Exercises for Chapter 2

1. In each of the four cases given, find a function $y(t)$ that solves the following equation:

   (a) $\frac{dy}{dt} = 5y$

   (b) $\frac{dy}{dt} = -3y$

   (c) $\frac{dy}{dt} = 12y$

   (d) $\frac{dy}{dt} = -1.5y$

2. In each of the four cases in Exercise 1, find a solution with

   (a) $y(0) = 1$

   (b) $y(1) = 1$

(c) $y(-1) = 1$

(d) $y(-1) = -1$

3. Which of the four solutions to Exercise 2a is largest when $t = 10$? Which is smallest when $t = 10$.

4. Find the first-order Taylor's approximation at $x_0 = 0$ for the functions

   (a) $f(x) = \sin x$

   (b) $f(x) = e^x$

   (c) $f(x) = x/(1 + x^2)$

   (d) $f(x) = e^x \sin x$

   (e) $f(x) = \sin(e^x)$

   [Remember that the first-order Taylor's expansion for a function $f(x)$ at a point x_0 is defined to be the simplest function, $g(x)$, whose value at x_0 is the same as that of f and whose first derivative at x_0 is also the same as f's first derivative at x_0.]

5. For a certain microorganism, birth is by budding off a fully formed copy of itself. Suppose that under reasonable favorable laboratory conditions (plenty of food and no predation), such births occur on average four times per day, and an individual lives, on the average, one day. Write a differential equation for the population, $p(t)$, of the microorganism as a function of time. Then find the solution given that at time zero, the population numbered 1000.

6. Design a hypothetical experiment with snails that would verify or rule out each of the two assumptions behind the simple model I present for snail handedness in the commentary on Reading 2.2.

3
Introduction
to Differential Equations*

Consider the case of fish in a pond. Suppose we started with a few fish in a pond. You might expect the fish population to increase with exponential speed at first, but would you expect the fish population to increase without bound? This is hardly likely. Rather, you might expect the fish population to level off and then fluctuate around some fixed population size.

We want to construct a mathematical model for the growth rate of the fish population. In particular, if $p(t)$ denotes the function of t that gives the fish population at time t, it is not unreasonable to *postulate* that $p(t)$ should satisfy a differential equation whose schematic form is

$$\frac{d}{dt}p = h(p)p. \tag{1}$$

Here, $h(p)$ is a function of the unknown p (to be determined) that plays the role of the relative growth rate (h = birth rate − death rate) for fish when the population is p.

In this regard, remark that previously, we considered a differential equation as in (1) where h was a constant (such as $h = 0.105$ or $h = -2.23$). However, it is unlikely that the relative growth rate is a constant, for when there are more fish, there is less food and so the death rate will be higher and the birth rate lower. The point is that on biological grounds, we expect the relative growth rate h to depend on the

---

*This chapter roughly paraphrases Harvard University notes by Otto Bretscher and Robin Gottlieb.

population size, p. When p is small, we expect h to be roughly constant, but in any event, h should turn negative when p gets large. In particular, here is a simple *model* for the function h:

$$h(p) = k - bp, \tag{2}$$

where k and b are positive constants to be determined. Note that this particular function has the property that when b is small (i.e., when $p \ll k/b$), then h is approximately equal to the constant k. Likewise, when p is large (i.e., when $p > k/b$), then h is negative.

The resulting equation,

$$\frac{d}{dt}p = kp - bp^2, \tag{3}$$

is called the *logistic equation*. [This equation was first introduced by the Belgian biologist Verhulst to predict the populations of Belgium and France. See Verhulst, P. F., "Notice sur la loi que la population suit dans son accroissement," Corr. Math. et Phys. 10 (1838) pp 113–120.]

For the sake of definiteness, consider (3) with $k = 0.5$ and $b = 0.001$, so

$$\frac{d}{dt}p = 0.001(500 - p)p. \tag{4}$$

It is important to stress that (4) or (3) is no more nor less than a mathematical model, and therefore its relevance to any given population problem is a question that can only be verified by *experiment*. The point is that (4) or (3) is a purely theoretical prediction (based on some simple intuition about nature) for the behavior of a population (fish, in this case). Any given theoretical prediction may or may not be correct. Only observations can decide whether (4) is useful.

3.1   The Behavior of the Logistic Equation

Equation (4) is not an "antiderivative problem." Rather, it is an equation that relates the derivative of p to p itself. Although (4) can be solved in closed form, this will not interest us. Closed form solutions to differential equations are extremely rare, and so we will use (4) as a testing ground for techniques that can be used on any differential equation of the form

$$\frac{d}{dt}p = f(p), \tag{5}$$

where $f(\cdot)$ is some function that is specified in advance.

In studying (4), there are two graphs that are useful. The first graph is that of the function $y = f(p) = 0.001(500 - p)p$. The graph looks as depicted in Figure 3.1.

Note that this is an inverted parabola that crosses the p-axis at the origin and at $p = 500$. Its maximum is at $p = 250$, and equals 62.5. For p between 0 and 500, the graph is above the p-axis, and for p greater than 500, the graph is below the p-axis.

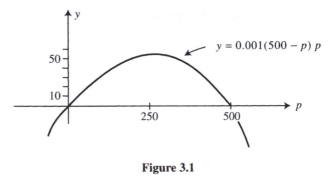

$$y = 0.001(500 - p) p$$

Figure 3.1

If the differential equation (4) is to hold, this means that $\frac{d}{dt} p$ is *positive* whenever p is between 0 and 500, and $\frac{d}{dt} p$ is *negative* when p is greater than 500. If p is exactly equal to 500, then $\frac{d}{dt} p = 0$, and p is a constant.

If you think about what was just said, we learned a great deal about the behavior of solutions to (4) without ever solving it exactly. We know that if, at some time t_0, the population obeys $p(t_0) > 500$, then p will decrease toward 500 for times $t > t_0$. On the other hand, if $p(t_0) < 500$ and not zero, then p will increase toward 500 for times $t > t_0$. Finally, if $p(t_0) = 500$, then $p(t) = 500$ for all times $t > t_0$. Figure 3.2 is a sketch of the various possibilities for p versus t.

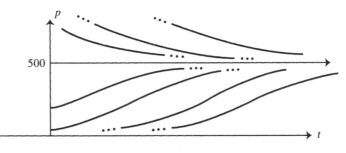

Figure 3.2

We say that $p = 500$ is an equilibrium value, since this constant function is a solution to the differential equation in (4). Likewise, $p = 0$ for all time is also a solution, so 0 is another equilibrium value.

Also, we learn that $p = 500$ is the carrying capacity of our pond, since over the long run, we expect roughly 500 fish in the pond. For example, this tells the owner of the pond that it doesn't make much sense to stock the pond with 1000 fish, since over the long run, half will die off. In fact, to save money, the owner can, in principle, stock the pond with a few fish, and if she is willing to wait long enough, she will find (if our model is correct) roughly 500 fish in the pond.

3.2   Another Example: Fishing Yields

Suppose now that the number of fish in the pond has stabilized to roughly 500, and the owner has decided to take a fish per week from the pond. Thus, 52 fish per year will be removed from the pond. We can ask whether this modest rate of fishing will deplete the stock, or whether this modest rate is sustainable. In fact, we can ask: What is the largest sustainable yield from this pond? That is, what is the maximum number of fish that can be caught per year without depleting the pond?

Consider the case where the owner catches 52 fish per year. To model this situation, we consider that $\frac{d}{dt}p$ will be given by (4) but with the modification that 52 fish are lost per year. That is, if time is measured in years, then

$$\frac{d}{dt}p = 0.001(500 - p)p - 52. \tag{6}$$

If we analyze this equation as we did (4), we should start by graphing the function $f(p) = 0.001(500 - p)p - 52$ (Figure 3.3).

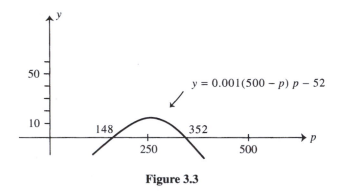

Figure 3.3

This is the same graph as in (5) except shifted down by 52: It is a downward pointing parabola with maximum $62.5 - 52 = 10.5$. Note that the graph still crosses the p-axis in two places, $p = 148$ and $p = 352$. The graph is positive when $148 < p < 352$ and the graph is negative when $p < 148$ or when $p > 352$.

If the differential equation in (6) is assumed to hold, then Figure 3.3 asserts that if p at time t_0 is between 148 and 352, then p will increase to 352 for times $t > t_0$. If $p(t_0) > 352$, then p will decrease to 352 for times $t > t_0$. And, if $p(t_0) < 148$, then p will decrease to zero for times $t > t_0$. Finally, if $p(t_0) = 148$ or if $p(t_0) = 352$, then p will stay at these values for all time. The graph in Figure 3.4 describes these various possibilities for p versus t.

In particular, we see that if $p(t_0) = 500$, then the fish stock will decline but will not decline to zero; rather it will stabilize at the new value of 352. Thus, a catch of 52 fish per year is sustainable. Is a catch of 75 fish per year sustainable? Is any catch that is greater than 62 fish per year sustainable? (Remember that the parabola in Figure 3.1 has maximum equal to 62.5.)

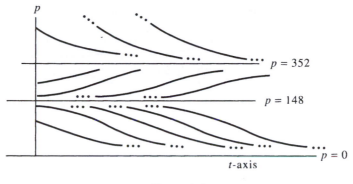

Figure 3.4

3.3   Brief Synopsis

In dealing with a differential equation $\frac{d}{dt}p = f(p)$, we are looking at two graphs: We first graph $y = \frac{d}{dt}p$ versus p [or, equivalently, $y = f(p)$ versus p] (Figure 3.5).

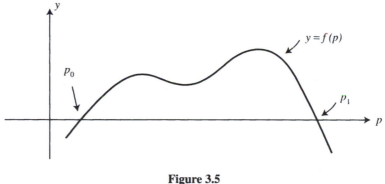

Figure 3.5

From the preceding graphs, we can tell whether $\frac{d}{dt}p$ is positive, negative, or zero for a given starting value at $t = t_0$. Then, for $t > t_0$, we can use Figure 3.5 to determine whether p increases or decreases. For example, if $f(p)$ is given by Figure 3.5, then sample solution trajectories, p versus t, have the form shown in Figure 3.6.

3.4   Determinism

It is crucial to realize that the equation $\frac{d}{dt}p = f(p)$ has a *unique* solution for each given starting value for p. That is, given the starting time t_0 and the starting point p_0, there is a *unique* solution $p(t)$ to the equation that has $p(t_0) = p_0$. It is in this sense

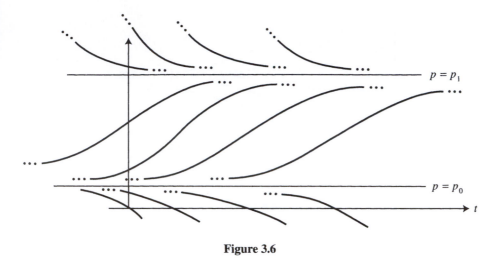

Figure 3.6

that equations of this type are *predictive*. You give the initial value for p, and the equation will tell you what happens for all subsequent times.

(Note that this uniqueness property of solutions to the equation implies that solution curves as drawn in Figures 3.2, 3.4, or 3.6 never intersect each other. If any two solution curves intersected, then at that point, there would be more than one solution trajectory to choose from.)

3.5   Global Temperature

Let $T(t)$ be the average temperature of the earth's surface at time t. In recent years, scientists have tried to understand the ways in which T changes. The rate of change, $\frac{d}{dt}T$, was thought to be influenced mainly by two factors: the radiation that reaches the earth from the sun, and the radiation that leaves the earth. To model this hypothesis, we write

$$\frac{d}{dt}T = f_{\text{in}} - f_{\text{out}}, \tag{7}$$

where f_{in} is the contribution from the incoming solar energy and where f_{out} is the contribution from the outgoing radiation. Both f_{in} and f_{out} depend on T. In particular, f_{out} increases with T since a hot body radiates more than a cool one. On the other hand, f_{in} also should increase with T, since at low temperatures, a greater fraction of the earth is covered with ice, which reflects sunlight and thus reduces the net radiation that reaches the surface. (This naive analysis does not take into account the effects of cloud cover. The cloud cover should also influence the behavior of f_{in} and f_{out}; more cloud cover should decrease the former and increase the latter. However, the temperature dependence of cloud cover is complicated.)

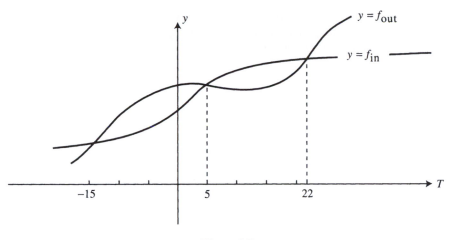

Figure 3.7

A simple model has f_{in} and f_{out} given by the graphs in Figure 3.7.

Note that $\frac{d}{dt}T$ is the difference between the two graphs, so is positive where $f_{in} > f_{out}$ and negative where $f_{in} < f_{out}$. From the graph, we see that there are three equilibrium points, roughly at $T = 22$, $T = 5$, and $T = -15$. The trajectories (T versus t) for various starting values can be read off the graph above and are sketched in Figure 3.8.

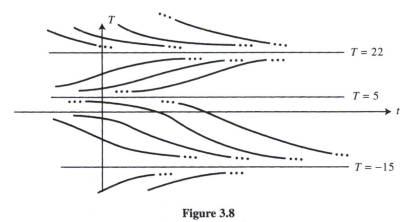

Figure 3.8

Note that the T versus t trajectories converge either to the $T = 22$ or the $T = -15$ solution. That is, if $T(t_0) > 5$, then $T(t)$ will converge to the $T = 22$ constant solution as time increases. However, if $T(t_0) < 5$, then $T(t)$ will converge to the $T = -15$ solution as t increases. This suggests that the earth's climate has two "equilibrium states," the warm state (which we are in now) with average temperature $T = 22$, and a cold state (the Ice Ages?) with average temperature $T = -15$. (This last discussion paraphrases material from Berger, A., ed., *Climatic Variations and Variability*, D. Reidel, Dordrecht.)

3.6   Lessons

Here are some key ideas from this chapter:

- The logistic equation is commonly used to model population growth in a resource-limited environment.

- The qualitive behavior of solutions $p(t)$ to the logistic equation, and to the generic equation $\frac{d}{dt}p = f(p)$, can be obtained from information in the graph of the function f.

- The points where the graph crosses the p-axis correspond to constant solutions to the differential equation. Meanwhile, if $p(t)$ is a time dependent solution and $f(p(t)) > 0$, then $p(t)$ moves to the right on the p-axis as time increases. Conversely, if $f(p(t)) < 0$, then $p(t)$ moves to the left.

- An equation of the form $\frac{d}{dt}p = f(p)$ is completely predictive. Choose any starting value for p and there is precisely one solution that starts at your chosen starting time with your chosen starting value.

READINGS FOR CHAPTER 3

READING 3.1

Biologists Sort the Lessons of Fisheries Collapse

Commentary: The purpose of this article is to illustrate the complicated issues that arise when biology collides with economics. The point of the article is that many ocean fisheries are in dire straights, that these dire straights were more or less predicted by scientists, but that the scientific predictions were not well received by the fishing industry. Scientists argued that if there is uncertainty in the predictions (which there almost always is), then one should err on the side of the resource—fish. Thus, cut back on fishing to decrease mortality. Meanwhile, the industry has argued that you should err on the side of economics and the fishing industry's short-term interests. The industry's voice tends to carry more weight on the relevant councils.

Biologists Sort the Lessons of Fisheries Collapse

Science, **264** (1994) 1252–1253.

Bob Holmes

The recent drumbeat of news reports about the collapse of many North American fisheries came as no surprise to fisheries biologist Vaughn Anthony. Anthony, who is the chief scientific advisor for the New England region of the National Marine Fisheries Service (NMFS), has been tracking declining fish stocks off the southern New England coast for years. Now, he says, "any dumb fool knows there's no fish around." Bottom-dwelling fish such as cod, haddock, and flounder are at or near all-time lows, and the venerable fishing industry there is collapsing. And New England is not alone: NMFS estimates that fully 45% of the fishstocks whose status is known are now overfished, and populations of some species have plummeted to less than 10% of the optimum level—the level that yields the largest sustainable catch. Says Anthony, "The status of the stock is so bad now that [nobody argues] about it."

But there's plenty of argument about how this crisis developed. Fishers blame a regulatory bureaucracy that was slow to act, while regulators say they were looking out for the industry's economic interests or that they were blindsided by unpredictable population swings and efficient new fishing technologies. Regulators also complain that biologists' uncertain estimates of fish populations and acceptable fishing levels failed to offer a solid case for tight fishing restrictions.

All of these claims carry varying degrees of truth, depending on the fishery. Still, most observers say that scientists have been giving clear warnings of the decline for years. Says Carl Safina, marine conservation director of the National Audubon Society, "The bottom line is that in fisheries where people have paid attention to the scientific recommendations, there are still fish around. In fisheries where the scientists have routinely been ignored or the most optimistic gloss has been put on the data, we have declines."

That bottom line may now be heeded. One result of the current crisis may be to build support for more cautious catch limits based on population data, even if the data are limited. The National Research Council, in a report issued this month, is calling on Congress to revise the current law governing fisheries management, the Magnuson Act, to build more biology into the regulatory process. And at the same time, scientists are trying to play a savvier political game by stressing the long-term economic benefits of cautious fisheries management. If these efforts succeed in limiting the fishing pressure, biologists say, even New England's sadly depleted fish populations stand a good chance of recovery—given enough time.

If so, it will be a homegrown solution to what is largely a homegrown problem. The Magnuson Act, passed in 1976, claimed for the United States the exclusive right to manage fisheries within 200 miles of its coastline, where the vast majority of commercial fishing takes place. The act excluded most of the foreign fishing vessels that were fishing these waters intensively, but the ensuing windfall of fish led to a massive building up of the U.S. fishing fleet.

Overfishing was supposed to be checked by eight regional fishery management councils, with the help of scientific advisers on the council staffs and at NMFS offices. These advisers monitor fish abundance based on the commercial fish harvest itself and on data from research vessels that fish at random. Since most fish are highly mobile and patchily distributed, however, fisheries biologists admit they're often lucky if their population estimates for any given year are within 30% of reality. Nor can scientists always tell whether fishing rather than natural factors such as shortage of prey or climate change is the key factor in a population decline, says marine scientist Brian Rothschild of the University of Maryland at Solomons. Human activities other than fishing can also affect abundances. Most notoriously, salmon populations off California, Oregon, and Washington almost certainly owe much of their steep decline not to fishing but to destruction of their spawning streams by dams and logging.

Signs of trouble. Despite these gaps in their understanding of fish population biology, most fisheries scientists—and many in the fishing industry as well—agree that only rarely do they simply blow the call on how much fishing a population can stand. Scientists can tell that a population is in trouble when its mortality rate, calculated from the age distribution of the fish in the commercial harvest, shows a steady rise, says Andrew Rosenberg, an NMFS scientist. Other factors besides fishing may be contributing, but a cutback in fishing is the only remedy available in most cases. For now, the sorry state of many fish populations often makes the scientists' call an easy one. "No matter how badly you estimate [reproduction] or natural mortality, you still will come to the conclusion that you should reduce the fishing mortality rate," says Rosenberg.

But where the shortcomings in the science do make a difference is in the researchers' ability to influence policy. Many fisheries scientists have seen the current crisis coming for as much as a decade, say Rosenberg and his colleagues, but when they presented their data to the management councils, their penchant for speaking in terms of probabilities and confidence intervals often served them— and the fish—poorly. Joseph Brancaleone, a former fisherman who chairs the New England Fishery Management Council, for example, argues that uncertainties in the population estimates weakened the case for severe restrictions. Carl Paulsen, program director of the National Coalition for Marine Conservation, an environmental group, agrees that the uncertainty leaves plenty of room for policy disputes. "We've argued that if there's uncertainty, you should err on the side of the resource," says Paulsen. "Industry has argued that you should err on the side of economies and the fishing industry."

And industry's voice tended to carry weight with the councils, say Paulsen and others, in part because of a decision made by Congress when it established the councils in 1976. Because it intended them to draw heavily on the expertise of the fishing industry, Congress went so far as to exempt council members—most of whom are federal appointees— from federal conflict-of-interest rules. As a result, members often hold direct interests in the fisheries they regulate. Such conflicts don't always lead to overfishing, but some councils have consistently overridden scientists' recommendations, many scientists and conservationists charge. "You have people in the industry, with livelihoods at stake, being unable to take the hits necessary to rebuild the stocks," says Paulsen.

Among the worst offenders has been the New England council, says Safina, who calls it "incredibly irresponsible and stupid" for allowing persistent overfishing of cod and flounder stocks. Despite a decade of warnings from fisheries biologists, the council has stubbornly resisted setting direct limits on fish harvests. Only recently did the council agree to restrict harvests by gradually limiting the number of days each boat can fish. And that plan is still not fully implemented. For now, says NMFS's Vaughn Anthony, fishers still catch around 60% of the entire fish population each year—more than twice the sustainable level "There's no room for rebuilding here," says Anthony.

Brancaleone, the chairman of the New England council, notes that it did respond to scientists' concerns by trying to reduce harvests through other, indirect, means: imposing minimum net-mesh sizes (which let more young fish escape) and staking out no-fishing zones. He defends the council's slow phase-in of more stringent controls as necessary to protect fishers from the economic pain of overregulation. Besides, he says, the science doesn't show a clear need to move any faster. "The data that we have are so slim that we can't put a number on [the effect of the controls]. By the third or fourth year, we'll have the data that will tell us [whether further restrictions are needed]," he says.

But more aggressive management has paid off in other fisheries, say researchers. Even the most outspoken critics of fishery management, such as Safina, agree that the North Pacific management council has done a good job of following scientists' recommendations in setting strict catch limits. As a result, many Alaskan fish populations such as Pacific halibut and salmon are still in good shape. And a moratorium on striped bass fishing in the mid-Atlantic states during the 1980s has allowed that fishery to rebound strongly from historic lows early in that decade.

These successes are encouraging scientists to change their approach in the fight over fishing limits, Rosenberg says. Many are now becoming more outspoken in arguing for conservative catch limits even when the data are uncertain. They've also learned a political lesson, Rosenberg says—the value of involving more members of the fishing industry in the stock-assessment process, "so people don't think we're doing something dark and mysterious." In addition, he notes, NMFS has begun to include economists in its analysis groups to evaluate the economic effects of various management strategies, thereby bolstering its claims that, in the long term, tighter regulation will benefit fishers.

Such regulation, say fisheries experts, might take the form of limits on the overall catch or of quotas assigned to individual fishers, which could be bought or sold. However the fishing pressure is eased, examples such as the striped bass suggest that fish populations can recover from even severe overfishing. Most biologists are reluctant to venture a guess as to how fast, though, because the speed of recovery also depends on the lifespan and reproductive rate of the fish—and the environmental vagaries that affect them.

Yellowtail flounder, for example, only reproduce well in years with cold winters. "Now what can we do about cold winters? If we don't have cold winters for the next 10 years, there won't be any yellowtail," says fishing-boat owner Barbara Stevenson of Portland, Maine. For her and other beleaguered fishers looking for a better future, therefore, the key words appear to be restraint—and patience.

READING 3.2

New Study Provides Some Good News for Fisheries

Commentary: When fish numbers are low, standard modeling of the population $p(t)$ at time t predicts that there will be an exponential equation, $p' = ap$, where a is a constant that is roughly the difference between the birth rate and death rate. This equation is assumed to be valid when the fish numbers are low and the environment is not changing very much from year to year. However, it may be that such a model is wrong and that the derivative of p when p is small has some other behavior that would cause the fish stock to become extinct if it ever dropped below some minimum amount. To some extent, this must be true, for if there is just one fish left of some species, then the exponential equation is clearly wrong. This is to say that the exponential equation is clearly only an approximation to reality when p is small, and maybe not a very good one. If this equation is way off at small population size, biologists say that "depensation" is occurring. The question is this: Is there evidence of significant depensation in the world's fisheries? It is crucial to determine when depensation is important, for if there is significant depensation, then the fish population will continue to plummet (to extinction) even with a complete ban on fishing.

This article reports on a study of 128 fisheries whose purpose was to determine the extent to which there is an observable deviation from the exponential model at low fish populations.

New Study Provides Some Good News for Fisheries

Science, **269** (1995) 1043.
Marcia Barinaga

In the nineteenth century, biologists and fishermen alike believed that the ocean's fish were a boundless resource that could never be depleted. But the population crashes that have struck many fish stocks during this century have dispelled that notion. Indeed, some fish populations have plummeted so precipitously that fisheries biologists in recent years have wondered whether something besides fishing was accelerating that decline: Perhaps at low numbers the fish were unable to reproduce efficiently or were succumbing disproportionately to predators. Such a phenomenon, known as depensation, has been documented in some dwindling insect and marine mammal populations, including whales; if it were to occur in depleted fish stocks, they would be unlikely to recover, even if fishing were stopped.

Now, a study of data on 128 heavily exploited fish stocks, reported in the following article, concludes that depensation doesn't appear to be at work in most fishery declines. "Fish stocks don't collapse because of depensation," says Ransom Myers of the Northwest Atlantic Fisheries Centre in St. John's, Newfoundland, the lead author on the study. "Fish stocks collapse because of plain simple overfishing."

"This can be thought of as good news," says John Beddingron, a population biologist at Imperial College, London. "An implication of their work is that there is little evidence to indicate that fish stocks won't recover once you stop fishing them." Adds Alec MacCall, director of the Tiburon, California, laboratory of the National Marine Fisheries Service (NMFS): "It really places the burden right back on the fisheries manager. It says it's really up to you folks to get [the fish] back."

Myers and his colleagues arrived at their conclusion from a meta-analysis of statistics gathered over at least 15 years on a wide range of major commercially fished stocks. They analyzed the numbers of spawning fish in individual stocks as well as the numbers of "recruits," young fish that survive to adulthood. The ratio of recruits to the total number of fish reflects a population's success at reproduction and survival. A drop in that ratio at low population numbers would mean depensation had occurred. In a handful of fish populations, including herring on the Georges Bank off New England, and Pacific sardines, depensation could not be ruled out. But in the vast majority of the fish stocks, including the greatly reduced Georges Bank cod and flounder stocks and Pacific sockeye salmon, the researchers found no evidence of depensation.

News like that will provide support for fishery managers in their efforts to regulate fishing, says Andrew Rosenberg, Northeast regional director for the NMFS and an author on the paper. The reason: They can now assure the fishing industry that cutbacks in fishing should lead to recovery of overexploited stocks. "You are continually trying to tell people there is a real benefit from reducing fishing, not just on stocks that have collapsed, but on stocks that have been reduced to arguably low levels," says Rosenberg. With the current results in hand, he says, "I can stand up in the council meet-

ings and argue that you're going to have to cutback the catch very severely this year, but if you maintain that low harvest rate for several years, you are going to start to see an increase in catch."

If the study had instead shown depensation to be widespread, the outlook for fisheries would be bleak. Overfished stocks would be less likely to recover, and managers would have to be "even more cautious in regulating fisheries to prevent overharvesting in the first place," says Rosenberg. In contrast, says Gunnar Stefansson of the Marine Research Institute in Reykjavik, Iceland, the new finding suggests that "if you just reduce the fishing effort to a moderate level, then you should be able to allow the stock to regain ... a healthy stock size, yet also maintain some harvest."

The NMFS is putting that kind of policy into practice on the ailing cod and flounder fisheries of the Georges Bank, based in part on the Myers team's findings. The fisheries were taking 66% of the fish stock every year, says Rosenberg, and the goal is to reduce that to about 20% without closing the fishery. "As the stock begins to recover and move up that stock recruitment curve," he adds, "you would expect 20% of whatever is out there will start to become a substantial number in a couple of years' time."

At the same time, Rosenberg cautions that these findings should not be used to justify fishing until stocks are almost completely gone. "It is really a false argument to say because it will come back, we don't have to worry about it," says Rosenberg. For any depleted stock to recover its health, Rosenberg and his co-authors say, fishing must be reduced

The earlier that reduction is begun, the better. A badly damaged fish stock may recover eventually, but not without wreaking severe economic hardship on the industry it supports, says Myers. He cites as an example Canada's lack of action when Newfoundland's cod stock got dangerously low in the late 1980s. By 1992, the population had dropped to 1/100th of its original size, and the fishery was shut down, causing the loss of 35,000 fishery jobs. There is no sign of recovery yet.

Indeed, the Myers team's lack of evidence for depensation has not calmed all fears that it is occurring in Newfoundland or elsewhere where fish stocks have crashed. "My feeling about this is even if one can't see depensation in stock recruitment data, there is something going on out there," says Jeremy Collie of the Graduate School of Oceanography at the University of Rhode Island in Narragansett. Reflecting on the Newfoundland cod fishery, Alaska's herring fishery, and others that have been slow to bounce back after fishing was stopped, he says, "the nagging fear is that maybe it is more complicated," and that depensation is occurring that can't be detected by the model used by Myers and his colleagues.

But even if the potential for depensation does lurk undetected in some declining fish stocks, Collie says, "the prescription [would be] basically the same ... reduce fishing mortality drastically." The difference, then, is not so much in the treatment as in the prognosis for the ailing patient, and that is where Myers and his colleagues have provided cause for optimism.

READING 3.3

Population Dynamics of Exploited Fish Stocks at Low Population Levels

Commentary: The authors propose a generalization to the standard exponential growth equation that has an extra parameter that can be set anywhere between 0 and ∞. When the parameter is equal to 1, there is no depensation and when the parameter is greater than 1, there is depensation, with the depensation being stronger for larger values of this parameter. The authors then compare the predictions of their equation for various values of this parameter with data from some 128 fisheries to see what value of the parameter makes their equation give data that most closely resemble the real data. They find that for most of the fisheries, their parameter is very close to 1, thus indicating that the fish stocks have not fallen to values where depensation will occur.

Population Dynamics of Exploited Fish Stocks at Low Population Levels

Science, **269** (1995) 1106–1108.
R. A. Myers, N. J. Barrowman, J. A. Hutchings, A. A. Rosenberg

Models of population dynamics in which per capita reproductive success declines at low population levels (variously known as depensation, the Allee effect, and inverse density-dependence) predict that populations can have multiple equilibria and may suddenly shift from one equilibrium to another. If such depensatory dynamics exist, reduced mortality may be insufficient to allow recovery of a population after abundance has been severely reduced by harvesting. Estimates of spawner abundance and number of surviving progeny for 128 fish stocks indicated only 3 stocks with significant depensation. Estimates of the statistical power of the tests strengthen the conclusion that depensatory dynamics are not apparent for fish populations at the levels studied.

Many of the world's fisheries are heavily exploited, and a number of stocks have experienced severe declines in abundance, many of them very suddenly [1]. The causes of the sudden declines and the potential for recovery for a stock when fishing is reduced have remained undetermined. The existence of depensatory dynamics would affect this potential for recovery.

Ecosystems that exhibit multiple stable states typically include a highly nonlinear functional feeding response in which predators are saturated at high levels of prey [2]. Similar dynamics will result if reproductive success is reduced at low population densities because of the difficulty in finding mates; this is sometimes known as the Allee effect. Both predator saturation and the Allee effect

---

Author affiliations: R. A. Myers and N. J. Barrowman: Northwest Atlantic Fisheries Centre, Science Branch. J. A. Hutchings: Department of Biology, Dalhousie University. A. A. Rosenberg: National Marine Fisheries Service.

can result in a low per capita production of new recruits to a population if the number of reproducing animals is reduced to a low level, an effect known as depensation [3]. Although these arguments are appealing, no rigorous empirical studies have examined the hypothesis that natural populations exhibit depensatory dynamics.

We analyzed models with and without depensation, using maximum likelihood estimation assuming log-normally distributed variation about the mean, and compared their goodness of fit to the observed data (Table 1) [4]. This is equivalent to estimation using log-transformed data and assuming additive normal errors. The models used the Beverton-Holt spawner and recruitment function, modified to include depensatory recruitment [5], given by

$$R = \frac{aS^\delta}{1 + (S^\delta/K)}$$

where R is recruitment of new fish to the population; S is a metric of spawner abundance; and α, K, and δ are all positive parameters. The parameter δ controls the depensation in the recruitment curve. If δ equals 1, there is no depensation. Depensatory dynamics are characterized by $\delta > 1$ and a sigmoidally shaped recruitment curve, with an unstable equilibrium point at low spawner abundance (Figure 1). Our test used the likelihood ratio between the maximum likelihood fit of the model with δ as a free parameter and the model with δ fixed at 1.

Data on 128 fish stocks were extracted from the database prepared by Myers *et al.* [4]. Spawner and recruitment time series for each stock were obtained from assessments prepared to advise management on the harvest of marine and anadromous fishes. We selected from the database those stocks for which the time series encompassed at least 15 years.

For 9 of the 128 stocks, the model with δ as a free parameter gave a significantly better fit at the 0.05 level (Figure 2 and Table 1). In three cases, the estimate of δ was greater than 1 (Figure 3). Spring-spawning Icelandic herring constitute the only population we examined in which the fishery collapsed and remained commercially extinct. Strong environmental changes have been identified that likely affected this stock and may have been responsible for its demise [6]. The only other stocks with significant depensation are Pacific salmon stocks, which

have been driven to extremely low levels by fishing and habitat loss [7].

Although not statistically significant at the 0.05 level, two other stocks show some evidence of depensatory dynamics. Pacific sardines and Georges Bank herring were both driven to very low abundance by overexploitation [8]. In both cases, no recovery was observed for decades, but now both stocks are appearing to increase. Depensation in the population dynamics of these two important stocks cannot be entirely ruled out, given this historical pattern. Depensatory dynamics may arise by multispecies interactions, but if present, they will appear in single-species data provided that many data sets are examined.

Depensatory recruitment, appearing as an inflection in the spawner-to-recruitment relation at low spawner abundance, will be difficult to detect in many data sets because there may be few observations in this portion of the curve. Therefore, we used a statistical power analysis [9] to assess the probability of detecting depensation if it was actually present. For each data set, we estimated the parameters α and K of the Beverton-Holt model, with the additional constraint that the asymptotic recruitment could be no greater than the maximum observed recruitment. Next, we set δ equal to 2 and constructed a sigmoid Beverton-Holt model that matched the fitted model at the 50% asymptotic recruitment point and at the asymptote at infinite spawner abundance. We then generated pseudo-random recruitment from a log-normal distribution, with the shape parameter estimated from the fitted model and the mean given by the constructed curve at each of the observed spawner abundances. Finally, we performed the likelihood ratio test for depensation described above and repeated the procedure 100 times to estimate the statistical power.

Statistical power was greater than 0.95 for 26 stocks for $\delta = 2$ (Fig. 3). In each of these, large declines in abundance have occurred, providing data at reduced spawner abundances. If depensatory recruitment is a general phenomenon in fish populations through this observed range of decrease, we would have expected more than 3 of the 128 stocks examined to show significant depensation in the observed data. These results are robust to gamma instead of log-normally distributed residuals, reasonable estimation error of spawners, and serial correlation in recruitment [10]. It is possible that more complex behavior might be masked by

Table 1.   Results of the likelihood ratio tests for depensation and results from the power analyses using the modified Beverton-Holt function for the 26 stocks for which the estimated power (when the true $\delta = 2$) was found to be >0.95. The table lists the number of pairs of data points n, the estimated depensation parameter δ, the P value from the likelihood ratio test, and the estimated power (when the true $\delta = 2$).

Population	n	δ	P value	Power
Clupeiformes				
Clupeidae				
Herring (*Clupea harengus*)				
Central Baltic	16	1.12	0.89	1.00
Downs stock	65	0.56	<0.01	1.00
Iceland (spring spawners)	23	1.78	<0.01	1.00
Iceland (summer spawners)	43	0.58	0.02	0.99
North Sea	41	1.42	0.29	0.97
Pacific sardine (*Sardinops caerulea*)				
California	31	1.21	0.18	1.00
Engraulidae				
Peruvian anchoveta (*Engraulis ringens*)				
Northern-Central Stock Peru	19	1.86	0.12	1.00
Gadiformes				
Gadidae				
Cod (*Gadus morhua*)				
Labrador	28	0.60	0.04	0.98
Southeast Baltic	22	0.82	0.59	1.00
Celtic Sea	20	1.14	0.82	1.00
Kattegat	19	1.06	0.95	1.00
Haddock (*Melanogrammus aeglefinus*)				
Georges Bank	58	1.19	0.43	1.00
Northeast Arctic	39	0.95	0.88	0.98
Merlucciidae				
Silver hake (*Merluccius bilinearis*)				
Georges Bank	33	0.86	0.25	1.00
Mid-Atlantic Bight	33	1.00	1.00	1.00
Salmoniformes				
Salmonidae				
Pink salmon (*Oncorhynchus gorbuscha*)				
Sashin Creek, Little Port Walter, Alaska	25	1.35	0.03	1.00
Prince William Sound, Alaska	15	1.47	0.04	1.00
Central Alaska	25	0.67	0.04	1.00
Sockeye salmon (*Oncorhynchus nerka*)				
Adams Complex, British Columbia, Canada	38	0.99	0.92	1.00
Chilko River, Canada	38	1.07	0.82	1.00
Egegik, Alaska	32	0.80	0.62	1.00
Horsefly River, Canada	38	1.00	0.96	1.00
Kvichak River, Alaska	25	0.91	0.53	1.00
Skeena River, Canada	39	1.24	0.69	0.96
Stellako River, Canada	38	0.76	0.60	1.00
Early Stuart Complex, Canada	38	0.73	0.02	1.00

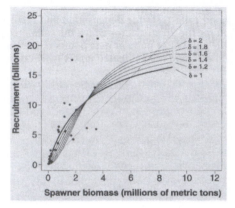

Fig. 1 Spawner recruitment data for the California sardine (dots) compared with theoretical modified Beverton-Holt spawner recruitment curves illustrating depensatory dynamics as determined by δ. The heavy solid line is the estimated Beverton-Holt curve ($\delta = 1$) for these data, whereas the light solid lines show the estimated curve modified to demonstrate depensation at levels $\delta = 1.2, 1.4, 1.6, 1.8,$ and 2.0. To keep the modified curves close to the estimated one, they were constrained to have the same asymptotic recruitment as the estimated curve and to pass through the same point of 50% asymptotic recruitment. The dashed line indicates the number of spawners produced by a given number of recruits if no fishing is assumed. Stable populations occur at the intersections of the solid and dashed lines.

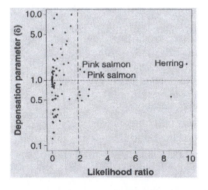

Fig. 2 Results of likelihood ratio tests for depensatory recruitment for 128 fish populations. The null hypothesis was that the depensation parameter δ was equal to 1. When the likelihood ratio (abscissa) is greater than $1/2\chi^2_{1,0.05}$ (the vertical dashed line), the null hypothesis is rejected with a type I error probability of 0.05. Depensation parameter estimates (ordinate) greater than 1 (the horizontal dashed line) indicate depensatory recruitment. Estimates less than 1 give spawner and recruitment relations with high slopes at the origin, indicating no apparent decline of recruitment at low spawner abundance. The populations with statistically significant depensation ($\delta > 1$) are spring-spawning Icelandic herring and pink salmon.

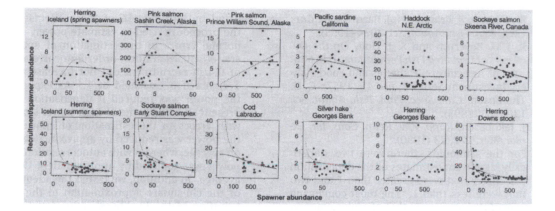

Fig. 3 Survival (recruits per spawner) versus spawner abundance (on a square root scale) for populations with significant depensation or high power to detect depensation. Evidence for depensation is decreased survival at low population levels (first three panels in the top row). Generally, survival was higher at low population levels. We computed survival by transforming recruitment and spawner abundance to the same units and then taking their ratio; thus, it is scaled in terms of replacement levels. When survival is less than 1, the population cannot replace itself (pink salmon from Sashin Creek were an exception; recruitment in that case was in terms of fry). Spawner units are in thousands of metric tons of spawners for marine fish and thousands of individuals for the salmon species. The solid curve is the fitted spawner and recruitment relation with δ fixed at 1. The dotted curve is the fitted relation with δ estimated as a free parameter. The curves were fit to the original recruitment and spawner data.

shortcomings in our approach.

Theoretical analyses and previous nonstatistical descriptions of depensatory recruitment for fish stocks [11] are not substantiated by our comparative analysis of the available data. None of the extant stocks of cod, plaice, hakes, or other commercially valuable species, many of which have been very heavily exploited, displayed depensatory dynamics in reproduction. The great majority of the populations show evidence of increased survival at lower population levels [12]. This analysis indicates that models with strongly reduced per capita reproductive success at the spawner abundance typical of currently surviving fish stocks are not generally applicable to fish population dynamics. The fish population collapses so far observed cannot be attributed to depensatory dynamics. The implication is that reductions in fishing mortality rates implemented by resource managers should enable currently remaining stocks to rebuild, unless environmental or ecosystem-level changes occur that alter the underlying dynamics of the stock. We conclude that the effects of overfishing are, at this point, still generally reversible.

REFERENCES

1. C. W. Clark, *The Optimal Management of Renewable Resources* (Wiley, New York, ed. 2, 1990), pp. 16–18; B. Dennis, *Nat. Resour. Model.* **3**, 481 (1989); C. W. Clark, *J. Cons. Int. Explor. Mer.* **36**, 7 (1974); J. A. Gulland, *ibid.* 37, 199 (1977); R. M. Peterman, *J. Fish. Res. Board Can.* **34**, 1130 (1977).

2. J. R. Beddington, in *Variability and Management of Large Marine Ecosystems*, K. Sherman and L. M. Alexander, Eds. (Selected Symposium 99, American Association for the Advancement of Science, Washington, DC, 1986); C. W. Clark, *Mathematical Bioeconomics* (Wiley-Interscience, New York, ed. 2, 1990); U.S. Department of Commerce, *Our Living Oceans* (NOAA Technical Memo NMFS-F/SPO-2, Washington, DC, 1992).

3. W. C. Allee, *The Social Life of Animals* (Norton, New York, 1938), p. 239; P. A. Larkin, R. F. Raleigh, N. J. Wilimovsky, *J. Fish. Res. Board Can.* **21**, 477 (1964); G. C. Varley, G. R. Gradwell, M. P. Hassell, *Insect Population Ecology: An Analytical Approach* (Blackwell, Oxford, 1973).

4. R. A. Myers, J. Bridson, N. J. Barrowman, *Can. Tech. Rep. Fish. Aquat. Sci.* **2024** (1995).
5. R. J. H. Beverton and S. J. Holt, *On the Dynamics of Exploited Fish Populations* (U.K. Ministry of Agriculture, Fisheries and Food Fishery Investment Series 2, no. 19, 1957); G. G. Thompson, *Can. Spec. Publ. Fish. Aquat. Sci.* **120**, 303 (1993).
6. J. Jakobssen, *Rapp. P.-V. Reun. Cons. Int. Explor. Mer.* **177**, 23 (1980).
7. For the salmonids, there were two statistically significant estimates of the depensation parameter that were greater than 1 and two that were less than 1. The most convincing case of depensation is that of pink salmon in Sashin Creek, AK, at a population abundance of less than 100 females; it is perhaps at this level that depensation is expected to occur for salmonids.
8. G. I. Murphy, *Proc. Calif. Acad. Sci.* **34**, 1 (1966); V. C. Anthony and G. Waring, *Rapp. P.-V. Reun. Cons. Int. Explor. Mer.* **177**, 72 (1980).
9. R. M. Peterman, *Can. J. Fish. Aquat. Sci.* **46**, 2 (1990).
10. We repeated the analyses assuming gamma-instead of log-normally distributed residuals; the results were almost identical. Robustness to estimation error in spawners and serial correlation in recruitment were investi-gated by introduction of these effects into the procedure used to estimate power. Log-normal errors in the estimation of spawners ($\sigma = 0.2$) and first-order autocorrelation of 0.4 in recruitment did not increase type 1 errors if δ was held at 1. As expected, the power was reduced if depensation was present ($\delta = 2$); for the 26 high-power stocks, with errors in the estimation of spawners, the power was reduced by approximately 3% on average, whereas with autocorrelation, the power was reduced by approximately 1% on average. In addition, we tested the adequacy of the chi-square approximation to the distribution of the likelihood ratio statistic by calculating the type 1 error rate when $\delta = 1$ (approximately 3% on average for the 26 high-power stocks).
11. F. Neave, *J. Fish. Res. Board Can.* **9**, 450 (1953); J. G. Hunter, *ibid.* 16, 835 (1959); O. Ulltang, *Rapp. P.-V. Reun. Cons. Int. Explor. Mer.* **177**, 489 (1980).
12. R. A. Myers and N. G. Cadigan, *Can. J. Fish. Aquat. Sci.* **50**, 8 (1993).
13. We thank the many fish population biologists who generously provided their data and the Canadian Department of Fisheries and Oceans Northern Cod Science Program for financial assistance.

Exercises for Chapter 3

The following are differential equations for an unknown function p of the variable t having the form $\frac{dp}{dt} = f(p)$. In each case, first graph $\frac{dp}{dt}$ versus p, and then use the resulting graph to sketch graphs on the t-p plane for solutions $p(t)$ for $t \geq 0$ that have the indicated values for $p(0)$.

1. $\frac{dp}{dt} = p(1 - p)$. Sketch graphs of $p(t)$ versus t for solutions with

    (a) $p(0) = -1$

    (b) $p(0) = 1/2$

    (c) $p(0) = 2$

2. $\frac{dp}{dt} = (p - 1)(p - 2)$. Sketch graphs of $p(t)$ versus t for solutions with

    (a) $p(0) = -1$

    (b) $p(0) = 0$

    (c) $p(0) = 3$

3. $\frac{dp}{dt} = e^p - 1$. Sketch graphs of $p(t)$ versus t for solutions with

    (a) $p(0) = -1$

    (b) $p(0) = 1$

    (c) $p(0) = 4$

4. $\frac{dp}{dt} = 1 - e^p$. Sketch graphs of $p(t)$ versus t for solutions with

    (a) $p(0) = -1$

    (b) $p(0) = 1$

    (c) $p(0) = 4$

4

Stability in a One-Component System

4.1 Initial Values of $x(t)$

You are trying to find a function of time, $x(t)$, that solves the equation $\frac{dx}{dt} = f(x)$. For example,

$$\frac{dx}{dt} = ax \tag{1}$$

has for its solution

$$x(t) = x(0)e^{at}. \tag{2}$$

Here, the value of x at time $t = 0$ is an arbitrary parameter. Each choice gives a different solution, and any choice is allowed.

Here is a second example:

$$\frac{dx}{dt} = ax^n, \tag{3}$$

whose general solution is

$$x(t) = x(0)\left(1 - a(n-1)x(0)^{n-1}t\right)^{-1/(n-1)}. \tag{4}$$

Here again, the value of x at time $t = 0$ [which is the parameter $x(0)$] is a parameter that you are free to specify. [You should check for yourself that the expression on the right side of (4) solves (3) for any choice of $x(0)$. Do this by differentiating the function of t on the right side of (4) and also by taking its nth power.]

Notice that if you determine an initial value for $x(0)$, say 5 or -1.33 or 22.9, then there is precisely one solution to either (1) or (3) whose value at $t = 0$ is equal to the predetermined $x(0)$.

It is no coincidence that the complete specification of a solution to (1) or (3) requires that you specify the value of $x(t)$ at $t = 0$. This is characteristic of any differential equation of the form

$$\frac{dx}{dt} = f(x) \tag{5}$$

for any choice of function f. That is, the solutions to (5) have one free parameter to specify, that being the value of the solution $x(t)$ at $t = 0$. For every such choice of $x(0)$, there is a unique solution to (5) that takes that value at $t = 0$.

This uniqueness can be understood in the following way: Starting at $t = 0$ at $x(0)$, (5) tells you the direction and speed at which to move, namely $f(x(0))$. After time Δt small, you are at a new position,

$$x(\Delta t) \approx x(0) + f(x(0))\Delta t, \tag{6}$$

and (5) again tells you the direction and speed at which to move, namely $f(x(\Delta t))$. After time Δt has passed again, you are at a new position,

$$x(2\Delta t) \approx x(0) + f(x(\Delta t))\Delta t + f(x(0))\,\Delta t, \tag{7}$$

and (5) again tells you the direction and speed to move. Continuing thus forever, you will have constructed an approximate solution to (5) starting at your chosen point $x(0)$. As you take Δt smaller, your approximate solution becomes more nearly a true solution. In some technical sense, in the limit of vanishingly small Δt, you will have constructed a solution to (5) that starts at $x(0)$ at $t = 0$.

4.2   Stability

Here is a one-component system: Suppose x is a function of t that measures the amount of some substance of interest. [Later, I sometimes use the term *trajectory* to mean the function $x(t)$, as I often think of t as time and view the correspondence $t \to x(t)$ as giving a movie of a glowing point as it traverses some part of the x-axis.] Suppose that a model for the behavior of x is given by

$$\frac{dx}{dt} = f(x), \tag{8}$$

where f is some function of x. Of interest here is the behavior of trajectories $x(t)$ that solve (8). Here are some important facts to keep in mind:

Fact 1: Given some value for x at time 0, say x_0, then there is precisely one solution to (8), $x(t)$, that obeys $x(0) = x_0$.

Fact 2: Likewise, given some value, say x_0, for x at some specific time t_0, there is precisely one solution $x(t)$ that obeys $x(t_0) = x_0$.

Here is a new term: An *equilibrium* point for (8) is a value for the starting value of the trajectory for which the ensuing solution is constant in time. That is, an equilibrium point for (8) is a point x_0 with the property that the unique solution to (8) with $x(0) = x_0$ obeys $x(t) = x_0$ for all t.

Equilibrium points are important for the following reason: It is an observational fact that if you take many systems in nature and wait for a long time, then eventually, all motion essentially stops. That is, many (although not all!) systems in nature tend to equilibrium points if you are willing to wait long enough.

The criterion for an equilibrium point is that

$$\frac{dx}{dt} = 0 \quad \text{for all } t. \tag{9}$$

Since (8) is assumed to hold, the equilibrium points are characterised as those values of x where $f(x) = 0$. Graph the function f, and the equilibrium points are the places where the graph crosses the x-axis. See Figure 4.1.

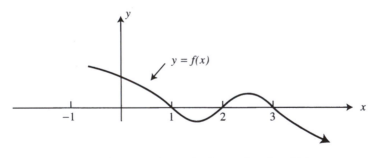

Figure 4.1 Example graph of some function $f(x)$

An equilibrium point x_0 for (8) is called *stable* if there is some segment (perhaps very short, perhaps long) of the x-axis that has x_0 inside it and also has the following special property:

Every solution $x(t)$ that starts in the segment gets continually closer to x_0 as time increases, and limits at x_0 as t tends to infinity.

If an equilibrium point is not stable, then it is called *unstable*. Unstable equilibrium points are almost never seen in nature. This is because a tiny perturbation of the system will move the system away from the equilibrium point. If the equilibrium point is stable, then the system will move back to the equilibrium point. If the system is unstable, then the system will move *away*. Since nature is full of tiny perturbations, you almost never view a system at an unstable equilibrium point.

As (8) is assumed to hold, then we can characterize an equilibrium point x_0 as stable or unstable in terms of the behavior of the function f near to x_0. An equilibrium point x is stable if $f(x) \leq 0$ for all x near x_0 but larger than x_0 and if $f(x) \geq 0$ for all x near x_0, but smaller than x_0. That is, x_0 is a stable equilibrium point if the function f decreases as we move past x_0 from left (smaller than x_0) to right (larger than x_0). In the preceding example, there are two stable equilibrium points, $x = 1$ and $x = 3$.

This is true for the following reason: If x_0 is supposed to be stable, then you want the following to happen: If you start at a point that is slightly *greater* than x_0; then you don't want the value of x to increase along the trajectory. (If it did, you would be moving away from x_0, and x_0 wouldn't be stable.) This requires $\frac{dx}{dt}$ [which is equal to $f(x)$] to be *negative* whenever x is slightly greater than x_0. Likewise, if you start at a point that is slightly less than x_0, then you don't want the trajectory to further decrease the x value, as then you would be moving away from x_0. Therefore, you want $\frac{dx}{dt}$ [which is $f(x)$] to be *positive* at all points x that are slightly less than x_0.

Modeling Example: Lake evaporation: Suppose that water flows into a lake at a constant rate, r, and that water evaporates from a lake at a rate that is proportional to (volume)$^{2/3}$. If the lake volume is constant, and if the evaporation rate is known, can you determine r? [Note that the evaporation rate is proportional to the area, and the area is proportional to (volume)$^{2/3}$ when the lake lies in a bowl that is shaped like an inverted cone. You might try to calculate this constant.]

Solution: Let $v(t)$ = volume at time t. Then $\frac{dv}{dt} = r - av^{2/3}$. By assumption the constant a is known. Set $f(v) = r - av^{2/3}$. There is one positive solution to $f(v) = 0$, namely $v_0 = (r/a)^{3/2}$. This is a stable equilibrium point. If it is assumed that a and v_0 can be measured, then r can be determined: $r = av_0^{2/3}$.

4.3   Lessons

Here are some key points from this chapter:

- An equation of the form $\frac{dx}{dt} = f(x)$ for a function, x, of the variable t is predictive in the sense that if you choose any value of x to start, then there is precisely one solution to this equation that starts at your chosen value.

- An equilibrium point for such an equation is a value, say x_0, for which the constant function $x(t) = x_0$ solves the equation. Thus, equilibrium points correspond to those values of x where $f(x) = 0$.

- An equilibrium point, x_0, is called stable when $f(x) > 0$ when x is slightly less than x_0 and also $f(x) < 0$ when x is slightly greater than x_0.

- Unstable equilibrium points are rarely seen in natural systems.

READINGS FOR CHAPTER 4

READING 4.1

Tempo and Mode of Speciation

Commentary: For our purposes, the point of this article is to illustrate how a function of time that can take whole number values (in this case, the number of species times 1000) can come to look very much like a continuous function when its values are large. This is the essence of Figure 2 of the article and this phenomenon is made explicit in the discussion in the article's final two paragraphs.

The fact is that discrete functions are commonly approximated by continuous ones. In particular, this is no more nor less than what one does when curve fitting to experimental data: You draw the best-fitting line (or parabola, or general curve) through your data points and pretend that the curve accurately interpolates your essentially finite data into continuously distributed data. An example of this sort of curve fitting can be seen in Figure 1 in the article *HIV-1 Dynamics in Vivo: Virion Clearance Rate, Infected Cell Lifespan, and Viral Generation Time*, A. Perelson, A. U. Neumann, M. Markowitz, J. M. Leonard and D. D. Ho, *Science* **271** (1996) 1582–1586, which you can see in Chapter 2's readings.

Tempo and Mode of Speciation

Science, 277 (1997) 1622–1623.
Michael L. Rosenzweig

How does one living species become two? And does the evolution of new species proceed fitfully or smoothly? If fitfully, what sort of cataclysms trigger progress? These are the basic questions that Klicka and Zink illuminate on page 1666 of this issue in a fascinating study of the timing of songbird speciation [1]. From this and other recent work, mixed messages emerge: Speciation serves more than one master. And its regularity depends on the scale at which it is measured.

Birds do not make very good fossils, so they contributed little to our understanding of the principles of evolution until the development of molecular tools such as mitochondrial DNA (mtDNA) sequencing. Because minor variations in mtDNA appear to have no consequences for fitness, mtDNA can be a neutral indicator of time and a valuable evolutionary clock. As more time passes, more random differences will accumulate between species. Now, knowing how to tell molecular evolutionary time, we can ask basic questions about bird evolution whose answers should apply more generally.

Evolutionists believe that speciation usually begins with the subdivision of species into isolated populations by a geographical barrier (allopatry). What better barrier than a continent-sized glacier knifing down through North America and slicing many of its major environments into eastern and western segments? Because many western and eastern species of birds are sisters, evolutionists attributed their speciation to geography and to its ally, the glacial scalpel.

Klicka and Zink show that we were wrong, at least about the glaciers. It makes one want to reexamine a whole class of geographical case histories-cases that emerged before molecular evidence became available. Were we also wrong to conclude from them that sympatric mechanisms of speciation—mechanisms that do not require geographical subdivision—are unimportant?

Author affiliation: Michael L. Rosenzweig: Department of Ecology and Evolutionary Biology, University of Arizona.

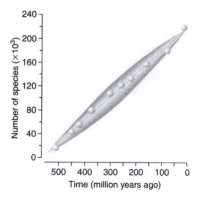

Fig. 1   Virtually constant accumulation of species throughout Phanerozoic time. The concave and convex curves show the limits to the curvature of the data given their rate variation [15]. Maximum rate: 611 species/million years; minimum rate: 292 species/million years.

Sympatric speciation differs most significantly from geographical speciation in that it is triggered by ecological interactions [2]. It targets ecological opportunities and produces new species with a high probability of survival. Evidence from fossils now strongly indicates that sympatric speciation was crucial in the radiation of innovative forms of life [3]. And new mathematical theory [4] has removed any remaining genetic qualms about it.

Nevertheless, geographical allopatric speciation usually dominates other mechanisms. To conclude otherwise would require us to explain away too many facts. For example, of the world's 18,818 species of fish, 36% live in fresh water although only 1% of the Earth's surface is fresh water, and fish productivity and population are considerably higher in marine continental shelves and upwelling currents. The myriad opportunities for fishes to form geographical isolates in fresh waters as compared with salt looks like unassailable evidence in favor of the dominance of geographical speciation.

Another fact: The number of species in a biological province such as the Neotropics depends on the area of the province. Larger provinces have more species, and the relation is virtually linear [2]. Dynamic theory based on geographical allopatric speciation predicts that correlation. Sympatric speciation does not.

The tempo of speciation also speaks in favor of a geographical mechanism. Sympatric speciation predicts that new species will emerge at an ever decreasing rate. But species form rather steadily during most periods. We see this in the fossil record [5], as well as in North American songbird DNA. And geographical speciation predicts it.

Yet, like a digital clock, genetic change is discrete at the finest scale. A digital clock seems to operate spasmodically at the scale of a second, but quite smoothly at that of an hour. Surely, at the scale of an individual base substitution, mtDNA change must resemble a "second hand." But the glacial hypothesis sought to extend fitfulness to a larger scale: 10^5 to 10^7 years.

The new DNA data from songbirds demonstrates that speciation events did not cluster as a result of the glaciation of North America. To a certain extent these events are distributed randomly throughout the time interval during which these pairs of species diverged. That is not unique. For instance, mammal fossils from the Meade Basin of Kansas show a similar scatter of speciation over a similar period [6].

But the bird DNA may also signal a real pulse of new species about 2.5 million years ago. One would ordinarily attribute the high value of the mode at 6% divergence to the workings of probability [7]. But a second set of 17 independent bird species shared this value almost exactly. Moreover, this pulse echoes a similar pulse in speciation events among antelopes, rodents, and hominids at about the same time in Africa [8]. Something suspicious is going on. This period coincides with the onset of a major global cooling off, a change that led eventually to the ice ages [9]. As a matter of fact, that very climate change may well have triggered our own speciation [10]. It certainly comes close to the time that hominids first began to show increased relative brain sizes [11]. Curious indeed. The geographical barrier of the glacier itself does not create a ripple in speciation rate, but the climate change that preceded it does.

Undoubtedly, the fossil record at scales of tens of millions of years shows a mixture of regularity and pulses. The pulses come in the form of mass extinction events that—perhaps periodically—sweep away 30% to 97% of the world's species and trigger prolonged episodes of replacement speciation [12]. During such times, I am not so sure that geographical speciation predominates.

Once we step back to the scale of hundreds of millions of years, however, the signal of pulses vanishes. The ups and downs of diversification disappear into a smoothness that almost defies belief. In the Figure 1, I plotted the adjusted numbers of marine invertebrate fossil species that first made their appearance during one of the 10 major time periods of the Phanerozoic [13] (adjustments correct for the sampling bias generated because we have less old rock than young rock). The ordinate is cumulative so that the slope of the curve gives the estimated rate of speciation. The data cluster closely about a straight line. That line—fit by regression—shows where the data would be if the rate of speciation had been absolutely constant.

So, like the proverbial fractal coastline, the rhythm of speciation looks jagged at very fine scales, takes on a smoother but still complex appearance at intermediate ones, and loses virtually all traces of irregularity when we look at the history of life at its grandest scale. This signifies the presence of worldwide steady states in diversity [2, 14]. Negative feedback loops hold its component process rates within bounds by acting through a single parameter—the extent of the area that the average species occupies. Because of competition and predation, high diversity leads to reduced average area. In turn, reduced area lowers speciation rates and elevates extinction rates. Variations in these rates may be extremely difficult to distinguish from noise.

REFERENCES

1. J. Klicka and R. M. Zink, *Science* **277**, 1666 (1997).

2. M. L. Rosenzweig, *Species Diversity in Space and Time* (Cambridge Univ. Press, Cambridge, 1995).

3. M. L. Rosenzweig and R. McCord, *Paleobiology* **17**, 202 (1991).

4. P. A. Johnson, F. C. Hoppensteadt, J. J. Smith, G. L. Bush, *Evol. Ecol.* **10**, 187 (1996).

5. M. L. Rosenzweig, in *Biodiversity Dynamics; Turnover of Populations, Taxa and Communities*, M. L. McKinney, Ed. (Columbia Univ. Press, New York, in press).

6. R. A. Martin and K. B. Fairbanks, *Evol. Ecol.*, in press.

7. J. K. McKee, *J. Theor. Biol.* **172**, 141 (1992).

8. E. S. Vrba, in *Macroevolutionary Patterns in the Plio-Pleistocene of Africa*, T. G. Brommage and F. Shrenk, Eds. (Oxford Univ. Press, Oxford, in press).

9. G. H. Denton, *ibid.*

10. E. S. Vrba, in *The Evolutionary History of Robust Australopithecines*, F. E. Grine, Ed. (Aldine Publishing, New York, 1988), pp. 405–426.

11. B. Wood and M. Collard, in [8].

12. J. J. Sepkoski, J. Geol. Soc. London 146, 7 (1989); D. Jablonski, *Science* **253**, 754 (1991); G. J. Retallack, *ibid.* **267**, 77 (1995).

13. Data for the Figure came from the following sources: Ages and durations from [W. Harland *et al.*, *A Geologic Time Scale*, 1989 (Cambridge Univ. Press, Cambridge, 1990)] with the help of J. J. Sepkoski Jr. to adjust the beginning times of the Cambrian and Ordovician periods. Rock areas and unadjusted diversities came from [D. M. Raup, *Paleobiology* **2**, 279 (1976a); J. J. Sepkoski Jr., *Acta Palaeontol. Pol.* **38**, 175 (1993)]. Raup counted 71,112 species entries in the zoological record from 1900 to 1970. He extrapolated 73,139 more from these 71,112. Of these 144,251 species, 7416 were insects, or uncertain, or lived in more than one period, or lived in the Pre-Cambrian; I excluded the latter. The remaining 136,835 species represent a substantial fraction of all known fossil forms. The adjusted number of species in a period is the number (of the 136,835) attributed to it divided by $A^{0.383}$, where A is the period's rock area divided by that of the Cenozoic.

14. P. W. Bretsky and S. W. Bretsky, *Lethaia* **9**, 223 (1976); W. D. Allmon, G. Rosenberg, R. W. Portell, K. S. Schindler, *Science* **260**, 1626 (1993).

15. First, arrange the data in order of declining accumulation rates, so that the "first" period has the highest rate and the "last" has the lowest. This imposes the largest possible negative second derivative on the data. Second, reverse the order (so that the one with the lowest rate is first). This imposes the largest possible positive second derivative on the data. The results are, in the first case, an exponent of 0.866 in a power fit, and in the second, an exponent of 1.204. The Figure 1 shows the polynomial approximations of these curves.

16. Title respectfully purloined from my late colleague, G. G. Simpson. I thank W. DiMichele, D. Jablonski, K. Flessa, M. McKinney, A. Miller, K. Niklas, S. Pimm, J. Sepkoski, and D. A. Thomson for stimulating ideas, advice, and feedback.

Exercises for Chapter 4

1. Suppose the function $x(t)$ evolves according to the following differential equation: $\frac{dx}{dt} = f(x)$. For each of the four candidate functions $f(x)$ graphed in (a) through (d), describe what happens to $x(t)$ as t gets large if $x(0) = 1$.

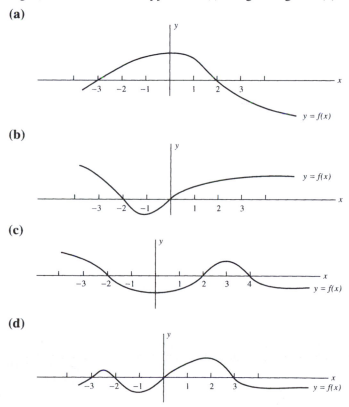

(a)

(b)

(c)

(d)

2. For each of the same four functions as in Exercise 1, describe what happens to $x(t)$ as t gets large if $x(0) = -4$.

3. Repeat Exercise 1, but now with $x(0) = -1$.

4. For each of the four functions from Exercise 1, list the equilibrium points and decide which are stable and which are unstable.

5. Find the first-order Taylor's expansion at $x_0 = 0$ for the following functions:
   - **(a)** $f(x) = \frac{1}{(1+2x)}$
   - **(b)** $f(x) = 3\cos(2x)$
   - **(c)** $f(x) = e^{4x}$

6. Suppose that someone proposed that the number $n(t)$ of AIDS viruses in a person evolved (approximately) in time as a solution to the equation $\frac{dn}{dt} = \alpha n - \nu n^2$, where α and ν are constants. How could a doctor measure α and ν?

5

Systems of First-Order Differential Equations*

Consider two types of animals, say left-curling snails and right-curling snails, that compete for the same resources. We will use $L(t)$ to denote the number of left-curling snails (in units of millions) at time t, and $R(t)$ to denote the corresponding number of right-curling snails. These two populations compete for the same resources. We might propose the following equations to govern the dynamics of the population numbers:

$$\frac{dR}{dt} = R - (R^2 + aRL)$$

$$\frac{dL}{dt} = L - (L^2 + aLR). \tag{1}$$

Here, a is a constant, a parameter that we can think of as a crude measure of the extent to which right- and left-handed snails interact with each other. Note that these equations treat right-curling and left-curling snails in an identical fashion. Thus, the model implicitly assumes that there is no biological curse inherent in left curling versus right curling.

Question: Does this simple model predict the observed phenomenon of near absence of left-curling shells as discussed by S. J. Gould in Reading 2.2?

Remarkably enough, this question can be answered without solving these equations explicitly. The method of doing so is called **phase plane analysis**, and it is the subject of this chapter and some of the subsequent chapters.

---

*Parts of this chapter paraphrase Harvard University notes by Otto Bretscher and Robin Gottlieb.

5.1   The Case $a > 1$

It turns out that the predictions of the model in (1) differ drastically depending on whether $a > 1$ or $a < 1$. To be concrete, consider first the case, where $a = 2$ and (1) reads

$$\frac{dR}{dt} = R - (R^2 + 2RL)$$

$$\frac{dL}{dt} = L - (L^2 + 2LR). \tag{2}$$

To see how phase plane analysis works, suppose initially at time $t = 0$, in Cambrian time, the number of right-curling snails was $R(0) = 510{,}000$ and that of left-curling snails was slightly less, say 500,000. We represent this by the point $(0.51, 0.50)$ in the (R, L) plane. A year later, there will be a new value of R and a new value of L; this new situation is represented by the point $(R(1), L(1))$ in the (R, L) plane. More generally, for each time t, the situation is represented by a point $(R(t), L(t))$ in the (R, L) plane. As t changes, this point may move, but in any event, it traces out a directed path in the (R, L) plane that is called the ***trajectory***. We are interested in the behavior of this trajectory as t gets very large—effectively, as t goes to infinity. Here, we will introduce some new notation and write $t \to \infty$ to mean "t goes to infinity."

Of course, we can never take t equal to infinity, but we can hope that as t gets very large, the trajectory approaches some limiting point or points in the (R, L) plane. In some situations, the trajectory may have a nice limit as t gets large, and in others, it most emphatically will not. (Note that the Cambrian period ended roughly half a billion years ago, so in the model at hand, *very large* means t on the order of 500 million.)

To discern the initial direction of the trajectory, consider substituting for R and L in (2) the values $(0.51, 0.50)$ at $t = 0$. This substitution on the right-hand side of (2) finds

$$\frac{dR}{dt} = -0.2601,$$

$$\frac{dL}{dt} = -0.26. \tag{3}$$

Both of these numbers are negative, so both R and L are decreasing initially. Thus, we know that the trajectory goes down and to the left in a plane with axes labeled R and L. Note that R, which starts somewhat larger than L, is decreasing slightly faster than L as drawn in Figure 5.1.

How can we analyze the trajectory in the long run? The fact that initially the trajectory moves down and to the left does not guarantee that this trend will continue, since both $\frac{dR}{dt}$ and $\frac{dL}{dt}$ change as movement proceeds along the trajectory.

The general direction of the trajectory is determined by the signs of $\frac{dR}{dt}$ and $\frac{dL}{dt}$. Before we find out where they are positive and negative, we determine the "borderline case," where $\frac{dR}{dt} = 0$ and where $\frac{dL}{dt} = 0$. For this purpose, we go back to (2) and read off that

$$\frac{dR}{dt} = 0 \text{ when } R(1 - R - 2L) = 0. \text{ Thus, } R = 0 \text{ or } L = (1 - R)/2.$$

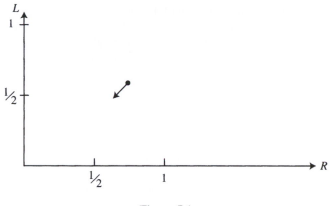

Figure 5.1

$$\frac{dR}{dt} = 0 \text{ when } L(1 - L - 2R) = 0. \text{ Thus, } L = 0 \text{ or } L = (1 - 2R). \tag{4}$$

The curves where $\frac{dR}{dt} = 0$ are called **R *null clines***, and those where $\frac{dL}{dt} = 0$ are called **L *null clines***. A picture of the null clines in the (R, L) plane is provided in Figure 5.2.

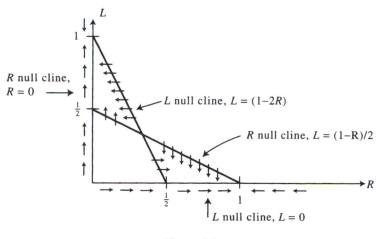

Figure 5.2

Note that the null clines where $\frac{dR}{dt} = 0$ are represented by vertical slash marks with arrows, and those where $\frac{dL}{dt} = 0$ by horizontal ones with arrows. There is a reason for this: As a trajectory hits an R null cline, then $\frac{dR}{dt} = 0$ (by definition) so the movement, at least instantaneously, is vertical only. There is no horizontal component to the direction of movement. The arrow indicates whether the vertical motion across this part of the null cline is up or down. The direction of the arrow can be found by computing the sign of the right-hand side of the second line of (2) at points on the null-cline segment. For example, at the point $(0.5, 0.25)$ on the R null cline, the left-hand side of the second line of (2) is negative. This means that the trajectory that crosses this null cline at $(0.5, 0.25)$ does so moving down. Note that as we moves along the R null cline, the direction of the arrow can change only where the R null cline intersects an L null cline, because these are precisely the places where the derivative of L is zero. Thus, we need only check the direction of motion for the R null clines at a limited number of points.

To summarize, the vertical slash lines with arrows on the R null cline indicate the direction of movement across the null cline. Likewise, the L null cline is marked by horizontal slash marks with arrows that show the direction of movement of a trajectory across this null cline. The motion here is horizontal because the vertical component of motion is zero due to the vanishing of $\frac{dL}{dt}$ on the L null cline. The arrow indicates whether $\frac{dR}{dt}$ is positive (so R is increasing) or negative (so R is decreasing) across this null cline. Here, the sign is determined by evaluating the right-hand side of the first line in (2) at the point in question. For example, at the point $(1/6, 2/3)$ on the L null cline, the right-hand side of the top line in (2) is negative. This means that the trajectory through this point moves from right to left. (Once again, this sign can only change where the L null cline intersects the R null cline.)

The null clines $\frac{dR}{dt} = 0$ and $\frac{dL}{dt} = 0$ intersect at the points $(0, 0)$, $(1, 0)$, $(0, 1)$, and $(1/3, 1/3)$. If the initial numbers of the two types of snails happen to be represented by one of these points, then the two population numbers remain consant since the derivatives of R and L are both zero there. These points are called the **equilibrium points** of the system. [For example, the trajectory that starts at $(1/3, 1/3)$ stays there for all time and so consists of only that point, rather than some path in the $R - L$ plane.]

Now we come back to the question of where $\frac{dR}{dt}$ and $\frac{dL}{dt}$ are positive and negative. We know that in the various regions marked (I), (II), (III), and (IV) in Figure 5.3, which are between the null clines, the signs of $\frac{dR}{dt}$ and $\frac{dL}{dt}$ do not change.

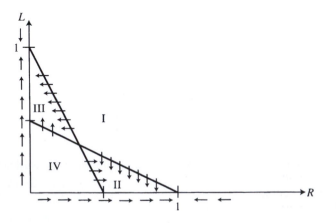

Figure 5.3

Thus, we can choose a sample point in each region to find these signs. For example, in the region marked I in Figure 5.3 we could pick the starting point $(0.51, 0.5)$ to find that both signs are negative. Thus, the general direction is therefore down and to the left throughout Region (I). For a second example, in Region (IV), we can choose the point $(0.1, 0.1)$ to find that

$$\frac{dR}{dt} = 0.1 - 0.03 = 0.7,$$
$$\frac{dL}{dt} = 0.1 - 0.03 = 0.7, \tag{5}$$

and so in Region (IV), the general direction is up and to the right. Applying the same method to all of the other regions and segments of null clines, we obtain Figure 5.4.

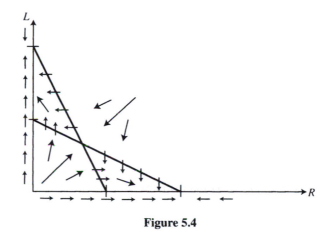

Figure 5.4

Based on this work, what can we say about the trajectory starting at $(0.51, 0.5)$? The trajectory initially moves down and to the left and will continue to do so until one of three things happen:

- L eventually becomes larger than R and the trajectory eventually crosses the L null cline where $L = (1 - 2R)$. The trajectory crosses this null cline moving horizontally and then, as it is in Region (III), it moves up and to the left. As it is moving up and to the left, it cannot recross the $L = (1 - 2R)$ null cline because the arrows on the slash marks for this null cline all point into Region (III). Thus, no trajectory can leave Region (III) across this null cline. Neither can it cross the $L = (1 - R)/2$ segment of the R null cline because the arrows on this null-cline segment all point to the left and thus also into Region (III). Hence, the trajectory is forced to continue in a manner that decreases R toward zero as it approaches the $R = 0$ portion of the R null cline.

- L stays less than R and the trajectory eventually crosses the R null-cline segment where $L = (1 - R)/2$. It crosses this segment vertically and then enters

Region (II), where it moves down and to the right. As it is moving down and to the right in this region, it cannot exit the region across either the $L = (1 - R)/2$ segment of the R null cline or the $L = (1 - 2R)$ segment of the L null cline. In either case, the arrows point into Region (II). Thus, the trajectory is forced to continue in a manner that decreases L towards zero as it approaches the the $L = 0$ portion of the L null cline.

- L and R approach equality along the trajectory as the trajectory approaches the $R = 1/3$ and $L = 1/3$ equilibrium point.

Thus, we see from the preceding that the evolution of the system unequivocally results in one of the following three scenarios at large time:

- Only left-curling snails survive and right-curling snails are extinct.

- Only right-curling snails survive and left-curling snails are extinct.

- There are essentially equal numbers of right- and left-curling snails.

In particular, this model has one outcome (the second point in the preceding list) that is reasonably close to the true distribution of snails.

5.2   The Case $a < 1$

To be concrete, let us take the case where $a = 1/2$ so that the model in (1) reads

$$\frac{dR}{dt} = R - (R^2 + RL/2),$$
$$\frac{dR}{dt} = L - (L^2 + LR/2). \tag{6}$$

As before, suppose that at time zero in the Cambrian, we have $(R, L) = (0.51, 0.5)$. We will analyze the subsequent evolution for this case in the same manner as with the case in Equation (2).

In this regard, the first step is to draw the perpendicular axis in the plane and label the horizontal axis R and the vertical axis L. Next, we draw the R null clines; these occur where the right-hand side of the top line in (6) vanishes—that is, where either

$$R = 0,$$
$$\text{or}\quad L = 2(1 - R). \tag{7}$$

With the R null clines drawn on the graph, we then draw the L null clines. The latter occur where the right-hand side of the second line of (7) vanishes. This happens where either

$$L = 0,$$
$$\text{or}\quad L = 1 - R/2. \tag{8}$$

The next step is to draw the vertical hash marks on the R null clines and the horizontal hash marks on the L null clines. Then, use the right-hand side of (6) to determine the direction of the arrows on the hash marks. The resulting picture is showin in Figure 5.5.

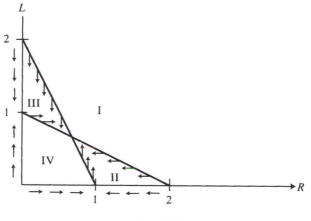

Figure 5.5

The intersections of the R null clines with the L null clines are the equilibrium points. These occur at $(0, 0)$, $(0, 1)$, $(1, 0)$, and $(2/3, 2/3)$.

The null clines break the first quadrant of the (R, L) plane into four regions, as labeled in Figure 5.6. Pick a point in each region and use its (R, L) values in the right hand side of (6) to determine the rough direction of the trajectories in that region. For example, the point $(1, 1)$ is in Region (I), and at this point, we find from (6) that

$$\frac{dR}{dt} = -1/2,$$
$$\frac{dL}{dt} = -1/2. \tag{9}$$

Thus, in Region (I), the trajectories move down and to the left. On the other hand, the point $(1/2, 1/2)$ is in Region (IV). Substituting $R = 1/2$ and $L = 1/2$ into the right-hand side of (6), we find that both of the derivatives in question are positive. Thus, in Region (IV), the trajectories all move up and to the right. The graph in Figure 5.6 indicates the behavior.

We are now ready to determine the long-time evolution of our model from the starting position where $R(0) = 0.51$ and $L(0) = 0.5$. We see that there are three possibilities:

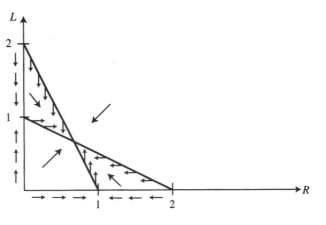

Figure 5.6

- The trajectory starts in Region (IV) and moves up and to the right initially. At some point, L becomes larger than R, and then sometime later, the trajectory enters Region (III) moving horizontally and to the right. In Region (III), the trajectory moves down and to the right, but it can never leave Region (III) since all of the hash mark arrows are pointing to this region. Thus, the trajectory is forced to approach ever closer the equilibrium point $(2/3, 2/3)$ as time evolves.

- The trajectory starts in Region (IV) and moves up and to the right initially. The value of R stays larger than L, and the trajectory eventually enters Region (II) by crossing the R null cline where $L = 2(1 - R)$ moving vertically. Once in Region (II), the trajectory moves up and to the left. The trajectory cannot exit Region (II) since all of the hash marks on the boundary of the region are pointing in. Thus, in this case also, the trajectory is forced to approach ever closer to the equilibrium point $(2/3, 2/3)$ as time evolves.

- The values of R and L approach equality as the trajectory advances up and to the right in Region (IV). Moreover, the trajectory stays in Region (IV), but approaches ever closer to the equilibrium point $(2/3, 2/3)$.

Thus, we see that in the case where $a = 1/2$ at least, all the preceding scenarios have the same result at large time: The initial value of $R = 0.51$ and $L = 0.5$ evolves in time toward the equilibrium point where R and L are equal with value $2/3$. As R and L are definitely not equal in real life, we can see that the case $a = 1/2$ for the model in Equation (1) can be discarded.

In fact, the $a = 1/2$ behavior is characteristic of all $a < 1$ versions of the model in Equation (1). Although there are quantitative differences with respect to the precise value for R and L of the limiting equilibrium point, the trajectory nonetheless approaches an equilibrium point where $R = L$. Thus, the case $a < 1$ in the model in Equation (1) does not match real-world data.

On the other hand, the case $a > 1$ is very much like that for $a = 2$, and in this case, the conclusions in (11) still hold. In particular, in the $a > 1$ case, one of the possibilities comes pretty close to the real-life situation. However, before we pat ourselves on the back, we should stop to ask whether $a > 1$ is a reasonable assumption. Indeed, this case, where $a > 1$, presents a simplified model for the situation where right-curling shells are more tolerant of the presence of other right-curling shells than other left-curling shells. Is this a reasonable assumption? What do you think?

Here are some possible scenarios: Perhaps shells cannot distinguish the curling direction of their neighbors. Alternately, suppose that shells do detect the curling direction of their neighbors, but are less tolerant of like-curling neighbors rather than more tolerant. Indeed, suppose that like-curling shells breed only with each other. Furthermore, suppose that snails come in males and females. (Do they?) Then right-curling males might fight right-curling males for dominance, but tolerate right-curling females; and likewise, right-curling females might fight right-curling females and tolerate right-curling males. Meanwhile, both ignore left curlers as these do not represent competition for breeding success, only competition for food. [Here is where a model suggests directions for field research and experiments. The point is that the question of whether the constant a in (1) is greater than 1 may, in principle, be verified by field research.]

5.3   The Lotka-Volterra Equation, a Predator-Prey Model

Austrian biophysicist Alfred Lotka and Italian mathematician Vito Volterra separately wrote down and analyzed a system of differential equations that model the interaction of predator and prey species. We give the example of the predator being foxes and the prey being hares. Let $F(t)$ denote the number of foxes at time t, and let $H(t)$ denote the number of hares. The model assumes that rates of change of $F(t)$ and $H(t)$ obey the equation

$$\frac{dH}{dt} = (a - bH - cF)H,$$
$$\frac{dF}{dt} = (-d + eH)F, \tag{10}$$

where a, b, c, d, and e are positive constants that we might hope to determine from field research data.

To help see the significance of these constants to the model, consider first writing the first line in (10) as

$$\frac{dH}{dt} = \alpha H, \tag{11}$$

where $\alpha = a - bH - cF$ is the net birth-death rate for hares when H is the number of hares and F is the number of foxes. Notice that when $F = 0$, so there are no foxes, then (11) is exactly the logistics equation that we studied previously. Thus, we can identify a as an intrinsic growth rate of hare in an ideal environment and we can identify a/b with the carrying capacity of the environment in the absence of foxes. Thus, we could, in principle, measure a and b by raising hares in a sufficiently large enclosure that is fenced to keep all foxes at bay. Meanwhile, if we let F be nonzero, we see that the term $-cFH$ in (10) models the predatory effects of foxes on the hares. Note that this effect increases with increasing number of foxes (as we might expect) and also with increasing number of hares (which is debatable). We might try to determine the constant c by measuring the birth versus death rate of hares in a patch of the environment that is not fenced to preclude fox predation.

In the second line of (10) we can identify the quantity $-d + eH$ as a net birth-death rate for foxes. Here, we see that when $H = 0$, the fox equation reads $F' = -dF$, with solution $F(t) = F(0)e^{-dt}$. This decreases to zero as t increases which is expected: Without hares, the foxes will starve. (Of course, this assumes that there is no alternative source of prey. Such an assumption may not be tenable in any given environment.) The term $+eHF$ in the second line of (10) models the positive effect of hares on the birth rate of foxes. That is, if hares are present, the foxes eat well and the birth rate increases, while the death rate decreases. So hares have a positive effect on the rate of change of F. The measurement of d and e can also be made (perhaps) by raising foxes in an enclosed environment where there food supply is controlled and their birth and death rates as a function of food supply are monitored.

To simplify the subsequent story, I will now choose the constants a, b, c, d, and e that appear in (10) so that the equations read

$$\frac{dH}{dt} = (2 - H - F)H,$$
$$\frac{dF}{dt} = (-1 + H)F. \tag{12}$$

As in the previous examples, the analysis of these equations for H and F starts with the drawing of a plane with axis labled H (say the horizontal axis) and F (say the vertical axis). We next draw the H null clines. From the right side of the first line in (12), we see that these occur where

$$H = 0 \quad \text{or} \quad F = 2 - H. \tag{13}$$

Mark these H null clines with vertical slash marks to indicate that the trajectories cross these lines moving vertically.

The next step is to draw the F null clines. From the right-hand side of the second line of (12), we see that these occur where $F = 0$ or $H = 1$. Mark the F null clines with horizontal slash marks to indicate that the trajectories cross them moving horizontally. The resulting (H, F) plane looks like Figure 5.7.

As remarked previously, the equilibrium points are the points where the H null clines cross the F null clines. In this example, they are $(0, 0)$, $(2, 0)$, and $(1, 1)$. Note that if a trajectory starts at an equilibrium point, it stays there forever since both the derivatives of H and of F vanish at such a point. (That is why they are called equilibrium points.) Note that neither $(1, 0)$ nor $(0, 2)$ are equilibrium points. Indeed, neither is the intersection of an H null cline with an F null cline as the former is the intersection of two H null clines and the latter is the intersection of two F null clines.

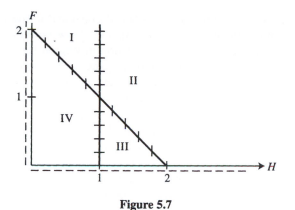

Figure 5.7

As indicated in Figure 5.7, the null clines break the plane into four regions. (The fact that all of our examples have four regions is a coincidence. There could be, in general, any number of regions.) By choosing a point in each region in turn, we can decide on the general direction of motion in that region by plugging into the right-hand side of (12) the H and F values for the chosen point. For example, the point $(3, 1)$ is in Region (II) and we see from the right-hand side of (12) that at this point

$$\frac{dH}{dt} = -6,$$

$$\frac{dF}{dt} = 2, \tag{14}$$

and so the motion in Region II is up and to the left. The general direction of motion in the remaining regions can be determined by a similar strategy. The resulting directions are marked on the H-F plane as in Figure 5.8.

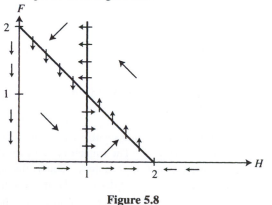

Figure 5.8

Figure 5.8 also has arrows exhibited on the null-cline slash marks to indicate the direction of motion across these null clines. These arrows can be determined as follows: We note that in Region (I) the motion is down and to the left, while in Region (II), it is up and to the left, so on the null cline between these two regions, the motion must be to the left. Likewise, the motion in Region (I) is down and to the left, while in Region (IV), it is down and to the right, so the motion on the null cline between these regions is down. On the other hand, the motion in Region (III) is up and to the right and as that in Region (IV) is down and to the right, the motion on the null cline between them must be purely to the right. The direction of motion on the null cline between Regions (II) and (III) is determined by a similar analysis.

To determine the direction of motion across the part of the boundary of Region (I) on the F-axis, we note that the motion in Region (I) is down and to the left and on this null cline, the direction is purely vertical, so it must be down. In Region (IV), the motion is down and to the right, so the motion on the F-axis part of the boundary of Region (IV) (which is vertical) must also be down, while that on the H-axis part of Region (IV) (which is horizontal) must be to the right. The directions of the arrows on the other parts of the H-axis are determined by a similar analysis.

With the H-F plane completely marked, we are now ready to consider the qualitative properties of hare-fox evolution as predicted by our model.

In particular, suppose we start at the point $(3, 1)$ in Region (II). The motion here is up and to the left, and so the trajectory takes off in this direction. This up and to the left motion persists until the F null cline at $H = 1$ is crossed from right to left and the trajectory enters Region (I). Here, the motion is down and to the right. This motion proceeds until the trajectory hits the H null cline where $F = 2 - H$. The trajectory crosses this null cline pointing down and continues into Region (IV).

At this point, the careful reader might wonder why the trajectory has no collision with the F-axis. The reason is quite simple: Motion on the F-axis is straight down, and this precludes a trajectory from hitting this axis from the side. Indeed, if a trajectory were ever to hit the F-axis, it would move straight down it. Then, if we imagine filming the action and running the film backward, we would see the trajectory move up the F-axis. As the time derivative of H on the F-axis is exactly zero, we would never see the trajectory leave this axis and thus it couldn't have hit the F-axis to begin with unless it started there. There is a general principle at work here:

- A vertical H null cline cannot be crossed since motion on this line is purely vertical.

- A horizontal F null cline cannot be crossed since motion on this line is purely horizontal.

In any event, once in Region (IV), the trajectory moves now down and to the right until it crosses the F null cline at $H = 1$. (The trajectory can't hit the H-axis unless it starts there since the time derivative of F is zero there. This is the second point above.) The trajectory then crosses the $H = 1$ part of the F null cline moving from left to right and enters Region (III) and then moves up and to the right. This motion persists until the H null cline where $H = 2 - F$ is crossed, this time moving up. The trajectory then reenters Region (II) and begins to cycle around again.

Figure 5.9 contains a rough sketch of the trajectory as determined so far.

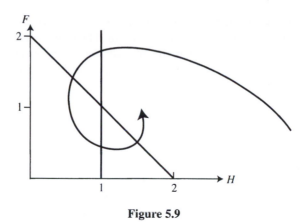

Figure 5.9

Thus, we see that the trajectory circles around the equilibrium point at $(1, 1)$. However, at this point, we do not have the tools to decide between the following two possibilities:

- The trajectory approaches a closed loop trajectory that encircles the equilibrium point $(1, 1)$. The latter would describe a cyclic oscillation of the predator and prey numbers.

- The trajectory spirals slowly into the equilibrium point $(1, 1)$. Note that (approximately) cyclic behavior in natural predator-prey populations is not uncommon.

5.4   Lessons

Here are some key points from this chapter:

- Information can be obtained from a differential equation as in (1) or (10) without having to solve the equation.

- Study the phase plane analysis for the examples of (1) and (10). In particular, familiarize yourself with the drawing and marking of null clines and equilibrium points in these examples, and study how they are used to discern the general direction of movement of a solution on the phase plane.

READINGS FOR CHAPTER 5

READING 5.1

Left Snails and Right Minds

Commentary: We are returning to this article from Chapter 2 (Reading 2.2; see page 23) to try to model the interaction between left-curling and right-curling snails of the same species. Let $L(t)$ and $R(t)$ denote their respective populations after time t. Here is a model: $\frac{dL}{dt} = L - L^2 - aRL$ and $\frac{dR}{dt} = R - R^2 - aLR$. Here, a is a positive constant that measures the relative interaction between right- and left-handed snails.

The case $a > 1$ holds when left and right curlers interfere more with each other than with their own kind. The case $a < 1$ is the opposite. The analysis finds the $a > 1$ case more like the real world. But only experiments with real snails can decide the correct value of a. Or, for that matter, only experiments can decide if such a simple-minded model is even reasonable.

Exercises for Chapter 5

1. Outline an experiment that can determine the value of the constant a, that appears in Equation (1) of Chapter 5.

2. Outline experiments that can determine the value of the constants a, b, c, d, and e that appear in Equation (10) of Chapter 5.

6

Phase Plane Analysis

This chapter continues our study of multicomponent differential equations. The discussion starts with a (simplified) model application: Consider the spread of a viral epidemic through an isolated population. Suppose that the following assumptions are valid:

1. Individuals are infected at a rate proportional to the product of the number of infected and susceptible individuals. The constant of proportionality is 0.05 per day.

2. The length of the incubation period is negligible. That is, infected individuals are immediately infectious.

3. On the average, an infected individual dies after 10 days.

4. Initially, no one is sick.

5. Infected individuals do not give birth, but susceptible individuals have a birth rate of 0.0003 per individual per year. Newborns are susceptible.

Here is a system of differential equations that can be used to model the course of the epidemic: Let $x(t)$ denote the number of susceptible people at time t. Let $y(t)$ denote the number of infected people at time t. Then the rates of change of x and y are given by

$$\frac{d}{dt}x = -0.05xy + 0.0003x \tag{1a}$$

$$\frac{d}{dt}y = +0.05xy - 0.1y. \qquad (1b)$$

(The first equation asserts that in a unit time interval, any given infected individual will infect any given susceptible individual with 5% probability; and furthermore, new susceptibles appear due to births. The second equation asserts that any given susceptible individual becomes infected by any given infected individual with probability 5% in a unit of time, but infected individuals disappear because 10% of them die on any given day.)

Our goal for this chapter will be to develop tools that will enable us to analyze (1) and similar equations.

6.1   Quick Observations

Here are some quick observations about (1):

- Given any choice, say (x_0, y_0), for the time $t = 0$ values of $x(t)$ and $y(t)$, there is a *unique* pair $(x(t), y(t))$ that solves (1) and has $x_0 = x(0)$ and $y_0 = y(0)$. That is, Equation (1) completely describes how $x(t)$ and $y(t)$ evolve in time given that you have decided on their starting values.

- In the first line of (1), if, at some time t, the value for $y(t)$ is one or more, then the time derivative of $x(t)$ is negative. Thus, the number of susceptible individuals will decrease steadily as long as there is at least one infected individual.

- In the second equation of (1), if, at some time t, the value of $x(t)$ is three or more, then the number of infected individuals will increase steadily.

- These last points imply the following: If there are one or more infected individuals to start with, the susceptible population number will decrease to zero. Then, when there is one or less susceptible left, the number of infected individuals will start to decrease too, due to death.

6.2   A Different Model

It is instructive to change (1) to read

$$\frac{d}{dt}x = -\lambda xy + 0.0003x, \qquad (2a)$$

$$\frac{d}{dt}y = +\lambda xy - 0.1y, \qquad (2b)$$

where λ is a constant that is a measure of the relative infectivity of the disease. We saw in the previous analysis that if λ is relatively large, then the disease will "burn itself out" in the sense that the number of susceptible individuals and also the number of infected individuals drops to zero. That is, high infectivity (relative to the birth rate of 0.0003 that appears in the first equation) makes everyone sick.

Consider the analogous case when λ is small, say 10^{-7}. Now, the behavior is complicated. Indeed, the time derivative of x is positive unless $y > 3000$. And the time derivative of y is negative unless $x > 1,000,000$. So a small number of infected individuals make for a disease that never reaches epidemic proportions. It simply infects a few individuals and then disappears (reminiscent of recent Ebola virus outbreaks in Africa). The point here is that the average time of infectivity (namely, 10 days) is too short for the disease to spread with such a low rate of infectivity. If there are but a few individuals initially infected, then there isn't time for them to infect their neighbors before they recover or die (provided they have fewer than 1 million neighbors). If the infectivity time were larger, say, 3 years (as with AIDS), then the second equation in (2) would be replaced by

$$\frac{d}{dt}y = +\lambda xy - 10^{-3}y, \tag{3}$$

and in this case, the disease would spread if the number of susceptible individuals was roughly 1000 or more. (Note that these models assume that the infected indivuals are up and about spreading germs. Quarantining infected individuals will drastically change the model equations. Can you guess how?)

6.3   The Existence of Solutions and Their Uniqueness

So the general problem is that of predicting the long-term behavior of a system of equations, as in Equation (2), or, more generally,

$$\frac{d}{dt}x = f(x, y) \quad \text{and} \quad \frac{d}{dt}y = g(x, y), \tag{4}$$

where f and g are given functions of two variables.

An absolutely crucial thing to remember about Equation (4) is that it is completely *predictive*. The point is that if you pick any place to start in the x-y plane at some given time, then there is precisely one solution to (4) that starts at your chosen place. This is to say that there *is* a solution that starts where you want it, and moreover, there is just one such solution.

An iron-clad proof of the preceding assertion can be found based on the following heuristic lines of reasoning: If you know the starting point at time zero, then (4) tells you the direction and speed of movement for t near zero. That is, for Δt small, (4) tells you that the values of $x(t)$ and $y(t)$ at $t = \Delta t$ are approximately $x_0 + f(x_0, y_0)\Delta t$ and $y_0 + g(x_0, y_0)\Delta t$, respectively. Here, I have introduced x_0 and y_0 to denote the x- and y-coordinates of the starting point. [Note that the observation that $x(\Delta t) - x_0$ is almost $f(x_0, y_0)\Delta t$ is, with the help of (4), just the definition of the derivative from one-variable calculus. Indeed, all I am saying is that $x(\Delta t) - x(0)$ is almost $\left(\frac{d}{dt}x\right)_{t=0}\Delta t$ when Δt is small.]

In any event, once you know the coordinates $x(\Delta t)$ and $y(\Delta t)$, then you can use (4) again (with the definition from one-variable calculus of $\frac{d}{dt}x$ and $\frac{d}{dt}y$) to tell you (approximately) the values for $x(2\Delta t)$ and $y(2\Delta t)$. Indeed, for small Δt, these differ only slightly from $x_0 + f[x(\Delta t), y(\Delta t)]\Delta t + f(x_0, y_0)\,\Delta t$ and $y_0 + g[x(\Delta t), y(\Delta t)]\Delta t + g(x_0, y_0)\,\Delta t$, respectively. Then, you can iterate the preceding to find x and y at $t = 3\Delta t, 4\Delta t, \ldots$, etc.

6.4   Generalities

Imagine a movie of the xy-plane with the position $(x(t), y(t))$ of the solution to (4) represented by a moving point in the plane. Then (4) tells you how the x- and y-coordinates of the moving point change with time. In particular, you can develop a sort of heuristic picture by first drawing the (x, y) plane and indicating the regions where $f = 0$, $f > 0$, and $f < 0$. Do likewise for the function g. Then, if you are in a region where, for example, $f > 0$ and also $g > 0$, you know that the path of $(x(t), y(t))$ moves up and to the right as the movie progresses because both $x(t)$ and $y(t)$ are increasing where $f > 0$ and $g > 0$. Similar analysis tells you the rough direction of motion for $(x(t), y(t))$, where $f > 0$ and $g < 0$ (down and to the right), and where $f < 0$ and where $g > 0$ or $g < 0$. This sort of analysis is called **phase plane analysis**.

6.5   Summary of Phase Plane Analysis

Here is a summary of phase plane analysis for a differential equation of the form in Equation (4) where f and g are two given functions on the xy-plane.

6.5.1 General Strategy

- Pick a starting point in the plane; there is a unique solution $\mathbf{v}(t) = \begin{pmatrix} x(t) \\ y(t) \end{pmatrix}$ to (4) that sits at your chosen starting point at $t = 0$.

- Think of $\mathbf{v}(t)$ as tracing out a path (trajectory) in the xy-plane as t increases. A goal is to predict the behavior of this path.

- The phase plane analysis described next is designed to help you predict the trajectory.

Here are the six steps for the phase plane analysis:

Step 1: Draw the curves where $f(x, y) = 0$. These are called the x null clines for the following reason: When $\mathbf{v}(t)$ lies on one of these null clines, then $\frac{dx}{dt} = 0$. Draw vertical slash marks on the x null clines to remind yourself that a trajectory that crosses such a null cline can only do so if it is moving purely in the vertical direction at the instant of crossing.

Step 2: Likewise, draw the curves where $g(x, y) = 0$. These are called the y null clines because when $\mathbf{v}(t)$ happens to sit on one, then $\frac{dy}{dt} = 0$. Draw horizontal slash marks on these null clines to remind yourself that a trajectory that crosses a y null cline does so by moving purely in the horizontal direction at the instant of crossing.

Step 3: Label the points where the x null clines intersect the y null clines. If $\mathbf{v}(t)$ is ever at one of these points, then both $\frac{dx}{dt}$ and $\frac{dy}{dt}$ vanish. This means that the trajectory stays at such a point for all time. These intersection points of x null clines and y null clines are called *equilibrium points*. If the system that is described by Equation (4) is going to settle into a steady state, then $\mathbf{v}(t)$ will have to approach one of the equilibrium points as t gets large.

Step 4: Label the regions of the xy-plane where $\frac{dx}{dt} < 0$ and where $\frac{dx}{dt} > 0$. (Note that these regions are always separated by x null clines.) Likewise, label the regions where $\frac{dy}{dt}$ is positive and negative.

Step 5: Go back and put arrows on the vertical hash marks of the x null clines. These arrows indicate whether motion across the null cline is up or down. The arrows are up on the parts of the x null cline in the $\frac{dy}{dt} > 0$ regions, and down on those parts of the x null cline in the $\frac{dy}{dt} < 0$ regions. Likewise, draw arrows on the horizontal slash marks that decorate the y null clines. The arrows are right pointing on the parts of the y null cline in the $\frac{dx}{dt} > 0$ regions and left pointing on the parts in the $\frac{dx}{dt} < 0$ regions.

Step 6: With the preceding completed, the analysis proceeds by observing that if the trajectory $\mathbf{v}(t)$ lies in a region where

(a) $\frac{dx}{dt} > 0$ and $\frac{dy}{dt} > 0$, then both $x(t)$ and $y(t)$ are increasing, so the trajectory must be moving up and to the right on the xy-plane.

(b) $\frac{dx}{dt} > 0$ and $\frac{dy}{dt} < 0$, then $x(t)$ is increasing but $y(t)$ is decreasing so the trajectory moves down and to the right.

(c) $\frac{dx}{dt} < 0$ and $\frac{dy}{dt} > 0$, then $x(t)$ is decreasing and $y(t)$ is increasing, so the trajectory moves up and to the left.

(d) $\frac{dx}{dt} < 0$ and also $\frac{dy}{dt} < 0$, then $x(t)$ and $y(t)$ are both decreasing, so the trajectory moves down and to the left.

Note: This sort of analysis is not quantitative (it is hard to get real numbers out of it), but it is a very powerful tool for analyzing the long-time evolution of a solution to an equation such as given in (4). However, there now exist good computer programs that will trace the trajectories in the xy-plane of solutions to differential equations like that in Equation (4).

6.6   Phase Plane Analysis for the Epidemic Model

As an example of phase plane analysis, consider the example in Equation (2) where $\lambda = 10^{-6}$. These equations read

$$\frac{d}{dt}x = -10^{-6}xy + 0.0003x, \tag{5a}$$

$$\frac{d}{dt}y = +10^{-6}xy - 0.1y. \tag{5b}$$

To begin, the x null clines are the curves in the plane where $\frac{d}{dt}x = 0$ along any trajectory. According to the first line in (5), this happens where $x \cdot (-10^{-6} \cdot y + 0.0003) = 0$. Thus, the x null clines consist of the two lines

$$x = 0 \quad \text{(the } y\text{-axis)}$$
$$y = 300 \quad \text{(a horizontal line)}. \tag{6}$$

See Figure 6.1.

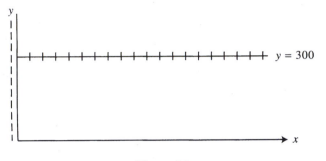

Figure 6.1

Note that trajectories are *vertical* when they hit the x null cline, because $\frac{d}{dt}x = 0$ there.

The y null clines occur where $y \cdot (10^{-6} \cdot x - 0.1) = 0$. Thus, the y null clines consist of the lines

$$y = 0 \text{ (the } x\text{-axis)}$$
$$x = 100,000 \text{ (a vertical line)}. \tag{7}$$

The y null clines are sketched in Figure 6.2. Note that trajectories are horizontal as they cross the y null clines because on these curves, the time derivative of y vanishes.

The equilibrium points are the points where the y null clines intersect the x null clines. These are the points $(0, 0)$ and $(300, 100,000)$. They are sketched in Figure 6.3. The regions where $\frac{d}{dt}x > 0$, $\frac{d}{dt}x < 0$, $\frac{d}{dt}y > 0$, and $\frac{d}{dt}y < 0$ are also labeled in Figure 6.3. From this information, you can determine the approximate course of the epidemic. (For example, if the initial condition is in the region where both $\frac{d}{dt}x > 0$ and $\frac{d}{dt}y > 0$, the trajectory for the epidemic will move, at least initially, up and to the right.)

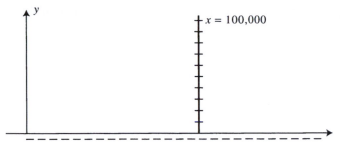

Figure 6.2

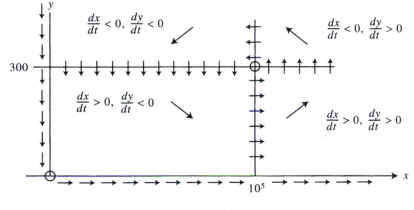

Figure 6.3

For example, suppose that the initial values for x and y are $x = 20,000$ and $y = 100$. In this region, $\frac{d}{dt}x > 0$ and $\frac{d}{dt}y < 0$, so x increases and y decreases. Note, however, that y can never actually hit zero in our model. This is because $y = 0$ is a y null cline, so on the $y = 0$ line, a trajectory has $\frac{d}{dt}y = 0$ and so must remain on this line for all time in the future and must have been on this line for all time in the past. Thus, as time increases, the number of infected individuals decreases, but according to the model, there are always some individuals who are infected. Meanwhile, the number of susceptible individuals increases steadily and eventually surpasses 100,000. After this point, $\frac{d}{dt}y > 0$, since there are enough susceptible individuals to make the epidemic "take off." In this region where $x > 100,000$ and $y > 0$ (but $y < 300$), both $\frac{d}{dt}x > 0$ and $\frac{d}{dt}y > 0$, so both increase. Eventually, y will increase past 300, after which $\frac{d}{dt}x$ turns negative. Thus, the number of infected individuals continues to increase, but the number of susceptible individuals drops steadily. Eventually, x decreases past 100,000, and then the number of infected individuals also starts to drop. Both x and y decrease until y decreases past 300 and then x starts to increase again.

Thus, the disease pattern seems cyclical. (We can't tell yet if there is a genuine cycle or if there is a spiraling into or out from the equilibrium point.)

6.7 Lessons

Here are key points from this chapter:

- Equation (4) is completely predictive; if you choose a starting point in the xy-plane, there is exactly one solution to (4) that starts at your chosen point.

- Learn well the six steps for doing phase plane analysis.

READINGS FOR CHAPTER 6

READING 6.1

Red Grouse and Their Predators

Commentary: This short correspondence provides a graph with discussion of the number versus time of red grouse shot on three moors in Scotland. These number versus time graphs are reasonably periodic (except at the latest dates), with roughly six year periods. However, the amplitudes at population maxima are not quite identical. The discussion is interesting in that the authors attribute the failure of the latest cycle on the Langholm moor to an unusually high raptor population on that moor. (For those who are not ornithologists, raptors and harriers are both hawk-like birds.)

Red Grouse and Their Predators

Science, **390** (1997) 547.
Simon Thirgood, Steve Redpath

Sir—We read with interest articles by May [1] and Mead [2] on the persecution of hen harriers and the impact of raptor predation on grouse-shooting bags. Some of Mead's comments, however, are misleading [2].

Mead suggests that the economic effects of harriers on grouse shooting are minimal. Unfortunately, that is not always the case. We recently completed six years of research on harrier and peregrine predation on grouse at Langholm in southwest Scotland. Raptors had bred freely on this moor since 1990, and female harrier numbers increased from two to twenty between 1992 and 1997. Peregrine

numbers increased from three to six pairs.

When raptor numbers were high, they removed 30% of breeding grouse in April and May and harriers removed 37% of the grouse chicks between June and August. Most of these losses appeared to be additional to other mortality, and we estimated that they reduced post-breeding numbers of grouse by 50%.

Historically, grouse bags at Langholm have shown a six-year cycle, peaking last in 1990, with 4,038 grouse shot (Fig. 1). Since 1990, grouse bags have declined, with 51 birds shot in 1997.

In contrast, grouse bags on two nearby moors,

Author affiliations: Simon Thirgood: Game Conservancy Trust. Steve Redpath: Institute of Terrestrial Ecology.

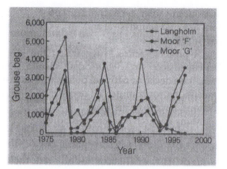

Fig. 1 Numbers of red grouse shot on Langholm moor during 1975–97 in comparison with numbers shot on nearby moors F and G during the same period. Grouse bags on all three moors fluctuated in synchrony from 1975 to 1993. After 1993, grouse bags on moors F and G increased while bags on Langholm moor continued to decline. Harrier and peregrine numbers on Langholm moor increased between 1992 and 1997 whereas the numbers of these raptors on moors F and G remained low.

with low raptor densities, having previously fluctuated in synchrony with Langholm moor, increased to high levels in 1997.

Increased predation by raptors at Langholm was considered the most likely cause for low grouse bags. Grouse management at Langholm cost £99,500 (US$168,000) in 1997 and, with grouse shooting producing £40 per bird, a bag of 2,487 grouse was required to balance costs. Clearly, if bags remain low the economic cost will be considerable.

Mead [2] suggests that more red grouse are killed on deer fences than are taken by harriers, citing work in Highland Scotland.

There are several flaws in this argument. First, deer fences are uncommon in red grouse range outside the Highlands, and indeed on many Highland moors. For example, there is little fencing at Langholm and collisions account for fewer than 1% of all recorded deaths [3]. Second, Highland studies suggest that 11% of red grouse deaths are due to collisions but 48% are due to raptor predation. Third, fences pose fewer problems to red grouse than to woodland grouse as strikes are concentrated near young plantations [4] and red grouse are birds of open moorland.

Finally, how can conflicts between raptors and grouse shooting be resolved [1]? As suggested by

Mead, predation patterns observed at Langholm will not apply everywhere. Our data suggest that, in the absence of persecution, harrier numbers will be related to densities of prey other than grouse [3]. In the long term, reducing the amount of grassland on moors may reduce the numbers of songbirds and voles, leading to reductions in harrier density and reducing their impact on grouse populations.

In the short term, however, raptor–grouse problems may require more active intervention in the form of supplementary feeding or raptor translocation. Such measures require the cooperation of conservation and shooting interests. The future of raptors, grouse, the moorland habitats they share and the rural communities they support depends upon it.

REFERENCES

1. May, R. M. *Nature* **389**, 330–331 (1997).
2. Mead, C. *Nature* **389**, 780 (1997).
3. Redpath, S. M. & Thirgood, S. J. *Birds of Prey and Red Grouse* (HMSO, 1997).
4. Baines, D. & Summers, R. W. *J. Appl. Ecol.* **34**, 941–948 (1997).
5. Hudson, P. J. *Grouse in Space and Time* (Game Conservancy Trust, 1992).
6. Mead, C. *The Times*, **7** and **23** November 1997.

READING 6.2

Wolves, Moose and Tree Rings on Isle Royale

Commentary: This article discusses a three-level ecosystem in which wolves prey on moose that eat balsam fir. For the present purposes, the article is used to illustrate the fluctuating (perhaps cyclic) populations of wolves, moose, and fir on Isle Royale. All three populations are seen to fluctuate somewhat out of phase with each other. However, it is by no means clear that the fluctuations provide an example in nature of an ecosystem whose population numbers oscillate with a regular period. In any event, the behavior of population numbers versus time of these mutually interacting populations is evidently quite complicated. A verifiable explanation for the time behavior of the population numbers is desirable. (Are there other arguments for the data besides those of the authors?)

Wolves, Moose, and Tree Rings on Isle Royale

Science, **266** (1994) 1555–1558.
B. E. McLaren, R. O. Peterson

Investigation of tree growth in Isle Royale National Park in Michigan revealed the influence of herbivores and carnivores on plants in an intimately linked food chain. Plant growth rates were regulated by cycles in animal density and responded to annual changes in primary productivity only when released from herbivory by wolf predation. Isle Royale's dendrochronology complements a rich literature on food chain control in aquatic systems, which often supports a trophic cascade model. This study provides evidence of top-down control in a forested ecosystem.

Terrestrial food chains of length three—plants, herbivores, and carnivores—are found throughout the temperate zone of the Northern Hemisphere, yet their establishment, pervasiveness, and stability are enigmatic subjects of debate among community ecologists [1]. According to one hypothesis, depletion of green plants by herbivores occurs only in exceptional circumstances because carnivores usually control herbivores [2]. However, systems in which increases in the density of a species at one trophic level accompany increases at higher, dependent trophic levels support a counterargument [3]. The top-down (trophic cascade) model predicts that changes in density at one trophic level are caused

by opposite changes in the next higher trophic level and that such inverse correlations cascade down a food chain. Accordingly, effects such as changes in primary productivity (the energy flow to plants) become noticeable only when higher, masking trophic levels are removed. After removal of carnivores from a three-level system, the control of density relationships is passed down the chain, to herbivores. The bottom-up model predicts that positive correlations occur between density changes at all trophic levels and especially between adjacent trophic levels, that changes in primary productivity affect higher trophic levels, and that extinction of the top trophic level does not change density patterns in lower levels.

We investigated food chain control in a large mammal system through observation of three trophic levels—a living system that is unlike two-level models of ungulate dynamics, which simulate the herbivore-plant interaction but are insensitive to parameter changes in a carnivore equation [4], and is also unlike models for ungulates that ignore the effects of vegetation change [5]. Tree-ring analyses were used to characterize the interaction between an herbivore population and its winter forage in a system with an apparent cycle between preda-

---

Author affiliations: McLaren, B. E., Peterson, R. O.: Michigan Technological University, School of Forestry and Wood Products, Houghton, MI 49931, USA.

Fig. 1 Isle Royale National Park, showing the location of sampled balsam fir trees. Samples were collected randomly from intersection points of township grid lines within two broad regions of high understory fir density (shaded area) [25]. Filled circles represent the locations of west-end trees, collected from a more established forest with a larger hardwood component; open circles represent east-end trees collected from boreal forests that were burned extensively in 19th-century fires. The two regions are separated by a deciduous forest that originated from an extensive fire in 1936 and by upland hardwood forest at higher elevations. The locations of two other sample sites are marked RH, Rock Harbor, and SS, Siskiwit Swamp.

tor and prey: the wolves (*Canis lupus*) and moose (*Alces alces*) in Isle Royale National Park, Michigan, the largest island (544 km^2) in Lake Superior [6, 7]. Ring width in balsam fir (*Abies balsamea*), a tree that makes up 59% of winter moose diet[8], provided an index of the herbivore food base, even though it is not optimal forage for moose. We assumed that the annual wood accrual for fir was proportional to its foliar biomass, which is an approximate measure of standing forage crop.

Balsam fir covers a large area of Isle Royale (Fig. 1), although its relative abundance in the overstory has declined since the arrival of moose early in the 1900s, from 46% in 1848, to 13% in 1978, to ~5% today [9]. The decline is attributed to the effect of moose herbivory; forests on small nearby islands that are less accessible to moose still have a large fir component [10]. The longevity of fir in the understory and its large contribution to moose diet on Isle Royale facilitated the present study [11]. Moreover, balsam fir exhibits an intriguing response to moose herbivory on Isle Royale, where areas of extreme suppression of fir growth on the west end of the island contrast with little-affected areas on the east end [12] (Fig. 2). We believe that the general response of fir is comparable to that of other forage species on Isle Royale.

The Isle Royale food chain appears to be a tightly linked, three-trophic-level system dominated by top-down control (Fig. 3). Over three decades, balsam fir trees from widely separated areas of the island displayed cyclic intervals of ring growth sup-

pression that accompanied elevated moose densities. Annual variation in ring width during suppression was very low and did not show any correlation with a climate series calculated for Isle Royale, which contradicts earlier proposals that variation in natural populations arises from the physical systems that surround them [13].

A common harmonic pattern of period 16 to 18 years in east- and west-end tree-ring chronologies [14] echoed similar periodicity in the moose population [15]. However, minima in the foliar biomass of understory balsam fir lagged behind moose population maxima by 1 to 2 years at the west end of the island and by 5 years at the east end [16]. That these two populations of fir should be out of phase in their response to herbivore density changes was actually consistent with the prediction of classical predator-prey models, given plant populations with unequal intrinsic rates of increase [17]. The west-end hardwood forest has usually had higher actual evapotranspiration (AET) associated with warmer, early summer temperatures than has the east-end boreal forest [18]; consequently, we expected higher plant growth rates on the west end of Isle Royale.

Herbivore density, in turn, was largely determined by predation. The wolf population has fluctuated with changes in the number of moose older than 9 years in the prey population, prompting consequent changes in herbivore density by influencing calf survival rates [6, 7, 19]. When old moose (and wolves) were relatively scarce, calf survival

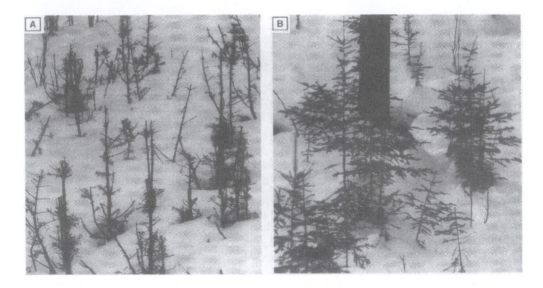

Fig. 2 Patterns of balsam fir growth suppression (**A**) and release (**B**) are evident throughout Isle Royale[12]. We explain the difference in terms of herbivory and forest disturbance processes.

was high and moose numbers grew, which then led to depletion of balsam fir forage. Subsequent decline in the moose population was closely linked to wolf increase, itself fueled by aging of the moose population. Vegetation response followed moose response (Fig. 3), so it was not directly responsible for herbivore fluctuations. In summary, because wolf maxima precede moose minima, and moose maxima are similarly followed by fir minima, and because the vegetation dynamics appear to be more intimately linked to the wolf-moose interaction than to seasonal weather patterns, we believe that we have evidence for top-down control in a nonaquatic three-trophic-level system [20].

A strong supporting argument presents itself in the recent wolf decline on Isle Royale, when the wolf population reached an unprecedented low [21]. The moose population accordingly reached a new, very high level, accompanied by strong suppression of balsam fir growth, in 1988 to 1991. This was especially apparent in the west-end sample (Fig. 3C). Without the context of a three-trophic-level interaction, fir decline and recovery cannot be explained; however, current disintegration of the system into a two-trophic-level interaction establishes even lower minima for fir growth than the suppressed level of the 1970s, which is more in agreement with the top-

down than the bottom-up model.

Nevertheless, release of balsam fir from herbivory by moose does occur on Isle Royale when large-scale forest disturbances allow bottom-up influences to prevail. Additional samples of balsam fir showed extremes in the outcome of the herbivore-plant interaction and helped explain the broadly contrasting pattern of suppression on the two ends of the island (Figs. 2 and 3). An east-end subsample of trees (RH), collected from an open-canopy area of past forest disturbance, increased in growth after a period of high wolf predation of moose in the late 1970s. A west-end subsample (SS), collected from a closed-canopy hardwood forest heavily browsed by moose, showed increasingly suppressed growth, particularly in the current decade of high moose density (Fig. 4). This high suppression accompanies lower fir density in many similar areas on the west end; such a pattern may explain the more tightly coupled cycle between moose and fir there. Whereas ring-width patterns before 1980 were coordinated in both samples and matched those in the extensive samples (for example, release in the 1960s, suppression in the late 1970s), the recent period of release, especially in the RH sample, was immediately associated with closer correlation with the primary productivity in-

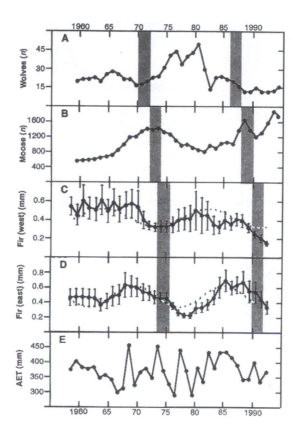

Fig. 3 The trophic system on Isle Royale, reconstructed for 1958 to 1994. (**A**) The number of wolves on the island from winter aerial counts. (**B**) Moose population size, reconstructed from collected skeletal remains (1959 to 1981) and from winter aerial counts (1982 to 1994). (**C**) Mean ring-width index[26] for eight balsam fir trees sampled in 1992 from the west end of Isle Royale (Fig. 1). Vertical bars are ±1 SEM from the arithmetic mean. Dashed line is the best-fit harmonic function[14]. (**D**) Mean ring-width index for eight trees from the east end of the island, as in (**C**). (**E**) Actual evapotranspiration from April to October, calculated for a weather station about 20 km from Isle Royale. The AET is the amount of water available to plants during the growing season as a function of both rainfall and temperature and has a close relation to primary productivity [24]. The shaded areas highlight intervals of forage suppression that we believe are closely tied to periods of elevated moose density, which in turn follow periods of low wolf density (note the lags between trophic levels). These intervals have no correspondence to AET.

dex (Table 1). Thus, only when moose density was relatively low, which allows the link to the wolf-moose interaction to be removed or weakened by disturbance, did understory balsam fir grow in a fashion consistent with the bottom-up model. We believe that external processes such as fire and large windstorms can override usual top-down effects in a forest, but that this effect occurs primarily

when the density of the next higher trophic level is low.

The uniqueness of the Isle Royale study is that it complements studies of top-down effects in aquatic systems [1]. The establishment of bottom-up control in a terrestrial system when higher trophic levels are removed also parallels results from aquatic studies [22] and follows analogous predictionsfor

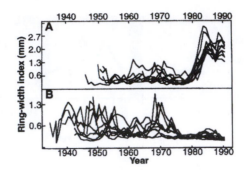

Fig. 4 Individual ring-width chronologies [26] for balsam fir trees collected from Isle Royale in 1992. (**A**) Ten trees, 26 to 48 years of age, from location RH (Fig. 1), in which fir height exceeded the reach of moose (3 m) in the late 1970s. (**B**) Nine trees, 48 to 60 years of age, from location SS (Fig. 1), for which total height at the time of sampling ranged from 1.08 to 1.57 m.

Table 1. Correlated annual changes in primary productivity and ring-width indices for balsam fir. The AET column shows the direction of change in AET [24] between a census year and the previous year (Fig. 3E). The RH and SS columns show the number of trees [out of 10 in the RH sample and 9 in the SS sample (Fig. 4)] for which the difference in ring width for these 2 years is in the same direction. Whereas ring-width release from 1979 to 1983 in the RH sample has no correlation with annual climate fluctuations, in the interval from 1984 to 1990, annual ring growth more closely tracks climate.

Year	AET	RH	SS
1979	−	3	2
1980	+	10	5
1981	−	0	3
1982	+	8	3
1983	−	1	3
1984	+	10	7
1985	+	8	6
1986	−	7	7
1987	−	8	7
1988	−	9	6
1989	+	8	2
1990	+	8	2
1991	−	10	5

ungulate populations without predators [5]. Furthermore, this study supports the top-down argument that was originally applied to the Isle Royale system [23].

REFERENCES

1. M. D. Hunger and P. W. Price, *Ecology* **73**, 724 (1992).
2. N. G. Hairston, F. E. Smith, L. B. Slobodkin, *Am. Nat.* **94**, 421 (1960).
3. S. L. Pimm, *The Balance of Nature* (Univ. of Chicago Press, Chicago, IL, 1992), chap. 14.
4. We considered the model in G. Caughley, *Analysis of Vertebrate Populations* (Wiley, New York, 1977), pp. 130–132.
5. A. T. Bergerud, *Trends Ecol. Evol.* **3**, 68 (1988); *Can. J. Zool.* **64**, 1515 (1986).
6. R. O. Peterson, R. E. Page, K. M. Dodge, *Science* **224**, 1350 (1984).
7. R. O. Peterson and R. E. Page, *J. Mammal.* **69**, 89 (1988); *Acta Zool. Fenn.* **174**, 251 (1983).
8. K. L. Risenhoover, thesis, Michigan Technological University (1987), p. 49.
9. R. A. Janke, D. McKaig, R. Raymond, *For. Sci.* **24**, 115 (1978); B. E. McLaren, unpublished

data (figures are for the west end of Isle Royale).

10. J. D. Snyder and R. A. Janke, *Am. Midl. Nat.* **95**, 79 (1976).

11. Suppression of balsam fir growth on Isle Royale by moose foraging occurs because of continuous removal of upper branches in winter. Trees remain alive in the forest understory for several decades, because lateral branches are protected from browsing by snow cover and because balsam fir is tolerant of shade. Trees in this study (except those represented in Fig. 4A) ranged from 34 to 68 years of age but came from a consistent height interval of 1 to 3 m at the time of sampling. The age variation made no contribution to variation in growth rates.

12. T. A. Brandner, R. O. Peterson, K. L. Risenhoover, *Ecology* **71**, 155 (1991).

13. G. Caughley and A. Gunn, *Oikos* **67**, 47 (1993); S. L. Pimm and A. Redfearn, *Nature* **334**, 613 (1984); M. H. Williamson, *The Analysis of Biological Populations* (Arnold, London, 1982), chap. 5.

14. A harmonic function fitted with an ordinary least squares approach, period 17.8 years, matched the west-end chronology with $R^2 = 0.38$; a function with period 16.1 years matched the east-end chronology with $R^2 = 0.70$. Following M. G. Bulmer [*J. Anim. Ecol.* **43**, 701 (1978)], functions incorporating both a harmonic term and a first-order autoregressive term were matched to the series, with improved fits of $R^2 = 0.68$ and $R^2 = 0.81$, respectively.

15. The cycle period inferred from empirical consideration of the three-trophic-level system is shorter than a period cited earlier, which was derived from a body-mass regression [6]. However, the confidence interval for the allometric Equation also included the shorter period of 16 to 18 years.

16. Ring-width suppression followed height-growth suppression in both samples by about 3 years.

17. G. Caughley, *Oecologia* **54**, 309 (1976). The plant with the inferior growth rate lags behind herbivore density more than does the plant with the superior growth rate.

18. B. E. McLaren, unpublished data; R. M. Linn, thesis, Duke University (1957).

19. R. O. Peterson, unpublished data.

20. Our result is new primarily because studies of long-term processes in terrestrial food chains are rare, especially studies of those in which top carnivores still exist. A. R. E. Sinclair and colleagues [*Am. Nat.* **141**, 173 (1993)] recently described the well-known snowshoe hare (*Lepus americana*) cycle in the Yukon Territory, Canada, as generating cyclic suppression of white spruce (*Picea glauca*). These authors, however, favored an explanation in meteorological driving forces. Other studies [for example, L. B. Keith, A. W. Todd, C. J. Brand, R. S. Adamcik, D. H. Rusch, *Int. Congr. Game Biol.* **13**, 151 (1976); L. B. Keith and L. A. Windberg, *Wildl. Monogr.* **58**, 70 (1978)] have considered plant-herbivore and herbivore-carnivore relations for the snowshoe hare, but in different study areas. The third trophic level in the fox-prey system studied by E. R. Lindstroem and colleagues [*Ecology* **75**, 1042 (1994)] was a parasite.

21. J. M. Thurber and R. O. Peterson, *J. Mammal.* **74**, 879 (1993).

22. D. J. McQueen, M. R. S. Johannes, J. R. Post, T. J. Stewart, D. R. S. Lean, *Ecol. Monogr.* **59**, 289 (1989); S. R. Carpenter *et al.*, *Ecology* **68**, 1863 (1987); M. Lynch and J. Shapiro, *Limnol. Oceanogr.* **26**, 86 (1981); J. L. Brooks and S. I. Dodson, *Science* **150**, 28 (1965).

23. L. Oksanen, S. D. Fretwell, J. Arruda, P. Niemela, *Am. Nat.* **118**, 240 (1981); for other contrasting published ideas concerning predator-prey cycles on Isle Royale, see [6] and J. Pastor and R. J. Naiman, *Am. Nat.* **139**, 690 (1992).

24. M. L. Rosenzweig, *Am. Nat.* **102**, 67 (1968). The index was calculated with the use of daily records from Thunder Bay, Ontario.

25. Understory fir density exceeds 250 ha^{-1} over about 70% of Isle Royale's land area and exceeds 10,000 ha^{-1} in many shoreline forests. Values were obtained by averaging density estimates for stems ≤ 2 m in 0.01-ha plots at the corners of each (square-mile) township section.

26. The ring-width index is the ratio of increments calculated as a summation of volume differences for a series of stacked conic sections representing the stems of sampled trees, divided by the cambial surface area at the beginning of each growing season for each year. An aggregate index, comprising ring-width measurements ($\pm 10^{-2}$ mm) averaged for four radii at 5- to 10-cm increments

along stems, is presented. This intensity of sampling permits an accurate height-growth reconstruction of the trees and measurement of the wood volume increment throughout the trees' stems.

27. We acknowledge the National Park Service (NPS) for permission to sample trees; V. G. Smith, University of Toronto, for the design of the ring-width index; R. J. Miller, Ontario Ministry of Natural Resources, for the development of analytic software and loan of measurement equipment; Environment Canada for supplying weather records; and the following for financial support: National Geographic Society, NPS Earthwatch, the Boone and Crockett Club, and NSF grant DEB-9317401. W. C. Kerfoot and F. H. Wagner kindly reviewed this manuscript and offered many helpful suggestions.

Exercises for Chapter 6

1. Suppose the function $x(t)$ evolves according to the differential equation $\frac{dx}{dt} = f(x)$. For each of the four choices for the function $f(x)$, describe what happens to $x(t)$ as t gets large if $x(0) = 1$.

    (a) $f(x) = (x+1)(x-2)$

    (b) $f(x) = (x+3)x(x-3)$

    (c) $f(x) = (x+1)x$

    (d) $f(x) = (3-x)(x-4)$

2. For each of the same four functions in Exercise 1, describe what happens to $x(t)$ as t gets large if $x(0) = -4$.

3. For each of the four functions from Exercise 1, list the equilibrium points and decide which are stable and which are unstable.

4. Find the first-order Taylor's expansion at $x_0 = 0$ for the following functions:

    (a) $f(x) = xe^{-6x}$

    (b) $f(x) = 2\cos(4x)$

    (c) $f(x) = 4x + x^2$

    (d) $f(x) = 3e^{12x}$

5. Consider the system of equations $\frac{dx}{dt} = y - x^2$, $\frac{dy}{dt} = x - 1$.

    (a) Sketch the null clines and label the equilibrium points.

    (b) Label (using arrows if you like) where $\frac{dx}{dt}$ and $\frac{dy}{dt}$ are positive and negative.

    (c) If $x(0) = 2$ and $y(0) = 1$, roughly, what happens to $x(t)$ and $y(t)$ as t tends to ∞? (Does the trajectory move arbitrarily far from the origin? If so, in what general direction? If not, what does the trajectory do?)

(d) If $x(0) = 0$ and $y(0) = 0$, what happens to $x(t)$ and $y(t)$ for small but positive t? [Is $x(t)$ greater or less than $x(0)$ if t is small and positive; also, is $y(t) > y(0)$ or not?]

6. Consider the system of equations $\frac{dx}{dt} = 2y$, $\frac{dy}{dt} = x^2 + y^2 - 1$.

(a) Sketch the null clines and label the equilibrium points.

(b) Label (using arrows if you like) where $\frac{dx}{dt}$ and $\frac{dy}{dt}$ are positive and negative.

(c) If $x(0) = 0$ and $y(0) = 2$, what happens to $x(t)$ and $y(t)$ as t tends to ∞?

(d) If $x(0) = 0$ and $y(0) = 1$, what happens to $x(t)$ and $y(t)$ for small but positive t?

7. Consider the system of equations $\frac{dx}{dt} = 2x(y^2 + 1)$, $\frac{dy}{dt} = e^x - y$.

(a) Sketch the null clines and label the equilibrium points.

(b) Label (using arrows if you like) where $\frac{dx}{dt}$ and $\frac{dy}{dt}$ are positive and negative.

(c) If $x(0) = 0$ and $y(0) = 2$, what happens to $x(t)$ and $y(t)$ as t tends to ∞?

(d) If $x(0) = 2$ and $y(0) = 0$, what happens to $x(t)$ and $y(t)$ for small but positive t?

7

Introduction to Vectors

This chapter serves as a digression of sorts to introduce some very useful shorthand notation for dealing with differential equations for more than one function. You should keep in mind the specific epidemic model of the previous chapter:

$$\frac{d}{dt}x = -0.05xy + 0.0003x$$

$$\frac{d}{dt}y = 0.05xy - 0.1y. \tag{1}$$

7.1   The General Equation

It proves useful to introduce some notational shorthand when thinking about a system of equations like the epidemic equation in (1), or in the more general case that follows, an equation that specifies the rate of change of a pair of functions of time, $x(t)$ and $y(t)$, in terms of their values at time t:

$$\frac{d}{dt}x = f(x, y),$$

$$\frac{d}{dt}y = g(x, y). \tag{2}$$

As the behavior of x determines that of y and vice versa, it proves convenient to think of the pair $x(t)$ and $y(t)$ as constituting two parts of a single object. That is, when we are thinking about $x(t)$ and $y(t)$ together, we will write them as a pair

$$\mathbf{v}(t) = \begin{pmatrix} x(t) \\ y(t) \end{pmatrix}. \tag{3}$$

Likewise, we will think of f and g as a pair,

$$\mathbf{F}(x, y) = \begin{pmatrix} f(x, y) \\ g(x, y) \end{pmatrix}. \tag{4}$$

Then Equation (2) becomes $\frac{d}{dt}\mathbf{v} = \mathbf{F}$. This notation is more condensed compared with (2).

I will justify the notation in three ways: First, it is much easier to write the shorthand than to write all of (2), even with fancy math word processing software. Second, the shorthand focuses on $x(t)$ and $y(t)$ as a sort of inseparable pair. Indeed, the differential equation in (2) intertwines their fate: Changing x changes y and vice versa; thus you should view these two functions as just two components of the entity I call $\mathbf{v}$. Likewise, the functions f and g (of the variables x and y) should be viewed as two components of a single entity that I denote as $\mathbf{F}$. Finally, using a single symbol to represent the pair of functions, as in (3) and (4), is standard notation in the literature, so you will most probably see it no matter what.

7.2 Definition of Vectors

In general, a *vector* (rightly, a two-dimensional vector) is a pair of numbers,

$$\mathbf{v} = \begin{pmatrix} a \\ b \end{pmatrix}. \tag{5}$$

The set of real numbers will be denoted by $\mathbb{R}$, and the set of two-dimensional vectors will be denoted by $\mathbb{R}^2$. (For the time being, I will use the boldface type to denote vectors so we can distinguish them from ordinary numbers, but after a while, I will surely succumb to laziness and forget, in which case it should be clear from the context of the equation in question that I am referring to a vector. In any event, notation of this sort is *not* uniform across the sciences, so it is not a good idea to become too attached to any one notational scheme.)

The numbers a and b in (5) are called the **components** of the vector $\mathbf{v}$; the top entry (a here) is called the first component, and the bottom entry (b here) is called the second component. Here are some examples of vectors: $\begin{pmatrix} 12 \\ 0 \end{pmatrix} \begin{pmatrix} -6 \\ 1.1 \end{pmatrix} \begin{pmatrix} 3.3 \\ -52 \end{pmatrix} \begin{pmatrix} 6.7 \\ 2.2 \end{pmatrix}$.

7.3   Philosophical Remarks

Notice that a point in the xy-plane is clearly a vector, and conversely, a vector [such as $\mathbf{v}$ in (5)] will define a point in the xy-plane by setting $x = a$ and $y = b$. But occasions will arise where it is convenient to think abstractly about a vector, without identifying it with a point in any particular coordinate plane. For example, at a fixed time t, $\mathbf{v}(t)$ in (3) is a pair of numbers, a vector, but it is also, obviously, a point in the xy-plane. But I can differentiate $\mathbf{v}(t)$ to get

$$\frac{d}{dt}\mathbf{v} = \begin{pmatrix} \frac{dx}{dt} \\ \frac{dy}{dt} \end{pmatrix}, \tag{6}$$

which gives a vector when evaluated at time t. However, the vector in (6) should never be thought of as residing in *the same* xy-plane as the vector in (3). [This is a fairly sophisticated concept, that the vectors in (6) and in (3) are in "different" xy-planes. In fact, this concept is so sophisticated that it isn't mentioned in most standard multi-variable calculus courses.]

Confusion can be avoided if you think of a vector as a pair of numbers, a point of the abstract space that we will call $\mathbb{R}^2$.

7.4   Vectors as Functions

As illustrated previously, we can (and will) consider vectors that are functions of time [such as $\mathbf{v}$ in (3) or its derivative, $\frac{d}{dt}\mathbf{v}$ in (6)], and we will consider vectors that are functions of other parameters [such as $\mathbf{F}$ in (4) being a function of the variables x and y].

Given a vector function, $\mathbf{v}$, of a parameter, say t, then the derivative of $\mathbf{v}$, written $\frac{d}{dt}\mathbf{v}$, is the vector function of t whose first component is the derivative of the first component of $\mathbf{v}$ and whose second component is the derivative of the second component of $\mathbf{v}$.

An antiderivative of a vector function of time $\mathbf{v}$ is a vector function of time, $\mathbf{w}$ that obeys $\frac{d}{dt}\mathbf{w} = \mathbf{v}$. Define the integral of $\mathbf{v}$,

$$\int_a^b \mathbf{v}(t)\,dt,$$

to be the vector whose first component is the integral of the first component of $\mathbf{v}$ and whose second component is the integral of $\mathbf{v}$'s second component.

7.5   Functions of Vectors

If f is a function of two variables, say $f = f(x, y)$, and if we are thinking of the pair x, y as a vector, $\mathbf{v}(t) = \begin{pmatrix} x(t) \\ y(t) \end{pmatrix}$, then we will write $f = f(\mathbf{v})$. That is, we will think of

f as a function of a vector. As a perverse example, the vector $\mathbf{F}$ in (4) is a function of the pair x, y, so we may think of $\mathbf{F} = \mathbf{F}(\mathbf{v})$ as a vector that is a function of a vector.

Don't forget, this is all simply notation at this point. However, this notation proves extremely convenient and so it is worth knowing and using.

7.6   Adding Vectors

You can add two vectors together by the following rule: Suppose $\mathbf{v} = \left(\begin{smallmatrix} a \\ b \end{smallmatrix}\right)$ and $\mathbf{w} = \left(\begin{smallmatrix} c \\ d \end{smallmatrix}\right)$ then $\mathbf{v} + \mathbf{w} = \left(\begin{smallmatrix} a+c \\ b+d \end{smallmatrix}\right)$. That is, the top components of $\mathbf{v}$ and $\mathbf{w}$ are added, as are their bottom components, to get the respective components of $\mathbf{v} + \mathbf{w}$.

7.7   Multiplying Vectors by Numbers

You can multiply a vector by a real number. Suppose that $\mathbf{v} = \left(\begin{smallmatrix} a \\ b \end{smallmatrix}\right)$ and that r is a real number. Then $r\mathbf{v} = \left(\begin{smallmatrix} ra \\ rb \end{smallmatrix}\right)$. (In many books, a real number is called a *scalar.*

7.8   The Length of a Vector

The length of the vector $\mathbf{v} = \left(\begin{smallmatrix} a \\ b \end{smallmatrix}\right)$ is defined to be the real number $|\mathbf{v}| = (a^2 + b^2)^{1/2}$. This is the length of the hypotenuse of a right triangle with side lengths $|a|$ and $|b|$. Note that multiplying $\mathbf{v}$ by a number, r, gives a vector whose length is rescaled by the factor $|r|$. Presumably, this is why r is called a scalar.

7.9   Dot Product

You can't multiply a vector with two components by another to get a third, but you can define a "multiplication" rule for vectors that produces a single number from a pair of vectors. The rule is as follows:

If $\mathbf{v} = \left(\begin{smallmatrix} a \\ b \end{smallmatrix}\right)$ and $\mathbf{w} = \left(\begin{smallmatrix} c \\ d \end{smallmatrix}\right)$, then the *dot product* of $\mathbf{v}$ with $\mathbf{w}$ is the number

$$\mathbf{v} \cdot \mathbf{w} = ac + bd. \tag{7}$$

Note that $\mathbf{v} \cdot \mathbf{w} = \mathbf{w} \cdot \mathbf{v}$ and that $\mathbf{v} \cdot \mathbf{v} = |\mathbf{v}|^2$.

7.10   Three-Component Vectors

Here is a parenthetical remark for those of you who plan to take some physics courses (I won't refer to the following material in the subsequent chapters): In college-level physics courses, you will see lots of vectors with three components, such as

$$\mathbf{v} = \begin{pmatrix} a \\ b \\ c \end{pmatrix}. \tag{8}$$

For example, these are profitably used to represent the positions and velocities of particles moving in space. If you have two such three-component vectors, say $\mathbf{v}$ and

$$\mathbf{w} = \begin{pmatrix} e \\ f \\ g \end{pmatrix}, \tag{9}$$

there is a dot product that associates a number to the pair. Here,

$$\mathbf{v} \cdot \mathbf{w} = ae + bf + cg. \tag{10}$$

That is, to compute $\mathbf{v} \cdot \mathbf{w}$, multiply each component of $\mathbf{v}$ by the corresponding component of $\mathbf{w}$ and add the three resulting numbers. (The obvious generalization of this rule defines the dot product between two vectors with four components, or two vectors with the same number of components, no matter how large.)

However, a pair of vectors with three components can also be multiplied by a different rule to get another three-component vector. (The number 3, of components, is very special. Unlike the dot product, this new product only makes sense for 3-component vectors.) This alternate multiplication rule is called the ***cross product*** and is written as $\mathbf{v} \times \mathbf{w}$. Here is the definition:

$$\mathbf{v} \times \mathbf{w} = \begin{pmatrix} bg - fc \\ ce - ag \\ af - eb \end{pmatrix}. \tag{11}$$

7.11   Lessons

This chapter introduced the following concepts:

- Vector notation

- Adding vectors and multiplying them by numbers

- Vectors whose components are functions of one or two variables

- Finding the length of a vector and taking the dot product between a pair of vectors

READINGS FOR CHAPTER 7

READING 7.1

Power Laws Governing Epidemics in Isolated Populations

Commentary: The purpose of studying this article is simply to illustrate some of the issues that epidemiologists grapple with. Here, the authors consider measles outbreaks in isolated populations (for example, in the Faroe Islands). They observe from the data that the number of epidemics with greater than s cases [denoted by $N(s)$] gives a function of s that is reasonably approximated by the power law relation $N(s) = \alpha s^{-1-b}$, where α and b are constants. (They find that $b = 0.29$ for the Faroe Islands.) Likewise, the authors introduce another function, $N(t)$, that is the number of epidemics which last for time greater than t. They find that this function is also modeled by a power law relation, $N(t) = \beta t^{-1-c}$, where β and c are constants. Here, $c = 0.8$ for the Faroe Islands. We might predict a relationship between the constants b and c based on the infectivity of measles and the manner in which it is spread. The authors present a model that gives an account of this relation. (A function, $f(x)$, of a variable x is said to obey a "power law" when it has the form $f(x) = ax^b$ where a and b are constants.)

Power Laws Governing Epidemics in Isolated Populations

Nature, **381** (1996) 600–602.
C. J. Rhodes, R. M. Anderson

Temporal changes in the incidence of measles virus infection within large urban communities in the developed world have been the focus of much discussion in the context of the identification and analysis of nonlinear and chaotic patterns in biological time series [1–11]. In contrast, the measles records for small isolated island populations are highly irregular, because of frequent fade-outs of infection [12–14], and traditional analysis [15] does not yield useful insight. Here we use measurements of the distribution of epidemic sizes and duration to show that regularities in the dynamics of such systems do become apparent. Specifically, these biological systems are characterized by well-defined power laws in a manner reminiscent of other nonlinear, spatially extended dynamical systems in the physical sciences [16–19]. We further show that the observed power-law exponents are well described by a simple lattice-based model which reflects the social interaction between individual hosts.

Isolated island populations have often provided a valuable arena for the study of evolutionary and ecological processes because the subject of investigation is confined to a well-defined region of space that is well insulated from the influence of external factors. In the context of population studies, it is often possible in such settings to better observe effects that are intrinsic to the community without having to quantify the effect of peripheral populations.

The Faroe Islands (population 25,000) have extensive detailed measles-case returns (Fig. 1A, top). It is considered to be an accurate epidemiological data set because the population is small and localized. Also, because of the comparative rarity of measles outbreaks, few measles cases escape notice [20]. In the 58-year interval, there are 43 distinct epidemic events. An epidemic event is defined as the presence of a finite number of cases recorded in a sequence of consecutive months bounded by an absence of cases. An epidemic event has a du-

Author affiliations: C. J. Rhodes, R. M. Anderson: Centre for the Epidemiology of Infectious Disease, Department of Zoology, University of Oxford.

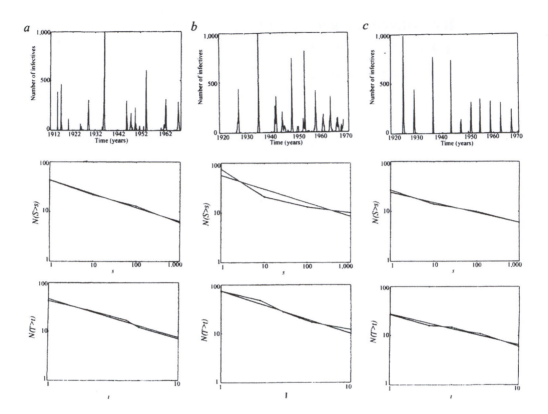

Fig. 1 A, Faroe Islands; B, Bornholm; C, Reykjavik (scaled). These are all examples of Bartlett's type III measles epidemic dynamics [12]. Top, monthly measles-case returns for each of the three communities. Between 1920 and 1970 the population of Reykjavik grew from 20,000 to 100,000, so we have scaled the incidence data per 25,000 of population. Middle, the distribution of epidemic event sizes. For the Faroes, $b \approx 0.28$; for Bornholm, $b \approx 0.28$; and for Reykjavik, $b \approx 0.21$. Bottom, distribution of epidemic event durations. For the Faroe Island data, $c \approx 0.8$; for Bornholm, $c \approx 0.85$; and for Reykjavik, $c \approx 0.62$. Clearly, scaling the measles incidence data will affect the epidemic sizes but leave the epidemic durations unscaled, hence this exponent is somewhat different from the Faroe and Bornholm cases. For each population, the exponents are unaffected by the specific choice of binning by decade.

ration, t, where $t = \tau_{\text{end}} - \tau_{\text{start}}$, τ_{start} is the month when cases in an event first appear, and τ_{end} is the next month when there are no more cases present. Similarly, an epidemic event has a size, s, where $s = \sum_{\tau_{\text{start}}}^{\tau_{\text{end}}} C(\tau)$ and $C(\tau)$ is the number of recorded cases of measles in the month τ.

From Fig. 1A, top, it is apparent that there is no obvious discernible pattern to the dynamics; there is wide variation in both the duration and size of the epidemic events, reminiscent of the sort of dynamics observed in the study of earthquakes. In that case, the Gutenberg–Richter [21] law acts as

an organizing principle connecting the frequency and magnitude of larger and smaller earthquakes. Performing the same analysis here, counting the number of epidemic events of size > s, there is a power-law dependence of the form $\log N(> s) = a - b \log(s)$. Thus, the number of epidemic events of size s scales as $N(s) \propto s^{-1-b}$ where $b \approx 0.28$. The same analysis for the epidemic event durations indicates that the number of epidemics of duration t scales as $N(t) \propto t^{-1-c}$, where, from Fig. 1A, bottom, $c \approx 0.8$. The presence of these scaling relations provides a useful way of estimating the like-

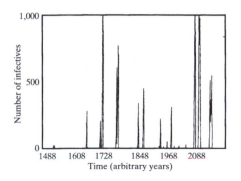

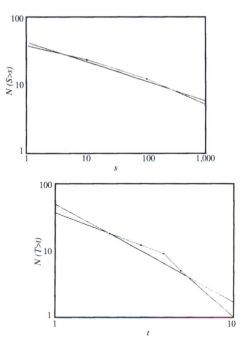

Fig. 2 Top left, time series of infectives for the model simulation. Each site in the $L \times L$ lattice is in one of three states: empty, occupied by a susceptible, or occupied by an infective. The lattice is updated synchronously at each time-step using the following rules: (1) susceptibles who are on nearest-neighbour sites to an infective become infective themselves; (2) infectives become inactive and the site they occupy becomes empty; (3) susceptibles are introduced onto empty lattice sites with a probability μ; (4) a new infective can arise when a susceptible is spontaneously infected with a probability ν. This effectively corresponds to an immigration term, whereby our lattice population is subject to infrequent infection from external sources. Periodic boundary conditions are used. These rules define a simple spatial $S - I$ model. It is the simplest possible model of epidemic spread on a lattice and is, at best, a caricature of the known epidemiological processes. The lattice is 250×250, with $\mu = 0.000026$ and $\nu = \mu/300$. Mean population density on the lattice is 25,000. At equilibrium, with an average life expectancy of 70 years, we expect ~ 1 new susceptible to appear each day. The 43 epidemic events in 58 years sets a lower limit of $\nu/\mu \approx 1/400$, although we use $\nu/\mu \approx 1/300$ for the simulations. Birth of new susceptibles and immigration of infectives onto the lattice are modelled as Poisson processes, with a mean of 1 new susceptible appearing on the lattice at each time-step and a mean of one infective immigrating every 300 time-steps. Thus 1 time-step can be equated with one day. The x-axis is labelled for 696 arbitrary months (58 years). Top right, distribution of epidemic event sizes for the simulation data, $b \simeq 0.29$. Bottom, distribution of epidemic event durations for the simulation data, $c \simeq 1.5$. The exponent is severely affected by the poor prediction of long epidemics > 10 months, but comparison with Fig. 1A, bottom, shows the simulation makes a good prediction of the distribution of epidemics of up to 5 months duration for the Faroes.

lihood, in a given interval of time, of epidemics of size s and duration t. Small short epidemics occur more often than large long epidemics, although their occurrence is connected and governed by the scaling exponents b and c. The exponents also apply to subportions of the data so, for example, we can use the epidemic distribution measured from the first half of the data set to predict the likely distribution of epidemic events in the second half.

We have also applied this analysis to two other subendemic populations with extensive and accurate measles records, Bornholm and Reykjavik. Fig-

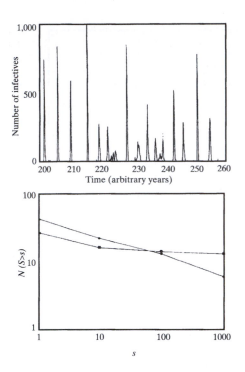

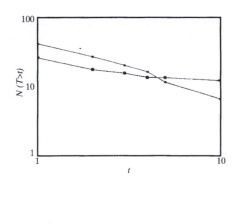

Fig. 3 Top left, time series of monthly number of infectives; bottom left, distribution of epidemic sizes (squares); and top right, epidemic duration (squares) for a stochastic Monte Carlo implementation of an SEIR model of measles in the Faroe Island population. The scaling for the Faroe Island (circles) data from Fig. 1 is shown for comparison. A constant population of 25,000 is used with a life expectancy of 50 years. The incubation and infectious intervals of measles are taken to be 7 days. R_0 is chosen to be 11. All births are into the susceptible class, with natural mortality occurring from each of the four classes. There is a Poisson influx of infectives into the community, on average every 300 days, as in the lattice simulation. The transients of the system are run off before data are recorded. No seasonality in the contact rate is used in these graphs, but it is found that inclusion of a sinusoidal seasonal driving term in the contact rate does not improve agreement between the model and the observed scaling.

ure 1*B*, *C* indicates that the exponents, *b* and *c*, for all three populations are remarkably similar.

Mechanisms leading to the emergence of power laws of this sort are still not fully understood, although recently many spatially extended interacting model systems have been shown to exhibit similar behaviour. We use one such lattice-based model previously used in discussion of 'forest-fire' dynamics [18, 22, 23], to provide a simple model for this phenomenon.

The dynamics of the model in Fig. 2 closely resemble those seen in the Faroe data. Exponents for the model data are similar to the actual epidemiological data; $b \simeq 0.29$ and $c \simeq 1.5$, with the simula-

tion underestimating the number of long epidemics of more than 10 months duration. The connectedness of the spatial distribution of the population on the lattice seems to reflect the social networks that exist in real communities [24].

For comparison, we illustrate the dynamics generated by a conventional stochastic SEIR model [3, 7, 8] of the Faroe Islands. Conventional measles parameters are used and the population is modelled as a single homogeneously mixing community, with the same low-level immigration rate of external infectives. As before, we calculate the distribution of epidemic sizes and durations (Fig. 3). The resulting times series of infectives looks similar to the observed island dynamics, but this calculation overestimates the number of larger epidemics and is not able to fit the observed data as well as the lattice model. The assumption of homogeneous mixing of susceptibles with infectives that underlie the SEIR equations appear to break down for small populations with infrequent epidemics. The replenishment of susceptibles onto vacant lattice sites and the nearest-neighbour contact spread of infection are central to reproducing the observed epidemic dynamics.

Our results suggest the existence of scaling laws for the size and duration of epidemics in the isolated island measles data sets. This places the dynamics of these epidemics in the same class as other spatially extended nonlinear dynamical systems, where scaling is also observed. In practice, this facilitates a form of prediction in which we can calculate the frequency of occurrence of epidemics of given size and duration. A simple spatial model generates almost the same exponents as are seen in our data analysis. The power-law phenomenon is also likely to be of relevance in the study of infrequent outbreaks of measles in highly vaccinated communities and in isolated rural populations in developing countries. The methods discussed here are completely general and could be applied to any other times series of communicable disease outbreaks in a small population.

REFERENCES

1. May, R. M. *Nature* **261**, 459-467 (1967).
2. May, R. M. *Proc. R. Soc. Lond.* A **413**, 27-44 (1987).
3. Grenfell, B. T. *J. R. Statist. Soc. B* **54**, 383-398 (1992).
4. Grenfell, B. T., Kleczkowski, A., Ellner, S. & Bolker, B. M. *Phil. Trans. R. Soc. Lond.* A **348**, 515-530 (1994).
5. Bolker, B. M. & Grenfell, B. T. *Proc. R. Soc. Lond.* B **251**, 75-81 (1993).
6. Grenfell, B. T., Bolker, B. M. & Kleczkowski, A. *Proc. R. Soc. Lond.* B **259**, 97-103 (1995).
7. Olsen, L. F., Truty, G. L. & Schaffer, W. M. *Theor. Pop. Biol.* **33**, 344-370 (1988).
8. Olsen, L. F. & Schaffer, W. M. *Science* **249**, 499-504 (1990).
9. Sugihara, G. & May, R. M. *Nature* **344**, 734-741 (1990).
10. Sugihara, G., Grenfell, B. & May, R. M. *Phil. Trans. R. Soc. Lond.* B **330**, 235-251 (1990).
11. Tidd, C. W., Olsen, L. F. & Schaffer, W. M. *Proc. R. Soc. Lond.* B **254**, 257-273 (1993).
12. Bartlett, M. S. *J. R. statist. Soc. A* **120**, 48-70 (1957).
13. Bartlett, M. S. *J. R. statist. Soc. A* **123**, 37-44 (1960).
14. Black, F. L. *J. theor. Biol.* **11**, 207-211 (1966).
15. Anderson, R. M. & May, R. M. *Infectious Diseases of Humans, Dynamics and Control* (Oxford Univ. Press, 1991).
16. Bak, P., Tang, C. & Weisenfeld, K. S. *Phys. Rev. A* **38**, 364-374 (1988).
17. Chen, K., Bak, P. & Obukhov, S. P. *Phys. Rev. A* **43**, 620-635 (1991).
18. Clar, S., Drossel, B. & Schwabl, F. *Phys. Rev. E* **50**, 1009-1018 (1994).
19. Somette, A. & Somette, D. *Europhys. Lett.* **9**, 197-202 (1989).
20. Cliff, A. D., Haggett, P. & Smallman-Raynor, M. *Measles. An Historical Geography of a Major Human Viral Disease from Global Expansion to Local Retreat 1840-1990* (Blackwell, Oxford, 1993).
21. Gutenberg, B. & Richter, C. F. *Ann. Geoffs.* **9**, 1-10 (1956).
22. Bak, P., Chen, K. & Tang, C. *Phys. Lett. A* **147**, 297-300 (1990).
23. Drossel, B. & Schwabl, F. *Phys. Rev. Lett.* **69**, 1629-1632 (1992).
24. Ferguson, N., May, R. M. & Anderson, R. M. in *Spatial Ecology: The Role of Space in Population Dynamics and Interspecific Interactions* (eds Tilman, D. & Karelva, P.) (Princeton Monographs in Population Biology, in press).

READING 7.2

Allometry and Simple Epidemic Models for Microparasites

Commentary: My purpose in presenting this article is to illustrate that the simple epidemic models that are discussed in this course are very much in use today. Note in particular the equations at the beginning of the last paragraph of the first page of this article. In the article, $S(t)$ denotes the number of susceptibles at time t, while $I(t)$ denotes the number of infected at time t. The equations read

$$\frac{d}{dt}S = (\nu - \mu)(1 - S/K)S - \lambda(I)S,$$

$$\frac{d}{dt}I = \lambda(I)S - (\mu + \alpha)I.$$

Here, K is the carrying capacity in the absence of pathogen, ν is the inherent birth rate, μ the inherent death rate, and α the increase in the death rate due to the disease. Meanwhile, $\lambda(I)$ is the function of I that characterizes how the disease is transmitted. The authors consider the two cases $\lambda(I) = \beta I$ and $\lambda(I) = \beta I/(1 + S)$, where β is a constant. The point of the article is to describe certain ideas for estimating this constant β.

Allometry and Simple Epidemic Models for Microparasites

Science, **379** (1996) 720–722.
Giulio A. De Leo, Andy P. Dobson

Simple mathematical models for microparasites offer a useful way to examine the population dynamics of different viral and bacterial pathogens. One constraint in applying these models in free-living host populations is the paucity of data with which to estimate transmission rates. Here we recast a standard epidemiological model by setting the birth and death rates of the host population and its density as simple allometric functions of host body weight. We then use standard threshold theorems for the model in order to estimate the minimum rate of transmission for the parasite to establish itself in a mammalian host population. Transmission rates that produce different comparable values of the parasites' basic reproductive number, R_0, are themselves allometric functions of host body size. We have extended the model to show that hosts having different body sizes suffer epidemic outbreaks whose frequency scales with body size. The expected epidemic periods for pathogens in different mammalian populations correspond to cycles observed in free-living populations.

The basic microparasite model takes the form [1] $dS/dt = (\nu - \mu)(1 - S/K)S - \lambda(I)S$, and $dI/dt = \lambda(I)S - (\mu + \alpha)I$, where S and I are the density of

susceptible and infected individuals respectively, K the carrying capacity in absence of the pathogen [2, 3], μ the host natural mortality rate, ν the maximum birth rate, α the disease-induced mortality, and $\lambda(I)$ the 'force of infection', that is, the rate at which animals become infected. The actual relationship between λ and I depends upon the type of interaction among infected and susceptibles. We examine two main types of transmission: density-dependent, in which the probability of a susceptible becoming infected is proportional to the density of infectives through the transmission rate β, namely $\lambda(I) = \beta I$; and frequency-dependent, in which the same probability is a function of the proportion of infectives, namely $\lambda(I) = \beta I/N$ (where $N = I + S$). Density-dependent transmission is usually assumed for wild-animal diseases [4, 5], but frequency dependence may characterize sexually and vectorially transmitted pathogens [6, 7]. It may also apply to species that are strongly territorial.

Following the introduction of a single infective into a population of K susceptibles, the disease will only spread provided $dI/dt > 0$. For this to occur, β must exceed a critical value β_{min}, where $\beta_{min} = (\mu + \alpha)/K$ for density-dependent transmission, and $\beta_{min} = \mu + \alpha$ for frequency-dependent

---

Author affiliations: Giulio A. De Leo, Andy P. Dobson: Ecology and Evolutionary Biology, Princeton University.

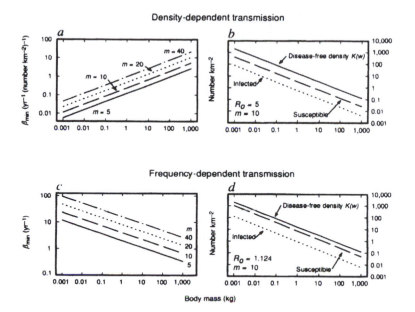

Fig. 1 Allometric relationships of $\beta_{\min}$ for density-dependent transmission (in a, $\beta_{\min} = 0.0247mw^{0.44}$) and frequency-dependent transmission (in c, $\beta_{\min} = 0.4mw^{-0.26}$). Host demographic parameters are: $K = 16.2w^{-0.70}$ (number km^{-2}), $r = \gamma - \mu = 0.6w^{-0.27}$ (years^{-1}), $\mu = 0.4w^{-0.26}$ (years^{-1}); m indicates the diseased induced mortality factor for infected individuals. If infection spreads ($\beta > \beta_{\min}$), host density will drop below the natural carrying capacity. Densities of susceptibles and infectives at equilibrium are inversely related to w (b and d). In contrast, the fraction of infected hosts and the degree of depression, $d = N_{eq}(W)/K(w)$, below the carrying capacity are independent of w. To derive the relationships in b and d, transmission rates have been chosen to produce comparable values of R_0 for mammals of different size, namely $\beta(w) = 5\beta_{\min}(w)$, and thus $R_0 = K\beta/(\alpha + \mu) = 5$ in the case of density-dependent transmission, and $\beta(w) = 1.124\beta_{\min}(w)$, and thus $R_0 = \beta/(\alpha + \mu) = 1.124$ in the case of frequency-dependent transmission. In the latter case, the host population is always driven to extinction when $R_0 \geq 1.22$. The allometric exponents of $I_{eq}(W)$ and $S_{eq}(w)$ are, to a large extent, determined by the carrying capacity $K(w)$ of the disease-free population, and are largely insensitive to the actual values of R_0 and α (provided $R_0 > 1$). Although transmission parameters combine several different epidemiological, environmental and social factors, the scaling of $\beta_{\min}$ is basically due to the following reasons: $\beta_{\min}$ increases with w in the density-dependent case because the number of encounters per unit time decreases with body size as a consequence of the lower population density, thus higher transmission rates are required for the disease to spread. In contrast, when transmission is frequency-dependent, the probability of successfully transmitting the disease does not depend upon density, but is proportional to the time spent in the infective class, $(\alpha + \mu)^{-1}$; as larger individuals have a longer infectious period, $\beta_{\min}$ decreases with w.

transmission. These thresholds are directly related to the basic reproductive number R_0, defined as the average number of secondary cases generated by one primary case in a susceptible population [5, 8]. The disease spreads in the population if $R_0 \geq 1$.

These threshold theorems are very useful in investigating disease dynamics and control policies like culling and vaccination, but their use is hampered in practice because direct measurements of

β are difficult, if not impossible, to obtain without extensive field data [1, 9]. In contrast, estimates of the basic host demography (ν, μ, and K) are available for many species. Extensive comparative studies [10–14] relate ν, μ and K to body size and show that they scale with height w as simple allometric relationships.

We can thus derive a similar allometric relationship for the transmission rate and predict the

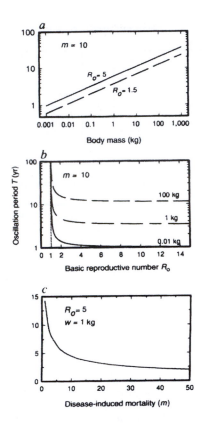

Fig. 2 The oscillation period T (in years) of the host population as a function of the body mass w (in a), the basic reproductive number $R_0 = K\beta/(\alpha + \mu)$ (in b) and the disease-induced mortality factor m (in c) in the case of density-dependent transmission. Note that for a pathogen that produces a 5–20-fold reduction of life expectancy of a 1-kg animal, the oscillation period ranges between 3 and 7 years, whereas T is practically constant for $R_0 > 3$. Scaling of the oscillation period is basically determined by the scaling of the intrinsic growth rate [16]. The stability properties and the transient dynamics to equilibrium are different when transmission is frequency-dependent: when $w < 10kg$, the system approaches equilibrium without oscillations, whereas for $w > 10kg$, damped oscillations with a very long period (> 60 years) may occur. In any case, no significant relationship can be detected between body size and equilibrium eigenvalues.

threshold values of β for species over a wide range of body sizes. To generalize this calculation we rescale the disease-induced mortality rate α with respect to the average lifespan of the species. By assuming that α is $(m - 1)$ times the natural mortality μ, more (or less) virulent pathogens will be characterized by larger (or smaller) values of m, as this produces an m-fold reduction of the $(\mu + \alpha)^{-1}$ life expectancy of infected individuals. .

After replacing $v(w)$, $\mu(w)$, $K(w)$ and $\alpha(w)$ into the equations for β_{min}, we find that it scales allometrically with body size. β_{min} increases with w when transmission is density-dependent and decreases when transmission is frequency-dependent (Fig. 1). With density-dependent transmission, the model exhibits damped oscillations towards equilibrium densities for any body size (Fig. 1), provided $\beta(w) > \beta_{min}(w)$. The oscillation period is allometric (Fig. 2), with an exponent of 0.260. Peterson *et al.* [15] analysed population cycles of 41 species of birds and mammals: they found that populations fluctuate with a period related to body mass

Model	Basic reproductive number	Threshold criterion for transmission rate	Allometric relationship $\beta(w)$
a, SI model with nonlinear density-dependent fecundity $S' = (v - \mu)[1 - (N/K)^C]S - \beta SI \quad N = I + Sc > 1$ $I' = \beta SI - (\alpha + \mu)I \qquad\qquad \alpha = (m - 1)\mu$	$R_0 = \dfrac{K\beta}{\alpha + \mu}$	$\beta \geq \beta_{\min} = \dfrac{\alpha + \mu}{K}$	$\beta_{\min} = 2.47\,10^{-2}\,m w^{0.44}$
b, SI model with density-dependent adult mortality $S' = (v - \mu)[1 - N/K]S - \beta SI \quad N = I + S$ $I' = \beta SI - (\alpha + \mu + \gamma N)I \quad\ \gamma = (v - \mu)/K$	$R_0 = \dfrac{K\beta}{\alpha + v}$	$\beta \geq \beta_{\min} = \dfrac{\alpha + v}{K}$	$\beta_{\min} = (2.47m + 8.02)10^{-2} w^{0.44}$
c, SIR: susceptible, infective and recovered $S' = (v - \mu)[1 - F/K]S - \beta SI \quad F = S + R$ $I' = \beta SI - (\alpha + \mu + \rho)I \qquad \gamma = (v - \mu)/K$ $R' = \rho I - \mu R$	$R_0 = \dfrac{K\beta}{\alpha + m\mu + \rho}$	$\beta \geq \beta_{\min} = \dfrac{\alpha + \mu + \rho}{K}$	$\beta_{\min} = 2.47\,10^{-2}(m + n)w^{0.44}$ where $n \equiv \rho/\mu$
d, SIR: susceptible, exposed and recovered $S' = (v - \mu)[1 - F/K]S - \beta SI \quad F = S + E + I$ $E' = \beta SI - (\sigma + \mu)E$ $I' = \sigma E - (\alpha + \mu)I$	$R_0 = \dfrac{\sigma \beta K}{(\sigma + \mu)(\alpha + \mu)}$	$\beta \geq \beta_{\min} = (1 + \frac{\mu}{\sigma})\dfrac{\alpha + \mu}{K}$	$\beta_{\min} \approx 2.47\,10^{-2}\,m w^{0.44}$ when $\sigma \gg \mu$
e, SIR: susceptible, exposed, infective and recovered $S' = (v - \mu)[1 - F/K]S - \beta SI \quad F = S + E + R$ $E' = \beta SI - (\sigma + \mu)E$ $I' = \sigma E - (\alpha + \mu + \rho)I$ $R' = \rho I - \mu R$	$R_0 = \dfrac{\sigma \beta K}{(\sigma + \mu)(\alpha + \mu + \rho)}$	$\beta \geq \beta_{\min} = (1 + \frac{\mu}{\sigma})\dfrac{\alpha + \mu + \rho}{K}$	$\beta_{\min} \approx 2.47\,10^{-2}\,m(m + n)w^{0.44}$ where $n \equiv \rho/\mu$ and $\sigma \gg \mu$

Table 1. Threshold conditions on β for five epidemiological models assuming different courses of infection and the density dependence of the host v and μ are the maximum fecundity and minimum mortality, respectively. The allometric relationship $\beta(w)$ between transmission and body size does not change with respect to the basic SI model when infected individuals can reproduce and the density-dependent reduction in host fecundity is not linear (model a). If density affects adult mortality rather than fecundity (model b), the threshold condition becomes $\beta \geq \beta_{\min} = (v + \alpha)/K$. As fecundity has roughly the same allometric coefficient as that of host mortality, $\beta_{\min}$ scales with body mass as in model a. Furthermore, if the disease-induced mortality factor m equals 20, $\beta_{\min}$ is only about 10% larger than in model a for the same m. The introduction of an immune class (model c) or a latent class (models d and e) slightly complicates the formula for the β threshold, but does not modify the allometry of $\beta_{\min}$ with respect to the previous models. In fact, the average time spent in the latent class, $1/\sigma$, is usually much smaller than the average life expectancy $1/\mu$, therefore $\mu/\sigma \approx 0$. Also, the rate of recovery ρ for the immune class can reasonably be assumed to scale with body size as μ. This explains why the exponent of the $w - \beta$ relationship is the same as usual. Although the allometric property of the threshold criterion is basically insensitive to variation in the model structure, the actual dynamics may change considerably as models with a latent class may show complex dynamics (such as limit cycles and chaos) if the basic reproductive number is sufficiently high, whereas the simple SI model cannot, unless exogenous seasonal factors are explicitly introduced.

as $8.51 w^{-0.263}$. Although our finding matches their data very well, other factors such as spatial and age structure, predation, plant–herbivore relationships and seasonality may be important in explaining the observed patterns. Oscillations are not detectable in the frequency-dependent case, which actually applies only to a minority of wild-animal diseases.

This model does not incorporate many features of the population at the epizoological level, such as different forms of density-dependent growth and the presence of latent and/or recovered classes. But several logical embellishments do not significantly modify the relationship between the body size of the host and the characteristic transmission rate: allometry holds for a wide class of epidemio-logical models (Table 1). These relations should not be interpreted as deterministic laws giving the exact transmission rate for any species, however, because other important details must be considered. For instance, there is considerable variability in the relationship between population density and body size [14]. Several studies [10, 13] have shown that carnivore populations are significantly less dense than those of herbivores and frugivores, so the relationships we described should be modified to include these differences. Also, the two components of the transmission rate should be included: production of infective agents and actual transmission of those agents from infective to susceptible. Large body size may allow greater production of infective viral particles—this would boost the transmission rate in the absence of any compensatory decrease in the actual transmission rate of pathogens.

The allometric relationships shown in Table 1 are a good indication of the general trends underlying the available data and provide a useful initial estimate of β for pathogens of mammals over a wide range of body sizes and population densities. The results in Fig. 1a, c offer an important insight into whether emergent diseases such as Hantavirus and Ebola virus could become established in the human population. If pathogens use mammals as their normal hosts and transmission is density-dependent, then outbreaks of these pathogens should die out in all but the most crowded human populations; in contrast, if transmission is frequency-dependent, as in the case of human immunodeficiency virus, they should be able to become established.

REFERENCES

1. Anderson, R. M. & May, R. M. *Infectious Diseases of Humans: Dynamics and Control* (Oxford Univ. Press, 1992).
2. Anderson, R. M. & Trewhella, W. *Phil. Trans. R. Soc. Lond.* B **310**, 327–381 (1985).
3. Fowler, C. W. *Ecology* **62**, 602–610 (1981).
4. Kermack, W. O. & McKendrick, A. G. *Proc. R. Soc. Lond.* A **115**, 700–721 (1927).
5. Anderson, R. M. & May, R. M. *Nature* **280**, 361–367 (1979).
6. Antonovics, J., Iwasa, J. Y. & Hassell, M. P. *Am. Nat.* **145**, 661–675 (1995).
7. Thrall, P. H., Biere, A. & Uyenoyama, M. K. *Am. Nat.* **145**, 43–62 (1995).
8. Anderson, R. M. & May, R. M. *Phil. Trans. R. Soc. Lond.* B **291**, 451–524 (1981).
9. Grenfall, B. T. & Dobson, A. P. *Infectious Diseases in Natural Populations* (Cambridge Univ. Press, 1995).
10. Peters, R. H. *The Ecological Implications of Body Size* (Cambridge Univ. Press, 1983).
11. Calder, W. A. *Size, Functions and Life History* (Harvard Univ. Press, Boston, 1984).
12. Schmidt-Nielsen, K. *Scaling: Why is Animal Size so Important?* (Cambridge Univ. Press, 1984).
13. Charnov, E. L. *Life History Invariants* (Oxford Univ. Press, 1993).
14. Silva, M. & Downing, J. A. *Am. Nat.* **145**, 704–727 (1995).
15. Peterson, R. O., Page, R. E. & Dodge, K. M. *Science* **224**, 1350–1352 (1984).
16. Mollison, D. *Nature* **310**, 224–225 (1984).

READING 7.3

Scope of the AIDS Epidemic in the United States

Commentary: The purpose of presenting this article is to illustrate the sorts of issues that epidemiologists must consider when dealing with a complicated disease such as AIDS. In particular, the article observes that the incidence of AIDS cases has slowed, approaching a plateau. The question is, How should this fact be interpreted? When AIDS first appeared, there was a large susceptible population with a broad range of ages, so the virus spread rapidly. As time went on, older susceptibles either became infected or else removed themselves from risk of infection. Thus, we expect to see such a plateau. However, for the purposes of predicting the future course of the epidemic, it is crucial to determine the rate of infectivity for young susceptibles, for there is a continuous supply of such individuals. However, this data is hard to come by for AIDS since the incubation period from initial infection to clinical symptoms can be very long. With this as background, the article uses epidemic models to estimate the number of infectives as a function of time and of age at infection.

Scope of the AIDS Epidemic in the United States

Science, **270** (1995) 1372–1375.
Philip S. Rosenberg

Two-dimensional deconvolution techniques are used here to reconstruct age-specific human immunodeficiency virus (HIV) infection rates in the United States from surveillance data on acquired immunodeficiency syndrome (AIDS). This approach suggests that 630,000 to 897,000 adults and adolescents in the United States were living with HIV infection as of January 1993, including 107,000 to 150,000 women. The estimated incidence of HIV infection declined markedly over time among white males, especially those older than 30 years.   In contrast, HIV incidence appears to have remained relatively constant among women and minorities. As of January 1993, prevalence was highest among young adults in their late twenties and thirties and among minorities.   An estimated 3 percent of black men and 1 percent of black women in their thirties were living with HIV infection as of that date. If infection rates remain at these levels, HIV must be considered as endemic in the United States.

Through the use of deconvolution methods known as backcalculation, the national AIDS database compiled by the Centers for Disease Control and Prevention (CDC) and the distribution of the incubation period between infection with HIV and diagnosis with AIDS can be used to reconstruct the historical incidence of HIV infection that best accounts for the observed epidemic of AIDS cases [1]. In recent years, the incidence of AIDS in the United States has slowed, with rates that are approaching a plateau [2]. This trend likely reflects both reduced infection rates since the mid-1980s and widespread use of prophylactic therapies, which delay the onset of AIDS [3]. Although the plateau in national AIDS cases would appear to be a favorable sign, optimism must be tempered by an appreciation of the dynamic nature of the epidemic. The HIV virus entered the United States during the late 1970s and spread rapidly during the early to mid-1980s. During this early period, there were large susceptible populations at risk over a broad range of ages. As the epidemic matured, one would expect that new entrants to at-risk populations—homosexual men, injection drug users, and high-risk heterosexuals—would tend to be young.   Hence, it is plausible that the epidemic would stabilize with fewer infections occurring in recent years compared to the mid-1980s. For purposes of epidemic monitoring, therefore, a key question is whether the incidence rate among young adults has declined in comparison to the rate among persons of the same age in the past.

Although there are only limited data available to address this question by direct observation, extensions of the backcalculation approach [4, 5] allow one to estimate the age-specific incidence of infection from the age-specific incidence of AIDS. To avoid potential biases resulting from CDC's expansion of the AIDS case definition in 1993 [2], this approach was applied to cases among adults and adolescents who were diagnosed up to 1 January 1993 and who met the previous 1987 case definition [6]. Incidence counts were adjusted for delays in reporting and estimates were inflated by 18% to reflect cases that will never be reported [7].

Although the overall rate of increase slowed after 1987, AIDS incidence trends differ according to birth cohort (Fig. 1A). Incidence among persons born before 1960 increased during the early to mid-1980s and then approached a plateau during 1991 to 1992, but AIDS incidence among persons born after 1960 was very low until 1986 and has increased steadily since then. This qualitative difference is apparent among men and women in each racial and ethnic group (Fig. 1B).

The infection rate function $v(s, a)$ specifies the number of infections per year at calendar time s among persons aged a years. Estimates of $v(s, a)$ were derived from age-specific AIDS incidence data on the basis of the incubation distribution. The fundamental convolution equation [4, 5] is given by

$$E(Y_{t,k}) = \int_{T_0}^{t} v(s, k - t + s) f(t - s | k - s, s) ds \quad (1)$$

Author affiliation: Philip S. Rosenberg: National Cancer Institute.

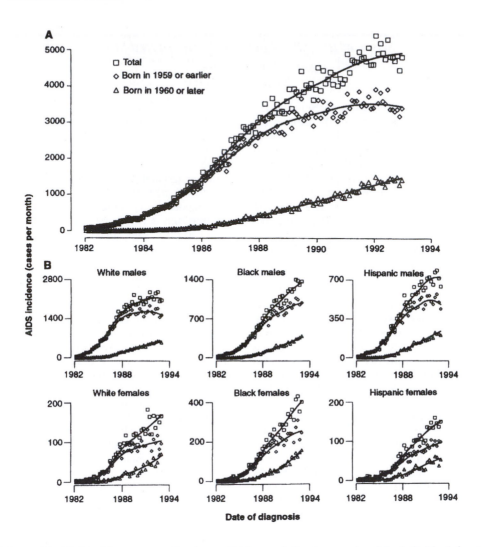

Fig. 1 (A) Monthly AIDS incidence counts for the total U.S. population (squares) and for individuals born in 1959 or earlier (diamonds) and in 1960 or later (triangles). (B) AIDS incidence for white, black, and Hispanic men and women. Locally weighted regression smoothing [19] with a bandwidth of 40% of the time axis was used to highlight trends in incidence (solid curves). The date axes mark 1 January of the years indicated.

where $E(Y_{t,k})$ is the expected number of cases occurring at calendar time t among persons aged k years at diagnosis, T_0 is the assumed start date of the epidemic, and $f(t|a, s)$ is the incubation period density function for persons aged a years at infection who were infected at time s. The incubation distribution varies by age to reflect that younger age is associated with slower progression [8] and

by time to account for the increasing use of prophylactic therapies since 1987 [1, 3-5]. Given observed AIDS incidence data $Y_{t,k}$ and an estimate of the family of incubation distributions $f(t|a, s)$ derived from cohort studies, $v(s, a)$ is estimated by maximum likelihood under the assumption that the observed counts followed a Poisson distribution [9].

The infection rate function was modeled with

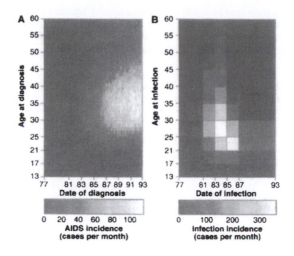

Fig. 2 (A) Monthly AIDS incidence counts among white males for January 1982 through December 1992 by single year of age (ages 13 to 59 are shown; ages 13 to 74 were used in the analyses). (B) Backcalculated number of infections per month by single year of age, calculated with the fast incubation distribution; results were qualitatively similar when the slow incubation distribution was used. Backcalculation models were fitted to the observed data after grouping the monthly periods into quarters and the single years of age into 17 age-at-diagnosis categories to reduce the computational burden. Appropriate expectations were calculated for the counts in these cells by integration of Eq. 1 [9]. Rectangular regions in (B) correspond to areas of assumed constant infection intensity. Results were similar in a sensitivity analysis with 10 knots on the time axis based on cutpoints set to 1 January of calendar years 1977, 1981 through 1988, and 1993.

bivariate step functions that had relatively broad steps in the time dimension and narrow steps in the age dimension. This approach yielded a good fit to the observed data without making strongly parametric assumptions about the infection rates [9]. The low hazard of AIDS in the first few years after infection makes it difficult to backcalculate recent HIV incidence [1]. For this reason, a 6-year step was used to cover the 1987 to 1992 period. Use of a long last step implies that estimated incidence for the recent past reflects a broad period average, and no inferences can be made about trends within this period.

Conservative estimates of incidence were derived from a Weibull model of the incubation distribution with a 9-year median time-to-AIDS for individuals aged 30 years at infection (the "fast" model) [10]. Larger estimates were derived from a "slow" model that incorporated a Weibull hazard with a change point to a linear slope [11] and with a median of 9.8 years to AIDS. A modest therapy effect on the incubation distribution was used to reflect that the efficacy of treatment appears to attenu-

ate within about 1 year [12]. Plausible ranges of estimates [1,9] were calculated by subtracting two standard errors from point estimates derived from the fast incubation distribution and by adding two standard errors to point estimates derived from the slow incubation distribution. Given estimates of the numbers of incident HIV infections derived by backcalculation, surveillance data on AIDS mortality and census data on population size were used to estimate the age-specific incidence of HIV infection per 100,000 person-years at risk and to estimate the percentage of the population that was alive with HIV infection as of 1 January 1993 [13].

In white males, AIDS cases have been concentrated among men in their thirties since the onset of the epidemic, but in recent years, AIDS incidence has been relatively high among men in their late twenties (Fig. 2A). In each age group, the number of infections per year increased and then decreased over time, with peaks in the 1983 to 1986 period (Fig. 2B). Although all age groups showed a decline in the number of new infections over time, the reductions were substantially greater for men older

Table 1.   Prevalence of HIV-1 infection in the United States as of 1 January 1993, estimated by backcalculation from AIDS incidence data [7] as described in the legends to Figs. 2 through 4. Point estimates are based on the fast model of the incubation distribution and may be conservative [10]. Plausible ranges, given in parentheses, reflect some uncertainty about the incubation distribution.

Group	Alive with HIV-1 infection (in thousands)	HIV-1–positive, ages 18 to 59 (%)
	Males	
White	255 (248–370)	0.49(0.48–0.70)
Black	184 (176–236)	2.29(2.20–2.91)
Hispanic	97 (91–131)	1.44(1.37–1.87)
Total	544 (523–747)	0.78(0.77–1.04)
	Females	
White	25 (23–34)	0.05(0.04–0.06)
Black	67 (63–82)	0.74(0.69–0.90)
Hispanic	24 (21–32)	0.34(0.31–0.45)
Total	117 (107–150)	0.16(0.15–0.20)
Grand total	660(630–897)	0.47(0.46–0.62)

than 30 years compared to younger men.

Estimates of the mean annual number of infections per 100,000 person-years at risk among white, black, and Hispanic men and women for 1981 to 1986 and 1987 to 1992 show a rapid rise in infection incidence beginning in the late teens and early twenties and peak rates among individuals in their mid- to late twenties (Fig. 3). Only among white males was there a marked decline over time. Rates may also have declined among younger Hispanic men. Black females showed increasing infection rates.

Among white males, age-specific HIV prevalence as of 1 January 1993 peaked at about 1% among individuals in their early thirties (Fig. 4). Prevalence rates were high among minority men over a broad range of ages, 30 to 44 years. About 3% of black males and 1.5% of Hispanic males in this age range were living with HIV infection. Prevalence was about 1% for black females and 0.5% for Hispanic females in their late twenties and early thirties. Prevalence was lowest in white women.

Between the start of the epidemic and 1 January 1993, an estimated 857,000 to 1.1 million Americans had become infected with HIV but 227,000 had died of AIDS. The resulting plausible range for persons living with HIV infection as of 1 January 1993 was 630,000 to 897,000 (Table 1). The total includes an estimated 107,000 to 150,000 HIV-infected women, yielding an estimated male-to-female prevalence ratio for that date of about 4.6 to 1.

The backcalculation estimate of 630,000 to 897,000 prevalent infections in the entire population as of January 1993 is lower than the previous Public Health Service (PHS) estimate of around 1 million infected [14], a figure that has been widely used for planning purposes. The difference is less striking when one considers that the plausible range for the PHS estimate was 800,000 to 1.2 million [14], a range that overlaps with previous ranges derived by backcalculation [1, 15] and the present study. Multiplying the incidence rates in Fig. 3 by the corresponding population sizes indicates that an average of 40,000 to 80,000 new infections occurred each year from 1987 to 1992, a range similar to previous estimates [1, 14]. However, this estimate provides no information about trends during the 6-year period and is subject to even greater uncertainty than the prevalence estimates.

The highest national prevalence rates were seen among minorities, and in particular among young black men. These estimates are derived from the large numbers of AIDS cases that have already been diagnosed. It is sobering to consider that 1 of every 50 black men in the United States aged 18 to 59 may be infected (Table 1), but a similar high rate in black men aged 18 to 59 of 1.9% was found in the Third National Household and Nutrition Examination Survey [16], a study that may have yielded conservative estimates because of nonresponse bias. The prevalence estimate derived here for women of 107,000 to 150,000 is also consistent with prevalence rates in the national Survey in Childbearing Women [17],

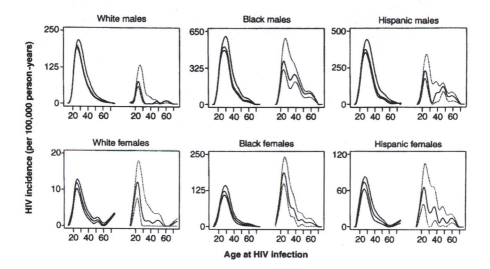

Fig. 3 Estimates of the number of infections per 100,000 person-years at risk by single year of age among white, black, and Hispanic men and women. Age-specific numbers of infections were calculated for each demographic group and for both models of the incubation distribution, as described in the legend to Fig. 2. For each model, the average infection rate per year was calculated for calendar periods 1981 through 1986 (left subplot) and 1987 through 1992 (right subplot), and a spline interpolant was drawn through the midpoints of the 11 age steps to highlight trends. Solid curves show point estimates based on the fast incubation distribution, and dotted curves show plausible ranges.

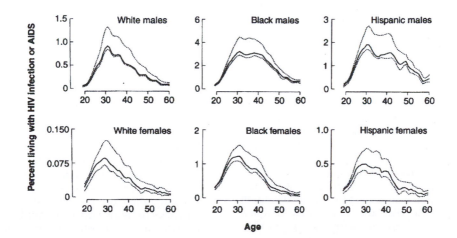

Fig. 4 Estimates of HIV prevalence as of 1 January 1993 by single year of age among white, black, and Hispanic men and women [13]. Solid curves show point estimates based on the fast incubation distribution and dotted curves show plausible ranges. Summary prevalences for persons aged 18 to 59 years are shown in Table 1.

which suggested that 130,000 women in the United States were infected as of 1992 [18].

It is important to recognize that backcalculated estimates are based on modeling rather than direct data and are very uncertain. The plausible ranges shown in Figs. 3 and 4 and in Table 1 account for a substantial amount, but not all, of the uncertainty about the incubation distribution and for random error from model fitting. These ranges do not account for uncertainty about the adjusted AIDS incidence counts or about the inflation factor used to account for cases that will never be reported, nor do they consider the choice of model of the infection rate function.

However, a central observation is that AIDS incidence increased much more rapidly in recent years among younger individuals born in 1960 or later than among older individuals (Fig. 1). Any proposed model must explain these qualitatively different epidemic curves. If infection rates remain at the levels indicated by these models, HIV must be considered an endemic infection affecting successive cohorts of young adults.

REFERENCES

1. R. Brookmeyer, *Science* **253**, 37 (1991).
2. Centers for Disease Control, *Morb. Mortal. Wkly. Rep.* **43** 826 (1994). This plateau was obscured after January 1993 by the expansion of case definition to include HIV-infected persons with severe CD4 depletion [Centers for Disease Control, *ibid.* **41** (no. RR-17) (1992)].
3. M. H. Gail, P. S. Rosenberg, J. J. Goedert, *J. AIDS* **3**, 296 (1990).
4. N. G. Becker and I. C. Marschner, *Biometrika* **80**, 165 (1993); A. Verdecchia *et al.*, *Int. J. Epidemiol.* **23**, 1027 (1994).
5. P. S. Rosenberg, *Stat. Med.* **13**, 1975 (1994).
6. Centers for Disease Control, *Morb. Mortal. Wkly. Rep.* **36** (suppl. 1S) (1987).
7. The analysis was based on all cases in the United States and its territories, including Puerto Rico, who were older than 13 years at diagnosis and who were reported to CDC as of March 1994. The AIDS incidence and mortality data for individuals who met the 1987 case definition [6] were adjusted for reporting delays with standard CDC adjustment weights. Incidence counts were inflated by 18%, an approximate figure, to account for unreported cases and for infected individuals who died before meeting the case definition. Adjusted totals of 336,000 incident cases of AIDS and 227,000 AIDS deaths as of 1 January 1993 were stratified by month of diagnosis, single year of age at diagnosis, gender, and race and ethnicity (non-Hispanic whites, non-Hispanic blacks, Hispanics, and others). Use of the 1987 definition maintains consistency between the AIDS data and available estimates of the incubation distribution.
8. M. E. Eyster *et al.*, *Ann. Intern. Med.* **107**, 1 (1987); A. R. Moss *et al.*, *Br. Med. J.* **296**, 745 (1988); A. B. Mariotto *et al.*, *Am. J. Epidemiol.* **135**, 428 (1992); P. S. Rosenberg, J. J. Goedert, R. J. Biggar, *AIDS* **8**, 803 (1994).
9. Technical details of the fitting procedures are described elsewhere [5].
10. This fast model may yield conservative estimates because the natural history hazard function may level off around 7 years after infection [J. C. M. Hendriks et al., *AIDS* **7**, 231 (1993)]. The Weibull hazard, in contrast, continues to increase. To reflect that progression rates appear to be slower with infection at younger ages, the age-specific hazard of progression for the fast model was assumed to increase (decrease) by the factor 1.042 (0.960) per year increase (decrease) in the age at infection compared to a person aged 30 years at infection, as derived in [9].
11. The estimated linear slope effect was 0.01 per year at 2.6 years after infection and the corresponding age effect was 1.037 [9]. Both the fast and slow estimates of the incubation distribution were derived from a large cohort study of HIV-infected homosexual men [R. J. Biggar and the International Registry of Seroconverters, *AIDS* **4**, 1059 (1990)].
12. Therapy was assumed to reduce the hazard of AIDS among treated individuals by the factor 0.50 for persons infected for about 5 years, and then to wear off [15]. The estimated cumulative proportion in treatment before AIDS was 40% for white males and 20% for other demographic groups. Therapy was assumed to have been introduced in April 1987 among white men and women and in April 1988 among other groups. These parameters are broadly consistent with clinical trial results [Concorde Coordinating Committee, *Lancet* **343**, 871 (1994)] and with estimates of the extent of therapy use in different groups [N. M.

H. Graham *et al.*, *J. AIDS* **4**, 267 (1991); P. S. Rosenberg *et al.*, *ibid.*, p. 392; W. Lang *et al.*, *ibid.* **6**, 191 (1993)].

13. To estimate age-specific prevalence at time T, note that individuals aged k years represent the surviving members of the cohort born at time $T - k$. Prevalence is estimated as the cumulative probability that a member of this cohort becomes infected as of T minus the cumulative probability that a member dies of AIDS as of T. These quantities can be estimated with actuarial formulas applied to census and AIDS mortality data and the results from the backcalculations.

14. Centers for Disease Control, *Morb. Mortal. Wkly. Rep.* **39**, 110 (1990).

15. P. S. Rosenberg, M. H. Gail, R. J. Carroll, *Stat. Med.* **11**, 1633 (1992).
16. G. M. McQuillan *et al.*, *J. AIDS* **7**, 1195 (1994).
17. M. Gwinn *et al.*, *J. Am. Med. Assoc.* **265**, 1704 (1991).
18. J. M. Karon, personal communication. This estimate is based on an analysis of unpublished CDC data.
19. W. S. Cleveland, *J. Am. Stat. Assoc.* **74**, 829 (1979).
20. I gratefully acknowledge M. Morgan for providing national AIDS surveillance data, T. Green for calculating the reporting delays, J. Smith of Information Management Systems for data management, and J. M. Karon and R. J. Biggar for many helpful discussions.

Exercises for Chapter 7

Part 1: Phase Plane Exercises

1. Do phase plane analysis for the following system of differential equations:

$$\frac{d}{dt}x = 3y - xy$$
$$\frac{d}{dt}y = 4x - xy.$$

That is, draw the xy-plane, draw the x and y null clines with the appropriate slash marks, give a rough direction for the trajectories in the regions between the null clines, and put arrows on the null-cline slash marks to indicate the direction of motion.

When you have finished with these tasks, explain what happens to a trajectory that starts at time zero at the point $(1/10, 1/10)$.

2. Do phase plane analysis for the following system of differential equations:

$$\frac{d}{dt}x = y - \cos(x)$$
$$\frac{d}{dt}y = xy - \pi y.$$

(Here, $\pi = 3.14\ldots.$) That is, draw the xy-plane, draw the x and y null clines with the appropriate slash marks, give a rough direction for the trajectories in the regions between the null clines, and put arrows on the null-cline slash marks to indicate the direction of motion.

When you have finished with these tasks, explain what happens to a trajectory that starts at time zero at the point $(0, 1/2)$.

Part 2: Vector Exercises

1. Compute $\mathbf{v} + \mathbf{w}$ and $\mathbf{v} - \mathbf{w}$ when $\mathbf{v}$ and $\mathbf{w}$ equal:

   (a) $\mathbf{v} = \left(\begin{smallmatrix} 5 \\ 5 \end{smallmatrix}\right)$ and $\mathbf{w} = \left(\begin{smallmatrix} -3 \\ 2 \end{smallmatrix}\right)$

   (b) $\mathbf{v} = \left(\begin{smallmatrix} 3 \\ 1 \end{smallmatrix}\right)$ and $\mathbf{w} = \left(\begin{smallmatrix} -10 \\ -31 \end{smallmatrix}\right)$

   (c) $\mathbf{v} = \left(\begin{smallmatrix} 2 \\ 0 \end{smallmatrix}\right)$ and $\mathbf{w} = \left(\begin{smallmatrix} 6 \\ -1 \end{smallmatrix}\right)$

   (d) $\mathbf{v} = \left(\begin{smallmatrix} 0 \\ 1 \end{smallmatrix}\right)$ and $\mathbf{w} = \left(\begin{smallmatrix} -1 \\ 0 \end{smallmatrix}\right)$

2. Compute $r\mathbf{v}$ when r and $\mathbf{v}$ are given by

   (a) $r = 3$ and $\mathbf{v} = \left(\begin{smallmatrix} -3 \\ 2 \end{smallmatrix}\right)$

   (b) $r = 1$ and $\mathbf{v} = \left(\begin{smallmatrix} 2 \\ 0 \end{smallmatrix}\right)$

   (c) $r = 0$ and $\mathbf{v} = \left(\begin{smallmatrix} -10 \\ -31 \end{smallmatrix}\right)$

   (d) $r = -2$ and $\mathbf{v} = \left(\begin{smallmatrix} 3 \\ 1 \end{smallmatrix}\right)$

3. Compute the length of the vectors $\mathbf{v}$ in Exercise 2.

4. Compute $\mathbf{v} \cdot \mathbf{w}$ for the vectors $\mathbf{v}$ and $\mathbf{w}$, in Exercise 1.

5. If $\mathbf{v}(t)$ is the vector function of time given, compute $\frac{d}{dt}\mathbf{v}$:

   (a) $\mathbf{v}(t) = \left(\begin{smallmatrix} \cos(2t) \\ 3e^{-2t} \end{smallmatrix}\right)$

   (b) $\mathbf{v}(t) = \left(\begin{smallmatrix} e^{2t} \\ t^2 \end{smallmatrix}\right)$

   (c) $\mathbf{v}(t) = \left(\begin{smallmatrix} 3 \\ 1 \end{smallmatrix}\right)$

   (d) $\mathbf{v}(t) = \left(\begin{smallmatrix} -\sin(t) \\ 2t \end{smallmatrix}\right)$

6. Compute an antiderivative for the vectors $\mathbf{v}(t)$ in Exercise 5.

8

Equilibrium in Two-Component, Linear Systems

Consider the pair of functions $x(t)$ and $y(t)$ of time t that satisfy the following differential equation:

$$\frac{d}{dt}x = ax + by,$$

$$\frac{d}{dt}y = cx + dy. \tag{1}$$

Here, a, b, c, and d are all constant, real numbers.

8.1 Uniqueness of Solutions

Here is one very important fact about (1): Given that a starting time t and starting points $\left(\begin{smallmatrix} x_{\text{start}} \\ y_{\text{start}} \end{smallmatrix}\right)$ have been specified, there is one and only one solution to (1) with $x(0) = x_{\text{start}}$ and with $y(0) = y_{\text{start}}$. Thus, an equation as in (1) is *predictive*. It also follows that the general solution to (1) has *two* free parameters: namely, you can freely specify where to start, and the equation takes over and determines where $x(t)$ and $y(t)$ proceed next.

[The argument here is basically the same as that for the one-component equations we studied first. Equation (1) tells you where to *go* in the xy-plane at time t if you know where you are at time t. Thus, if you know where you are at time 0, or at any given time t_0, then (1) tells you where you will be at all other times.]

8.2   Equilibrium Solutions

An *equilibrium* solution is a solution to (1) where $\mathbf{v}(t) = \begin{pmatrix} x(t) \\ y(t) \end{pmatrix}$ is a constant vector (i.e., where $x = x_0$ and $y = y_0$ are independent of t). For an equilibrium solution, the time derivative $\frac{d}{dt}\mathbf{v} = 0$. This means that

$$ax_0 + by_0 = 0,$$
$$cx_0 + dy_0 = 0. \qquad (2)$$

Unless a and b both vanish, or c and d both vanish, each of the preceding equations is an equation for a line through the origin in the xy-plane, and the equilibrium solution is the point (if it exists) where these two lines intersect. (Remember that these lines are called null clines.)

   If the two lines have different slopes (so b/a is not equal to d/c), then the two lines will intersect at the origin. The diagram in Figure 8.1 shows where $\frac{d}{dt}x$ and $\frac{d}{dt}y$ are positive and negative on the xy-plane. This diagram is drawn for the particular case where

$$\mathbf{A} = \begin{pmatrix} a & b \\ c & d \end{pmatrix} = \begin{pmatrix} 1 & 1 \\ 1 & -1 \end{pmatrix}. \qquad (3)$$

Figure 8.1 shows the lines defined by (2) in the case where a, b, c, d have the values given in (3). Notice that in Figure 8.1, I have drawn the regions where $\frac{d}{dt}x$ and $\frac{d}{dt}y$ are individually positive and negative.

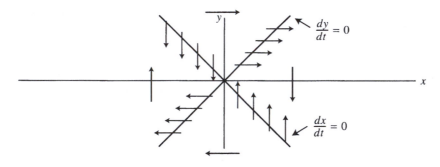

Figure 8.1

   In Equation (3), I have introduced the array, or *matrix*, $\mathbf{A}$. I did this because it is convenient to think of the four numbers a, b, c, and d that appear in (1) as forming a matrix as in (3). The point is that the properties of solutions to (1) depend on the choices for a, b, c, d, and to keep this fact in mind, we think of these four parameters as one object, the matrix $\mathbf{A}$. (In mathematically oriented literature, the term matrix is used to denote any rectangular array of numbers. If the matrix has n elements in each row and column, where n can be 1, 2, 3, $\ldots$, then one refers to an $n \times n$ matrix. In these chapters, the matrices will be almost always 2×2, and you should assume this unless the evidence to the contrary is incontrovertible.)

8.3   Determinants

As I remarked previously, the lines that are defined by the two equations in (2) are not parallel when $ad - cb$ is not zero. When this quantity is zero, then they are parallel (assuming that both a and b and also both c and d are not identically zero). The number given by $ad - cb$ occurs over and over in subsequent discussions, and so it has been given a name, the **determinant** of **A**, written det(**A**). That is, given an array (or *matrix*) **A**, we have introduced the number

$$\det(\mathbf{A}) = ad - bc, \tag{4}$$

which is the product of the top left entry and the bottom right entry *minus* the product of the bottom left entry times the top right entry. (There is also a notion of determinant for an $n \times n$ matrix for n larger than 2, but the discussion of determinants in the general case won't be given.)

8.4   Stability

The constant solution $\mathbf{0} = \left(\begin{smallmatrix} 0 \\ 0 \end{smallmatrix}\right)$ to Equation (1) is said to be **stable** when *all* trajectories that start in some region with **0** inside move closer to **0** as time increases and limit to zero as time gets arbitrarily large. If **0** is not stable, it will be called *unstable*. Thus, an equilibrium point is unstable when there is at least *one* trajectory that doesn't move toward **0** as time increases. The point is that there may be some trajectories that do move toward the unstable equilibrium point, but not all of them do. An unstable solution is unlikely to be seen in nature.

The criteria for stability of the equilibrium point in (4) can be stated in terms of the matrix **A** as follows: Reintroduce $\det(\mathbf{A}) = (ad - bc)$, and introduce the quantity $a + d$, which is called the **trace** of **A** and written tr(**A**). Then **A** is stable if and only if *both*

$$\text{tr}(\mathbf{A}) < 0,$$
$$\det(\mathbf{A}) > 0. \tag{5}$$

For future reference in the subsequent chapters, it is absolutely imperative that you remember that (5) is the criteria for the stability of the equilibrium solution to (1).

8.5   An Equation for $x(t)$

Here is how to verify the preceding assertion: Consider a solution $\mathbf{v}(t) = \left(\begin{smallmatrix} x(t) \\ y(t) \end{smallmatrix}\right)$ to (1) that is close to the zero vector, $\mathbf{0} = \left(\begin{smallmatrix} 0 \\ 0 \end{smallmatrix}\right)$, at some time $t = t_0$. We want to check that (5) will imply that $\mathbf{v}(t)$ limits to **0** as $t \to \infty$. The trick is to solve Equation (1) for $\mathbf{v}(t)$ in closed form. The exact solution will tell us whether $\mathbf{v}(t)$ must stay small for all t, or whether $\mathbf{v}(t)$ can grow as t gets large. I present the exact solution. The derivation itself is of independent interest but would take this chapter where I don't plan to go. Indeed, your memory of the criterion in (5) for stability is more important in these chapters than the details of the derivation. In any event, I will leave the derivation out and simply refer you to any standard introductory book on linear algebra. [However, it requires nothing more than a certain level of stamina to take the time derivative of the solution offered here and verify that it does indeed satisfy (1).]

There are two cases to consider. The simplest case is when both b and c vanish. In this case, the first line in (1) does not involve y, and the second line does not involve x. So we may apply our criteria for stability from Chapter 4 to each line of (1) separately to learn that $\mathbf{0}$ is a stable equilibrium point if and only if both a and d are negative. This is equivalent to (5) when $b, c = 0$.

The answer in the general case is given next, but a digression is needed first to set the stage. To begin the digression, remember that once you choose a starting point for the trajectory, then there is a unique solution $\mathbf{v}(t)$ to Equation (1) that starts at your chosen point. Let's agree to denote your starting point by

$$\mathbf{v}_0 = \begin{pmatrix} x(0) \\ y(0) \end{pmatrix}. \tag{6}$$

Now I need to introduce a second vector,

$$\mathbf{w}_0 = \begin{pmatrix} \frac{1}{2}(a-d)x(0) + by(0) \\ cx(0) + \frac{1}{2}(d-a)y(0) \end{pmatrix}. \tag{7}$$

Finally, I need to introduce the number

$$\Delta = \frac{1}{4}\operatorname{tr}(\mathbf{A})^2 - \det(\mathbf{A}). \tag{8}$$

The precise form of the general solution to (1) depends on Δ. There are three cases to consider.

Case 1: If $\Delta > 0$, then the general form for $\mathbf{v}(t)$ is

$$\mathbf{v}(t) = 2^{-1}e^{\operatorname{tr}(\mathbf{A})t/2}\left(e^{\sqrt{\Delta}t}(\mathbf{v}_0 + \Delta^{-1/2}\mathbf{w}_0) + e^{-\sqrt{\Delta}t}(\mathbf{v}_0 - \Delta^{-1/2}\mathbf{w}_0)\right). \tag{9}$$

Remember, $\mathbf{v}_0$ is yours to choose. Choose any $\mathbf{v}_0$ and then (9) describes the trajectory that is traced out by the solution $\mathbf{v}(t)$ to (1) that starts at your chosen $\mathbf{v}_0$ at $t = 0$.

Case 2: If $\Delta = 0$, then the general form for $\mathbf{v}(t)$ is

$$\mathbf{v}(t) = e^{\operatorname{tr}(\mathbf{A})t/2}(\mathbf{v}_0 + t\mathbf{w}_0). \tag{10}$$

Once again, $\mathbf{v}_0$ is yours to choose.

Case 3: If $\Delta < 0$, then the general form for $\mathbf{v}(t)$ is

$$\mathbf{v}(t) = 2^{-1}e^{\operatorname{tr}(\mathbf{A})t/2}\left(\cos(|\Delta|^{1/2}t)\mathbf{v}_0 + |\Delta|^{-1/2}\sin(|\Delta|^{1/2}t)\mathbf{w}_0\right). \tag{11}$$

As before, $\mathbf{v}_0$ is yours to choose.

To finish the story, note that in each of the three cases, there are solutions $\begin{pmatrix} x(t) \\ y(t) \end{pmatrix}$ that do not approach $\mathbf{0}$ as time increases when the conditions in (5) are violated. [You have to choose $x(0)$ and $y(0)$ correctly in (6) for this to happen. Actually, this happens for the generic choice—you have to choose $x(0)$ and $y(0)$ in a very special way in order for it not to happen.] When the conditions in (5) hold, then *all* solutions will approach $\mathbf{0}$ as time increases.

8.6   Lessons

Here are some key points to take from this chapter:

- Learn the condition for the stability of the equilibrium point to (1).

- Remember the definition of the matrix in (3), and remember the definition of its trace and determinant. Then learn the criteria in (5) for stability.

READINGS FOR CHAPTER 8

READING 8.1

Better Protection of the Ozone Layer

Commentary: The purpose of this article is only to point out that even the very simple Equation (1) in Chapter 8 has its applications. In this case, the applications are to the equations for the production of chlorine in the stratosphere. In particular, the first two lines of Equation (1) in this article constitute a two-component differential equation that is exactly of the form of Equation (1) in Chapter 8. Indeed, setting $x = B_T$ and $y = B_S$, the constants (a, b, c, d) in Equation (1) of Chapter 8 are given by the constants in Equation (1) of this article by the following rule: $a = -1/L_T - f/\tau_1$, $b = 1/\tau_1$, $c = -1/L_S - 1/\tau_1$, and $d = f/\tau_1$. Notice that the article's equations (2a, b) give an exact solution for certain starting values for $B_T(0)$ and $B_S(0)$. Compare with Equation (9) in Section 8.5 of this chapter.

Better Protection of the Ozone Layer

Nature, **367** (1994) 505–508.
Malcolm K. W. Ko, Nien-Dak Sze, Michael J. Prather

How can we extend the Montreal Protocol to other ozone-depleting chemicals, such as fuel from the Space Shuttle and pharmaceuticals, when the life cycles of these compounds and the scales of the industries are different?

International agreements have been enacted to protect the ozone layer by regulating the release of chlorine- and bromine-bearing chemicals such as

---

Author affiliations: Malcolm K. W. Ko and Nien-Dak Sze: Atmospheric and Environmental Research Inc., Cambridge, Massachusetts. Michael J. Prather: Department of Earth System Science, University of California, Irvine.

the chlorofluorocarbons and the halons. One of the criteria by which such chemicals are assessed and regulated is the ozone depletion potential (ODP) [1–5]. But because more and more chemicals are turning out to be ozone-depleting, we believe that a more refined approach is needed for effective and equitable control. Here, we address ways to extend international agreements to include other contemporary ozone-depleting chemicals whose applications and life cycles are very different from synthetic halocarbons used in the industries already singled out for concern, such as refrigeration, insulation and fire-fighting applications.

During the past decade, an international regime has emerged for limiting the consumption of chemicals believed to carry chlorine (Cl) and bromine (Br) into the upper atmosphere (stratosphere) (see box). The Convention on the Protection of the Ozone Layer (Vienna, 1985) was followed in 1987 by the Montreal Protocol, subsequently amended in 1990 (London amendment) and in 1992 (Copenhagen amendment). The governments that have signed these treaties are committed to phase out the production and consumption of CFCs (chlorofluorocarbons) and halons (brominated hydrocarbons), and to limit the production of their substitutes (for example hydrochlorofluorocarbons, HCFCs) in the expectation that the stratospheric concentration of chlorine and bromine can thereby be reduced. Ultimately, the goal is to reverse the observed downward trend of global ozone, and to limit the possible consequential damage to the biosphere from increased ultraviolet radiation. The adherence of governments to these agreements, which have the status in international law of formal treaties, is mirrored in national legislation such as the US Clean Air Act (1992).

The evolving protocols have given precedence to controlling synthetic CFCs and halons used in applications such as electronic and metal cleaning, foam blowing and refrigeration (CFCs); and fire extinguishers (halons). These chemicals are released in the lower atmosphere (the troposphere), but are sufficiently long-lived that they can be transported to the stratosphere where most of them are broken down by ultraviolet radiation, producing highly reactive Cl and Br radicals that are chiefly responsible for the catalytic destruction of stratospheric ozone. Because the timescale of mixing in the troposphere is less than the residence time of these halocarbons, the effect on ozone (as measured by the ODP) does not depend exactly on where, when and how they are released.

But there are other more direct and effective means by which chlorine can enter the stratosphere. These include solid-fuel rocket motors in the Space Shuttle launches, which deposit chlorine directly in the stratosphere. Prudence, as well as consistency, requires that these sources should also be evaluated under the same criteria (for example, ODPs) to determine their contributions, if any, to ozone depletion. Before the Copenhagen meeting, the scientific community was asked to quantify the impact of other atmospheric emissions such as solid rocket motors [6], stratospheric aircraft [7] and use of methyl bromide (CH_3Br) [8] in agricultural activities. But no ODPs were calculated for solid rocket motors and stratospheric aircraft. Based on the findings on CH_3Br, a freeze on its production beginning in 1995 was adopted in the Copenhagen amendment.

What follows is a demonstration that some potential ozone-depleting substances have life cycles that differ qualitatively as well as quantitatively from the chemicals now controlled by the agreements. We discuss what factors should be considered when developing a strategy for control of compounds whose applications and life cycles are very different from the CFCs. Specifically, we look for ways to ensure that the resulting strategy is practical (can it be applied easily?), effective (does it omit chemicals whose net impact may be comparable to individual HCFCs or CFCs?), and equitable (does it ban or impose excessive penalty on uses that have relatively small effects on the ozone layer?). We will use three examples to illustrate these points: chlorine deposited by the solid-fuel rockets from the Space Shuttle launches; CH_3Br used in soil fumigation; and brominated compounds in pharmaceutical use.

Ozone depletion potentials

First defined for CFCs a decade ago, the ODP is an index measuring the time-integrated ozone depletion caused by specific quantity of a chemical relative to that caused by the same quantity of the chlorofluorocarbon CFC-11 (the fully substituted methane $CFCl_3$). The definition presumes the chemical is ultimately released into the atmosphere. Policy makers and industry representatives have since asked scientists to extend and calculate ODPs for HCFCs [2] and the halons [3, 5]. To-

Chronology of ozone-protection agreements

The Vienna Convention for the Protection of the Ozone Layer was agreed March 1985 and entered into force in September 1988. It was set up with the intention of preventing any further damage to the ozone layer by "recognising the possibility that world-wide emissions and use of fully halogenated chlorofluorocarbons (CFCs) and other chlorine-containing substances can significantly deplete and otherwise modify the ozone layer, leading to potentially adverse effects on human health, crops, marine life, materials and climate, and recognising at the same time the need to further assess possible modifications and their potentially adverse effects."

It was also agreed that negotiations would continue on the development of a protocol to control equitably global production, emissions and use of CFCs. From this, the Montreal Protocol, an international agreement to phase out ozone-depleting substances, was agreed in 1987.

- 1987. Montreal Protocol on Substances that Deplete the Ozone Layer. It was agreed that global action was needed to phase out CFCs and halons.

- 1989. First meeting of the parties to the Protocol in Helsinki. Declaration adopted calling for CFCs and halons to be phased out by 2000.

- 1990. Second meeting of the parties of the Protocol in London. Agreement to phase out CFCs, halons and carbon tetrachloride by 2000 and methyl chloroform by 2005. Financial mechanism set up to assist developing countries.

- 1991. New European regulation 594/91 came into force. CFCs to be phased out within the European Communities by 1997.

Second meeting of the parties to the Vienna Convention and third meeting of the parties to the Montreal Protocol.

- 1992. Fourth meeting of the parties to the Montreal Protocol in Copenhagen. Parties agreed to bring forward phase-out dates for CFCs, carbon tetrachloride and methyl chloroform to 1996; halons to be phased out by 1994. Controls agreed for methyl bromide and HCFCs.

- 1993. Fifth meeting of the parties to the Montreal Protocol. It was agreed that there would be no essential uses for halons in 1994. Multilateral fund replenished.

Third meeting of the parties to the Vienna Convention.
Source: UK Department of the Environment.

---

tal chlorine loading of the atmosphere [9] has also been used to assess the global ozone loss caused by these chemicals, either separately or in combination for specific emission predictions. The amount of chlorine in the stratosphere not still tied up in the parent halocarbon is defined as the stratospheric chlorine loading [10]. To understand the relation between ODP and chlorine loading, it is necessary to examine the life cycles of the halocarbons.

The life cycles of the halocarbons are illustrated in the figure. These halocarbons are, in general, relatively long-lived and thus well mixed in the troposphere. They cycle between the troposphere and stratosphere where a portion of the compound is dissociated during repeated excursions into the stratosphere and the troposphere. The Cl and Br atoms formed in the stratosphere participate in catalytic cycles removing ozone until, on average after

3 years, they are transported back, to the troposphere. The Cl and Br atoms in the troposphere, either produced in situ by degradation of the halocarbons or transported from the stratosphere, are assumed to be removed quickly by rain out or surface deposition. For CFCs, almost all the chlorine atoms are produced in the stratosphere. The HCFCs react with hydroxyl radicals (OH) and thus are destroyed readily (on a timescale of a few years to decades) in the troposphere. As a result, they contribute less to the stratosphere chlorine loading on a per-molecule-emitted basis.

Numerical simulations using photochemical models indicate that the time-integrated stratospheric chlorine loading is a good proxy for cumulative ozone loss [9,10]. We have used a simple two-box model (with tropospheric and stratospheric compartments) to calculate, as a function of time after release, the amounts of chlorine still tied up in the undissociated halocarbons in the troposphere (B_T) and stratosphere (B_S), as well as the quantity of free chlorine in the stratosphere $(C,$ stratospheric chlorine loading). The differential equations describing the evolution over time of the three reservoirs are

$$\frac{dB_T}{dt} = -\frac{B_T}{L_T} - \frac{B_T f - B_S}{\tau_t}$$

$$\frac{dB_S}{dt} = -\frac{B_S}{L_S} - \frac{B_S - B_T f}{\tau_t} \qquad (1)$$

$$\frac{dC}{dt} = -\frac{C}{\tau_t} + \frac{B_S}{L_S}$$

where L_T and L_S are the tropospheric and stratospheric lifetimes for chemical loss of the halocarbons and τ_1 is the turnover time for replacing stratospheric air by tropospheric air. In calculating the exchange flux, the tropospheric burden is scaled by a factor $f = 0.15/0.85$ (assuming here that 15% of the atmospheric mass is in the stratosphere) so that the flux is proportional to difference in the mixing ratios. We select $\tau_t = 3$ yr, consistent with Holton's [11] derivation of 2.5 yr for the turnover of air above 100 mbar. Based on two-dimensional model results [12], we select $L_T = 1,000$ yr and $L_S = 5$ yr for CFC-11.

Assuming $B_T(0) = 1$ kg chlorine in the form of CFC-11, $B_S(0) = C_{\text{F-11}}(0) = 0$, the burdens (in kg chlorine) for year t are

$$B_T = 0.930 \exp[-t/46.5]$$
$$+ 0.070 \exp[-t/1.75] \qquad (2a)$$
$$B_S(t) = 0.107(\exp[-t/46.5]$$
$$- \exp[-t/1.75]) \qquad (2b)$$
$$C_{\text{F-11}}(t) = 0.068 \exp[-t/46.5]$$
$$+ 0.090 \exp[-t/1.75]$$
$$- 0.158 \exp[-t/3] \qquad (2c)$$

The integrated chlorine loading (p.p.t.v.-year) for emission of 1 kton $(1 \times 10^6$ kg) of CFC-11 is obtained by integrating equation $(2c)$ and multiplying by 0.81.

$$IC_{\text{F-11}}(t) = 2.33(1 + 0.16 \exp[-t/3]$$
$$- 1.11 \exp[-t/46.5]$$
$$- 0.05 \exp[-t/1.75])\text{p.p.t.v.-year} \qquad (3)$$

The ODP of a compound X can be approximated as the ratio $IC_X(t)/IC_{\text{F-11}}(t)$, where $IC_x(t)$ is the integrated chlorine loading due to 1 kiloton release of X calculated using appropriate L_T and L_S. The steady state ODP corresponds to the ratio in the limit as $t \to \infty$.

Regulating halocarbons

The Protocol to date has focused on limiting the production of halocarbons used in traditional applications. The list of products containing controlled species (annex D of the Protocol) specifies automobile and truck air-conditioning units; domestic and commercial refrigeration equipment; aerosol products (except medical aerosols); portable fire extinguishers; insulation boards, panels and pipe covers and pre-polymers. The list of controlled species (annexes A–C) is mainly limited to synthetic fluorochlorocarbons or bromocarbons with no known natural sources.

Although the Protocol recognizes the ODP as a measure of the environmental danger, the lists of controlled species in the annex are not grouped by ODP values. Rather, the species were classified according to chemical type, by use and by when they were included in the Protocol. However, the ODP values are used explicitly in the following way. Instead of imposing production limits on an individual chemical, the control is applied to groups of chemicals. Within each group, the production of each chemical can be traded after appropriate weighting by its ODP value. The Protocol states that "each Party shall, for each group of substances ... determine its calculated levels of Production by (i) multiplying its annual *production* of each controlled substance by the ozone depleting potential specified ... ; (ii) adding together, for each such Group, the resulting figures." Under the Copenhagen amendment, the annual production limit on the transition substances (HCFCs) are specified as the 1989 production plus 3.1% of the ODP weighted annual production of the CFCs in the same year.

Although it is not explicitly stated, a reading of the protocols and the US Clean Air Act suggests that their formulations are based on considerations of both ODP and chlorine loading from the current emission rates. There is agreement to phase out production of the CFCs (annex A group I and annex B group I) and CCl_4 (annex B group II) by 1996; and halons (annex A group II) by this year. These compounds all have ODPs of about 0.5 or larger.

The atmospheric ($1/e$ folding) lifetimes of these compounds are 30 years or longer (with the exception of halon-1211). Once introduced into the atmosphere, these chemicals will continue to release Cl/Br atoms into the stratosphere over their respective lifetimes. Thus, even if their production is stopped immediately, it will take many decades (until the chemicals are purged from the atmosphere) before the chlorine concentration can return to the level of about 2,000 p.p.t.v. which existed before the ozone hole [10].

In the Copenhagen (1992) amendment, there is a proposal to phase out the substitute HCFCs (annex C) by the year 2030, and to impose stringent limitations on production rates during the interim period. The HCFCs, with atmospheric lifetimes of about 15 years or less, are more rapidly purged from the atmosphere and thus allow for more rapid decay of stratospheric chlorine once use of the chemicals is phased out. One rationale for controlling HCFCs with ODPs as small as 0.02 is that the short-term relative impact of such species is much larger than suggested by the steady-state ODP [4].

Methyl chloroform (CH_3CCl_3, annex B group III), a compound that fits loosely with the definition of HCFC (but without fluorine), has an atmospheric lifetime of about 6 years and an ODP of 0.12. It is due to be phased out in 1996. This decision was presumably made because the very large production rate contributes significantly to the chlorine loading of the current atmosphere. Chemicals in another class—methylene chloride (CH_2Cl_2), perchloroethylene (C_2Cl_4) and trichloroethylene (C_2HCl_3)—have annual productions comparable to that of CH_3CCl_3, but have not yet been regulated, presumably because they have much shorter lifetimes, smaller ODPs and give rise to smaller chlorine loading.

Some countries have formulated unilateral regulations based mainly on the calculated values of ODPs. Under the US Clean Air Act, any chemical with an ODP larger than 0.2 can be classified as a class I substance. Unless they are in current production, such chemicals may not be manufactured. Heavy fines (up to $25,000 per violation per day) can be imposed for any intentional venting of these materials; and indeed the Environmental Protection Agency is currently seeking large fines from 28 individuals and businesses for such violations [13]. This ODP cut-off is applied independently of the amount released, the argument being

that the combined effects from a large variety of ozone-depleting chemicals, even though produced in small quantities, will add up and must be regulated as a collective industry. The ODP cutoff is also applied regardless of how the chemicals will be used. As illustrated by the examples that follow, this approach may be problematic when extended to control chemicals outside the traditional CFC industries because the ODP may depend sensitively on end-use, which in turn determines the life cycle or fate of the chemicals.

ODP of solid rocket fuel

Most scientific studies and early regulations have focused on chemicals whose atmospheric life cycles and participation in ozone depletion are similar to that of CFC-11. The chlorine loading from Space Shuttle launches requires a somewhat different approach. The solid rocket fuel is composed of 16% by weight of aluminum, 70% of ammonium perchlorate (NH_4ClO_4), and 14% of a polymer matrix. Combustion gives rise to the exhaust of HCl into the atmosphere. Each launch emits approximately 200 tons (200,000 kg) Cl out of a total fuel load of 1,700 tons (750 tons of liquid fuel and 950 tons of solid fuel). About one-third of the solid rocket motor exhaust, or 68 tons of Cl, is deposited in the stratosphere (defined here as altitudes above 15 km) per launch (see figure). The Shuttle emissions are removed from the stratosphere with a residence time of 3 years. The average integrated chlorine loading in the stratosphere per kton NH_4ClO_4 used as fuel is given by

$$IC_{SS}(t) = 0.318(1.0\exp[-t/3]) \text{ p.p.t.v.-year} \quad (4)$$

The behaviour of the chlorine loading is more comparable to that of HCFCs, as the atmospheric recovery is relatively rapid following cessation of emissions.

The time-dependent ODP is approximately $IC_{ss}(t)/IC_{F-11}(t)$. The ratio of the lead coefficients gives the steady-state ODP value of 0.14. Note that the calculated ODP remains above 0.2 for 50 years. As is also obvious from the equations, the time-dependent ODP is very large for a short time after emission because the impact of the Space Shuttle is immediate whereas CFC-11 (and other HCFCs) take years to reach the stratosphere.

One ambiguity in using equation (4) to calculate the ODP for the rocket fuel lies in what is meant by "emission of 1 kton". If we restrict ourselves

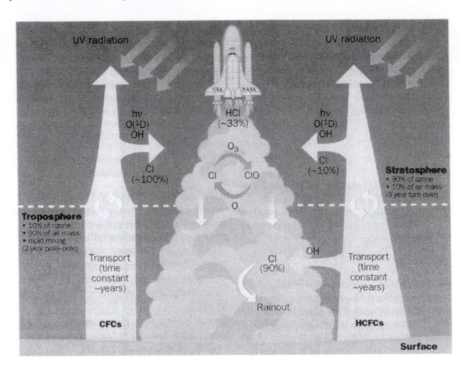

Fig. 1 Life cycles of CFCs, HCFCs and solid rocket fuel (NH₄ClO₄). Chlorine can either be carried into the stratosphere by atmospheric transport of CFCs, HFCs and by Space Shuttle exhaust. The main difference between CFCs and HCFCs is that most (~ 90%) HCFCs are removed in the troposphere by hydroxyl radicals (OH). In the case of Space Shuttle, approximately one-third of the fuel (NH₄ClO₄) is burnt in the atmosphere, producing HCl.

to the emission of stratospheric HCl (the primary form of chlorine in the exhaust), then the coefficient in equation (4) is larger and the steady state ODP becomes 1.3. On the other hand, if we consider the emissions to be the total fuel load (liquid plus solid), then the coefficient in equation (4) scales down and the ODP becomes 0.05. There is no precedent for using a similar definition to reduce the ODP of a CFC used as an aerosol propellant (when the ODP is defined as the original ODP discounted by the weight fraction of the CFC in the spray-can product).

How do the total anticipated emissions from the Space Shuttle launches compare with chlorine loading expected from the substitute HCFCs to be used in refrigeration, or against the CFC background expected over the next decades? Previous assessments [6,14] showed that a launch rate of one Shuttle per month over several years would increase stratospheric chlorine loading by about 3 p.p.t.v.

(as much as 10 p.p.t.v. at northern mid latitudes). This could be compared with a chlorine loading of about 3,000 p.p.t.v. generated by the CFCs emitted over the past several decades. Because the chlorine loading is expected to decrease, we believe that it is more appropriate to use another measure. A 10 p.p.t.v. contribution is comparable to a loading of 30 p.p.t.v. expected from a continuous annual release of 50 kton of a typical HCFC (1 Cl atom, 10-year lifetime, atomic weight 100), and of the same order as the contribution from the annual use of CFCs and HCFCs by some developing countries.

Application-sensitive chemicals

Before the establishment of recovery practice for recycling or destruction, the applications of the CFCs (apart from use as feedstock to produce other chemicals) are such that, except for a time delay of up to a few decades (the so-called banking time), all the material produced is eventually released to the atmosphere (very little is destroyed during applications). The Protocol recognizes this and explicitly exempts the amount used in feedstock, and the amount recaptured for recycling and destruction, from being counted as production in the basket approach within each group. In extending the control strategy to a wider class of compounds, however, we find that some usage may result in having only part of the applied material released to the atmosphere because the compound is partially destroyed in the application. For example, when used as a soil fumigant, about 10 to 50% of the CH_3Br undergoes prompt chemical transformation in the soil and is not released to the atmosphere [8]. These values are based on several model calculations and limited measurements. If a large fraction of the compound were destroyed (without emitting Cl or Br into the atmosphere) during its application and a portion of the chemical emitted to the atmosphere were destroyed by exchange with the Earth's surface or the ocean, should we extend the existing definition of ODP which currently only focuses on the atmospheric portion of a compound's life cycle? By defining the ODP over an extended life cycle, we will be able to include whatever effects are associated with the differences in life cycles and thus provide a more meaningful measure of relative ozone depletion. In any case, the amount destroyed during any application should be quantified and counted as 'amount destroyed' by approved technologies.

Small-use chemicals

Although the Protocol contains provisions to exempt small medical use of existing CFCs, there is no uniform guideline for assessing new drugs. Many specialized pharmaceuticals contain bromine. Because a Br atom is about 40 times more efficient than a Cl atom in removing ozone [3], the ODPs of these compounds are about the same as the CECs even if their lifetimes are many times shorter. Although these compounds do pose a short-term risk to the ozone layer, they are expected to have very small annual production (a few tons). How should production of such unique, beneficial drugs be controlled while protecting the ozone layer?

Proposed strategy

We have discussed here several ideas that have been considered implicitly in formulating the control strategy in the Protocol to limit the chlorine loading expected from long-lived halocarbons used in traditional industries. As the list of potential ozone-depleting substances grows, the protection of the ozone layer may require extension of the Protocol to the release of Cl and Br in the stratosphere in connection with other human activities. To this end, it is necessary to follow an approach that relies on our scientific understanding of stratospheric ozone and the life cycles of the compound in question. The aim is to provide sound, long-term protection for the ozone layer, while having a less capricious impact on technological development. The guidelines and the rationale follow.

- Register all industrially produced chlorine- and bromine-bearing compounds that can potentially put reactive chlorine and bromine species into the stratosphere, and determine their ODPs taking into account the product life cycles. We need to identify all potential sources of stratospheric chlorine and bromine.

- Ban or place stringent limits on the production of those compounds with long atmospheric lifetimes (lifetime threshold to be determined). Using only short-lived compounds will allow for a reasonably rapid recovery of the atmosphere should unforeseen factors, such as volcanoes or other unanticipated global changes, suddenly enhance chlorine-driven ozone depletion.

- Set a predetermined goal for the recovery of stratospheric ozone, and thus a schedule for stratospheric chlorine levels (or chlorine plus bromine equivalent measure). Define a market basket approach for the use of all Cl- and Br-bearing chemicals so that the ODP-weighted emissions meet this schedule.

The idea of an ODP-weighted basket is not new. It is part of the protocol in calculating the production within each group of chemicals and in specifying the limit for substitutes. What is different here is the extension to other, non-CFC sources of stratospheric pollution by using an ODP-weighted emission basket that can be tied directly to stratospheric chlorine loading. For the CFCs and HCFCs, the ODP-weighted emission gives results equivalent to those for the ODP-weighted production after exempting feedstock, and recapture for recycling and destruction. However, the particular use of a compound becomes important when evaluating nonvolatile chlorinated chemicals, such as the NH_4ClO_4 in Space Shuttle fuel, which are not expected to put reactive chlorine into the stratosphere unless they are used in highly specific ways. For example, static tests of the shuttle engine on the ground should not affect stratospheric ozone. In such cases, the distinction between production and emission becomes important. Extending the existing definition of ODP over a product life cycle could help reconcile this difficulty.

Future prospects
The broad outline proposed here aims to impose taxes or to place caps on the different end-uses of a compound in which part of the chemical is destroyed. It allows for limited use of any Cl- and Br-bearing compounds considered beneficial to society whose values and costs are to be determined by market forces or by national priorities.

The global limit on ODP-weighted emissions can be agreed by parties to the Protocol and each country can choose the manner in which it limits combined emissions of chloro- and bromocarbons (by taxation, by free-market auction of permits, or by explicit caps for specific compounds). The permit trading being used for controlling SO_2 emissions in the United States [15] may be a viable approach to setting the pollution fees involved in releasing ozone-depleting chemicals. Chlorine-free rocket fuels are being developed as alternatives to NH_4ClO_4 [16]. The pollution fee system could help to translate the environmental pay-off into more concrete terms.

The rationale for the control of other ozone-depleting substances unrelated to the chlorine and bromine chemical cycles (for example, nitrogen fertilizer, which produces nitrous oxide as a degradation product) is far more complicated and, we believe, needs to be understood further before being regulated.

REFERENCES

1. Wuebbles, D. J. *J. geophys. Res.* **88**, 1433–1443 (1983).
2. Fisher, D. A. *et al.*, *Nature* **344**, 508–512 (1990).
3. Solomon, S. *et al.*, *J. geophys. Res.* **97**, 825–842 (1992).
4. Solomon, S. & Albritton, D. L. *Nature* **357**, 33–37 (1992).
5. *Scientific Assessment of Ozone Depletion: 1991* Ch. 6 Global Ozone Research and Monitoring Project Rap. No. 25 (World Meteorological Organization, Geneva, 1992).
6. *Scientific Assessment of Ozone Depletion: 1991* Ch. 10 Global Ozone Research and Monitoring Project Rep. No. 25 (World Meteorological Organization, Geneva, 1992).
7. *Scientific Assessment of Ozone Depletion: 1991* Ch. 9 Global Ozone Research and Monitoring Project Rep. No. 25 (World Meteorological Organization, Geneva, 1992).
8. *Methyl Bromide: Its Atmospheric Science, Technology and Economics* Montreal Protocol Assessment Supplement (United Nations Environment Programme, Nairobi, 1992).
9. Prather, M. J. & Watson. R. T. *Nature* **244**, 729–734 (1990).
10. *Scientific Assessment of Ozone Depletion: 1991* Ch. 8 Global Ozone Research and Monitoring Project Rep. No. 25 (World Meteorological Organization, Geneva, 1992).
11. Holton, J. R. *J. atmos. Sci.* **47**, 392–395 (1990).
12. Ko, M.K.W. *et al.* *J. geophys. Res.* **96**, 7547–7552 (1991).
13. *Chem. Engng. News* p. 14 (16 August 1993).
14. Prather, M. J. *et al.* *J. geophys. Res.* **95**, 18583–18590 (1990).
15. *Science* **260**, 1884–1885 (1993).
16. *Chem. Engng. News* p. 18 (17 January 1994).

Exercises for Chapter 8

1. Draw the null clines (where $\frac{d}{dt}x = 0$ or $\frac{d}{dt}y = 0$) and find the equilibrium point for the equation

$$\frac{d}{dt}\begin{pmatrix} x \\ y \end{pmatrix} = \begin{pmatrix} 3x + 4y \\ y \end{pmatrix}.$$

2. Label the regions on the (x, y) plane where $\frac{d}{dt}x$ is positive and negative, and likewise for $\frac{d}{dt}y$ for the equation in the previous exercise.

3. Find the equilibrium points for the equation

$$\frac{d}{dt}\begin{pmatrix} x \\ y \end{pmatrix} = \begin{pmatrix} x - 2y \\ x + y \end{pmatrix}.$$

4. Label the regions on the (x, y) plane where $\frac{d}{dt}x$ is positive and negative, and likewise for $\frac{d}{dt}y$ for the equation in Exercise 3.

5. Answer the same questions as in Exercises 1 and 2 for the equation

$$\frac{d}{dt}\begin{pmatrix} x \\ y \end{pmatrix} = \begin{pmatrix} -x - y \\ x - y \end{pmatrix}.$$

6. In Exercise 1, if a solution starts at $\left(\begin{smallmatrix} 1 \\ 1 \end{smallmatrix}\right)$ at $t = 0$, what happens to the trajectory at large times? [Does it go very far from $\left(\begin{smallmatrix} 1 \\ 1 \end{smallmatrix}\right)$?] Here is a hint for answering this question: Go back to your drawing with the null clines and ask yourself: Can the trajectory $\left(\begin{smallmatrix} x(t) \\ y(t) \end{smallmatrix}\right)$ leave the region where $\frac{d}{dt}x$ and $\frac{d}{dt}y$ are both positive? If it stays in this region, ask yourself what will happen to it.

7. Decide whether the equilibrium points in Exercises 1, 3, and 5 are stable.

8. Answer in a few sentences: Why are unstable solutions rarely seen in nature?

9. Verify by substitution that $\mathbf{v}(t)$ from (9), (10), and (11) solve Equation (1) when the quantity Δ in Equation (8) is positive, zero, and negative, respectively.

9

Stability in Nonlinear Systems

As the title indicates, the subject of this chapter is stability for equilibrium solutions to the generic differential for a vector function of time $\mathbf{v}(t) = \begin{pmatrix} x(t) \\ y(t) \end{pmatrix}$. In particular, suppose that $\mathbf{v}$ obeys the equation

$$\frac{d}{dt}\begin{pmatrix} x(t) \\ y(t) \end{pmatrix} = \begin{pmatrix} f(x, y) \\ g(x, y) \end{pmatrix}, \tag{1}$$

where f and g are functions on the xy-plane. The equilibrium solutions are points of the form $\mathbf{v}_0 = \begin{pmatrix} x_0 \\ y_0 \end{pmatrix}$, where x_0 and y_0 are simultaneous solutions to the equations

$$f(x, y) = 0, \tag{2a}$$
$$g(x, y) = 0. \tag{2b}$$

As you should recall from previous chapters, the set of pairs $\begin{pmatrix} x \\ y \end{pmatrix}$ that solves the first Equation (2a) is called the x null cline and the set of pairs $\begin{pmatrix} x \\ y \end{pmatrix}$ that solve the second equation is the y null cline.

For example, consider an epidemic model for a city where new susceptibles enter at a constant rate β per day. The following model assumes that infected people either recover or die after some number of days; and if they recover, then they are immune. The variables are $S(t)$ ($=$ the number of susceptibles at time t) and $I(t)$ ($=$ the number of infected individuals at time t). The analog of Equation (1) for the functions S and I is

$$\frac{d}{dt}S = -\alpha SI + \beta$$

$$\frac{d}{dt}I = -\sigma I + \alpha S I.$$ (3)

The constant α is determined by the probability of contact between an infected person and a susceptible person, while $1/\sigma$ is equal to the mean time that a person is infectious. The constant β is equal to the rate at which new susceptibles enter the city.

This model has one equilibrium point,

$$S = \sigma/\alpha \quad \text{and} \quad I = \beta/\sigma.$$ (4)

The issue here is stability.

9.1   Definition of Stability

Consider an equilibrium point $\mathbf{v}_0 = \begin{pmatrix} x_0 \\ y_0 \end{pmatrix}$ for the equation in (1). Here is a working definition of stability: Call $\mathbf{v}_0$ stable if the following occurs: Any solution $\mathbf{v}(t) = \begin{pmatrix} x(t) \\ y(t) \end{pmatrix}$ to (1) which is sufficiently close to $\mathbf{v}_0$ at some time $t = t_0$ gets closer and closer to $\mathbf{v}_0$ for all time $t > t_0$, and gets arbitrarily close as t approaches infinity.

9.2   Criteron for Stability

Here is a condition that, when satisfied, implies that $\mathbf{v}_0$ is stable. To state the condition, introduce the following four numbers:

$$a = \left(\frac{d}{dx} f(x, y_0)\right)|_{x=x_0}$$ (5a)

$$b = \left(\frac{d}{dy} f(x_0, y)\right)|_{y=y_0}$$ (5b)

$$c = \left(\frac{d}{dx} g(x, y_0)\right)|_{x=x_0}$$ (5c)

$$d = \left(\frac{d}{dy} g(x_0, y)\right)|_{y=y_0}.$$ (5d)

You should think of these four numbers as forming a matrix,

$$\mathbf{D} = \begin{pmatrix} a & b \\ c & d \end{pmatrix}.$$ (6)

Here is the fundamental result:

Theorem.   The equilibrium point $\mathbf{v}_0$ is stable when both of the following hold:

(1) $ad - bc > 0$

(2) $a + d < 0$

If either of these fails, then $\mathbf{v}_0$ is not stable. Furthermore, if the preceding conditions hold, then the following alternate version of stability also holds: A trajectory that starts near the original equilibrium point, but evolves in time according to some slightly perturbed version of Equation (1) will, nonetheless, stay near to $\mathbf{v}_0$ for all time $t > 0$. [Think of the perturbed equations as being defined by Equation (1) but with f replaced by some slightly modified function $f'(x, y)$, and with g replaced by some slightly modified $g'(x, y)$.]

The preceding stability result is very important, and you should commit it to memory.

Just to review some terminology, an array as in (6) is called a matrix, or, more formally, a "2 by 2 matrix." (You could have larger arrays, with more entries.) The numbers a, b, c, and d that define the matrix $\mathbf{D}$ are called the **entries** of $\mathbf{D}$. Also, the particular combinations of the entries that appear in the theorem have names since they arise quite often in various contexts. The combination $ad - bc$ is called the **determinant** of $\mathbf{D}$ and is written det($\mathbf{D}$). Meanwhile, the combination $a + d$ is called the **trace** of $\mathbf{D}$ and is written tr($\mathbf{D}$). Thus, you should remember that the stability for criteria are

$$\det(\mathbf{D}) = ad - bc > 0 \tag{7a}$$
$$\text{trace}(\mathbf{D}) = a + d < 0. \tag{7b}$$

9.3   Partial Derivatives

Notice that the x-derivative in (5a) is taken on the function $f(x, y_0)$, a function only of x that is obtained from the two-variable function $f(x, y)$ by fixing the y-coordinate at y_0 and letting only the x-coordinate vary. Meanwhile, the derivative in (5b) is taken on the function $f(x_0, y)$, a function only of y that is obtained from the two-variable function $f(x, y)$ by fixing the x-coordinate at x_0 and letting only the y-coordinate vary.

In general, if $h(x, y)$ is a function of the two variables x and y, then the **partial derivative** of h in the x-direction, denoted by $\frac{\partial h}{\partial x}$, is a new function of x and y whose value at some point $\left(\begin{smallmatrix} x_0 \\ y_0 \end{smallmatrix}\right)$ is given by

$$\left(\frac{\partial h}{\partial x}\right)(x_0, y_0) = \left(\frac{d}{dx}h(x, y_0)\right)_{x=x_0}. \tag{8}$$

Likewise, the partial derivative of h in the y-direction, denoted by $\frac{\partial h}{\partial y}$, is a new function of x and y whose value at some point $\left(\begin{smallmatrix} x_0 \\ y_0 \end{smallmatrix}\right)$ is given by

$$\left(\frac{\partial h}{\partial y}\right)(x_0, y_0) = \left(\frac{d}{dy}h(x_0, y)\right)_{y=y_0}. \tag{9}$$

The point here is that a function, h, of two variables x and y, has two "derivatives," one on the x-direction, $\frac{\partial h}{\partial x}$, and one in the y-direction, $\frac{\partial h}{\partial y}$.

9.4 Applications of the Partial Derivative

Note that there is one important application of the partial derivative that you should recognize. It is a generalization of the chain rule. Remember that the chain rule states the following: Let h be a function of x and let x be a function of t. Then the t-derivative of $h(x(t))$ is given by

$$\frac{d}{dt} h(x(t), y(t)) = \left(\frac{d}{dx} h\right)(x(t))\left(\frac{d}{dt} x\right)(t). \tag{10}$$

The chain rule generalization states the following: Let h be a function of x and y, and let x and y be functions of t. Then the t-derivative of $h(x(t), y(t))$ is given by

$$\frac{d}{dt} h(x(t)) = \left(\frac{\partial h}{\partial x}\right)(x(t), y(t))\left(\frac{d}{dt} x\right)(t) + \left(\frac{\partial h}{\partial y}\right)(x(t), y(t))\left(\frac{d}{dt} y\right)(t). \tag{11}$$

9.5 Examples of Stability

Consider the example in (3) with the equilibrium point given in (4). The matrix $\mathbf{D}$ in this case is

$$\mathbf{D} = \begin{pmatrix} -\alpha\beta/\sigma & -\sigma \\ \alpha\beta/\sigma & 0 \end{pmatrix}. \tag{12}$$

Note that $\det(\mathbf{D}) = \alpha\beta > 0$ while $\text{trace}(\mathbf{D}) = -\alpha\beta/\sigma < 0$, so the equilibrium point is stable.

9.6 Heuristics

Here is why the matrix $\mathbf{D}$ in (6) is relevant to the stability question: At $\mathbf{v}_0 = \begin{pmatrix} x_0 \\ y_0 \end{pmatrix}$, both functions $f(x, y)$ and $g(x, y)$ vanish. If $\mathbf{v} = \begin{pmatrix} x \\ y \end{pmatrix}$ is sufficiently close to $\mathbf{v}_0$, then both $f(x, y)$ and $g(x, y)$ are small, but not necessarily zero. We can approximate the values of these functions by their first-order Taylor's approximations. Taylor's approximations were discussed earlier in Chapter 2, but here, things are more complicated because f and g are functions of x and y, not functions of a single variable. To understand the two-variable Taylor's approximation, we must first remember the Taylor's approximation for a function, say h, of a single variable, say t. If $h(t_0) = 0$, then the first-order Taylor's approximation for the function h near t_0 is the best approximation of h by a function of the form $k(t)$ of the form $\alpha + \beta t$. Here, α and β are constants to be determined. In this case, $k(t) = \left(\frac{dh}{dt}|_{t=t_0}\right)(t - t_0)$, which involves the derivative of h at $t = t_0$. Note that the idea here is to reverse the usual definition of the derivative: For $\Delta t = t - t_0$ very small, we have

$$\frac{\Delta h}{\Delta t} = \frac{h(t) - h(t_0)}{t - t_0} \approx \left(\frac{dh}{dt}\bigg|_{t=t_0}\right).$$

Since $h(t_0) = 0$ here by assumption, multiplication of both sides of this last equation by $\Delta t = t - t_0$ yields $h(t) \approx \left(\frac{dh}{dt}\big|_{t=t_0}\right)(t - t_0)$, which is the first-order Taylor's approximation.

The best approximation to a two-variable function $h(x, y)$, near a point $\mathbf{v}_0 = \left(\begin{smallmatrix} x_0 \\ y_0 \end{smallmatrix}\right)$ in the xy-plane by a function of the form $k(x, y) = \gamma + \alpha x + \beta y$ (with α, β, and γ, constant) is obtained using

$$k(x, y) = h(x_0, y_0) + \left(\frac{\partial h}{\partial x}\bigg|_{\substack{x=x_0 \\ y=y_0}}\right)(x - x_0) + \left(\frac{\partial h}{\partial y}\bigg|_{\substack{x=x_0 \\ y=y_0}}\right)(y - y_0). \qquad (13)$$

This function $k(x, y)$ is the best approximation to $h(x, y)$ near $\mathbf{v}_0$ among all functions of x and y that have the simple form $\gamma + \alpha x + \beta y$ with α, β, and γ constant. (A function of the form $\alpha x + \beta y$ with α and β both constants is called a ***linear*** function of x and y in mathematically oriented literature. Strangely enough, the graph, $z = \alpha x + \beta y$, of such a function is not a line in 3-space; rather, it is a plane.)

The function $k(x, y)$ in (13) is called the first-order Taylor's approximation for h near $\mathbf{v}_0$. Notice that it depends on the value of the original function h at $\mathbf{v}_0$ and also on the values at $\mathbf{v}_0$ of the partial derivatives of h in both the x- and y-directions. However, it sees no more of the possibly complicated behavior of h than these three numbers.

With the preceding understood, return now to the milieu of the equation in (1) and suppose that we are only interested in the behavior of a solution $\mathbf{v}(t)$ to this equation as long as $\mathbf{v}$ stays near the equilibrium point $\mathbf{v}_0$. At the equilibrium point, $f = 0$ so we can get a reasonable approximation for f by replacing it by its first-order Taylor's approximation:

$$k(x, y) = f(x_0, y_0) + \left(\frac{\partial f}{\partial x}\bigg|_{\substack{x=x_0 \\ y=y_0}}\right)(x - x_0) + \left(\frac{\partial f}{\partial y}\bigg|_{\substack{x=x_0 \\ y=y_0}}\right)(y - y_0). \qquad (14)$$

Likewise, we can get a reasonable approximation for g near $\mathbf{v}_0$ by replacing g by its first-order Taylor's approximation.

To the extent that replacing f by the function on the right side of (14) is legitimate, and likewise so is the analogous replacement of g, we learn that the behavior of a solution $\mathbf{v}(t)$ to Equation (1) near to the equilibrium point $\mathbf{v}_0$ can be analyzed by studying the behavior of a solution to the following simpler system of equations:

$$\frac{d}{dt}\begin{pmatrix} x(t) \\ y(t) \end{pmatrix} = \begin{pmatrix} a[x(t) - x_0] + b[y(t) - y_0] \\ c[x(t) - x_0] + d[y(t) - y_0] \end{pmatrix}, \qquad (15)$$

where a, b, c, and d are given by (5).

This last equation can be put into a more useful form if we remember that we are interested ultimately in whether or not $x(t) - x_0$ and $y(t) - y_0$ gets large as t gets large.

Since both x_0 and y_0 are time independent, their time derivatives vanish. Hence, (15) implies the following equation for the functions $x(t) - x_0$ and $y(t) - y_0$:

$$\frac{d}{dt}\begin{pmatrix} x - x_0 \\ y - y_0 \end{pmatrix} = \begin{pmatrix} a(x - x_0) + b(y - y_0) \\ c(x - x_0) + d(y - y_0) \end{pmatrix}. \tag{16}$$

The solutions of this last equation were given explicitly in the previous chapter. There, we saw that when (7) holds, all solutions have the property that $x(t) - x_0$ and also $y(t) - y_0$ tend to zero as t gets large. That is, the equilibrium point v_0 is stable for Equation (16).

Finally, remark that inasmuch as (16) is a good approximation for the equation of interest [which is Equation (1)] when $v(t)$ is close to v_0, we have learned that all solutions to (1) that start near enough to v_0 will tend to v_0 as t gets large provided that the conditions in (7) hold. [The precise notion of "near enough" is determined by how well f and g are approximated by their first-order Taylor's approximations as in (14). In any event, these approximations get better and better as the distance to v_0 decreases, so we can conclude that inside some distance, the approximation is good enough to allow us to replace Equation (1) with Equation (16) in order to deduce the behavior of $v(t)$.]

9.7   Lessons

Here are some key points from this chapter:

- Learn the stability criteria from Subsection 9.2. In particular, know how to compute the matrix in (6) for the equilibrium point in question and how to compute its determinant and trace. Remember that the equilibrium point is stable if (7) holds and unstable otherwise.

- The stability question is important because unstable solutions are rarely observed in nature.

- Know how to compute partial derivatives.

- Know how to use the two-variable version of the chain rule.

Exercises for Chapter 9

Sketch the null clines, find all equilibrium points, compute the matrix **D**, and determine stability for the following systems of equations.

1. $\dfrac{d}{dt}\begin{pmatrix} x \\ y \end{pmatrix} = \begin{pmatrix} xy - 2x \\ xy - 2y \end{pmatrix}$

2. $\dfrac{d}{dt}\begin{pmatrix} x \\ y \end{pmatrix} = \begin{pmatrix} x - x^2 - xy \\ y - 2xy - 2y^2 \end{pmatrix}$

3. $\dfrac{d}{dt}\begin{pmatrix} x \\ y \end{pmatrix} = \begin{pmatrix} xy - x \\ y - xy - y^2 \end{pmatrix}$

4. Compute the partial derivatives $\frac{\partial h}{\partial x}$ and $\frac{\partial h}{\partial y}$ when h is given by

   (a) $h(x, y) = xy^3 - x^2y^2$

   (b) $h(x, y) = \cos(xy)$

   (c) $h(x, y) = x \cos(y) - 1$

   (d) $h(x, y) = y \cos(x) - x^2$

5. Take $x(t) = \cos(t)$ and $y(t) = \sin(t)$. Then, in each of the four cases for $h(x, y)$ given in Exercise 4, do the following:

   (a) Write out $h(x(t), y(t))$.

   (b) Compute $\frac{d}{dt}h(x(t), y(t))$ directly.

   (c) Use the two-variable chain rule to compute $\frac{d}{dt}h(x(t), y(t))$.

10

Nonlinear Stability Revisited

This chapter is mostly a review of stability for a two-component differential equation, but there are some ruminations on partial derivatives at the chapter's end.

10.1 Review of the Stability Criteria for a One-Component Equation

Remember the stability criteria for an equilibrium solution to the one-component equation

$$\frac{d}{dt}x = f(x).\tag{1}$$

The equilibrium solution is $x(t) = x_0$, a constant that obeys $f(x_0) = 0$.

Remember that x is stable if $(\frac{df}{dx})_{x=x_0} < 0$. Indeed, if $(\frac{df}{dx})_{x=x_0} < 0$, then

$$
\begin{aligned}
\frac{d}{dt}x &< 0 \quad \text{whenever } x(t) \text{ is near to, but greater than } x_0. \\
\frac{d}{dt}x &> 0 \quad \text{whenever } x(t) \text{ is near to, but less than } x_0.
\end{aligned}
\tag{2}
$$

For example, Figure 10.1 is a graph of $f(x) = 3x - x^2 - 1$ with the stable and unstable equilibrium points marked.

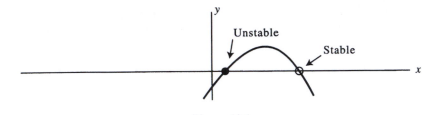

Figure 10.1

10.2   Stability for a Two-Component System

Consider the two-component system

$$\frac{d}{dt}\begin{pmatrix} x \\ y \end{pmatrix} = \begin{pmatrix} f(x, y) \\ g(x, y) \end{pmatrix}, \tag{3}$$

where f and g are functions of the points x and y in the plane. For example, let $x(t) =$ number of grass plants at time t in your yard back home, and let $y(t) =$ number of weeds at time t there. The broadleaf weeds compete with the grass for space and nutrients. We might expect that $\mathbf{v}(t) = \begin{pmatrix} x(t) \\ y(t) \end{pmatrix}$ would evolve with time according to an equation of the form

$$\frac{d}{dt}\begin{pmatrix} x \\ y \end{pmatrix} = \begin{pmatrix} \alpha x - \beta x^2 - \gamma xy \\ \sigma y - \delta y^2 - \rho xy \end{pmatrix}, \tag{4}$$

where $\alpha, \beta, \gamma, \sigma, \delta$, and ρ are positive constants that must be determined experimentally. (See Exercise 2 at the end of this chapter.)

The equilibrium points to (3) are the constant solutions that have $\begin{pmatrix} x(t) \\ y(t) \end{pmatrix} = \begin{pmatrix} x_0 \\ y_0 \end{pmatrix}$, where x_0 and y_0 solve $f(x_0, y_0) = 0$ and $g(x_0, y_0) = 0$.

In the example (4), take, for the sake of argument, $\alpha = 1, \beta = 1, \gamma = 2$; and take $\sigma = 2, \delta = 1$ and $\rho = 3$ so that (4) reads

$$\frac{d}{dt}\begin{pmatrix} x \\ y \end{pmatrix} = \begin{pmatrix} x - x^2 - 2xy \\ 2y - y^2 - 3xy \end{pmatrix}. \tag{5}$$

Here, the equilibrium solutions are

$$\begin{pmatrix} 0 \\ 0 \end{pmatrix}, \quad \begin{pmatrix} 1 \\ 0 \end{pmatrix}, \quad \begin{pmatrix} 0 \\ 2 \end{pmatrix}, \quad \begin{pmatrix} 0.6 \\ 0.2 \end{pmatrix}. \tag{6}$$

As was the case with (1), the stability of the equilibrium solutions in (3) can be determined by looking at the derivatives of the functions f and g at the point (x_0, y_0). However, in this case, each function has two derivatives.

Indeed, any function, $h(x, y)$, of two variables has two derivatives. The first, written $\frac{\partial h}{\partial x}$, is obtained by keeping y fixed at the value y_0 and considering h as a function of x only. The second, written $\frac{\partial h}{\partial y}$, is obtained by keeping x fixed at the value x_0 and considering h as a function of y only. Thus,

$$
\begin{aligned}
\left(\frac{\partial h}{\partial x} \right)_{x=x_0, y=y_0} &= \lim_{\Delta x \to 0} \frac{h(x_0 + \Delta x, y_0) - h(x_0, y_0)}{\Delta x}, \\
\left(\frac{\partial h}{\partial y} \right)_{x=x_0, y=y_0} &= \lim_{\Delta y \to 0} \frac{h(x_0, y_0 + \Delta y) - h(x_0, y_0)}{\Delta y}.
\end{aligned}
\tag{7}
$$

Returning to Equation (3), both f and g have two derivatives each at $\binom{x_0}{y_0}$, and so we might expect that the equilibrium criteria are determined by these four possible derivatives. As pointed out in the previous chapter, this is precisely the case. Indeed, introduce the matrix

$$
\mathbf{D} = \begin{pmatrix} \frac{\partial f}{\partial x} & \frac{\partial f}{\partial y} \\ \frac{\partial g}{\partial x} & \frac{\partial g}{\partial y} \end{pmatrix},
\tag{8}
$$

where all derivatives are evaluated at $x = x_0$ and $y = y_0$, and compute

$$
\text{trace}(\mathbf{D}) = \left(\frac{\partial f}{\partial x} \right)_{x=x_0, y=y_0} + \left(\frac{\partial g}{\partial y} \right)_{x=x_0, y=y_0},
\tag{9a}
$$

$$
\det(\mathbf{D}) = \left(\frac{\partial f}{\partial x} \right) \left(\frac{\partial g}{\partial y} \right)_{x=x_0, y=y_0} - \left(\frac{\partial f}{\partial y} \right) \left(\frac{\partial g}{\partial x} \right)_{x=x_0, y=y_0}.
\tag{9b}
$$

Then the solution $\binom{x_0}{y_0}$ is stable if

$$
\text{trace}(\mathbf{D}) < 0 \quad \text{and} \quad \det(\mathbf{D}) > 0.
\tag{10}
$$

On the other hand, the solution $\binom{x_0}{y_0}$ is unstable otherwise. Here, you should recall that the matrix $\mathbf{D}$ was used to define the simpler differential equation

$$
\frac{d}{dt} \begin{pmatrix} x \\ y \end{pmatrix} = \begin{pmatrix} \frac{\partial f}{\partial x}(x - x_0) + \frac{\partial f}{\partial y}(y - y_0) \\ \frac{\partial g}{\partial x}(x - x_0) + \frac{\partial g}{\partial y}(y - y_0) \end{pmatrix}.
\tag{11}
$$

Then a solution to this last equation that is close to $\binom{x_0}{y_0}$ is also close to a solution to (3), and vice versa.

10.3   The Example

Consider the competition model in (5). The stability criteria of (9) and (10) can be considered for each of the four possible equilibrium points. The results are as follows: Only the equilibrium point $\binom{0}{2}$ is stable; the others are unstable.

10.4   More on Partial Derivatives

Suppose we have a function, $h(x, y)$, of two variables. For example,

$$
\begin{aligned}
h &= x^2 y^3; \\
h &= x \cos(xy); \\
h &= \sin(x + y^2); \\
h &= xe^y.
\end{aligned}
\tag{12}
$$

We can then take the partial derivative of h in the x-direction to obtain a new function, and also in the y-direction to obtain (usually) a different function. For examples in (12), these new functions are

$$
\begin{aligned}
\frac{\partial h}{\partial x} &= 2xy^3 & \text{and} && \frac{\partial h}{\partial y} &= 3x^2 y^2 \\
\frac{\partial h}{\partial x} &= \cos(xy) - xy\sin(xy) & \text{and} && \frac{\partial h}{\partial y} &= -x^2 \sin(xy) \\
\frac{\partial h}{\partial x} &= \cos(x + y^2) & \text{and} && \frac{\partial h}{\partial y} &= 2y\cos(x + y^2) \\
\frac{\partial h}{\partial x} &= e^y & \text{and} && \frac{\partial h}{\partial y} &= xe^y.
\end{aligned}
\tag{13}
$$

Just as we can take two (or more) derivatives of a function of a single variable, we can take two (or more) derivatives of a function of more than one variable. For example, we can then take either the x or y partial derivatives of, say $\frac{\partial h}{\partial x}$, to obtain two additional functions. These are generally denoted by $\frac{\partial^2 h}{\partial x^2}$ and $\frac{\partial^2 h}{\partial x \partial y}$, respectively. Likewise, we can take the x or y partials of $\frac{\partial h}{\partial y}$ to obtain two more functions that are usually denoted by $\frac{\partial^2 h}{\partial y^2}$ and $\frac{\partial^2 h}{\partial y \partial x}$. Thus, the second derivatives of h really comprise an array of four functions. Arguing along the same lines, we obtain eight functions that are third derivatives of h and then 16 fourth derivative functions, etc.

However, in this regard, it is important to remember the following general rule:

When taking some number of x derivatives of a function and some number of y derivatives, the answer is insensitive *to the order in which these derivatives are taken. Thus, for example, we can take all of the x derivatives first and then all of the y derivatives, or vice versa, or we can alternate x and then y derivatives until all of one are used up, etc. The answer will not depend on the order, only on the overall number of x derivatives taken and y derivatives taken.* (14)

In the case of second derivatives, this general rule boils down to the following:

$$
\frac{\partial^2 h}{\partial y \partial x} = \frac{\partial^2 h}{\partial x \partial y}.
\tag{15}
$$

10.5 Integration

In one-variable calculus, you learned about differentiation, but also about integration. So you can ask whether there are integration formulas for functions of two (or more) variables. The answer is yes. However, since this kind of integration plays no real role in the subsequent chapters, I will only provide a brief discussion of this matter.

To begin, suppose that x and y are the variables and $h(x, y)$ is a function of them. For example, $h(x, y) = xy^2$. Then you can integrate the function h over any reasonable region in the xy-plane. The procedure will be messy for any complicated region, but things are absolutely straightforward if the region is rectangular. Indeed, suppose that the region in question is the rectangle, R, that is defined by the conditions

$$a \leq x \leq b,$$
$$c \leq y \leq d,$$

(16)

where $a < b$ and $c < d$ are some given numbers. Figure 10.2 is an example where $a = 0, b = 1, c = 0$ and $d = 2$.

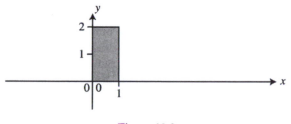

Figure 10.2

The integral of $h(x, y)$ over the rectangle in (16) is obtained by first computing, for *each* fixed y, the integral

$$\int_a^b h(x, y) \, dx.$$

(17)

When I say to keep y fixed, the point is that you should treat the appearance of y as you would that of any number or constant so that $h(x, y)$ is (temporarily) viewed as depending solely on x. You then compute the expression in (17) via the rules of one-variable calculus. However, you will usually get a different answer for this integral for each possible value of y, so the expression in (17) is a function of y. With this last point understood, integrate this function of y from c to d by the rules of one-variable calculus to get the final answer:

$$\int_c^d \left(\int_a^b h(x, y), dx \right) dy.$$

(18)

Alternately, you can do the y-integration from c to d first, treating x as you would any constant, to obtain a function of x, namely

$$\int_c^d h(x, y)\, dy. \tag{19}$$

Indeed, with x treated as a constant, the rules of one-variable calculus apply to the expression in (19) and these are the rules you use to do the computation. But the result depends (for a typical choice of h) on x. This function of x can then be integrated from a to b as you would any function of one variable:

$$\int_a^b \left(\int_c^d h(x, y)\, dy \right) dx. \tag{20}$$

The numbers obtained from (18) and (20) will be identical. The point is that the order in which the integration is done (x first and then y or vice versa) is irrelevant. Just remember that if you do the x first, the result is a function of y, and if you do the y first, the result is a function of x.

Here is one example: Consider the integral of the function $h(x, y) = xy^2$ over the rectangle where $0 \le x \le 1$ and $0 \le y \le 2$. Doing the x-integral first, you obtain

$$\int_0^1 xy^2\, dx = y \int_0^1 x\, dx = y^2(x^2/2)\Big|_0^1 = y^2/2. \tag{21}$$

Notice here how I took the y^2 outside the integral, since I am treating it as I would any fixed number when integrating over x. In any event, now I do the y integral of the function $y^2/2$ over the interval where $0 \le y \le 2$ to obtain

$$\int_0^2 y^2/2\, dy = (y^3/6)\Big|_0^2 = 4/3. \tag{22}$$

Alternately, I can integrate $h(x, y) = xy^2$ over the same rectangle by doing the y-integral first while treating x as a constant:

$$\int_0^2 xy^2\, dy = x \int_0^2 y^2\, dy = x(y^3/3)\Big|_0^2 = 8x/3. \tag{23}$$

Next, I complete the project by integrating $8x/3$ over the interval $0 \le x \le 1$ to obtain

$$\int_0^1 8x/3\, dx = 4/3(x^2)\Big|_0^1 = 4/3. \tag{24}$$

Notice that this is the same as (22), as it should be.

10.6   Lessons

Here are some key points from this chapter:

- Know the stability criteria for a differential equation of the form given in (3).

- Know how to compute the stability matrix in (8) at an equilibrium point; and know how to compute its trace and determinant.

- Know how to take partial second derivatives. In this regard, remember that when taking a mixed partial derivative using both the x-derivative and the y-derivative, the answer is insensitive to the order in which you take them.

- Know how to integrate a function of two variables, say x and y, over any rectangle in the xy-plane.

READINGS FOR CHAPTER 10

READING 10.1

Hopes for the Future: Restoration Ecology and Conservation Biology

Commentary: This article presents a mathematical model whose goal is to predict the ultimate distribution of land useage. Consider, in particular, Equations (1)–(4). Here, F is the area of pristine forest habitat, A is the area of agricultural land, U is the area of unused land, and P is the human population using the land. All are functions of time. Also, $s, d, b, a, r,$ and h are constants. The relevant equilibrium point is given by Equations (5)–(8). The authors assume that this equilbrium point is stable. In any event, they discuss the ramifications of what they find, and I leave it to you to judge whether their arguments are convincing.

Hopes for the Future: Restoration Ecology and Conservation Biology

Science, **277** (1997) 515–522.
Andy P. Dobson, A. D. Bradshaw, A. J. M. Baker

Conversion of natural habitats into agricultural and industrial landscapes, and ultimately into degraded land, is the major impact of humans on the natural environment, posing a great threat to biodiversity. The emerging discipline of restoration ecology provides a powerful suite of tools for speeding the recovery of degraded lands. In doing so, restoration ecology provides a crucial complement to the establishment of nature reserves as a way of increasing land for the preservation of biodiversity. An integrated understanding of how human population growth and changes in agricultural practice interact with natural recovery processes and restoration ecology

---

Author affiliations: A. P. Dobson: Department of Ecology and Evolutionary Biology, Princeton University. A. D. Bradshaw: School of Biological Sciences, University of Liverpool. A. J. M. Baker: Department of Animal and Plant Sciences, Sheffield University.

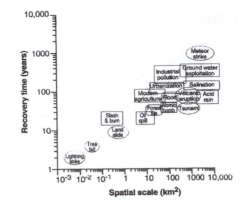

Fig. 1 The relation between the spatial scale of natural and anthropogenic disasters and their approximate expected time to recovery. Natural disasters are depicted in ellipses, and anthropogenic disasters are represented by rectangles. The data used to construct the figure were taken from a number of sources [6, 13, 32].

provides some hope for the future of the environment.

The impact of humans on the natural environment occurs at a variety of temporal and spatial scales. Industrial accidents, such as the *Exxon Valdez* oil spill or the melt-down at Chernobyl, often dominate the world's headlines and produce sudden dramatic ecological change over a large, but usually restricted, area of the landscape. Other changes—such as industrial pollution, deforestation, and conversion of natural habitats into agricultural and industrial land—occur chronically over large sections of each continent. All of these anthropogenic activities alter the habitat available for most other species and usually lead to a reduction in biodiversity.

Where catastrophic environmental changes occur, their major impact on biodiversity occurs instantaneously, although residual effects may last for several years. In contrast, the impacts of long-term habitat conversion may occur over a much longer time scale as individual species become threatened and eventually go extinct. Moreover, the disruptions in community structure and ecosystem function that occur as species are lost will exacerbate this accumulated extinction debt [1]. Yet the scale and magnitude of these disturbances is often comparable with the "natural" disasters from which ecosystems usually recover (Figure 1). In this review we describe how developments in restoration ecology and phytoremediation can be inte-

grated with conservation biology to speed the recovery of natural ecosystems from local and more widespread anthropogenic changes. From the perspective of conservation biology, it is essential that restoration is undertaken before substantial losses of biodiversity have occurred. It is also crucial that the cleanup of industrial accidents have a minimum impact on biodiversity. In both cases, many of the more innovative and cost-effective approaches to solving these problems rely on harnessing natural ecosystem processes that are mediated by the different components of biodiversity.

Habitat Conversion and Loss of Biodiversity

Habitat conversion is the major threat to biodiversity. In particular, tropical forests [2], along with temperate forests, savannas, and coastal marshes, are being converted into land for agriculture, private homes, shopping malls, and cities. The length of time that the habitat remains viable for agricultural use is determined by the duration of soil productivity, or the rate of accumulation of weeds and other pest and pathogen species. Similarly, in areas of industrial activity, such as mining, use of an area commonly persists only until the mineral resource is exhausted; where there is manufacturing, use often comes to an end when the industry becomes outdated.

Throughout human history, habitat conversion has taken place at different rates and on different spatial scales [3]. In Europe, the Middle East,

China, and Meso America, rates of habitat conversion were initially slow, and the deep alluvial soils underlying many of these areas have eroded slowly, permitting agriculture to persist for many centuries. In North America, habitat conversion has taken place more rapidly: Conversion has occurred in localized patches over the last 10,000 years, but the main changes have predominantly spread from east to west across the continent over the last 400 years (Figure 2A). Strikingly, habitat conversion in the tropics has occurred primarily during the second half of the 20th century (Figure 2B). This rapid rate of change has been caused by a number of economic and political forces driven by the large increases in the human population density in most tropical countries. In many places, widespread commercial logging of tropical forests has provided access into areas previously inaccessible to anyone other than the endemic groups who practiced low-level swidden agriculture. In the 1970s, the governments of Brazil and Indonesia provided tax incentives and other forms of political pressure that encouraged transmigration from areas of high human population density into areas that might be converted to agriculture [4]. In developed countries, the same process can be driven by the common desire of new industry to establish on virgin "green-field" sites.

Habitat conversion from forests to agriculture and then to degraded land is the single biggest factor in the present biological diversity crisis [5]. Data from various continents suggest that tropical forests (and temperate forests) are being destroyed at annual rates of between 1 and 4% of their current areas [2, 4, 6]. A significant additional effect is that habitat conversion to agriculture is occurring on land that only retains its utility as agricultural land for 3 to 5 years; it is then abandoned when invasion by weeds or erosion of the topsoil reduces its agricultural viability [4, 7]. These degraded areas will accumulate, because although natural colonization and succession will occur, the process can be slow and can produce considerably impoverished fauna and flora. Where widespread clearing has lead to the local extinction of previously common pioneer species and mid- and late-succession species, recovery occurs at a much slower pace. Models of habitat loss suggest that the process produces an "extinction debt," a pool of species that will eventually go extinct unless the habitat is repaired or restored [1]. Although populations of some of these species may be maintained as captive populations in zoos and botanical gardens for eventual reintroduction, only a limited number of species are likely to be saved in this way [8]. In contrast, restoration of the habitat before these extinctions occur may provide an important means of allowing a significant number of species to recover. This suggests that human efforts to aid habitat restoration will increasingly become a crucial aspect of the conservation of biodiversity.

Modeling habitat conversion. The basic dynamics of habitat conversion and recovery can be described by a simple mathematical model. This model examines the impact, at the landscape level, of habitat conversion driven by the agricultural needs produced by a growing human population. The structure of the model is similar to the compartmental susceptible-infectious-recovered (SIR) models used in epidemiology [9]. In this case, the compartments correspond to periods of time over which patches of converted natural habitat (such as tropical forest) can be used as agricultural (or industrial) land and the time that the resulting degraded land takes to recover to forest. The model also applies to areas used for industrial purposes that subsequently become obsolete, leading to the accumulation of derelict land, which then either recovers naturally or is reclaimed artificially.

The model considers the rate at which an original area of pristine forest habitat (of area F) is converted first to agricultural land (area A), which after a period of time $1/a$, becomes unused land (area U), which in turn recovers through natural succession or ecological restoration to become forest after a time interval $1/s$. Unused land may also be restored to agriculturally viable land after a time interval $1/b$. The basic parameters of the model can be readily estimated from current studies of tropical and temperate forests, and the framework is readily adaptable to other types of habitat. Rates of habitat conversion are assumed to be a simple function of the number of humans P using the land at any time. Initially this parameter is assumed to be constant; more detailed models can allow technological advances to lead to increases in the rates of habitat conversion. It is assumed that human population growth r occurs at a maximum rate of 4% and can be modeled as a simple logistic function with carrying capacity, given by the minimum amount of land h required to support an individual human. Technological and agricultural advances may also

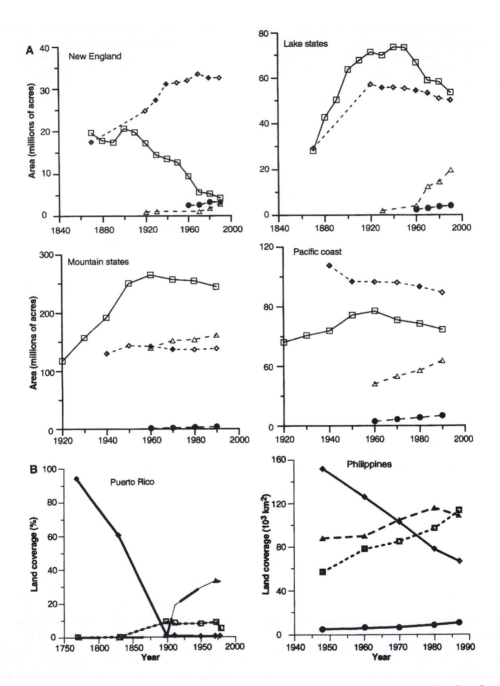

Fig. 2 (**A**) Changes in land use for four different regions of the United States, 1850 to 1990 (data from [49]). The changes generally occurred from east (New England) to west (Pacific Coast). (**B**) Observed patterns of habitat conversion in the tropics: Puerto Rico [50] and the Philippines [4]. Squares indicate farm area, diamonds are forest area, circles are urban areas, and triangles are miscellaneous (degraded) lands.

produce temporal changes in this parameter [10], but the initial assumption is that it remains constant. The basic form of the model consists of four coupled differential equations.

$$\frac{dF}{dt} = sU - dPF \tag{1}$$

$$\frac{dA}{dt} = dPF + bU - aA \tag{2}$$

$$\frac{dU}{dt} = aA - (b + s)U \tag{3}$$

$$\frac{dP}{dt} = rP\frac{A - hP}{A} \tag{4}$$

This framework assumes that we start with an initial area of forest F, which can be either the entire forest in a country or a patch of forest connected to other patches in a spatial array. The loss of biodiversity from the original habitat can be modeled in a number of different ways. In the simplest case, assume that the total number of species NS of any taxon living in the habitat of area A can be estimated using a simple power-law relation, $NS = cA^z$ [11] (c is a constant of proportionality, and z is the power-law constant). A number of studies have shown that this relation provides a useful estimate of the proportion of species that will be found in a habitat as its total area declines [12].

The equations provide a framework in which to compare the impact on natural habitats of agricultural expansions, ecological restoration, improvements in agricultural efficiency, and human population growth. It is important to note that although the equilibrium dynamics are unaffected by the rate of growth of the human population, the transient dynamics are strongly dependent on the rate of human population increase: The more rapidly the human population increases, the more rapidly the forest is degraded and the landscape becomes dominated by unused land (Figure 3A).

The equilibrium expressions for each of the model's variables (denoted by an asterisk) can provide simple insights into the factors that determine the proportion of each type of habitat that will be present once the system has reached a steady state.

$$A^* = hP^* \tag{5}$$

$$U^* = \frac{aA^*}{(s + b)} \tag{6}$$

$$F^* = \frac{ah}{d}\left(\frac{s}{s + b}\right) \tag{7}$$

The size of the human population at equilibrium depends on the initial total area F_0

$$P^* = \frac{\left(F_0 - \frac{ah}{d}\right)\left(\frac{s}{s+b}\right)}{h\left(\frac{a}{s+b} + 1\right)} \tag{8}$$

The equilibrium results illustrate the sensitivity of the resultant landscape to the rate processes, which determine the duration of time for which the land can be used for agriculture ($1/a$) and the length of time it takes to recover from degraded land to forest ($1/s$) (Figure 3B). The most striking result of this analysis is that as the length of time for which land can be used for agriculture increases, the less forest remains in the final landscape. This situation corresponds fairly closely to Europe and the American prairies today. It would also apply to successful industrial areas, where one industry is succeeded by another and derelict land is reused immediately. In contrast, when land can only be used for agriculture for a short time and human population density is low, it is possible that significant amounts of forest remain, a situation that corresponds to swidden agriculture in many tropical forests up until the early 20th century. In all cases, increases in the land available to agriculture will result from reductions in the recovery time from previous exploitation. It is also important to notice that the amount of natural habitat remaining at equilibrium (F^*) decreases both with the efficiency with which humans convert natural habitat to agricultural land (d) and with the efficiency of agricultural production ($1/h$). Moreover, further reductions in the natural habitat remaining at equilibrium are likely to occur if restoration produces new agricultural land, rather than "natural" habitats.

It is possible to build modifications into this basic framework. In many cases, tropical deforestation is not primarily the result of conversion for agriculture but is to open up logging roads for selective logging [4, 6, 13]. This phenomenon can readily be included into the model with the addition of another equation. Similarly, the model can include more subtle details of the way in which human population growth responds to changes in resources [14]. Alternatively, additional expressions for resources such as water may be included; because resources such as water are vital to agricultural activity and human welfare and may require

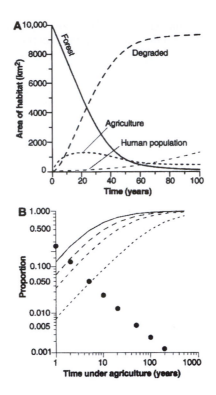

Fig. 3 (A) The transient dynamics of the model when an initial patch of 10,000 km² of forest are invaded by 50 people. Pristine forest declines and the area under agricultural land increases, as does unused and unusable land that is slowly recovering to forest. (B) The relation between the period of time for which land remains viable for agriculture and the proportion of habitat remaining in its pristine (or recovered) state (large dots). The contours illustrate the proportion of agricultural land in a landscape: They are drawn for a range of times that reflect different rates of recovery of degraded land back into forest (solid line, 5 years; long dashes, 10 years; intermediate dashes, 20 years; short dashes, 100 years).

large areas of natural habitat to act as watersheds, the inclusion of these resources may lead to significant increases in the amount of land that remains as natural habitat.

Habitat conversion and the loss of biological diversity. All of this work presents a fairly pessimistic outlook for the future of the biological diversity that inhabits tropical and temperate forests. Although it seems likely that small patches of forest will remain, these patches are likely to be of the order of the size of the small patches of pristine forests remaining in Europe and the eastern United States. If the land can support agriculture for prolonged periods of time, the bulk of the landscape is dominated by agricultural land. In contrast, where

fragile soils allow only short-term agriculture, the landscape becomes dominated by degraded land that is slowly recovering into forests. The same principles apply to land used for industrial areas.

This situation suggests two main approaches for the conservation of biodiversity. The simplest and most crucial is the setting aside of protected areas of land as parks and nature reserves. It is crucial that, within these, a full range and sufficient amount of "original" ecosystem is retained because they will provide the natural colonists for the regeneration of degraded patches. Mapping and geographical information systems will provide important methods for identifying key areas of habitat for endangered species [15].

Table 1.   The time scales for biological and physical processes involved in the development of ecosystems on a newly produced bare area.

Biological processes		Physical processes	
Time scale (years)	Process	Time scale (years)	Process
1–50	Immigration of appropriate plant species	1–1000	Accumulation of fine material by rock weathering or physical deposition
1–50	Establishment of appropriate plant species		
1–10	Accumulation of fine materials captured by plants	1–1000	Decomposition of soil minerals by weathering
1–100	Accumulation of nutrients by plants from soil minerals	1–100	Improvements of soil available water capacity
1–100	Accumulation of N by biological fixation and from atmospheric inputs	1–1000	Release of mineral nutrients from soil minerals
1–20	Immigration of soil flora and fauna supported by accumulating organic matter		
1–20	Changes in soil-structure and organic-matter turnover due to plant, soil microorganism, and animal activities		
1–20	Improvements in soil water-holding capacity due to changes in soil structure	10–10000	Leaching of mobile materials from surface to lower layers
10–1000	Reduction in toxicities due to accumulation of organic matter	100–10000	Formation of distinctive horizons in the soil profile

The second main approach is the restoration of degraded agricultural and industrial land, to reduce the pressure for further natural habitat conversion by providing lands that can be used for agricultural or recreational activities, and even to provide new land for nature reserves. In a world where development should be curtailed yet all agricultural and industrial development cannot be stopped, there is intense psychological and practical importance to growing new, or restored, ecosystems. Indeed, restoration can now be considered a critical element in managing the world's environment. It provides a powerful way of reducing the length of time for which habitat remains in the degraded and unused category of the simple model described above.

Lessons from Primary Succession

In the current climate of gloom over the future of our environment, people tend to think that all environmental damage is irreversible.   It is not. The functioning ecosystems in every part of Earth's surface, including such inhospitable loca-

tions as moraines left by retreating glaciers [16] and the dunes formed by the accretion of lake shore sand [17], originated in the natural processes of primary succession [18]. Primary succession is ecosystem development in situations where no previously developed soil exists. The processes involved fall into two groups, biological and physical (Table 1). Although the primary characteristics of habitats are physicochemical, the biological processes, particularly the processes of nutrient accumulation, are more important for the development of a habitat that can support a properly functioning ecosystem. This dependence on biology is especially true for nitrogen because in soils it is stored only in organic matter, from which it is released by slow decomposition. In temperate regions, where decomposition rates are 10% per year or less, the soil must accumulate at least 1000 kg of N per hectare to provide for the annual needs of an ecosystem (about 100 kg of N per hectare). Because the initial degraded material is likely to have very little nitrogen, accumulation is frequently the limiting factor controlling ecosystem development [19].

In primary successions, community development accompanies the development of the habitat. The establishment of different species can be determined by chance, the state of the habitat, and interactions between new species and those already present. These processes are recognized in the alternative models for the mechanism of succession of tolerance, facilitation, and inhibition [20]. In situations in which the environment is degraded by human activity, the processes of primary succession bring about ecosystem development in the same manner; examples include hard-rock quarries [21], ironstone banks [22], and kaolin mining wastes [23]. The early stages of succession, characterized by the poverty and openness of the ecosystem, help generate ecosystems rich in species sensitive to competition that cannot find living space in more developed ecosystems [24]. The processes of secondary succession (ecosystem development in situations where the original soil remains) also have a bearing on what may be achieved. However, secondary succession represents, ecologically, a simpler problem, most of which is included in primary succession.

Ecological Restoration

The problem with leaving restoration to natural processes is that they take time, measured in decades or centuries (Table 1); redevelopment of advanced communities may take a millennium or more. However, this long time scale is due to specific problems that, once identified, can be overcome by artificial interventions, which are most successful if they use or mimic natural processes. This process of identification and intervention is the essence of ecological restoration.

The substrate. The soil usually provides the major problems, but for each problem, there are immediate and long-term treatments (Table 2). For example, nitrogen deficiency can be overcome in the short term by the application of artificial or organic fertilizers, and in the long term by the introduction of nitrogen-fixing plant species. A range of herbaceous and woody species, which can accumulate over $100 \, \text{kg N ha}^{-1} \, \text{year}^{-1}$, can be used to raise the nitrogen capital of the soil: For instance, herbaceous legumes such as *Trifolium* and *Lespedeza* have been used temporarily in the restoration of pasture land on coal wastes in the United Kingdom

and Australia, and tree species such as *Casuarina* and *Acacia* are an integral part of the final ecosystem in the restoration of forests on metal and coal mine wastes in India. The use of nitrogen-fixing species requires good knowledge of their biology, both their soil preferences and their interactions with other species. Although many of the structural characteristics of soils take a long time to restore, the restoration of their biological properties can usually be brought about swiftly, so that proper ecosystem function can be achieved within 10 years.

These treatments can often be obviated if the original soil can be removed before disturbance and replaced afterward, as is now required for surface mining in most developed countries. But a considerable backlog of sites where this has not been done exists throughout the world, and many situations occur where it is still not required. In 1974 in the United States, the total area of degraded land resulting from surface mining alone was over 1,784,000 ha; in Britain, the total amount of officially recognized derelict land was over 43,000 ha. Recent surveys suggest little change because the restoration level has been matched by the accumulation of additional dereliction. Similarly, data from the Food and Agriculture Organization (FAO) on the conversion of forests into agricultural land in the tropics show that the loss of forests is more closely matched by increases in degraded land than by increases in agricultural land [25].

The community. Once soil characteristics have been restored, it is not difficult to restore a full suite of plant species to form the required vegetation. The common approach has been to choose (i) species important for restoring ecosystem function, together with (ii) species that are to be the main components of the final ecosystem, leaving (iii) the many plant and animal species that should make up the final biodiversity of the ecosystem to recolonize by their own efforts. It is now realized, from studies of chemical waste heaps [26] and the area covered by the eruption of Mount St. Helens [27], that the recolonization of the final biodiversity is likely to be slow and unreliable, because so many of the desired species are no longer available in the vicinity and migration over long distances can be limited.

There is considerable interest in devising means for enhancing natural immigration that results

Table 2.   Short-term and long-term approaches to soil problems in ecological restoration.

Category	Problem	Immediate treatment	Long-term treatment
		Physical	
Texture	Coarse	Organic matter or fines	Vegetation
	Fine	Organic matter	Vegetation
Structure	Compact	Rip or scarify	Vegetation
	Loose	Compact	Vegetation
Stability	Unstable	Stabilizer or nurse	Regrade or vegetation
Moisture	Wet	Drain	Drain
	Dry	Irrigate or mulch	Tolerant vegetation
		Nutrition	
Macronutrients	Nitrogen	Fertilizer	N-fixing species
	Others	Fertilizer and lime	Fertilizer and lime
Micronutrients	Deficient	Fertilizer	
Toxicity	Low	Lime	Lime or tolerant species
pH	High	Pyritic waste or organic matter	Weathering
Heavy metals	High	Organic matter or tolerant plants	Inert covering or bioremediation
Organic compounds	High	Inert covering	Microbial breakdown
Salinity	High	Weathering or irrigate	Weathering or tolerant species

from the natural processes of seed dispersal by en-couraging the movement of seed carriers, especially birds [28]. Where this approach is not possible, ar-tificial reintroduction must be used. More informa-tion is needed about the establishment character-istics and requirements of individual wild species, as is illustrated by current work on European heath communities: The establishment of heath species that require low amounts of nutrients and lots of light, such as *Calluna*, can be aided by the addition of fertilizer and the presence of a grass species that act as a "nurse" [29]. Much has been hypothesized about assembly rules for communities; ecological restoration provides the opportunity to test these ideas. It appears that, so long as the environmental requirements for the establishment of the individ-ual species can be met (and this may not be easy), species can be introduced together rather than se-quentially. This method will not, however, ap-ply in more highly structured communities—such as forests [30], especially high-diversity tropical forests [31]—where the age-determined elements of vegetation structure cannot be put back immedi-ately. The current challenge of ecological restora-tion is to steer development so that all of the sub-tleties of structure and function recover, allowing the full range of species to find their niches [32]. Nevertheless, there is always the possibility that the whole process may go in an undesired direction [31,

33]. Species that arrive first by chance may persist and dominate the ecosystem for many decades, es-pecially if they establish in large numbers. Manip-ulation of the initial composition of species may therefore be necessary.

Another gap in current knowledge is in deter-mining how many of the component species of the final ecosystem can be left to enter the commu-nity on their own. Mobile species such as birds can certainly do this, provided that the develop-ing ecosystem is suitably structured, as has been shown in gravel pit restoration [34]. Fish and wa-ter plants usually cannot colonize unless the pits are connected to a river system, but they can be added as individuals or in soil or dredged mud. Many woodland species are immobile [35], so they must be added as seed or plants. For artificially in-troduced species, it is important to use individuals from the locality, which are well adapted to the en-vironment. Evolutionary adaptation can take place specifically in relation to the conditions occurring in degraded habitats [36], suggesting that material from the degraded habitat itself is not necessar-ily most suitable for the final restored ecosystem. However, such adapted individuals will be valuable in the initial, degraded condition, and there is no reason such material will not adapt to subsequent improvements in the environment.

Phytoremediation. The most recalcitrant problem in restoration is the decontamination of soils polluted with heavy metals and certain organic compounds. Techniques in current use for metals are based on physicochemical extraction, such as acid leaching and electro-osmosis, or on in situ immobilization, for example, by vitrification, a thermal process in which metals are fused with the silica fraction of the soil to form an inert glass [37]. These methods require specialized equipment and trained operators; they are therefore costly and only appropriate for the decontamination of small areas where pressure of land use or development potential merits the outlay. Furthermore, they remove all biological activity from the treated medium and adversely affect its physical structure and suitability for plant growth during revegetation.

In the last few years, there has been much interest in the potential offered by plant uptake of heavy metals as a means of soil decontamination [38, 39]. The principle of phytoextraction (one form of phytoremediation) is simple and elegant in concept: The substrate is "cropped" for metals, which are progressively and selectively removed, leaving it in all other ways unaffected. The biomass is harvested, removed, and disposed of as hazardous waste or is incinerated at low temperature, allowing the recovery of the metals concentrated in the ash.

Some plants endemic to metalliferous soils are capable of accumulating exceptionally high concentrations of potentially phytotoxic metals such as Zn, Ni, Cd, Pb, Cu, and Co in their harvestable biomass. For these metals, some 400 hyperaccumulators have now been identified [40]. Concentrations of metals can reach several percent in the above-ground dry biomass of these plants. In Europe, a large proportion of hyperaccumulators are members of the Brassicaceae (a family that includes many important crops): Included are species of the genus *Alyssum* found on serpentine (naturally Ni- and Cr-rich) soils in southern Europe, which can accumulate concentrations of Ni in excess of 2% on a dry-weight basis, and some *Thlaspi* species from calamine (naturally Zn-, Cd-, and Pb-rich) soils, which can accumulate Zn to more than 5%, Cd to 0.2%, and Pb to 1%. Hyperaccumulators of Zn, Ni, Cu, and Co have also been discovered in the metallophyte floras of the tropics and subtropics. They include representatives of many families and range in growth form from annual herbs

to shrubs and trees. Baker *et al.* [41] used a range of Zn- and Ni-hyperaccumulator plants grown under intensive agronomic conditions to remove metals from agricultural soils contaminated in the rhizosphere (soil reachable by roots) through historical application of industrially contaminated sewage sludges. Uptake of Zn by the hyperaccumulator *Thlaspi caerulescens* was 30 kg ha^{-1}; subsequent provenance and agronomic trials elsewhere have pushed this value to well over 100 kg ha^{-1} and up to 2 kg ha^{-1} for Cd [39]. This approach to phytoextraction is still under development in both the United States (U.S. Department of Agriculture's Agricultural Research Service, Beltsville, Maryland) and the United Kingdom (Institute of Arable Crop Research, Rothamsted). A commercial incentive to this development is that it has been estimated that the harvesting of 1 ha of *Thlaspi* yields 20 tons of biomass, which could contain over $1000 in recoverable metals.

The difficulties of using hyperaccumulator plants for wide-scale soil decontamination center on their relatively low biomass and slow rate of growth. Faster growing selection lines of annual *Brassica* crops such as Indian mustard (*B. juncea*) have been used for Pb phytoextraction [42]. The lower concentration of metal in the *Brassica* shoots (compared to that in Pb hyperaccumulators) is more than compensated for by the greater biomass achievable and the possibility of growing multiple crops within a season.

Phytoextraction, either by exploitation of the unique physiological properties of hyperaccumulator plants or by means of high-biomass accumulator crops such as *B. juncea*, provides an attractive option for soil cleanup if the time scale is acceptable and the contamination is within the rhizosphere. Existing phytoextraction crops could also be improved through the use of conventional breeding techniques and genetic engineering approaches on high-biomass host plants. However, deep-seated contamination cannot be accessed by the roots; in these circumstances, plowing may be required. Bioavailability is also a problem, as metallic contaminants are usually present in insoluble and unavailable forms. The use of chemical amendments, such as chelating agents, to increase bioavailability in the rhizosphere has proven effective for Pb phytoextraction [43]. There is considerable scope for enhancing metal bioavailability locally within the rhizosphere while protecting against leaching and ground-water contamination. Phytotech (Monmouth Junction, New Jersey) successfully used Pb phytoextraction at two brownfield (industrially contaminated) sites in New Jersey, cleaning most of the contaminated areas to state industrial standards (1000 mg kg^{-1} Pb) within one summer of multiple cropping with *B. juncea* [44]. Phytoextraction also offers great possibilities for the decontamination of radionuclides such as ^{137}Cs and ^{90}Sr from contaminated soils and effluents and for the removal of excess Na, Se, and B from saline soils. Phytovolatilization (the use of plants to extract inorganic contaminants, which are then dispersed into the atmosphere by volatilization from aerial parts) of Se and Hg also provides considerable scope for commercial development [44].

A further variant of phytoremediation under development is phytostabilization, the use of plants to immobilize or stabilize metal contaminants in the soil [38]. Gary Pierzynski of Kansas State University and Jerry Schnoor of the University of Iowa have used plants successfully to stabilize soils and decrease the movement of metals to ground water at a smelter Superfund site in Dearing, Kansas. Nothing has grown on the waste piles since the site was abandoned in 1919; concentrations as high as 20,000 mg kg^{-1} Pb and 200,000 mg kg^{-1} Zn made it impossible for even weeds to invade the site. The researchers planted 3100 hybrid poplar trees on two acres at the site with soil amendments. Survival has exceeded expectations, and the site is now revegetated. More importantly, the primary risk to humans in the vicinity, that of windblown dust, has been reduced.

Bioremediation technology for organic compounds has been developing faster than that for inorganics, so it is tempting to believe that for every type of organic contaminant a suitable microbial organism or consortium can be found with the capacity to degrade the contaminant in situ, ultimately into CO_2 and water. However, the major leap from in vitro demonstration to successful field decontamination has proved too great in many promising initiatives, but commercial bioremediation systems are now available for many of the more environmentally hazardous organic contaminants. Phytoremediation of organic contaminants is also proving possible either directly through plant uptake or by degradation or stabilization in the rhizosphere [45]. In an attempt to obtain a better understanding of what occurs when plants are moved

from the bench to the field, researchers at the University of Washington constructed a series of field test plots to study the interactions between a variety of trees and contaminated ground-water streams. This pilot scale study, using a series of artificial aquifers, allows them to compare accumulation, metabolism, and transpiration to a known level of exposure to trichloroethylene or carbon tetrachloride. This type of data may be crucial to the acceptance of this technology by more conservative remediation personnel. On the basis of conclusions drawn from laboratory data and results from the pilot site, the University of Washington group, in conjunction with the Oregon Department of Environmental Quality, are attempting to remediate an aquifer that was contaminated with 1,1,1-trichloroethane about 10 years ago. This site has been proposed for a detailed study of the efficiency of poplar trees in the remediation of a contaminated aquifer.

Outlook

Although human habitat conversion has generally been detrimental to most other species, restoration ecology is beginning to provide opportunities to reverse the trend and to create new habitats for biodiversity. Primary succession provides good evidence of the power of natural processes in recreating ecosystems without aid. When coupled with interventions aimed at treating any serious long-term problems that may occur, rapid improvements can be brought about. There are many examples of successful restoration [46]: A single outstanding, current case is the restoration of the 10,000 ha of barren land around the nickel smelters at Sudbury, Ontario [47]. With biologically based technologies such as phytoremediation, it will be possible to treat the most serious types of environmental damage in ways that not only restore a functional ecosystem but also recover resources that are valuable to industry. It is not yet clear the degree to which efforts should be aimed at true restoration (of what was there previously) as opposed to producing something new (replacing what was there originally), either modeled on other ecosystems or totally new artificial constructs. They are each possible alternatives (Figure 4). To retain the world's biodiversity with the use of such creative conservation techniques can be effective and is welcomed by many conservationists. Whereas the principal aim of conservation

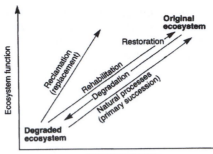

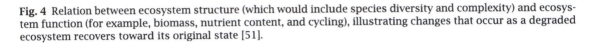

Fig. 4 Relation between ecosystem structure (which would include species diversity and complexity) and ecosystem function (for example, biomass, nutrient content, and cycling), illustrating changes that occur as a degraded ecosystem recovers toward its original state [51].

biology will be to protect current nature reserves and national parks, while identifying and protecting unprotected areas that are also naturally rich in biodiversity [15], there is also a need for restoration to become a standard part of the conservation biologist's armory.

Although it is obvious that wherever possible environmental damage should not be allowed to occur in the first place, human development and population growth mean that damage is inevitable. The demands on land use are such that large areas of land will continue to be converted into agricultural and industrial areas. Increasingly, these will be "marginal" lands that will only be viable for agriculture for a short period of time. Ironically, the poorer soils of these areas may support a higher level of biodiversity than the areas with deep rich soils that have already been occupied preferentially [48]. Rapidly restoring natural ecosystems that have only been transiently used for agriculture provides an important opportunity to ensure that development becomes sustainable. Most attempts at sustainable use of natural resources have focused at the population level, but we also need to consider the use of natural resources at the landscape level. Here the key to successful restoration will still rest on there being a potential pool of colonists that recolonize restored land. The only way to ensure this resource is to rigorously protect biodiversity in nature reserves and other natural habitats.

Ecological restoration will continue to provide important insights into the way that ecological communities are assembled and ecosystems func-

tion. There is a direct analogy with engineering: It is a relatively straightforward exercise to take apart an ecosystem or an automobile engine, yet quantifying the relative number of parts in an automobile engine (or an ecosystem) tells us little about how it functions. In contrast, reassembling the engine (or the ecosystem) will reveal a deeper level of understanding of how each of its components functions.

REFERENCES

1. D. Tilman, R. M. May, C. L. Lehman, M. A. Nowak, *Nature* **371**, 65 (1994).

2. W. V. Reid, in *Tropical Deforestation and Species Extinction*, T. C. Whitmore and J. A. Sayer, Eds. (Chapman and Hall, London, 1992), pp. 55–73; J. A. Sayer and T. C. Whitmore, *Biol. Conserv.* **55**, 199 (1991); T. C. Whitmore and J. A. Sayer, in *Tropical Deforestation and Species Extinction*, T. C. Whitmore and J. A. Sayer, Eds. (Chapman and Hall, London, 1992), pp. 1–14.

3. W. B. Meyer and B. L. Turner II, *Annu. Rev. Ecol. Syst.* **23**, 39 (1992); A. Goudie, *The Human Impact on the Natural Environment* (MIT Press, Cambridge, MA, 1990).

4. D. L. Skole, W. H. Chomentowski, W. A. Salas, A. D. Nobre, *Bioscience* **44**, 314 (1994); D. M. Kummer and B. L. Turner II, *ibid.*, p. 323.

5. W. V. Reid and K. R. Miller, *Keeping Options Alive: The Scientific Basis for Conserving Biodiversity* (World Resources Institute, Washington, DC, 1989); A. A. Burbridge and N. L. McKenzie, *Biol. Conserv.* **50**, 143 (1989); H. Koopowitz and H. Kaye, *Plant Extinction:*

A *Global Crisis* (Christopher Helm, London, 1983); R. M. May, J. H. Lawton, N. E. Stork, in *Extinction Rates*, J. H. Lawton and R. M. May, Eds. (Oxford Univ. Press, Oxford, 1995), pp. 1–24; T. D. Sisk, A. E. Launer, K. R. Switky, P. R. Ehrlich, *Bioscience* **44**, 592 (1994); A. T. Durning, *Worldwatch Pap.* **117**, 1 (1993); S. L. Pimm, G. J. Russell, J. L. Gittleman, T. M. Brooks, *Science* **269**, 347 (1995); R. T. Watson *et al.*, *Global Biodiversity Assessment: Summary for Policy-Makers* (Cambridge Univ. Press, Cambridge, 1995).

6. D. Skole and C. Tucker, *Science* **260**, 1905 (1993); R. F. W. Barnes, *Afr. J. Ecol.* **28**, 161 (1990).

7. C. Uhl and C. F. Jordan, *Ecology* **65**, 1476 (1984); C. Uhl and P. G. Murphy, *Trop. Ecol.* **22**, 219 (1981); C. Uhl, in *Biodiversity*, E. O. Wilson, Ed. (National Academy Press, Washington, DC, 1988), pp. 326–332.

8. W. G. Conway, *Int. Zoo Yearb.* **24/25**, 210 (1986); T. J. Foose, in *The Last Extinction*, L. Kaufman and K. Mallory, Eds. (MIT Press, Cambridge, MA, 1993), pp. 149–178); J. H. W. Gipps, *Beyond Captive Breeding: Re-introducing Endangered Mammals of the World* (Oxford Univ. Press, Oxford, 1991); A. M. Lyles and R. M. May, *Nature* **326**, 245 (1987); B. Griffith, J. M. Scott, J. W. Carpenter, C. Reed, *Science* **245**, 477 (1989).

9. R. M. Anderson and R. M. May, *Infectious Diseases of Humans: Dynamics and Control* (Oxford Univ. Press, Oxford, 1991); R. V. O'Neill, R. H. Gardner, M. G. Turner, W. H. Romme, *Landscape Ecol.* **7**, 19 (1992); P. Faeth, C. Cort, R. Livernash, *Evaluating the Carbon Sequestration Benefits of Forestry Projects in Developing Countries* (World Resources Institute, Washington, DC, 1994).

10. M. L. Primack, *J. Econ. Hist.* **22**, 484 (1962); D. R. Headrick, in *The Earth as Transformed by Human Action*, B. L. Turner II *et al.*, Eds. (Cambridge Univ. Press, Cambridge, 1990), pp. 55–67.

11. R. H. MacArthur and E. O. Wilson, *The Theory of Island Biogeography* (Princeton Univ. Press, Princeton, NJ, 1967); W. D. Newmark, *Biol. J. Linn. Soc.* **28**, 83 (1986); H. A. Gleason, *Ecology* **3**, 158 (1922); O. Arrhenius, *J. Ecol.* **9**, 95 (1921); D. Simberloff, *Tropical Deforestation and Species Extinction*, T. C. Whitmore and J. A. Sayer, Eds. (Chapman and Hall, London, 1992), pp. 75–90.

12. Studies of the extinction of woodland birds in the eastern United States indicate that this approach provides accurate estimates of the observed extinction rate [see S. L. Pimm and R. A. Askins, *Proc. Natl. Acad. Sci. U.S.A.* **92**, 9343 (1995)]. A recent review of habitat loss and fragmentation [H. Andren, *Oikos* **71**, 355 (1995)] illustrates that percolation theory provides important insights into the fragmentation process that is concomitant to habitat loss.

13. D. Pimentel *et al.*, *Oikos* **46**, 404 (1986); L. E. Sponsel, T. N. Headland, R. C. Bailey, *Tropical Deforestation: The Human Dimension* (Columbia Univ. Press, New York, 1996).

14. J. P. Holdren and P. R. Ehrlich, *Am. Sci.* **62**, 282 (1974); J. E. Cohen, *Science* **269**, 341 (1995); P. M. Vitousek, P. R. Ehrlich, A. H. Ehrlich, P. M. Matson, *Bioscience* **36**, 368 (1986); G. C. Daily and P. R. Ehrlich, *ibid.* **42**, 761 (1992).

15. R. L. Pressey, C. J. Humphries, C. R. Margules, R. I. Vane-Wright, P. H. Williams, *Trends Ecol. Evol.* **8**, 124 (1993); J. M. Scott, B. Csuti, J. D. Jacobi, J. E. Estes, *Bioscience* **37**, 782 (1987); J. M. Scott et al., *Wildl. Monogr.* **123**, 1 (1993); A. P. Dobson, J. P. Rodriguez, W. M. Roberts, D. S. Wilcove, *Science* **275**, 550 (1997).

16. R. L. Crocker and J. Major, *J. Ecol.* **43**, 427 (1955).

17. J. S. Olsen, *Bot. Gaz.* **199**, 125 (1958).

18. J. Miles and D. W. H. Walton, Eds., *Primary Succession on Land* (Blackwell, Oxford, 1993).

19. R. H. Marrs and A. D. Bradshaw, in *ibid.*, pp. 221–235; A. D. Bradshaw, *J. Ecol.* **20**, 151 (1938).

20. J. H. Connell and R. D. Slatyer, *Am. Nat.* **111**, 1119 (1977).

21. B. N. K. Davis, *Biol. Conserv.* **10**, 249 (1976).

22. G. A. Leisman, *Ecol. Monogr.* **27**, 221 (1957).

23. W. S. Dancer, J. F. Handley, A. D. Bradshaw, *Plant Soil* **48**, 153 (1977).

24. D. Ratcliffe, *Proc. R. Soc. London Ser. B* **339**, 355 (1974).

25. R. A. Houghton [*Bioscience* **44**, 305 (1994)] quotes FAO data for the period 1980 to 1985, which suggest that in the tropical regions of America, Africa, and Asia, forests are being lost at rates of 5.0, 2.8, and 2.1 million hectares per year, respectively. Although agricultural lands are increasing at 3.3, 0.3, and 0.9 million hectares per year, degraded lands are increasing at rates of 1.7, 2.6, and 1.3 million hectares per year.

26. H. J. Ash, R. P. Gemmell, A. D. Bradshaw, *J. Appl. Ecol.* **31**, 74 (1994).

27. J. F. Franklin, J. A. MacMahon, F. J. Swanson, J. R. Sedell, *Natl. Geogr. Res.* **1**, 198 (1985); J. F. Franklin, P. M. Frenzen, F. J. Swanson, in *Rehabilitating Damaged Ecosystems*, J. Cairns Jr., Ed. (CRC Press, Boca Raton, FL, 1995), pp. 288-332.

28. G. R. Robinson, S. N. Handel, *Conserv. Biol.* **7**, 271 (1993).

29. Environmental Advisory Unit, *Heathland Restoration: A Handbook of Techniques* (British Gas, Southampton, UK, 1988).

30. W. C. Ashby, in *Restoration Ecology*, W. R. Jordan, M. E. Gilpin, J. D. Aber, Eds. (Cambridge Univ. Press, Cambridge, 1987), pp. 89-97.

31. D. H. Janzen, *Conservation Biology: The Science of Scarcity and Diversity*, M. E. Soule, Ed. (Sinauer, Northampton, MA, 1986), pp. 286-303; D. H. Janzen, *Science* **239**, 243 (1988); D. H. Janzen, *Trends Ecol. Evol.* **9**, 365 (1994).

32. W. R. Jordan, M. E. Gilpin, J. D. Aber. *Restoration Ecology* (Cambridge Univ. Press, Cambridge, 1987).

33. H.-K. Luh, S. L. Pimm, *J. Anim. Ecol.* **62**, 749 (1993).

34. J. Andrews, D. Kinsman, *Gravel Pit Restoration for Wildlife* (Royal Society for Protection of Birds, Sandy Beds, UK, 1990).

35. G. F. Peterken. *Biol. Conserv.* **6**, 239 (1974).

36. T. McNeilly, in *Restoration Ecology*, W. R. Jordan, M. E. Gilpin, J. D. Aber, Eds. (Cambridge Univ. Press, Cambridge, 1987), pp. 271-283.

37. U.S. Army Toxic and Hazardous Materials Agency, Interim Tech. Rep. AMXTH-TE-CR-86101 (1987).

38. D. E. Salt *et al.*, *Biotechnology* **13**, 468 (1995); S. D. Cunningham and D. W. Ow, *Plant Physiol.* **110**, 715 (1996); D. Comis, *J. Soil Water Conserv.* **51**, 184 (1996).

39. R. L. Chaney *et al.*, *Curr. Opin. Biotechnol.* **8**, 279 (1997).

40. A. J. M. Baker and R. R. Brooks, *Biorecovery* **1**, 81 (1989); R. D. Reeves, A. J. M. Baker, R. R. Brooks, *Min. Environ. Manage.* **3**, 4 (September 1995).

41. A. J. M. Baker, S. P. McGrath, C. M. D. Sidoli, R. D. Reeves, *Resour. Conserv. Recycling* **11**, 41 (1994).

42. P. B. A. N. Kumar, V. Dushenkov, H. Motto, I. Raskin, *Environ. Sci. Technol.* **29**, 1232 (1995).

43. M. J. Blaylock *et al.*, *ibid.* **31**, 860 (1997).

44. M. E. Watanabe, *ibid.*, p. 182A.

45. S. D. Cunningham, T. A. Anderson, A. P. Schwab, F. C. Hsu, *Adv. Agron.* **56**, 55 (1996).

46. A. D. Bradshaw and M. J. Chadwick, *The Restoration of Land* (Univ. of California Press, Berkeley, CA, 1980); M. K. Wali, Ed., *Ecosystem Rehabilitation*, (SPB Academic, The Hague, 1992); J. M. Gunn, *Restoration and Recovery of an Industrial Region* (Springer, New York, 1995).

47. K. Winterhalder, *Environ. Rev.* **4**, 185 (1996).

48. M. Huston, *Science* **262**, 1676 (1993).

49. U.S. Bureau of the Census, *Statistical Abstract of the United States* (1952); *ibid.* (1962); *ibid.* (1972); *ibid.* (1982); *ibid.* (1990); *ibid.* (1993). [Context Link]

50. A. R. Barash, *Biol. Conserv.* **39**, 97 (1987).

51. A. D. Bradshaw, *Restoration Ecology and Sustainable Development*, K. Urbanska and N. R. Webb, Eds. (Cambridge Univ. Press, Cambridge, 1997).

52. A.P.D. would like to thank G. DeLeo, A. M. Lyles, S. J. Ryan, J. P. Rodriguez, and D. Wilcove for comments on the manuscript and G. Dorner for extracting the data on habitat conversion in the United States. A.P.D.'s work is supported by a grant from the Charles Stewart Mott Foundation to the Environmental Defense Fund.

READING 10.2

Effects of Disturbance on River Food Webs

Commentary: Here is another article concerning population biology in which the equilibrium points of a differential equation model play a central role. The article models the food chain in a river environment. There are more than two levels in the chain. (Such a chain is called multitrophic.) I direct your attention to Equation (1) of the article for the differential equations. As can be seen, these constitute a four-component generalization of the two-component equations that we have been analyzing. The authors look for the stable equilibrium points and analyze the relative populations as a function of some of the parameters. (The phrase "Solved at steady state" from the article means simply that the equilibrium solutions were found.) The authors then check some of their predictions with experimental data.

Effects of Disturbance on River Food Webs

Science, **273** (1996) 1558–1561.
J. Timothy Wootton, Michael S. Parker, Mary E. Power

A multitrophic model integrating the effects of flooding disturbance and food web interactions in rivers predicted that removing floods would cause increases of predator-resistant grazing insects, which would divert energy away from the food chain leading to predatory fish. Experimental manipulations of predator-resistant grazers and top predators, and large-scale comparisons of regulated and unregulated rivers, verified the model predictions. Thus, multitrophic models can successfully synthesize a variety of ecological processes, and conservation programs may benefit by taking a food web perspective instead of concentrating on a single species.

Although conservation programs typically concentrate on the direct impacts of environmental change on a single species, ecological experiments and theory demonstrate that species are affected in complex ways by other species, ecosystem productivity, and disturbance regimes [1–4]. Therefore, to understand and predict the consequences of impacts on the environment, ecologists must shift from an autecological perspective to consideration of the interaction of multiple causal factors. For example, changes in climate, land use, and water regulation or diversion all may alter the flooding regime of rivers. How do changes in flooding disturbance affect species in river food webs? Successfully predicting the answer to this question requires a framework that can synthesize the direct effect of disturbance-induced mortality, as well as the indirect compensatory or reinforcing effects of interactions among various species in an ecological community.

Multitrophic dynamic models of species interactions provide a potentially useful synthetic theoretical framework to simultaneously examine a variety of ecological processes [1,2]. Within this framework, species interactions are modeled explicitly, and the dynamics of limiting resources can be treated as species at the bottom of the food web. Disturbance can be incorporated by adding density-independent mortality terms to the dynamics of each species. In previous studies of rivers in northern California, we have shown that such a framework can predict the consequences of removing top predators from a food web and of varying productivity at the base of the food web [1,3]. Here, we consider whether the effects of disturbance can also be incorporated into such a theoretical framework.

Our observations of free-flowing rivers in northern California during the droughts of 1990–1992 and 1994, when scouring floods typical of the winter rainy season were absent or reduced, indicated that flood disturbance can have important effects on river food webs. In the absence of floods, we observed a marked increase in a predator-resistant caddisfly, *Dicosmoecus gilvipes* [5]. The combined effect of large size and robust protective cases effectively eliminates predation on this species by most fish and invertebrate predators, but these same traits can make it susceptible to flood mortality [6]. Its heavy protective case restricts *Dicosmoecus* to the river bottom, and its large size renders most interstitial spaces ineffective as refuges from mortality. As a result, *Dicosmoecus* has difficulty avoiding rolling rocks during scouring

---

Author affiliations: J. T. Wootton: Department of Ecology and Evolution, University of Chicago. M. S. Parker: Department of Biology, Southern Oregon State College. M. E. Power: Department of Integrative Biology, University of California.

floods. Moreover, attaining large size requires longer aquatic larval periods than those of smaller grazers such as midges (Chironomidae) or mayflies (Ephemeroptera), which increases the time during which *Dicosmoecus* is at risk of flood-induced mortality. Two important observations further support this inference of a tradeoff between resistance to flood disturbance and resistance to predation: *Dicosmoecus* densities in April 1992 showed a 77% reduction [from 82.3 ± 65.4 (SD) to 18.8 ± 23.6 individuals per square meter] after a brief spate, and no consumption of *Dicosmoecus* occurred in feeding trials using a range of predators [7].

These observations indicated that a reasonable multitrophic model of this river system might consist of two linked food chains, one going from algae to predator-susceptible grazers to predatory fish and insects and the other going from algae to predator-resistant grazers. Such a situation can be modeled as

$$\frac{dA}{dt} = b_a A L e^{-c_x A} - c_h H A - c_d D A - m_a A$$
$$\frac{dH}{dt} = b_h c_h A L H - c_p P H - m_h H$$
$$\frac{dP}{dt} = b_p c_p H P - m_p P$$
$$\frac{dD}{dt} = b_d c_d A D - m_d D5 \qquad (1)$$

where A, H, D, and P are the abundances of algae, predator-susceptible herbivores, predators, and predator-resistant grazers (*Dicosmoecus*), respectively; L is the amount of incident light available for algae to convert into new offspring by photosynthesis [1]; b_x (where x is a, h, d, or p) is the conversion efficiency of consumed resource into individuals of species x; c_x is the per capita consumption rate of resources by species x; e is the base of the natural logarithm; and m_x is the density-independent mortality (due to disturbance or other causes) experienced by species x.

Solved at steady state, this multitrophic model makes predictions about the average consequences of reducing disturbance (decreasing m_x), given that disturbance increases the mortality of predator-resistant grazers more than that of predator-susceptible grazers. In short, disturbance reduction is predicted to diminish energy flow in the food chain leading to predatory fish, diverting most of the ecosystem energy to predator-resistant grazers. First, reduction of flooding disturbance should in-

crease predator-resistant grazers because of a direct reduction in mortality. Second, removal of disturbance should indirectly decrease algal abundance because of increases in grazing pressure. Third, reduced flooding should indirectly decrease predator abundance because less energy becomes available to the longer food chain. Finally, predator-susceptible grazers should neither increase nor decrease, despite the change in the abundance of their competitors, because decreases in population growth rate arising from lower algal crops are ultimately offset by lower predator populations. This last prediction has interesting implications for recent experimental investigations focusing on whether pairs of species compete [8], because it indicates that the existence of competition may be underestimated when interactions among the larger community are ignored.

To test the predictions of the model, we experimentally manipulated the abundance of *Dicosmoecus* in mesocosms placed in the South Fork Eel River on the Angelo Coast Range Preserve, Mendocino County, California [1, 3, 6]. We constructed instream channels (1.56 m by 1.17 m by 0.78 m) of wooden frames, clear plastic sides, and ends and bottom of plastic screen (6-mm mesh). Twenty-four channels were placed in blocks of four in six similar reaches in the river, and river gravel was added to each channel to a depth of 5 cm. We also added eight ceramic tiles (7.5 by 7.5 cm) to the bottom of each channel to serve as uniform sampling substrates [1]. On 24 June 1992, we introduced 120 *Dicosmoecus* into each of two randomly chosen channels in each block. Additionally, we crossed each treatment with an additional treatment, the presence or absence of three juvenile steelhead (*Oncorhynchus mykiss*, 40 to 80 mm standard length), to test several other model predictions: The addition of steelhead (which increases mortality of small fish and predatory invertebrates) should reduce small predators, increase predator-susceptible grazing insects, and reduce algal cover. We censused the channels on 24 July 1992, just before *Dicosmoecus* diapause [9].

As predicted, manipulating *Dicosmoecus* and steelhead caused significant changes in community structure [multivariate analysis of variance (MANOVA), $P < 0.0002$ for *Dicosmoecus* treatment, $P < 0.00005$ for steelhead treatment, $P > 0.45$ for interaction of *Dicosmoecus* and steelhead, and $P > 0.1$ for location effect] (Figure 1). Adding *Dicosmoe-*

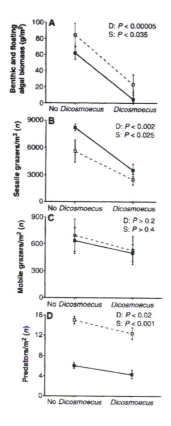

Fig. 1 Mean (±1 SE) block-adjusted responses of (**A**) algal biomass, (**B**) sessile predator-susceptible grazers, (**C**) mobile predator-susceptible grazers, and (**D**) small predators to the manipulation of the predator-resistant grazer *Dicosmoecus gilvipes* (D) and the top predator *Oncorhynchus mykiss* (steelhead; S). Lines indicate treatments with steelhead present (—) and absent (- - -). Probabilities were generated from one-sided *t* tests used to test the directional predictions of the model.

cus caused significant ($P < 0.05$) declines in algal biomass (−83%), predator abundance (−23%), and sessile grazer abundance (midges and the caddisfly *Tinodes* sp.; −56%), but not in mobile grazer abundance; whereas adding steelhead caused significant declines in predator abundance (−62%) and algal biomass (−54%), as well as significant increases in sessile grazer abundance (43%), but no change in mobile grazer abundance. The trophic level–specific predictions of the model were supported for all changes except the decline in sessile insect abundance in the presence of *Dicosmoecus*.

The unanticipated response of sessile insects to *Dicosmoecus* appeared to arise from two factors. First, *Dicosmoecus* appeared to prey on sessile insect larvae, and *Dicosmoecus* feeding activity destroyed protective tubes and prevented establishment of sessile insects on the substrate. Second, the increase in algae in the absence of *Dicosmoecus* provided a physical substrate in which insects could hide [10]. Although adding *Dicosmoecus* predation on midges into the model does not result in the decline of other grazers, destroying protective tubes and reducing algal cover modifies the interaction between predators and their prey [11], which increases predation intensity on grazers. A model that includes this type of indirect interaction predicts all of the patterns arising in the experimental results.

To determine whether the predictions of the

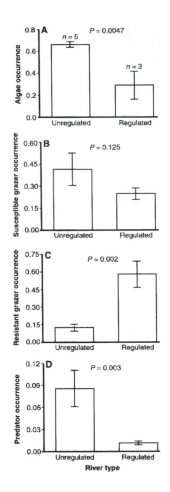

Fig. 2 Average (±1 SE) occurrence (percentage of transect sites with a given trophic group) during the summer growing season of (**A**) visibly conspicuous algae (filamentous diatoms, *Oscillatoria, Rivularia, Nostoc, Cladophora, Oedogonium,* and *Zygnematales*), (**B**) predator-susceptible grazers, (**C**) predator-resistant grazers, and (**D**) predators in unregulated (with flood disturbance) and regulated rivers (flood disturbance greatly reduced) in northern California. Probabilities were generated from one-sided *t* tests used to test the directional predictions of the model.

model could be generalized beyond our experimental channels, we compared surveys of the biota of unregulated (regularly flooding) and regulated (flooding prevented) rivers in northern California [12]. As predicted by the model and experimental results, the regulated rivers we surveyed exhibited a significantly higher occurrence of predator-resistant grazers (369%), a significantly lower occurrence of algae (−58%) and predators (−87%), and a slightly, but not significantly, lower occurrence of

predator-susceptible grazers (−41%), relative to the rivers with natural flows we surveyed (Figure 2).

Our results have several implications for both basic and applied issues. From a basic perspective, our results demonstrate that multitrophic models can provide a useful framework to synthesize multispecies interactions and disturbance regimes, and thus may provide one approach for understanding the role of disturbance in complex natural ecosystems. From an applied perspective, our results emphasize the need to shift from a single-population perspective to a community or food web perspective when developing strategies for resource management or conservation. For example, there has been an alarming decline in salmonid populations along the Pacific coast of North America, and the regulation of rivers through damming has been strongly implicated as one major cause [13, 14]. Managers trying to remedy this situation have concentrated on the possible direct effects of dams on salmonids, for example, in preventing spawning migrations or grinding up juveniles in turbines, and have consequently recommended autecological solutions. By taking a food-web approach, our results demonstrate that such strategies may be insufficient because, in the absence of scouring floods, the food web beneath the fish collapses. Therefore, although mitigation of direct effects of dams on fish may be important, alternative approaches such as pulsed water releases to create riverbed scouring may be required to restore food web integrity, as well as to address other impacts related to river geomorphology [14, 15].

REFERENCES

1. J. T. Wootton, M. E. Power, *Proc. Natl. Acad. Sci. U.S.A.* **90**, 1384 (1992).
2. M. L. Rosenzweig, *Science* **171**, 385 (1971); L. Oksanen, S. D. Fretwell, J. Arruda, P. Niemela, *Am. Nat.* **118**, 240 (1981); S. L. Pimm, *Food Webs* (Chapman and Hall, London, 1982); D. Tilman, *Am. Nat.* **129**, 769 (1987); S. L. Pimm, R. L. Kitching, *Oikos* **50**, 302 (1987); P. A. Abrams, *Am. Nat.* **141**, 351 (1993).
3. M. E. Power, *Science* **250**, 811 (1990).
4. R. T. Paine, *Am. Nat.* **100**, 65 (1966); J. H. Connell, Science 199, 1302 (1978); D. L. DeAngelis, *Dynamics of Nutrient Cycling and Food Webs* (Chapman and Hall, London, 1992); J. T. Wootton, *Annu. Rev. Ecol. Syst.* **25**, 443 (1994).
5. M. E. Power, M. S. Parker, J. T. Wootton, in *Food Webs: Integration of Patterns and Dynamics*, G. A. Polis and K. O. Winemiller, Eds. (Chapman and Hall, New York, 1995), pp. 286–297.
6. M. E. Power, *Arch. Hydrobiol.* **125**, 385 (1992).
7. *Dicosmoecus* exposed to seven predator species in 12.6-liter tubs (six replicates per predator species; 10 grazers and one predator per tub) experienced no mortality over 24 hours, whereas a representative grazer, the mayfly *Centroptilum*, experienced

50.7 ± 24.5% (SD) mortality due to predation pooled across all replicates of all predator species.

8. J. H. Connell, *Am. Nat.* **122**, 661 (1983); T. W. Schoener, *ibid*, p. 240.

9. We visually counted all fish (*Hesperoleucas symmetricus* and *Gasterosteus aculeatus*) and odonates (primarily *Aeshna californica and Archilestes californica*) in the entire channel and counted all grazing insects [predominantly *Centroptilum, Nixe*, and *Paroleptophlebia* (Ephemeroptera); *Mysticides, Lepidosoma, Tinodes*, and *Gumaga* (Trichoptera); *Ferrissia* and *Physella* (Pulmonata); and *Eubrianix* (Psephenidae)], except for midge (Chironomidae) larvae, on the top and bottom of each tile. Midges were counted on only one tile because of their high abundance. We collected benthic algae (largely a mixture of Cladophora glomerata, Nostoc spp., and Epithemia spp.) from one randomly selected tile in each treatment by scraping the tile with a razor blade and measuring ash-free dry weight in the laboratory. All floating algal mats in a channel were collected with an aquarium net, spun in a salad spinner for a standard 50 turns to remove excess water, and weighed. Subsamples of floating algae were collected, dried, weighed, and analyzed through measurement of ash-free dry weight to calibrate wet mass with ash-free dry weight. We used a blocked, two-way MANOVA to test for community-wide differences among treatments, and then tested for the specific differences predicted a priori by the model, using one-tailed paired t tests.

10. L. B. Crowder and W. E. Cooper, *Ecology* **63**, 1802 (1982); M. E. Power, *Oikos* **58**, 67 (1990).

11. J. T. Wootton, *Am. Nat.* **141**, 71 (1993).

12. Methods are described in [6]. Data were taken from survey locations described in [6] and from surveys of reaches of the Mad River (regulated) and Van Duzen River (unregulated) in Six Rivers National Forest, CA, conducted throughout the summer of 1994. All variables except algal occurrence were log-transformed before analysis to stabilize the variance.

13. M. Waldichuk, *Can. Bull. Fish. Aquat. Sci.* **226**, 295 (1993); A. G. Maule, C. B. Schreck, C. S. Bradford, B. A. Barton, *Trans. Am. Fish. Soc.* **117**, 245 (1988); H. L. Raymond, *N. Am. J. Fish. Manage.* **8**, 1 (1988).

14. R. W. Nelson, J. R. Dwyer, W. E. Greenberg, *Environ. Manage.* **11**, 479 (1987).

15. F. K. Ligon, W. E. Dietrich, W. J. Trust. *Bioscience* **45**, 183 (1995); M. E. Power, A. Sun, G. Parker, W. E. Dietrich, J. T. Wootton, *ibid.*, p. 159; C. S. Holling and G. K. Meffe, *Conserv. Biol.* **10**, 328 (1996).

16. We thank B. Amerson, C. Bailey, J. Chase, M. Eskridge, D. Gordon, N. Guthrie, S. Kupfer-burg, S. Lane, M. Liu, J. Lyons, J. Marks, S. McGuire, K. Meier, E. Noonberg, C. Pfister, M. Pizer, W. Roberts, M. Salzer, A. Sun, C. Wang, and J. Wootton for field assistance and P. Steel for logistical support. Funded in part by NSF, the California State Water Resources Center, the Miller Institute for Basic Research, and the University of Chicago Block Fund.

Exercises for Chapter 10

In the following exercises, all equation numbers refer to Chapter 10.

1. Compute the matrix $\mathbf{D}$ for each of the four equilibrium points for Equation (5). Compute $\det(D)$ and $\operatorname{trace}(D)$ at each of these four equilibrium points and then verify the assertions above concerning the stability of these equilibrium points.

2. As remarked, the α, β, γ, σ, δ, and ρ that appear in (4) must all be determined by experiment. For each of these constants, suggest an experiment (say, in a greenhouse) that can be used to determine the given constants value. (You needn't use more than a sentence or two to describe each experiment. Try to translate the abstract differential equation to the real world.)

3. Compute the partials $\frac{\partial^2 h}{\partial x^2}$ and $\frac{\partial^2 h}{\partial x \partial y}$ and also $\frac{\partial^2 h}{\partial y^2}$ and $\frac{\partial^2 h}{\partial y \partial x}$ for the examples of h in (12). Use the results to verify that the equality in (15) holds.

4. Integrate the function $h(x, y)$ in the cases given over the indicated rectangle.

    (a) $h(x, y) = 1$ over the rectangle where $0 \leq x \leq 1$ and $-1 \leq y \leq 2$

    (b) $h(x, y) = x$ over the rectangle where $-1 \leq x \leq 1$ and $0 \leq y \leq 1$

    (c) $h(x, y) = x + y$ over the rectangle where $0 \leq x \leq 1$ and $0 \leq y \leq 1$

    (d) $h(x, y) = xy$ over the rectangle where $-2 \leq x \leq 1$ and $2 \leq y \leq 3$

    (e) $h(x, y) = \cos(xy)$ over the rectangle where $0 \leq x \leq 1$ and $0 \leq y \leq 1$

11

Matrix Notation

The purpose of this chapter is to introduce some concise and often used notation for the following differential equation for functions $x(t)$ and $y(t)$ of the variable t:

$$\frac{d}{dt}x = ax + by,$$
$$\frac{d}{dt}y = cx + dy. \tag{1}$$

Here, a, b, c, and d are constants. In vector notation, this is the equation for the vector function of time $\mathbf{v}(t) = \begin{pmatrix} x(t) \\ y(t) \end{pmatrix}$ that obeys the equation

$$\frac{d}{dt}\mathbf{v} = \mathbf{f}(\mathbf{v}), \tag{2}$$

where $\mathbf{f}(\mathbf{v}) = \begin{pmatrix} ax + by \\ cx + dy \end{pmatrix}$ is a vector whose entries are functions of the vector $\mathbf{v}$.

11.1   Multiplying Matrices Against Vectors

The important thing to note is that the equation in (1) or, equivalently, that in (2) depends on the numbers a, b, c, and d. As I stressed in previous chapters, it proves convenient to think of these four numbers as forming the matrix

$$\mathbf{M} = \begin{pmatrix} a & b \\ c & d \end{pmatrix}. \tag{3}$$

(Properly, this $\mathbf{M}$ is a 2×2 matrix since it is a square array with each row having two entries and with each column having two entries. We can imagine $p \times q$ arrays with p equal to the number of rows and with q equal to the number of columns.)

The point here is that the right-hand side of Equation (1) provides a formula that combines the entries of a vector (here, x and y) with those of the matrix in (3) to obtain a new vector. This formula can be viewed as one that provides a generalization of multiplication of two numbers to produce a new number; here, a matrix is "multiplied" against a vector to produce a new vector. To be precise, use $\mathbf{v}$ to denote that the vector is $\mathbf{v} = \begin{pmatrix} x \\ y \end{pmatrix}$ and then the multiplication rule gives (by its very definition)

$$\mathbf{Mv} = \begin{pmatrix} ax + by \\ cx + dy \end{pmatrix}. \tag{4}$$

The fact is that matrices were invented to simplify the notational complexity of vector transformations that send a vector $\mathbf{v}$, as previously, to the new vector $\begin{pmatrix} ax + by \\ cx + dy \end{pmatrix}$.

Note that when multiplying a matrix times a vector, the notation here puts the matrix first and the vector second. This is the convention, one that will be followed strictly. (A mathematician will get confused if you put the vector before the matrix.)

In any event, with $\mathbf{Mv}$ understood as in (4), then (1) has the following concise form:

$$\frac{d}{dt}\mathbf{v} = \mathbf{Mv}. \tag{5}$$

11.2   Manipulating Matrices

The notion in (4) of multiplying a matrix against a vector is not the only reasonable, algebraic operation that can be performed with matrices. In particular, the remainder of this chapter describes some of the more common ones that you might run into.

11.3   Adding Matrices Together

A pair of matrices (both 2×2, or both having the same number of rows as each other and likewise the same number of columns) can be added to each other to get a third matrix. For example, if $\mathbf{M}$ is given by (3) and if $\mathbf{M}' = \begin{pmatrix} a' & b' \\ c' & d' \end{pmatrix}$, then

$$\mathbf{M} + \mathbf{M}' = \begin{pmatrix} a + a' & b + b' \\ c + c' & d + d' \end{pmatrix}. \tag{6}$$

That is, matrices are added by adding individually each entry.

11.4 Multiplying Matrices by a Number

A matrix can also be multiplied by a real number, r. For example,

$$r\mathbf{M} = \begin{pmatrix} ra & rb \\ rc & rd \end{pmatrix}. \tag{7}$$

That is, multiplying a number against a matrix simply multiplies each entry by that number. Note that $r(\mathbf{M} + \mathbf{M}') = r\mathbf{M} + r\mathbf{M}'$.

11.5 Multiplying Matrices by Matrices

There is also a way to multiply a pair of matrices together to obtain a third matrix. Here is how:

$$\mathbf{M}\mathbf{M}' = \begin{pmatrix} aa' + bc' & ab' + bd' \\ ca' + dc' & cb' + dd' \end{pmatrix}. \tag{8}$$

It is important to remember that $\mathbf{M}\mathbf{M}' \neq \mathbf{M}'\mathbf{M}$ in general. Consider, for example, the case where $\mathbf{M} = \begin{pmatrix} 0 & 1 \\ 0 & 0 \end{pmatrix}$ and $\mathbf{M}' = \begin{pmatrix} 0 & 0 \\ 1 & 0 \end{pmatrix}$. In this case,

$$\mathbf{M}\mathbf{M}' = \begin{pmatrix} 1 & 0 \\ 0 & 0 \end{pmatrix} \quad \text{and} \quad \mathbf{M}'\mathbf{M} = \begin{pmatrix} 0 & 0 \\ 0 & 1 \end{pmatrix}.$$

11.6 Diagonal Matrices

A matrix $\mathbf{M}$ is called *diagonal* when $\mathbf{M} = \begin{pmatrix} a & 0 \\ 0 & d \end{pmatrix}$.

11.7 Determinants and Traces

Remember that the determinant of the matrix $\mathbf{M}$ is

$$\det(\mathbf{M}) = ad - bc. \tag{9}$$

Meanwhile, the trace of the matrix $\mathbf{M}$ is

$$\mathrm{tr}(\mathbf{M}) = a + d. \tag{10}$$

11.8   Eigenvectors and Eigenvalues

When **M** is a matrix, a vector **w** is said to be an *eigenvector* of **M** if the result of multiplying **M** against **w** gives a vector that is a real number multiple of **w**. This is to say that **w** is an eigenvector for **M** when

$$\mathbf{Mw} = r\mathbf{w}, \qquad (11)$$

where r is some real number. Moreover, if (11) holds, then the number r is called an *eigenvalue*. (You might notice that if **w** is an eigenvector, so is any nonzero, real number multiple of **w**; but these all have the same eigenvalue. Of course, the vector **0** with all entries 0 is an eigenvector of every matrix, so the convention is to ignore this trivial case.)

If you find an eigenvector with its eigenvalue, then you have also found a solution to Equation (1) [or (5)]. Indeed, if **w** obeys (11), then

$$\mathbf{v}(t) = \alpha e^{rt}\mathbf{w} \qquad (12)$$

obeys (5) for any constant, real number α. This is to say that the components of $\mathbf{v}(t)$ obey (1).

For example, diagonal matrix $\mathbf{M} = \begin{pmatrix} a & 0 \\ 0 & d \end{pmatrix}$ has two non-proportional eigenvectors, namely $\mathbf{w}_1 = \begin{pmatrix} 1 \\ 0 \end{pmatrix}$ and $\mathbf{w}_2 = \begin{pmatrix} 0 \\ 1 \end{pmatrix}$. The corresponding eigenvalues are a and d, respectively. Thus, $\mathbf{v}(t) = e^{at}\mathbf{w}_1$ and $\mathbf{v}(t) = e^{dt}\mathbf{w}_2$ both obey (5) and (1).

Eigenvectors and eigenvalues often enter discussions of differential equations having the form of (5) because if you find them, then you know some solutions to (5). Most standard linear algebra texts devote a good deal of space to the problem of finding eigenvectors and eigenvalues of matrices, but I will say no more about them here. (However, these terms do enter again in some later chapters, but in a very different context and with a different, though not unrelated, meaning.)

11.9   Lessons

Here are some key points from this chapter:

- Remember what a matrix is.

- Keep in mind how to add matrices, multiply them by numbers, multiply them against vectors, and multiply them against each other.

- Know how to compute determinants and traces.

Exercises for Chapter 11

1. Compute **Mv** for

   (a) $\mathbf{M} = \begin{pmatrix} 1 & 2 \\ 0 & 1 \end{pmatrix}$ and $\mathbf{v} = \begin{pmatrix} 1 \\ 1 \end{pmatrix}$

   (b) $\mathbf{M} = \begin{pmatrix} 3 & 2 \\ -5 & 4 \end{pmatrix}$ and $\mathbf{v} = \begin{pmatrix} 1 \\ 3 \end{pmatrix}$

   (c) $\mathbf{M} = \begin{pmatrix} 0.5 & 0.2 \\ 8 & 0.4 \end{pmatrix}$ and $\mathbf{v} = \begin{pmatrix} 0 \\ 1 \end{pmatrix}$

2. Find **MM'** and **M'M** when

   (a) $\mathbf{M} = \begin{pmatrix} 1 & 2 \\ 0 & 1 \end{pmatrix}$ and $\mathbf{M}' = \begin{pmatrix} 2 & 5 \\ 1 & 6 \end{pmatrix}$

   (b) $\mathbf{M} = \begin{pmatrix} 0 & 9 \\ 3 & -4 \end{pmatrix}$ and $\mathbf{M}' = \begin{pmatrix} 2 & 5 \\ 1 & 1 \end{pmatrix}$

   (c) $\mathbf{M} = \begin{pmatrix} 2 & 5 \\ 1 & 6 \end{pmatrix}$ and $\mathbf{M}' = \begin{pmatrix} 0 & 9 \\ 3 & -4 \end{pmatrix}$

3. Compute both determinant and trace for **M** in Exercises 2(a), 2(b), and 2(c).

4. Verify that $\mathbf{v} = \begin{pmatrix} 1 \\ 0 \end{pmatrix}$ is an eigenvector for the matrix in Exercise 1(a). Find its eigenvalue.

12

Remarks About Australian Predators

You should read Reading 12.1 (page 181, from *Natural History*, by Tim Flannery) before reading this chapter. The predator situation in Australia as described in Flannery's article might be modeled by the following differential equation for $k(t)$ = number of kangaroos at time t (the prey) and $p(t)$ = number of predators at time t:

$$\frac{d}{dt}\begin{pmatrix} k \\ p \end{pmatrix} = \begin{pmatrix} \alpha k - \beta k^2 - \gamma kp \\ -\sigma p + \lambda kp \end{pmatrix},\tag{1}$$

where α, β, γ, σ, and λ are constants.

12.1   The Phase Plane

The phase plane analysis starts with the k null clines. These occur where

$$k = 0 \quad \text{or where} \quad p = (\alpha - \beta k)/\gamma.\tag{2}$$

The p null clines occur where

$$p = 0 \quad \text{or where} \quad k = \sigma/\lambda.\tag{3}$$

Note that the only biologically relevant part of the phase diagram is the region where $k \geq 0$ and $p \geq 0$. The diagram in Figure 12.1 is sketched for the following two cases: Case 1 has $\alpha/\beta < \sigma/\lambda$. Case 2 has $\alpha/\beta > \sigma/\lambda$.

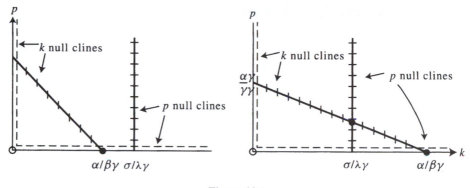

Figure 12.1

The equilibrium points in the two cases are circled. They are as follows:

$$k = 0, \ p = 0 \tag{4a}$$
$$k = \alpha/\beta, \ p = 0 \tag{4b}$$
$$k = \beta/\lambda, \ p = \alpha/\gamma - \beta\sigma/(\lambda\gamma). \tag{4c}$$

12.2   The Large Mammal Case

Notice that the third equilibrium point has no biological significance ($p < 0$) when

$$\alpha/\beta < \sigma/\lambda. \tag{5}$$

Assume for the moment that (5) holds. The surprise is that this condition is precisely the condition that makes the equilibrium ($k = \alpha/\beta$, $p = 0$) stable. Indeed, the matrix of partial derivatives for this equilibrium point is

$$\mathbf{D} = \begin{pmatrix} -\alpha & -\gamma\alpha/\beta \\ 0 & -\sigma + \lambda\alpha/\beta \end{pmatrix}. \tag{6}$$

Since $\alpha > 0$, the determinant of $\mathbf{D}$ will be positive if and only if $-\sigma + \lambda\alpha/\beta < 0$, which is the same condition as in (5). This will also ensure that the trace of $\mathbf{D}$ is negative. Thus, when (5) holds, the equilibrium point $k = \alpha/\beta$ and $p = 0$ is stable. Thus, if Flannery's ideas in his *Natural History* article are correct, then the case where (5) holds must be the model for the case where the predator is a large mammal. Indeed, Flannery proposes that the only stable equilibrium when the predator is a large mammal has zero predators, and this is precisely what happens when (5) holds.

12.3   The Large Reptile Case

Now suppose that

$$\alpha/\beta > \sigma/\lambda. \tag{7}$$

In this case, the preceding discussion shows that the equilibrium point $k = \alpha/\beta$, $p = 0$ is now unstable. However, now the third equilibrium point, $[k = \beta/\lambda, \; p = \alpha/\gamma - \beta\sigma/(\lambda\gamma)]$ in Equation (4), is a reasonable alternative. To test its stability, consider the matrix of derivatives $\mathbf{D}$ again. Here,

$$\mathbf{D} = \begin{pmatrix} -\beta\sigma/\lambda & -\gamma\sigma/\lambda \\ (\lambda\alpha - \beta\sigma)/\gamma & 0 \end{pmatrix}. \tag{8}$$

In this case, trace($\mathbf{D}$) < 0 and det($\mathbf{D}$) $= \sigma(\lambda\alpha - \beta\sigma)/\lambda$, which is positive under the assumption in (7). Thus, under the assumption in (7), the third equilibrium point in Equation (4) is now stable.

   This last equilibrium point has a positive number of prey and a positive number of predators. Thus, if Flannery's proposal is correct, then the case where (7) holds must correspond to the case where the predators are large reptiles; for here we see that there is a stable equilibrium point with nonzero numbers of predators and prey.

12.4   Comments

Do the preceding conclusions contradict the thesis in the *Natural History* article? The coefficients α and β that appear in (5) and (7) are presumably determined by the species of prey, and so do not change with change of predator from large mammal to large reptile. With α/β fixed, the ratio σ/λ determines whether (5) or (7) holds. Large σ/λ (relative to the fixed value of α/β) means that (5) holds and small σ/λ means that (7) holds.

   Note that a species of long-lived animal will have small σ, while a short-lived animal will have large σ. A species of animal that efficiently utilizes food to make babies will have large λ, but a relatively inefficient animal will have small λ.

   Thus, the thesis of Flannery's *Natural History* article can be tested to some degree by measuring the ratio σ/λ for various mammalian predators and for various reptilian predators.

12.5   Lessons

Here are some key points from this chapter:

- Change the parameter values in a differential equation and the equilibrium points will generally move. Moreover, as an equilibrium point moves in response to a change of the parameter values, it can change from being stable to being unstable and vice versa.

- Mathematical modeling can sometimes be used to pinpoint key issues for experimental investigation. In particular, by varying parameters in a model, you may learn that some parameters are relevant to the given biological issue and some are not.

READINGS FOR CHAPTER 12

READING 12.1

The Case of the Missing Meat Eaters

Commentary: This article observes that Meganesia (Australia, New Guinea, and nearby islands) has historically lacked large mammalian predators. In fact, the dominant predators before the advent of humans have been rather large lizards. (Moreover, Australia has a particularly diverse and interesting snake population.) The author suggests a cause that relates ultimately to the low productivity of the soil. As a result (says the author), large herbivores are both rare and thinly spread. He suggests that such a distribution favors reptilian carnivores over mammalian ones. A mathematical model can be proposed, as in this chapter, that suggests certain experiments that can corroborate the author's thesis.

The Case of the Missing Meat Eaters

Natural History, June 1993, pages 22–24.
Tim Flannery

During ice ages, when sea level is low, Australia, Tasmania, New Guinea, and their smaller neighbors coalesce into a single great island. Dubbed Meganesia by scientists, this landmass covers an area of almost 4 million square miles and is a single geological entity, carried across the Southern Hemisphere by plate tectonics. Meganesian plants and animals thus share a common biological heritage, and even when the sea carves their great island homeland into discrete pieces, the flora and fauna retain their affinities. In addition to abundance of marsupial, or pouched, mammals and a dearth of placental mammals, one of Meganesia's most striking features is its extraordinary lack of large mammalian carnivores.

This unusual situation is perhaps best illustrated by going back some 60,000 years, before the arrival of humans in Australia. At that time, Meganesia was home to approximately sixty species of mammals that weighed more than twenty pounds. Of these, not more than three were meat eaters, and all are now extinct. Only two other warmblooded carnivores weighed more than ten pounds, and they still survive: the Tasmanian devil and the spotted-tailed quoll. Each of these species fills, or filled, a somewhat different ecological niche.

The Tasmanian devil, a scavenger and bone cruncher that takes whatever prey comes its way, is perhaps best described as a miniature marsupial hyena. Also a marsupial, the spotted-tailed quoll is weasel-like or civetlike in both appearance and in its stealthy behavior. The now-extinct thylacine (which survived in Tasmania until 1936) was roughly the size and shape of a wolf and was Meganesia's only doglike marsupial carnivore. Also extinct, the marsupial lion, one of the few carnivores to have arisen from herbivorous ancestors, had large, slicing premolars. This led some nineteenth-century scientists to speculate that it was a vegetarian that fed mainly on melons. But the discovery of a well-preserved fossil paw revealed that it was equipped with a big hooded claw, and the marsupial lion is now believed to have been an adept predator. Despite its common name, it was closer in size to a leopard than a lion and may have been the marsupial equivalent of the medium-sized cats on other continents.

The marsupial giant rat kangaroo weighed a hefty ninety pounds and stood some four feet tall but had teeth similar to those of much smaller insectivores. This beast lived throughout eastern Australia—in woodland, grassy steppeland, and savanna—during the last ice age. Its ecology is enigmatic, however. It may well have been an omnivore, eating plants, scavenging carcasses, and opportunistically preying on bird eggs and small vertebrates. If such an interpretation of its diet and habits is correct, this primitive kangaroo may have filled a niche similar to that of some small bears.

In the entire Australasian region, therefore, the broad carnivore niches were filled by just one mammal species each—doglike, catlike, civetlike, scavenging, and, possibly, bearlike animals. In contrast, even today the United States (the lower forty-eight of which are roughly the size of Australia) is inhabited by three bear species, five kinds of dogs, six kinds of cats, six species of weasels and their relatives, as well as raccoons, ringtails, and coatis. And the region's abundance of carnivores pales when compared with its fauna during the Pleistocene, when dire wolves, various bears, jaguars, cheetahs, lions, and saber-tooths also roamed the continent. This diversity of mammalian carnivores is by no means exceptional; Europe, Asia, Africa, and South America either did or still do support similarly diverse carnivore guilds. In all these regions, the broad cat and dog niches are subdivided according to size, prey type, and habitat, allowing many species to coexist.

Biologist have long speculated on the cause of the imbalance in the Meganesian mammal fauna. One of the most important limitations known to affect carnivores is simply the size of the landmass they inhabit. Alhough Australia, which comprises the bulk of Meganesia, is indeed the smallest continent, it is still about 3 million square miles in area. Yet the Meganesian carnivore assemblage is not much richer than that of the island of Madagascar, which is only one-twentieth the size of Meganesia.

Another school of thought holds that marsupials, having relatively small brains, were unable to evolve into successful predators. A quick look at the fossil record of South America, however, disproves the hypothesis that a connection exists between brain size and predation skill. Many species of doglike marsupials, ranging in size from bearlike to civetlike, lived in South America during the Tertiary period, about 65 to 2 million years ago. A remarkable subfamily of carnivorous marsupials evolved into catlike animals, resembling North American saber-tooths, that were capable of killing the largest of prey. The group that includes the ancestors of the American opossum also produced large flesh eaters. While all of these beasts became extinct when placental carnivores arrived in South America over the past 5 million years, they thrived for many millions of years, preying mainly upon large placental herbivores.

Since there appears to be no intrinsic bar to carnivory in marsupials, perhaps the environment holds a clue to Meganesia's paucity of large carnivorous mammals. Meat eaters sit at the apex of a broad-based food pyramid and are thus the most vulnerable of life forms to disturbances in the food chain. For example, an area of grassland that supports billions of individual grasses may sustain only a few thousand large herbivores. These, in turn, may be able to support fewer than one hundred large carnivores. If the environment is poor, large herbivores will be rare and thinly spread, and a critical point may be reached where the density of prey is so low that a population of large meateaters cannot be sustained. If further impoverished, such an environment can no longer support any large carnivores.

Australia is notoriously infertile. An old continent with a stable geological history, it has experienced no widespread glaciation, mountain building, or volcanic activity—the forces that create new soil—over the past 50 million years. As a result of its quiet past, Australia is a land of old, thin, and leached soils. In the country's semiarid zone, for example, soils have about half the levels of nitrates and phosphates of equivalent soils elsewhere. The amount and quality of arable land is another good measure of productivity, and even the 10 percent of Australia's total land area that is considered arable is marginal when compared with other landmasses. Other indications of poor soil come from Australia's plants, which have developed a variety of strategies, including slow average growth rates, to cope with the lack of nutrients.

A contributing environmental factor is El Niño, or Southern Oscillation cycle, which influences rainfall with a periodicity of roughly a decade. In some years, Australia receives high levels of rainfall, and productivity peaks, as it did in 1990. But in El Niño years, such as 1992, rainfall is reduced and prolonged droughts are likely. On no other continent does the cycle have such an extreme impact. Its effects can readily be seen in the high degree of nomadism and nonseasonal breeding in many Australian animals, particularly birds. When such variability is superimposed on a system that is already marked by low productivity, top-order carnivores are subject to exceptional stress.

These climatic factors have so shaped the biology of the region that even areas of rain forest lack big carnivores. The largest area of rain forest in Meganesia is in New Guinea, which is even more noteworthy than Australia for its lack of meat-eating mammals. Here we have no evidence of indigenous large catlike or scavenging predators. Before human settlement, New Guinea supported some two hundred species of rather small mammalian herbivores and insectivores, but was home to just one large warmblooded carnivore, the thylacine. Today, apart from humans, the largest predator is the bronze quoll, a two-pound, civetlike species.

If large meat-eating mammals are disadvantaged in such a system, might animals that require less food and energy fare better? Reptiles eat far less than mammals do, having no need to create inner body heat. They can survive long periods of food shortage and can exist at higher population densities than mammals, relative to their prey. Cold-bloodedness thus becomes a great boon to survival. I believe this is what has happened in Meganesia, home to a remarkable array of carnivorous reptiles. Before the arrival of humans, the largest carnivores in the region were *Wonambi*, a 110-pound pythonlike snake with a 12-inch girth; a giant land crocodile known as *Quinkana*; and a goanna—a kind of monitor lizard—called *Megalania*. Weighing as much as a ton, and more than twenty feet long, *Megalania* would have dwarfed present-day reptiles. Its nearest living relative is the Komodo dragon, which lives on a few small Indonesian islands adjacent to Australia. Although it weighs only a fraction as much as *Magalania*, the Komodo dragon is capable of killing goats, calves,

and even humans. *Megalania* would have been powerful enough to subdue diprotodons, the rhino-sized marsupial plant eaters that were the largest of all Australian mammals. *Wonambi*, the snake, occupied a far different ecological niche. It lived much farther south than large snakes do today and its remains are often found in rocks and caves. Its head was enormous; its jaws filled with hundreds of tiny teeth. It may have fed upon wombat- and wallaby-sized mammals. The least well known of these reptiles is the 10-foot long, 500-pound crocodile *Quinkana*. It seems to have been quite independent of water, for its fossils have been found in caves that contain only remains of terrestrial species. *Quinkana* had a large, boxlike snout and compressed, serrated teeth. It may have competed with young *Megalania* for the now extinct kangaroos and smaller diprotodons.

Are the climatic patterns of the Pleistocene and more recent times an aberration in the history of Meganesia? Paleontological research suggests that during much of the "age of mammals," and certainly since about 20 million years ago, Meganesia has been relatively resource poor and lacking in mammalian carnivores. On the other end of the spectrum, leading into historic times, Meganesia has been colonized by humans and, more recently, by animals introduced by them. How have the predators among them fared? The number of humans in Meganesia since people first crossed the sea from Asia some 40,000 years ago remained small prior to European settlement. Adaptable and omnivorous, humans also became the top predators: their hunting prowess probably led to extinction of all terrestrial vertebrate species that exceeded them in size, including all of the land carnivores larger than the thylacine. The dog, known as the dingo, introduced some 3,500 years ago, apparently drove both the thylacine and Tasmanian devil to extinction on the mainland. The success of other smaller, introduced predators such as the fox has been detrimental to native predators such as quolls. Humans, dingoes, and foxes have not caused a net increase in the number of mammalian carnivores in Meganesia; they have simply replaced the few existing warmblooded carnivore species. But today, Australia, Tasmania, and New Guinea can still boast a rich supply of reptiles—ten species of goannas and a further ten species of pythons that weigh at least ten pounds. This remains a record number of sizable coldblooded carnivores.

The case of the Meganesian meat eaters opens up new areas for exploration while reinforcing the view that because of its unusual climatic conditions and lone isolation, Meganesia is truly a separate experiment in evolution.

READING 12.2

Uniting Two General Patterns in the Distribution of Species

Commentary: This article attempts to explain two empirically derived relationships that seem to hold in some generality in population biology. The first is the "species-area" relation, which predicts that the number of species in an isolated habitat of area A is proportional to some power of A. The second is called the "distribution-abundance" relation, which predicts that the local abundance of a species with wide distribution is greater than that with narrow distribution.

My point in presenting this article is simply to focus your attention on Equation (1) of the article, which is a differential equation for a set of functions $\{p_{ij}(t)\}$ that gives the population of species labeled i on the island labeled with j. Here, i runs from 1 to some specified whole number and likewise j. Notice that the variable $C_i(t)$, which appears on the right-hand side of Equation (1), is later taken to be equal to an i-dependent constant times the sum over j of p_{ij} times the area of island j. This makes Equation (1) into a multicomponent version of our generic two-component equation, $\frac{d}{dt}x = f(x, y)$ and $\frac{d}{dt}y = g(x, y)$.

Uniting Two General Patterns in the Distribution of Species

Science, **275** (1997) 397–400.
Ilkka Hanski, Mats Gyllenberg

Two patterns in the distribution of species have become firmly but independently established in ecology: the species-area curve, which describes how rapidly the number of species increases with area, and the positive relation between species' geographical distribution and average local abundance. There is no generally agreed explanation of either pattern, but for both the two main hypotheses are essentially the same: divergence of species along the ecological specialist-generalist continuum and colonization-extinction dynamics. A model is described that merges the two mechanisms, predicts both patterns, and thereby shows how the two general, but formerly disconnected, patterns are interrelated.

The species-area (SA) curve is one of the few universally accepted generalizations in community ecology [1–3], but ecologists have failed to agree on the mechanisms that produce this pattern [3]. According to the habitat heterogeneity hypothesis, large areas have more species than small ones because of their greater range of distinct resources, which facilitates the occurrence of ecological specialists [3]. As an alternative, MacArthur and Wilson [2] advanced the dynamic theory of island biogeography, which predicts that species richness increases with area owing to decreasing extinction rate with increasing area.

Another general pattern in the distribution of species has been well documented only during the

---

Author affiliations: Ilkka Hanski: Department of Ecology and Systematics, University of Helsinki. M. Gyllenberg: Department of Mathematics, University of Turku.

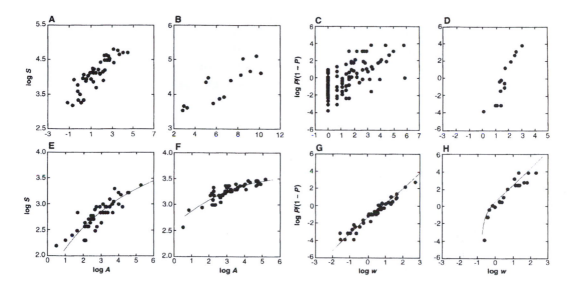

Fig. 1 Empirical and theoretical examples of approximately linear SA and DA curves. (**A** and **B**) The logarithm (base e) of species number against the logarithm of island area for (**A**) moths in a mainland-island system [28] and (**B**) birds in a metapopulation system [29]. (**C** and **D**) The respective DA curves, with the logarithm of $P/(1 - P)$ plotted against the logarithm of w [18]. The slope values and their standard errors (in parentheses) are (**A**) 0.29 (0.02), (**B**) 0.18 (0.03), (**C**) 0.83 (0.09), and (**D**) 3.16 (0.40). (**E** to **H**) Corresponding model-predicted results, in which the continuous lines give the expected values [19], and the dots give a stochastic realization obtained by assigning species to a set of 50 islands with their predicted incidences on these islands [15]. The following parameter values were used in all cases: $Q = 50$, $m_A = 3$, $\sigma_A = 1$, and $\sigma_w = 1.5$. To have comparable species numbers on the islands in the mainland-island and metapopulation models, we used the values of $c = 0.01$ in the former and $c = 0.00005$ in the latter, respectively.

past 15 years [4, 5]: species with wide distributions tend to be locally more abundant than species with narrow distributions. We call this relation the distribution-abundance (DA) curve. The two most widely recognized explanations of the DA curve are Brown's niche breadth hypothesis and metapopulation dynamics. According to Brown's hypothesis [5], generalist species, or species using ubiquitous resources [6], are both locally common and widely distributed, whereas specialists are constrained to have narrow distribution and tend to be locally uncommon. Metapopulation dynamic models predict that locally common species become widely distributed because of their low extinction rates and high colonization rates [7, 8]. High migration rates from existing large populations may additionally "rescue" small populations from extinction, in which case a wide distribution with many large populations tends to enhance average local abundance [7].

Surprisingly, although the two main hypotheses about the SA and DA curves are strikingly similar, the two patterns themselves have been studied without any reference to each other [9]. To bring conceptual unity to this area of ecology, we demonstrate that the SA and DA curves are both predicted by the same model, which furthermore merges the two "competing" hypotheses, namely, ecological specialization (habitat heterogeneity) and extinction-colonization dynamics.

To construct the model, consider a set of R islands [10] populated by a "pool" of Q species. The islands differ in area; we denote by m_A and σ_A^2 the mean and the variance of the logarithm of island areas (base e is used throughout this report). Likewise, the species differ in their abundances per unit area (density), with m_w and σ_w^2 denoting the mean and the variance of the logarithm of species densities [11]. By definition, the "carrying capacity" (equilibrium population size) of species i on

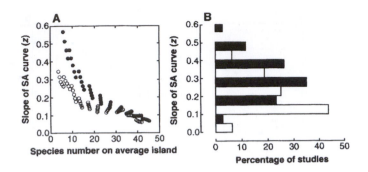

Fig. 2 (A) Plot of the slope of the SA curve against the species number on an averagesized island ($m_A = 3$) in the mainland-island (●) and metapopulation models (○), respectively. These results were obtained for the following parameter values: $R = 50$, $Q = 50$, $m_A = 3$, $\sigma_A = 1$, $\sigma_w = 1.5, 1.6, \ldots, 2.0$, and $\log c = -5, -4, \ldots, 5$ in the mainland-island model and $\log c = -15, -14, \ldots, -5$ in the metapopulation model. (B) Distribution of empirical z values, shown separately for data sets from archipelagoes [mainland-island situations (■), $n = 35$] and mainlands [classical metapopulations (open square), $n = 16$]. The difference between the two is significant at the 5% level. Data are from [30], omitting data sets in which areas covered less than three orders of magnitude and which hence can be expected to yield unreliable slope estimates.

island j is given by $K_{ij} = w_i A_j$, where w_i is the density of species i and A_j is the area of island j.

Following the standard approach to modeling metapopulation dynamics [7], we model changes in the probability $p_{ij}(t)$ of species i being present on island j at time t, in the absence of interspecific interactions, as

$$\frac{dp_{ij}}{dt} = C_i(t)[1 - p_{ij}] - \mu_{ij}p_{ij} \qquad (1)$$

where $C_i(t)$ is the colonization rate of empty islands and μ_{ij} is the extinction rate of extant populations. Empirical studies suggest that μ_{ij} is roughly proportional to $1/K_{ij}$ [12, 13], and we use this approximation below. The appropriate expression for $C_i(t)$ is different for two fundamentally different scenarios. In a mainland-island situation, the presence of species on islands is dependent on colonization from a permanent mainland community, where the density of species i is w_i. In this case, $C_i(t)$ is given by cw_i, where the value of parameter c decreases with increasing distance to the mainland. In contrast, in the classical metapopulation there is no external mainland, and the empty islands are colonized from other occupied islands [7, 14]. Here it is reasonable to assume that the colonization rate of empty islands is proportional to the pooled abundance of the species in the network of islands, and hence $C_i(t) = cw_i \sum p_{ij}(t)A_j$.

We focus in our analysis on the equilibrium value of $p_{ij}(t)$, which is called the incidence of species i on island j, J_{ij} [15]. The SA and DA curves are related to each other, because both curves are obtained by summing up the same set of incidences, but over different indices. The sum of the incidences across species gives the expected number of species on island j, S_j, whereas the sum of the incidences across islands gives the expected geographical distribution of species i, D_i. In the metapopulation literature, distribution is typically measured by the fraction of occupied islands, $P_i = D_i/R$ [7].

The most widely used statistical (descriptive) model of the SA curve is the power function model [1-3, 16], $S = kA^z$, which is generally used in the log-transformed form [3],

$$\log S = \log k + z \log A$$

This model has the obvious drawback of being unbounded, contrary to common sense and empirical results [13, 17]. However, ecologists have found that the logarithm of species number ($\log S$) generally increases roughly linearly with the logarithm of island area ($\log A$) for a large range of island areas (Figure 1, A and B). Empirical studies typically report the slope of the linear regression line, z.

The DA curve lacks a similar widely used statistical model. Here the empirical studies have been concerned with the demonstration that some sort

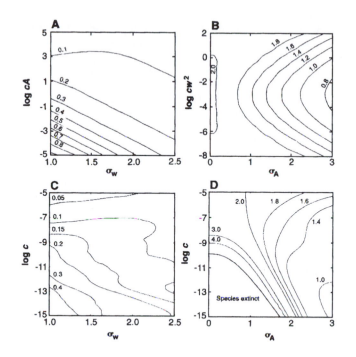

Fig. 3 Dependence of the model-predicted slopes of the SA and DA curves on parameter values. The contour lines join equal z values. (**A**) The slope of the SA curve in the mainland-island model as a function of cA and σ_w [19]. (**B**) The slope of the DA curve in the mainland-island model as a function of cw^2 and σ_A for $m_A = 3$ [19]. (**C**) The slope of the SA curve in the metapopulation model as a function of c and σ_w, based on numerical evaluation of the derivative at $m_A = 3$ [15]. The range of c values was selected to roughly match the values used in the respective mainland-island model. (**D**) The slope of the DA curve in the metapopulation model as a function of c and σ_A for $w = 1$, based on numerical evaluation of the derivative at $m_A = 3$ [15]. In the region marked "Species extinct" the value of $\log w$ is to the left of the vertical asymptote in Figure 1H [22], and the metapopulation goes extinct.

of positive relation exists [4, 5]. Considering a mainland-island situation, the rarest species on the mainland often do not occur on any of the islands ($P = 0$), whereas the commonest species are found on all ($P = 1$) or most islands. This observation suggests a logistic model for the DA curve, with P increasing from zero to 1 with increasing density, w, or, as we will assume here, with increasing logarithm of density, $\log w$:

$$P = \frac{1}{1 + e^{-a - b \log w}}$$

where a and b are two parameters. The logistic model can be linearized with the logit-transformation [18] (Figure 1, C and D).

We are now ready to examine the model-predicted SA and DA curves. For the mainland-

island situation, we have derived exact mathematical formulas for the slopes of the SA and DA curves [19]. The predicted SA curve is approximately linear for several orders of magnitude of island areas (Figure 1E). The slope of the curve is a function of two parameters, σ_w and cA [19], which can be estimated with empirical data [20]. The parameter combination cA, and therefore the value of the slope, is a function of the ratio of colonization to extinction probabilities [13, 21]. The slope of the model-predicted DA curve is also approximately linear for a large range of species' densities (Figure 1G), and it depends on three parameters: the mean (m_A) and the variance (σ_A) of island areas, and the parameter combination cw_i^2 [19].

For classical metapopulations without an external mainland, the slopes of the SA and DA curves can be calculated numerically [15]. The model generates approximately linear SA and DA curves (Figure 1, F and H), comparable to the respective curves for the mainland-island situation (Figure 1, E and G), except that the DA curve now has a vertical asymptote for small values of w [22] owing to an extinction threshold in classical metapopulation models [7].

Comparison of the mainland-island and metapopulation models (Figure 2A) yields distinctly smaller SA slopes for the metapopulation model for comparable parameter values. In contrast, the DA curve is expected to be steeper for classical metapopulations than for mainland-island systems (Figure 1, G and H, and Figure 3, B and D) owing to the low vertical asymptote in the former.

Despite its simplicity, the model makes several quantitative predictions. First, it predicts the slopes of the SA and DA curves on the basis of measurable ecological parameters: the moments of the species abundance and island area distributions, and species' incidences on islands (Figure 3). Parameter values that allow 20 to 80% of the species in the species pool to occur on an average-sized island generate z values ranging from 0.1 to 0.45 (Figure 2A), which is the typical range of empirical values (Figure 2B) [2, 3]. The model predicts that z increases with isolation, which is observed in empirical data [2, 3] (for exceptions, see below), and that z decreases with increasing variance of the species abundance distribution (Figure 3, A and C) [19].

Second, the model accounts for the empirically observed small z values on mainlands (Figure 2B) [2, 3] without resorting to an ad hoc assumption about "transient" species inflating the species number in small study areas [3]. In contrast, in the present model the small z values on mainland are due to the lack of rare species in the largest study areas; rare species go extinct from the entire network (Figure 1, H and D).

Third, the model explains the apparent exception of small z values in some very isolated archipelagoes [13]. The z value is predicted to be large for isolated islands if colonization occurs from the mainland, but small if colonization occurs among the islands, which may well happen in the most isolated archipelagoes.

Fourth, the model demonstrates that the exact shape of the species abundance distribution is not critical for realistic SA curves and z values [23], contrary to the suggestions of phenomenological

(sampling) models based on the canonical lognormal species abundance distribution [24]. What is critical is that there are some interspecific differences in abundance; previous dynamic models that ignored interspecific differences failed to predict realistic SA curves without making the implausible assumption of complete density compensation among competing species [13, 25]. Furthermore, interspecific differences in abundance must generate differences in species' distributions, as happens mechanistically in the present model. In contrast, the assumption by previous models of a random or nonrandom but fixed distribution of individuals in space [26] does not generate realistic SA curves and entirely fails to predict the effect of isolation on z.

In summary, not only can the SA and DA curves be predicted by the same model, the DA curve appears also to be a necessary ingredient of realistic SA curves. The model also predicts another widespread property of island communities, the nested-subset distribution of species on islands and habitat islands [27]. One open question that remains is to what extent the present results, based on a model of regional dynamics, can be extended to a continental scale.

REFERENCES

1. C. B. Williams, *Nature* **152**, 264 (1943); F. W. Preston, *Ecology* **29**, 254 (1948); R. P. McIntosh, *The Background of Ecology* (Cambridge Univ. Press, Cambridge, 1985).
2. R. H. MacArthur and E. O. Wilson, *The Theory of Island Biogeography* (Princeton Univ. Press, Princeton, NJ, 1967).
3. F. W. Preston, *Ecology* **41**, 611 (1960); E. F. Connor and E. D. McCoy, *Am. Nat.* **113**, 791 (1979); M. L. Rosenzweig, *Species Diversity in Space and Time* (Cambridge Univ. Press, Cambridge, 1995).
4. I. Hanski, *Oikos* **38**, 210 (1982); _____, J. Kouki, A. Halkka, in *Species Diversity in Ecological Communities: Historical and Geographical Perspectives*, R. E. Ricklefs and D. Schluter, Eds. (Univ. of Chicago Press, Chicago, 1993), pp. 108–116; K. J. Gaston and J. H. Lawton, *Oikos* **58**, 329 (1990); J. H. Lawton, *Trends Ecol. Evol.* **8**, 409 (1993); K. J. Gaston, *Rarity* (Chapman and Hall, London, 1994); _____, T. M. Blackburn, J. H. Lawton, in preparation.
5. J. H. Brown, *Am. Nat.* **124**, 255 (1984).
6. I. Hanski, *Oikos* **62**, 88 (1991).

7. M. Gyllenberg and I. Hanski, *Theor. Popul. Biol.* **42**, 35 (1992); I. Hanski and M. Gyllenberg, *Am. Nat.* **142**, 17 (1993).
8. S. Nee, R. D. Gregory, R. M. May, *Oikos* **62**, 83 (1991); L. A. Venier and L. Fahrig, *ibid.* **76**, 564 (1996).
9. To our knowledge, the only reference to a possible connection between the SA and DA curves is by I. Hanski, J. Kouki, and A. Halkka [in *Species Diversity in Ecological Communities: Historical and Geographical Perspectives,* R. E. Ricklefs and D. Schluter, Eds. (Univ. of Chicago Press, Chicago, 1993), p. 108].
10. We use the term "island" to refer both to true and habitat islands, and to what are called habitat patches, or fragments, in the metapopulation literature [7].
11. Without any loss of generality, we assume that m_w equals unity.
12. M. E. Gilpin and J. M. Diamond, *Proc. Natl. Acad. Sci. U.S.A.* **73**, 4130 (1976); I. Hanski, *Trends Ecol. Evol.* **9**, 131 (1994); *J. Anim. Ecol.* **63**, 151 (1994). In the latter two references, the more general relation $\mu_{ij} = e/K_{ij}^x$ is used, but we assume here for simplicity that $x = 1$, which is a good approximation for many species. The rate parameter e has been here absorbed in the unit of island area, to give the per-year extinction probability $1 - e^{-1/A}$ for species with $w = 1$. A scaling constant d is given by $d = A'/A$, where A' is island area in (say) kilometers squared. d may be calculated from knowledge of the per-year extinction probability and A'.
13. T. W. Schoener, in *16th International Ornithological Congress,* Canberra, Australia, 12 to 17 August 1974 (Australian Academy of Sciences, Canberra City, 1976), pp. 629–642.
14. Somewhat confusingly, the classical metapopulation scenario, where there is no external mainland, is referred to as the "mainland" regression in the species-area literature.
15. The equilibrium probability of species i occupying island j, p_{ij}^*, which is called the incidence, J_{ij}, is obtained from Equation 1 as

$$p_{ij}^* = J_{ij} = \frac{C_i^* w_i A_j}{C_i^* w_i A_j + 1} \qquad (2)$$

where C_i^* is the equilibrium value of $C_i(t)$. In the mainland-island model, where $C_i(t) =$

cw_i, we obtain

$$J_{ij} = \frac{cA_j w_i^2}{cA_j w_i^2 + 1} \qquad (3)$$

In the metapopulation model, the incidences can be calculated only numerically. Substituting Equation 2 into the expression $C_i^* = cw_i \sum J_{ij} A_j$, which gives the equilibrium value of $C_i(t)$ in the metapopulation model, we obtain

$$1 = cw_i^2 \sum_{j=1}^{R} \frac{A_j^2}{C_i^* w_i A_j + 1} \qquad (4)$$

from which C_i^* can be solved provided that $cw_i^2 \sum A_j^2 > 1$, which is a necessary and sufficient condition for species i to persist in the network of islands. The incidences can then be calculated from Equation 2.

16. O. Arrhenius, *J. Ecol.* **9**, 95 (1921); H. A. Gleason, *Ecology* **3**, 158 (1922); F. W. Preston, *ibid.* **41**, 611 (1960).
17. M. H. Buys, J. S. Maritz, J. J. A. Van Der Walt, *J. Veg. Sci.* **5**, 63 (1994); M. R. Williams, *Ecology* **76**, 2607 (1995).
18. The nonlinear logistic model can be linearized with the logit-transformation, $\log[P/(1 - P)] = a + b \log w$, which we apply throughout this report.
19. Assuming that Q (species number in the pool) is large and that $\log w$ is uniformly distributed with zero mean, we obtain after some calculation the expected number of species on island j as

$$S_j = \sum_i J_{ij} = \frac{Q}{4\sigma_w \sqrt{3}} \log \Gamma \qquad (5)$$

where (dropping the subscript j)

$$\Gamma = \frac{1 + cAe^{2\sigma_w \sqrt{3}}}{1 + cAe^{-2\sigma_w \sqrt{3}}}$$

The slope of the SA curve is then given by

$$\frac{\partial \log S}{\partial \log A} = \frac{1 - \Gamma^{-1}}{\Gamma \log \Gamma} \qquad (6)$$

The distribution of species i is given by

$$P_i = \frac{1}{R} \sum_j J_{ij} = 1 - \frac{1}{2\sigma_A \sqrt{3}} \log \frac{cw_i^2 + q_1}{cw_i^2 + q_2} \qquad (7)$$

and the slope of the DA curve is (dropping the subscripts)

$$\frac{\partial \log(P/[1-P])}{\partial \log w}$$

$$= \frac{2cw^2q}{(cw^2 + q_1)(cw^2 - q_2)P(1-P)} \quad (8)$$

where

$$q_1 = e^{-m_A + \sigma_A\sqrt{3}}, \quad q_2 = e^{-m_A - \sigma_A\sqrt{3}},$$

and

$$q = \frac{1}{2\sigma_A\sqrt{3}}[q_1 - q_2]$$

20. σ_w can be estimated as the standard deviation of the species abundance distribution on the mainland. In 10 examples of invertebrate and bird communities, empirical values ranged from 0.95 to 2.57 (mean = 1.63, SD= 0.62; details can be obtained from I. H. upon request). G. Sugihara [*Am. Nat.* **116**, 770 (1980)] has reported a wider range of values than reported here, but the data in his study do not always represent "local" communities (for example, birds in North America). Using Equation 3, one can express the parameter combination cA as $cA = w^{-2}J/(1-J)$, which suggests that, in principle, c can be estimated with data on island areas, species' abundances on mainland, and on their incidences on islands. To aid intuition about the cA values, consider an average species with $w_i = 1$. For such a species, $cA = J/(1-J)$, and hence the range of cA values in Figure 3A from 10^{-2} to 10^2 corresponds to the incidence on average-sized islands ranging from 0.01 to 0.99, which covers a very large range.

21. Denoting the per-year colonization and extinction probabilities by λ and μ, we obtain

$$cA = \frac{\lambda}{\mu}E[w^2]$$

where E denotes the expected value. That the slope of the SA curve depends on the ratio of colonization to extinction probabilities has been suggested by R. E. Ricklefs and G. W. Cox [*Am. Nat.* **106**, 195 (1972)] and M. P. Johnson and D. S. Simberloff [*J. Biogeogr.* **1**, 149 (1974)].

22. The vertical asymptote in Figure 1H is given by

$$\log w = -\frac{1}{2}\left(\log c + \log \sum_{j=1}^{R} A_j^2\right)$$

The other asymptote has a slope of 2.

23. We derived the slope values [19] also for back-to-back exponential and lognormal distributions of w. The expressions are more complicated in these cases, but the results are very similar.

24. F. W. Preston, *Ecology* **43**, 185 (1962); R. M. May, in *Ecology and Evolution of Communities*, M. L. Cody and J. M. Diamond, Eds. (Belknap, Cambridge, MA, 1975), p. 81.

25. C. Wissel and B. Maier, *J. Biogeogr.* **19**, 355 (1992).

26. B. D. Coleman, *Math. Biosci.* **54**, 191 (1981); H. Caswell and J. E. Cohen, in *Species Diversity in Ecological Communities: Historical and Geographical Perspectives*, R. E. Ricklefs and D. Schluter, Eds. (Univ. of Chicago Press, Chicago, 1993), p. 99.

27. B. D. Patterson, *Conserv. Biol.* **1**, 323 (1987).

28. M. Nieminen, *Oecologia* **108**, 643 (1996). The moth example includes 46 islands located about 15 km off the mainland in southwest Finland. Moths were trapped with sugar-bait traps in summer 1993. The parameter log w gives the abundance on mainland.

29. Z. Witkowski and P. Plonka, *Bull. Pol. Acad. Sci. Biol. Sci.* **32**, 241 (1984); W. B. Harms and P. Opdam, in *Changing Landscapes: An Ecological Perspective*, I. G. Zonneveld and R. T. T. Forman, Eds. (Springer-Verlag, New York, 1990), p. 73. The bird examples are from studies in Poland (Witkowski and Plonka) and The Netherlands (Harms and Opdam). "Islands" in this instance are habitat islands on mainland. The parameter log w in Figure 1H gives the estimated density of territories.

30. E. F. Connor and E. C. McCoy, *Am. Nat.* **113**, 791 (1979).

31. We thank F. Adler, J. Brown, M. Camara, E. Connor, M. Kuussaari, J. Lawton, A. Moilanen, and T. Schoener for comments on the manuscript. W. W. Murdoch and National Center for Ecological Analysis and Synthesis (NCEAS) in Santa Barbara are thanked for supporting this research.

Exercise for Chapter 12

1. With reference to the constants σ and λ that appear in Equation (1) at the start of this chapter, design an experiment that can measure the ratio σ/λ for large lizard and mammal predators. In this regard, try to think of an experiment that can be performed on captive animals, and then a second one for wild animals.

13

Introduction to Advection

In this chapter, we begin to consider equations that can be used to predict the behavior of quantities that depend on both time and space. That is, we wish to predict the values of some function $u(t, x)$ that is a function of t time and x space. Here is a simple example for the sort of analysis that lies ahead: Suppose that the wind is blowing from west to east at a constant velocity of 3 meters per second. Suppose that an explosion in a chemical warehouse has pumped particulate pollution into the air. Suppose that the particles fall out of the air at a constant percentage rate r. What do we need to know to predict the particle concentration east and west of the explosion as a function of time and distance from the explosion?

13.1 What Comes In Must Go Out

Let $u(t, x)$ be the function that gives the density (number of particles per meter) of particulate matter in the air at time t and point x along the west/east line through the warehouse. (Make the origin at the warehouse.) The following considerations lead to an equation for u: First, fix a point x and a small distance Δx. The amount of particulate matter in the region between x and $x + \Delta x$ at time t is given (approximately) by

$$u(t, x)\Delta x. \tag{1}$$

(This approximation becomes more and more accurate as Δx shrinks toward zero.)

192

The rate of change of (1) with respect to time is

$$\frac{d}{dt}(u(t, x)\Delta x) = q(t, x) - q(t, x + \Delta x) - k(t, x)\Delta x, \tag{2}$$

where

1. $q(t, x)$ is the number of particles per second that pass x from left to right minus the number or particles per second that pass x from right to left.

2. $q(t, x + \Delta x)$ is the number of particles per second that pass $x + \Delta x$ from left to right minus the number of particles per second that pass $x + \Delta x$ from right to left.

3. $k(t, x)\Delta x$ is the approximate number of particles that are created in the region between x and $x + \Delta x$ at time t minus the number that are destroyed in this same region at time t. [In the example below in Equation (2), $k(t, x) = -ru(t, x)$, but we can imagine more complicated terms here.]   (3)

It is important to realize that Equation (2) expresses nothing more than the tautology that the rate of change of the number of particles in the region between x and $x + \Delta x$ is given by

1. Adding the number of particles that enter across x and subtracting the number that leave across x [the term $q(t, x)$].

2. Subtracting the number that leave across $x + \Delta x$ and adding the number that enter across $x + \Delta x$ [the term $q(t, x + \Delta x)$].

3. Adding the number that are created and subtracting the number that are destroyed in the region between x and $x + \Delta x$ [this is given by the term $k(t, x)\Delta x$].

If we divide both sides of (2) by Δx and take the limit as Δx tends to zero, we obtain

$$\frac{\partial}{\partial t}u(t, x) = -\frac{\partial}{\partial x}q(t, x) + k(t, x). \tag{4}$$

[Remember from Chapter 9 that we have *defined*

$$\frac{\partial}{\partial x}q(t, x) = \lim_{\Delta x \to 0} \frac{q(t, x + \Delta x) - q(t, x)}{\Delta x}, \tag{5}$$

which explains the first term on the right-hand side of Equation (4).]

It is important for you to pause here to digest the fact that Equation (4), though perhaps very elegant, does no more than cryptically summarize the bookkeeping discussion in the paragraph that precedes it.

13.2   The Form of q

In particular, Equation (4) is not useful unless we can find a reasonable form for the functions $q(t, x)$ and $k(t, x)$. In the example of the explosion, we have already decided to take $k(t, x) = -ru(t, x)$ to account for the fact that the particles are lost at a constant rate r.

In our explosion example, there is also a relatively simple form for $q(t, x)$ if we assume that the motion of the particles is entirely due to their being pushed along by the wind. In this case, the number of particles that pass the point x from left to right at time t is given by $3u(t, x)$ particles per second; and the number of particles that pass from right to left is equal to zero. That is, if the wind is blowing from left to right at a speed of 3 meters per second, then the number of particles per second that pass the point x at time t is given by the density (in units of particles per meter) times the wind speed (in units of meters per second). To summarize, under the preceding assumptions, we should take

$$q(t, x) = 3u(t, x). \tag{6}$$

Please note the assumption we used to derive (6): The particles are moving only because of the motion of the wind. This is a reasonably valid assumption if the particles are heavy, but if the particles are very light, then we expect some random dissipative motion even without the wind blowing. (We will see in a subsequent chapter how to model the dissipative case.)

13.3   The Advection Equation

If we plug $k = -ru$ and $q = 3u$ into Equation (4), we find that the function $u(t, x)$ is predicted to be a solution to the *advection* equation

$$\frac{\partial}{\partial t}u(t, x) = -3\frac{\partial}{\partial x}u(t, x) - ru(t, x). \tag{7}$$

So, with the preceding assumption about the form for $q(t, x)$, the density function $u(t, x)$ (which is what we are interested in) is *constrained* in the sense that its partial derivatives in the t- and the x-directions are related according to (7). Thus, if we know the solutions to (7), then we know something about the form of our mysterious function u. [Equation (7) is our first example of what is often called a **partial differential equation**.]

Fact: Every solution to this equation can be written as

$$u(t, x) = e^{-rt} f(x - 3t), \tag{8}$$

where $f(x - 3t)$ means the following: Take any function f of one variable, say s. Then create the function of $f(x - 3t)$ of the variables t and x that is obtained by evaluating your original function f at the point $s = x - 3t$.

Note that there is a huge set of solutions to (7), for I can choose any function f to use in (8).

Here is how to prove that (8) solves (7): Simply compute the partial derivatives. As there are a number of them, this is lengthy process. In any event, here is how it goes:

1. $\frac{\partial}{\partial t} u(t, x)$: Compute this by holding x fixed (as if it were a constant number) to obtain a function just of t alone. Take the usual $\frac{d}{dt}$ of this function of t alone. Here is the result:

$$\frac{\partial}{\partial t} u(t, x) = \frac{\partial}{\partial t} \left(e^{-rt} f(x - 3t) \right) = \left(\frac{d}{dt} e^{-rt} \right) f(x - 3t) + e^{-rt} \frac{\partial}{\partial t} f(x - 3t)$$

$$= -r e^{-rt} f(x - 3t) - e^{-rt} 3 \left(\frac{d}{ds} f \right) (x - 3t). \tag{9}$$

(You have to use the chain rule to get the last term right.)

2. Compute $\frac{\partial}{\partial x} (e^{-rt} f(x - 3t))$ by holding t fixed to consider $e^{-rt} f(x - 3t)$ as a function only of x; then take the usual $\frac{d}{dx}$ of this function. Using the chain rule, we find

$$\frac{\partial}{\partial x} \left(e^{-rt} f(x - 3t) \right) = e^{-rt} \left(\frac{d}{ds} f \right) (x - 3t). \tag{10}$$

3. Now compare Equations (9) and (10) with (7) to see that (7) holds.

13.4   Initial Conditions

There are infinitely many solutions of (7), one for each choice of function $f(s)$ in (8). How do we determine the precise function $f(s)$ to use in (8)? Here is where the *initial conditions* enter. The term *initial condition* signifies the value of $u(t, x)$ as x varies but t is fixed at some predetermined time (say $t = 0$). We shall also meet the term *boundary condition*, which signifies the value of $u(t, x)$ as t varies while x is fixed at some predetermined value.

For example, suppose that the value of u at $t = 0$ has been determined a priori to be given as a function of x. Call this new function $g(x)$. That is, suppose that knowledge of $g(x)$, which is the time zero density, *has been given*. Then there is one and only one solution to (7) with $u(0, x) = g(x)$. This is the solution to (7) where the function f at the point s is given by taking $f(s) = g(s)$. That is,

$$u(t, x) = e^{-rt} g(x - 3t) \tag{11}$$

is the *only* solution to (7) that *obeys* the initial condition

$$u(0, x) = g(x). \tag{12}$$

Fact: The point here is that (7) predicts the values for u at all times t and at all points x after the values for u at time 0 and at all points x have been specified.

In the preceding explosion scenario, we can imagine that a satellite photo has been taken at time 0 (a few minutes after the explosion) and that the particle density $g(x)$ has been determined as a function of x at time 0 from the satellite photograph.

The point here is that the values $u(t_0, x)$ as x varies at fixed time $t = t_0$ completely determine the solution $u(t, x)$ to (7).

It also turns out that knowledge of the values $u(t, x_0)$ as t varies at fixed x_0 is sufficient to uniquely determine a solution $u(t, x)$ to (7).

Fact: In particular, (6) is predictive when the value of $u(t, x)$ at $x = 0$ and all times t has been specified.

Consider, for example, that $u(t, 0)$ has been a priori determined to be given by the function $h(t)$. Then there is one and only one solution to (7) at which $u(t, 0) = h(t)$ for all times t. This solution is given in the form of (8) by taking f to be $f(s) = e^{-rs/3}h(-s/3)$. That is, the unique solution to (7) that also obeys $u(t, 0) = h(t)$ is given by

$$u(t, x) = e^{-rt}e^{-r(x-3t)/3}h((3t - x)/3). \tag{13}$$

In the explosion example, if you measure the particle density as a function of time *at the origin*, then Equation (7) gives you the density as a function of time at *all* times t and at *all* points x. Note the tremendous savings of labor here, for to determine the particle density at all time t and at all points x, you need only place one particle detector at the origin and record the density there as a function of time. This will give you the function $h(t)$. The mathematics then determines $u(t, x)$ for you at any t and any x via Equation (13).

13.5   Traveling Waves

To get a feeling for how these solutions look, consider Equation (7) in the case where $r = 0$. As asserted previously, the general solution has the form $u(t, x) = f(x - 3t)$, where $f(\cdot)$ can be any function of one-variable and $u(t, x)$ is obtained from f by evaluating the latter at the point $x - 3t$. What does this say about u? Among other things, it says that the value of u at a point (t, x) is the same as that of u at time 0, but not at x, rather at $3t$ units to the left of x. Stated differently, if you were to walk to the right (increasing x) at the same speed as the moving air, speed 3, so your x-coordinate increases by the amount $3t$ after time t, then you would not see any change in the value of u. In this sense, the solution $u(t, x)$ to the $r = 0$ version of Equation (7) describes a concentration of particles that moves at speed 3 to the right but otherwise maintains its shape. Likewise, the versions of Equation (7) with $r \neq 0$ describe concentrations of particles that move at speed 3 to the right and either decrease ($r < 0$) or increase ($r > 0$) in time as they move. Some of the exercises for Chapter 13 provide practice visualizing this traveling wave phenomenon.

13.6   Lessons

Here are some key points from this chapter:

- There is a tautological differential equation, (4), that describes the time and space dependence of the density of particles moving in a fluid. Here, $q(t, x)$ takes into account the net flow of particles past x at time t, while $k(t, x)$ takes into account the net number of particles created and destroyed at x at time t.

- In the case where particle motion is due to the constant velocity flow of the ambient fluid, then $q = -cu$, where $c > 0$ if the flow is from left to right on the x-axis, and otherwise $c \leq 0$. With this choice of q, (4) is called an advection equation.

- The advection equation in (7) is predictive in the sense that you can specify what you want for the function u at time $t = 0$ as a function of x and then there is a unique solution to the equation that is your specified function of x at $t = 0$.

- Alternately, you can specify what you want for u at a particular value of x (say $x = 0$) for all time, and there is a unique solution $u(t, x)$ to (7) with the specified behavior at your chosen x-value.

- The solution to (7) resembles a traveling wave.

READINGS FOR CHAPTER 13

READING 13.1

Malaria: Focus on Mosquito Genes

Commentary: This is a Research News article that reports on a promising avenue for controlling mosquito-borne diseases such as malaria: Insert into the mosquito genome a gene that makes the mosquito stomach inhospitable for the malaria parasite. If this gene is inserted in a highly mobile part of the genome, it should spread rapidly through the mosquito population, thus eventually extinguishing the malaria parasite. For our purposes, this idea is interesting inasmuch as it introduces spatial dependence into a population biology, time evolution problem. To elaborate, suppose that this idea is workable and that you have set a timetable for conquest of malaria in central Africa in, say, 50 years. Here is a key question: Where in Africa should resistant mosquitos be installed in order for the resistance to spread throughout the target zone in 50 years? In this regard, note that it is impossible (due to the size of the region) initially to seed the whole of central Africa with resistant mosquitos.

   Do you think that the preceding question can be answered with the help of an advection equation? My sense is that it cannot.

Malaria: Focus on Mosquito Genes

Science, **261** (1993) 546–548.

Peter Aldhous

Researchers battling malaria and some other insect-borne diseases hope to engineer strains of insects that cannot carry the human pathogens they now transmit.

In the early 1960s, malaria seemed finally to be a disease on the run. Field workers armed with DDT were winning the battle against the disease by attacking its transmission routes, or vectors: malaria-carrying mosquitoes. But by the end of the decade, the tide of the battle had turned as the mosquitoes evolved resistance to DDT and other insecticides. Malaria again seemed unstoppable. Indeed, the mosquitoes even evaded another of science's wonder weapons, the technique of swamping insect populations with sterile males to slow down their rate of reproduction. The wild mosquitoes simply bred too fast to make the technique workable. It was "like bailing out the ocean with a sieve," says vector biologist Frank Collins of the Centers for Disease Control and Prevention (CDC) in Atlanta.

After these setbacks, most researchers all but abandoned hope of stamping out insect borne diseases by focusing on the insects themselves and turned instead to developing drugs and vaccines—only to find themselves confronted by another wily foe, the malaria parasite, which developed drug resistance and evaded candidate vaccines (*Science*, **21** January 1990, p. 399). But now biologists are mounting new campaigns against malaria-carrying mosquitoes and other insect disease vectors. They have a powerful new weapon—genetic engineering—and a new strategy: Instead of trying to eradicate the insects, vector biologists hope to produce transgenic strains that are incapable of transmitting disease. Researchers are already searching for genes that can help insect vectors resist infection with human pathogens, and they are looking for ways to spread these genes through insect populations in the wild.

The stakes are high: Malaria alone kills more than a million people each year in Africa, where it is carried by the mosquito *Anopheles gambiae*, which has been notoriously difficult to control by conventional means. (Strategies involving insect control and antimalaria drugs have, however, been more effective elsewhere in the tropics, where malaria

is transmitted by other types of mosquito.) "The [*An. gambiae*] mosquito itself is the problem," says malariologist Louis Miller of the National Institute of Allergy and Infectious Diseases (NIAID).

This new focus on insect vector genetics owes much to the MacArthur Foundation—and, in Britain, the Wellcome Trust—which backed this approach in the late 1980s, before other agencies recognized its potential. The MacArthur program, in particular, invigorated the field, bringing in heavyweights from neighboring disciplines, such as *Drosophila* geneticist Fotis Kafatos, now director-general of the European Molecular Biology Laboratory. And this week sees a major conquest for the program, with the publication on page 605 of a linkage map of the X chromosome of *An. gambiae*, produced by Kafatos' lab at Harvard University and CDC's Collins. When complete, the map may lead researchers to genes that influence the interaction between mosquitoes and the protozoan malaria parasite it carries—and perhaps to key genes that can be manipulated to reduce malaria transmission.

Within a year, Kafatos expects to have achieved another victory by extending the linkage map to the other two *An. gambiae* chromosomes. Armed with this and a physical genome map, completed in Kafatos' Harvard University lab in 1991, biologists will be "in a position to map [*An. gambiae*] traits very effectively," he says. And the malaria mosquito is not the only mosquito to have a genome project all its own: In the current issue of the *Journal of Heredity*, a team led by Dave Severson of the University of Wisconsin describes a linkage map for *Aedes aegypti*—which transmits the dengue and yellow fever viruses.

Antimalaria mosquitoes

One of the first hints that the key to malaria control may lie in the mosquito genome came in the mid-1980s, when Collins bred a strain of *An. gambiae* that disables malaria ookinetes, the form of the parasite that crosses the mosquito gut wall. Once ookinetes get into the insect's body cavity, they attach to the gut and develop into the so-called oocysts, which in turn rupture after a cou-

---

Insect-Borne Viruses: Help From Plants

Biologists studying insect-borne viral diseases may have a head start over their colleagues working on diseases like malaria that are spread by protozoan parasites. The pathogenic human viruses transmitted by insects are similar genetically to those that infect plants. And because "a lot of work has been done on plant viruses," says virologist Stephen Higgs of Colorado State University, viral vector biologists are able to borrow ideas directly from genetic strategies that have been used successfully to protect plants from disease.

Higgs and his Colorado State colleagues are hoping to tackle viruses like dengue and yellow fever, which are transmitted by the mosquito *Aedes aegypti*. First, however, they are honing their techniques by working on the La Crosse virus, which causes an occasionally fatal form of encephalitis in the midwestern United States and is transmitted by the *Aedes triseriatus* mosquito. To introduce foreign genes in *Aedes* cells, the researchers use a modified Sindbis virus—a relatively harmless virus that infects a wide range of insects. The foreign genes are hooked up to a promoter sequence and spliced into a clone of Sindbis complementary DNA. Once inside *Aedes* cells, this cDNA produces infectious Sindbis virus and expresses the foreign genes.

Using this method, the Colorado State team has been studying suspected antiviral genes. "Antisense strategies seem to work very well," says molecular biologist Ken Olson. Taking a cue from studies with transgenic plants, the team has spliced a "back to front" version of the gene for the coat protein of the La Crosse virus into the Sindbis cDNA. The antisense cDNA produced by the recombinant Sindbis, says Olson, seems to interfere with the corresponding La Crosse virus mRNA and almost completely prevents replication of the La Crosse virus inside *Aedes* cells. The next step is to feed blood containing the recombinant Sindbis cDNA to live *A. triseriatus* mosquitoes to see if this reduces their ability to transmit La Crosse encephalitis.

Unfortunately, the Sindbis virus system does not carry the genes into the mosquito genome. So, just as in malaria vector biology, there is no immediate prospect of making a transgenic insect capable of spreading antiviral genes through wild insect populations. But when the transgenic technology becomes available, the Colorado State researchers hope to have an armory of antiviral genes at the ready.

---

ple of weeks to release large numbers of sporozoites, the stage of the malaria parasite that infects humans. Collins' malaria-resistant mosquitoes halt the process at an early stage by imprisoning ookinetes in a capsule made from the pigment melanin—a spectacular effect that is due to the enhancement of the mosquitoes' normal defenses against infection. Simple genetic modifications—such as attaching stronger promoter sequences to the genes responsible—could enhance this effect against malaria. But first Collins has to find those genes—and that is where the maps come in. He has already narrowed down the search to a region occupying about 10% of the *An. gambiae* genome and even has a candidate gene: a gene for a serine protease enzyme thought to be involved in the encapsulation response.

Collins and medical entomologist Susan Paskewitz of the University of Wisconsin have now cloned the serine protease gene from the malaria-resistant strain, and Collins aims to use the new genome maps to see if it maps to the same precise position as the enhanced encapsulation trait. Collins rates his chances of having stumbled across the correct gene on his first attempt as a "very long shot." But he is confident that the new maps will eventually lead him to the genes that underlie the strain's par-

asite resistance.

Other groups are plotting a different line of attack: Rather than searching for antimalaria genes in *An. gambiae* itself, they want to introduce foreign genes into the mosquito to make it resist the malaria parasite. Molecular geneticist Julian Crampton of the Liverpool School of Tropical Medicine and cell biologist Robert Sinden of London's Imperial College may have the ideal candidate: a mammalian immune system gene that produces an antibody against a malaria antigen carried by the ookinete. Crampton and Sinden's group are now cloning the gene for a fragment of the antibody that could be produced in insect cells, and which by itself disables the ookinete.

A particularly elegant part of the team's strategy is that they may be able to switch on the gene at just the right time and place. This possibility stems from work by a third arm of the collaboration, a group led by molecular biologist Andrea Crisanti of Rome's La Sapienza University. Crisanti's group has sequenced the genes for two of the mosquito's digestive enzymes, called trypsins, that are released into its midgut just as the malaria parasite is trying to cross the gut wall. The promoter sequences from these enzyme genes could act as a trigger if linked to the antibody gene, releas-

Bacteria May Provide Access to the Tsetse Fly

As geneticists search for ways to produce transgenic insects capable of spreading new genes through wild mosquito populations (see main text), tsetse fly biologists are trying another tactic. The tsetse fly *Glossina morsitans* not only transmits African trypanosomes, the protozoan parasites that cause sleeping sickness, but it also carries symbiotic bacteria, or symbionts, inside the cells of its gut and other tissues. This has led researchers to think about creating "pseudo-transgenic" tsetses by inserting genes into the symbionts rather than directly into the fly's own genome.

The possibility of using this approach was raised in 1987, when entomologist Ian Maudlin and biochemist Susan Welburn of the University of Bristol, England, cultured one of the species of bacterium found in the tsetse gut. And last year, researchers at Yale University brought it one step closer to reality when they genetically transformed the bacterial symbiont using a plasmid, a loop of DNA, carrying genes that confer resistance to antibiotics. The Bristol and Yale groups are now working to isolate genes that might confer resistance to trypanosomes. If such genes could be inserted into the symbiont, it may be possible to produce tsetses that are "immune" to trypanosomes.

Maudlin's group is focusing on a particular biochemical interaction among parasite, vector, and symbiont: Cells in the tsetse midgut secrete a lectin protein that can kill trypanosomes before they leave the gut. But this is inhibited by a sugar called N-acetyl-D-glucosamine, which is in turn produced when a chitinase enzyme secreted by the symbiont breaks down the tough chitin lining of the tsetse's gut wall. As a result, Maudlin has found that flies carrying large numbers of symbionts are more susceptible to trypanosomes. He reasons, however, that if he can find a way to remove the sugar or produce more lectin, he could block the trypanosome's life cycle. Maudlin's group is now working on ways to alter the symbiont, either by inserting the gene for the tsetse lectin or by adding a gene to make the symbiont mop up the lectin-inhibiting sugar.

The Yale tsetse group, led by molecular parasitologist Serap Aksoy, is taking a different tack. It is hoping to coax the symbiont into expressing a gene from a mammalian immune system. Angray Kang of the Scripps Research Institute in La Jolla, working in collaboration with Aksoy's group, aims to isolate the gene that produces an antibody fragment that attacks a trypanosome antigen called procyclin. The Yale group will then splice it into the symbiont to produce a transformed bacterium that should kill trypanosomes.

Even if both groups succeed in producing trypanosome-killing symbionts, however, they still face a tough problem: how to get the bacteria back into the fly. The normal symbionts would first have to be flushed out with an antibiotic, but this would also remove a second symbiotic bacterium carried inside specialized cells in the tsetse gut. This symbiont is believed to supply the tsetse with an essential nutrient—and because it has not yet been cultured there is currently no way to reintroduce it into the flies.

A second conundrum is how to spread engineered symbionts through wild populations. Tsetse symbionts are passed from female flies to their offspring, but this alone will not drive a transformed symbiont through a large wild population. The Yale researchers, however, believe that they may have found a solution. Using a probe that recognizes a specific genetic sequence, they have identified a third bacterium, called *Wolbachia pipientis*, in the ovaries of female tsetses. In other insect species, *Wolbachia* exerts an effect called cytoplasmic incompatibility: Female insects infected with *Wolbachia* can mate with both infected and uninfected males, but females lacking the bacterium produce viable offspring only with uninfected mates. As a result, *Wolbachia* can spread rapidly through an insect population—and some strains maximize their transmission at the expense of others, through similar incompatibility effects.

The Yale researchers do not yet know if *Wolbachia* will cause mating incompatibility in the tsetse. But if so, says Aksoy, it could be used to drive engineered symbionts through a population, as *Wolbachia* and an altered symbiont would be passed down together from females to their offspring. "You could couple the two bacterial systems to spread genes," she says.

That idea has sent a ripple of excitement through the malaria vector research community. Although the malaria mosquito *Anopheles gambiae* carries neither *Wolbachia* nor gut symbionts, entomologist Chris Curtis of the London School of Hygiene and Tropical Medicine is now collaborating with Scott O'Neill, the Yale group's *Wolbachia* expert, to see if *Wolbachia* can be introduced into *An. gambiae*. Curtis and O'Neill believe *Wolbachia* could be used to drive antimalaria genes through mosquito populations, if the genes were inserted into a vehicle that, like *Wolbachia*, is maternally inherited. "We're looking at small DNA viruses," says O'Neill. The leading candidate? A class of insect viruses called densoviruses, some of which are transmitted from mother to offspring.

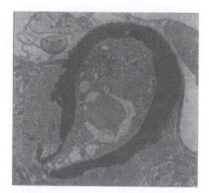

Locked up. Malaria ookinetes encapsulated in melanin in *An. gambiae* gut.

ing large quantities of antibody into the gut when the parasite is most vulnerable. "If it works," says Crampton, "you could use the same general strategy against any insect-borne disease."

Crossing the species barrier

Before Crampton's strategy can be tested, however, he must find a reliable way to insert foreign genes into the mosquito—a barrier that will eventually confront all the groups hoping to create transgenic insects. Some tsetse fly biologists are attempting to sidestep the problem by ignoring the fly's genome and modifying its symbiotic bacteria instead (see box, p. 200). Others are taking a more conventional approach: Molecular geneticist Anthony James of the University of California, Irvine, says his group has "preliminary, but promising" data indicating that it may be possible to use retroviruses to smuggle genes into mosquito chromosomes.

Most vector biologists are, however, currently pinning their hopes on a genetic manipulation technique used by *Drosophila* geneticists in which foreign genes are spliced into fragments of DNA known as transposable elements. Geneticists have found several of these elements that can readily be inserted into *Drosophila* chromosomes, and they can move around the genome of germline cells, multiplying as they do so—which means a transposable element can spread very rapidly through a population. For example, one such element, the *P* element, was first spotted in *Drosophila melanogaster* about 50 years ago, but it is now almost ubiquitous in the species. If a similar transposable element could be used to insert genes into *An. gambiae*, biologists could release relatively small numbers of the resulting transgenic insects, and the element would spread the genes through the wild population.

Given this potential, the race is on to find a transposable element that functions in *An. gambiae*. One leading candidate, called *minos*, was isolated from *Drosophila hydei* by geneticist Babis Savakis of the Institute for Molecular Biology and Biotechnology in Heraklion, Crete. *Minos* has already been introduced into *D. melanogaster* and Savakis is now collaborating with researchers at NIAID to see if it will function in *An. gambiae*.

Molecular entomologists Hugh Robertson and David Lampe of the University of Illinois, meanwhile, are focusing on a widespread family of transposable elements called *mariner*, which occur in the genome of many insects. *An. gambiae* possesses a *mariner* that is almost identical to those found in the horn fly and a species of lacewing, but these *mariners* all seem to have lost their ability to move. So Robertson and Lampe have taken the lacewing element and hooked it up to a more powerful promoter sequence, in the hope that this will kick it back into action.

The engineered *mariner* has yet to be put to the test, but Robertson is optimistic that it will function in a range of insect species—including *An. gambiae*. But even if this particular element does not live up to its promise, most insect geneticists believe that it is only a matter of time before researchers come up with elements that will function in *An. gambiae* and other disease-carrying insects. "My guess is that in virtually every organism you would be able to find elements that would work," says Kafatos.

Kafatos and his fellow vector biologists hope that by the time researchers have worked out systems to create transgenic insects, several candidate antimalaria genes will have been identified. The ultimate goal is to load several such genes together into a single strain to reduce the chances of the malaria parasite evolving resistance.

It will be a long campaign, however, and extensive safety testing will be required before large numbers of transgenic insects can be released into the wild. "I think we are talking about a 15- to 20-year timeframe," says Tore Godal, director of the World Health Organization's Tropical Disease Research program. But if the remaining technical hurdles and safety questions can be overcome, this new approach "might even lead to the eradication of malaria," says NIAID's Miller.

REFERENCES

1. L. Zheng, *et al.*, "Low-Resolution Genome Map of the Malaria Mosquito *Anopheles gambiae*," *Proc. Nat. Acad. Sci. U.S.* **88** 11187 (1 December 1991).

2. F. H. Collins, *et al.*, "Genetic Selection of a Plasmodium-Refractory Strain of the Malaria Vector *Anopheles gambiae*," *Science* **234**, 607 (31 October 1986).

3. D. W. Severson, *et al.*, "Linkage Map for *Aedes aegypti* Using Restriction Fragment Length Polymorphisms," *Journal of Heredity* **84**, (July/August 1993).

4. H-M. Müller, *et al.*, "Members of a Trypsin Gene Family in *Anopheles gambiae* Are Induced in the Gut by Blood Meal," *EMBO Journal* **12**, 2891 (July 1993).

5. A. A. James, "Mosquito Molecular Genetics: The Hands That Feed Bite Back," *Science* **257**, 37 (3 July 1992).

6. H. M. Robertson, "The *Mariner* Transposable Element is Widespread in Insects," *Nature* **362** 241 (18 March 1993).

7. S. Higgs, *et al.*, "Alphavirus Expression Systems: Applications to Mosquito Vector Studies," *Parasitology Today* (in press).

8. C. B. Beard, *et al.*, "Modification of Arthropod Vector Competence via Symbiotic Bacteria," *Parasitology Today* **9** 179 (May 1993).

READING 13.2

Research Community Swats Grasshopper Control Trial

Commentary: This is another article whose subject suggests the spatial dependence in a problem that involves the time dependence of a population. The particular issue here is that of introducing biological controls to deal with grasshopper infestations. Such controls might spread from the target area to affect essentially beneficial insects far from the scene of the crime. In particular, we must consider the time and space dependence of both the grasshopper population and that of its biological control. There are many famous examples of biological controls that got out of hand, the most recent being the inadvertent release of a rabbit-killing virus in Australia.

Research Community Swats Grasshopper Control Trial

Science, **260** (1993) 887.
Billy Goodman

As chemical pesticides have gained a bad reputation as indiscriminate killers of good and bad insects alike, agricultural scientists trying to protect crops have been pinning their hopes on biological controls—combating pests with pests of their own. The U.S. Department of Agriculture (USDA) is, how-

Author affiliation: Billy Goodman is a science writer based in Montclair, New Jersey.

ever, about to exterminate plans to test one such method this year, fearing the cure may be worse than the disease. Pests brought in to do the killing, they worry, could wipe out not just the target insects but beneficial species as well—just the problem that has given chemical pesticides a bad name.

The target in this case is grasshoppers. Once or twice a decade on western U.S. rangelands, grasshopper populations boom, sometimes reaching 100 hoppers per square meter. Large areas of grazing land are nearly stripped bare, and what the grasshoppers eat, cattle cannot. So ranchers and government range managers usually turn to broad-spectrum pesticides such as malathion.

In 1987, after the worst hopper outbreak in 50 years (55 million acres infested; widespread crop and forage damage), USDA scientists realized that chemical control was not only environmentally unwise, but simply wasn't working. They began evaluating alternatives and eventually turned to some out-of-town help: The scientists imported two grasshopper enemies, a wasp and a fungus, from Australia.

The Australians do their dirty work in different ways. The fungus, *Entomophaga praxibuli*, lurks as spores among range grasses and latches onto a passing grasshopper. It then penetrates the insect's exoskeleton, where it grows rapidly, digesting the tissue of its unfortunate host and killing it. The wasp, on the other hand, is an egg parasite. *Scelio parvicornis* lays its eggs in buried grasshopper egg pods. The hungry wasp larvae hatch inside a grasshopper egg and ravenously devour the developing insect.

The fungus was released on small test plots in North Dakota and Alaska in 1989, 1990, and 1991. The initial results were promising, and in 1992 the USDA's Agricultural Research Service (ARS) planned to enlarge the fungus program with additional releases. But these plans got derailed when a dispute arose over a separate proposal to release the Australian wasps, bringing the whole notion of plaguing grasshoppers with imported pests under fresh scrutiny.

The fuss began in 1991 when Richard Dysart, a research entomologist with the ARS in Sidney, Montana, applied to the USDA's Animal and Plant Health Inspection Service (APHIS) for a permit to release the wasps onto test plots in three states. "The action, if successful, would reduce the amount of pesticides poured out on the habitat," says Dysart. But

when University of Wyoming entomologist Jeffrey Lockwood learned of the planned release, he became alarmed. The wasp isn't particularly choosy about the grasshopper species it attacks. Yet only 10 to 15 of the more than 300 grasshopper species in the lower 48 states have the periodic population booms that turn them into real pests. The nontarget grasshoppers, Lockwood says, probably do no harm and in some cases may even do some good. For example, the species *Hesperotettix viridis* suppresses the poisonous snakeweed plant, which could spread unchecked if that particular grasshopper population plunged.

Lockwood and some colleagues took these objections to APHIS officials, prompting the service to do a more thorough assessment of both the wasp and fungus programs, beginning in 1992. The review, undertaken by a division of APHIS's Plant Protection and Quarantine section, has stalled the wasp release and halted further releases of the fungus. While results of the wasp assessment are not due until later this month, both Dysart and Gary Cunningham, the director of the ARS grasshopper control project, say the official in charge of the assessment told them 3 weeks ago that the wasp permit will be denied. The ARS entomologist in charge of the fungus program, Don Hostetter, has heard nothing about that evaluation, but since his funding is about to run out, even a positive reply will come too late. "I think this plan is dead in the water," he says.

ARS entomologist Raymond Carruthers, who helped initiate the fungus research, says that when the fungus was first released "there wasn't much concern about the impacts on nontarget species." Further research has shown that the fungus too has a fairly wide host range, encompassing not only some of the worst pests but many nontarget species as well. Both of these enemies do, however, appear to be restricted to grasshoppers, which makes them much more specific than chemical insecticides.

The fungus, Carruthers feels, as well as the wasp, have a built-in control mechanism that makes them safe to use: density dependence. They should only build up to large populations when there is a grasshopper outbreak. As the enemies cut short the outbreak, their own populations would crash. Rarer, nontarget grasshoppers would largely be spared from any attacks.

Lockwood, however, questions this logic. "Density dependence says that as the host declines, it becomes harder for an enemy to find," he says. "But if the enemy can switch hosts easily, it might continue to hammer away at the new host." He'd prefer the USDA to develop alternative techniques such as more specific native biological control agents and more surgical use of insecticides in baits.

Jerry Onsager, research leader of the USDA's Rangeland Insect Laboratory in Bozeman, Montana, agrees that it is impossible to predict all of the effects of a test release accurately. Yet he thinks the risks of harmful side effects from such a release are low and are worth it, because lab studies won't answer one crucial question: What wider environmental impacts will such a release have? "You either grit your teeth and take chances or spend the rest of your career doing cage studies, and in 30 years you won't necessarily know much," he says.

Lockwood, though, has a darker view. "Biological control agents are not bounded in space and time," he says. "Their impact will not decline as a function of time or distance from the release, if the releases are successful. If they're a bad idea, we can't get them back."

READING 13.3

Hosts and Parasitoids in Space;
Insect Parasitoid Species Respond to Forest Structure at Different
Spatial Scales

Commentary: These two articles illustrate recent work that incorporates space as well as time dependence in population biology modeling. The first article is a review and perspective on the second. The articles discuss a study of the interaction between a moth and its parasites in the case where the forest habitat is fragmented into isolated stands of trees. They use statistical analysis in an attempt to correlate data consisting of observed moth and parasite abundance with parameters such as the density of the moth and parasites in previous generations and the amount of forest cover available in the given stand of trees.

Hosts and Parasitoids in Space

Nature, **386** (1997) 660–661.
H. C. J. Godfray, M. P. Hassell

Parasitoids are insects whose larvae develop on or in the body of other insects, eventually killing them (Fig. 1). Their population dynamics have been studied intensively, both because of their economic importance in controlling pests, but also because they provide a relatively simple system for investigating dynamic processes common to all consumer–resource interactions.

In recent years, theorists have been exploring how an explicit consideration of space affects the dynamic interactions between hosts and parasitoids, a move that reflects an increasing re-alization of the importance of spatial processes in all branches of population dynamics. But although the mathematical models reveal a fascinating menagerie of dynamic patterns, experimental studies have lagged behind, chiefly because of the great logistical problems involved. However, a series of studies, including that of Roland and Taylor (to follow) have demonstrated the rewards of taking a spatial approach.

Roland and Taylor studied a moth called *Malacosoma disstria*, the forest tent caterpillar, which

---

Author affiliations: H. C. J. Godfray and M. P. Hassell: Department of Biology and the NERC Centre for Population Biology, Imperial College at Silwood Park.

Fig. 1 Parasitoid attack—here a brood of the braconid wasp *Cotesia glomeratus* emerges from the caterpillar of a small white butterfly, *Pieris rapae*, before undergoing pupation.

feeds in aspen woodland in central Canada. The moth's caterpillars live gregariously in silken webs, and can become common enough to defoliate large areas of forest. Although the population dynamics are not wholly understood, it is thought that outbreaks are suppressed by parasitoids of the moth, particularly certain flies of the families Tachinidae and Sarcophagidae.

Before human intervention, aspen forests occurred over wide areas, but today they have been fragmented into stands of varying size. Roland and Taylor asked how fragmentation and host density influence the pattern of attack on *M. disstria* by the different species of parasitoid. They sampled populations of the moth and parasitoid at 127 sites within a 25 × 25 km grid, and repeated the exercise at 109 sites within a smaller 800 × 800 m grid. Employing statistical modelling techniques, they then sought to explain observed rates of parasitism by each of four species of parasitoid using two classes of explanatory variable: first, the density of the host in the current and previous generations; second, a series of measures of forest fragmentation consisting of estimates of forest cover over successively largely circular areas centred at the sampling site (the estimates being obtained from satellite photographs involving GIS (Geographical Information System) technology).

Parasitism by the three largest parasitoids showed significant positive dependence on host density while the smallest species showed negative density dependence. The extent of forest fragmentation influenced parasitism by all flies. But

the spatial scale at which parasitism and fragmentation were most highly correlated differed between species, with larger species being influenced by fragmentation measured over the greatest area. Moreover, whereas the three largest parasitoids showed reduced rates of attack in fragmented forests, parasitism by the smallest species was increased in small woods.

This study is significant for two reasons. First, it illustrates spatial patterns of positive and negative density dependence on a scale never previously studied. There is some controversy over exactly how density dependence in parasitoid attack influences host dynamics, and data such as these are invaluable in guiding new theory. Second, it shows how habitat change can alter host-parasitoid population dynamics. In a fragmented forest, the larger species that are probably most important in regulating the host perform poorly, something that might explain the longer outbreaks of *M. disstria* observed in such woodland. Further studies will need to try to disentangle the joint effects of the four parasitoids on host dynamics (here treated independently), as well as developing techniques for dealing with possible statistical non-independence of samples collected from a spatial grid.

In a highly fragmented landscape, individual populations may be destined to become extinct in a relatively short period of time. However, the ensemble of populations may persist if they are loosely coupled by migration and if colonization is able to balance extinction. Such a population structure is known as a classical, or "blinking lights," metapopulation.

One of the best examples of a classical metapopulation is provided by the work of Hanski and colleagues on *Melitaea cinxia*, the Glanville fritillary butterfly, in the Aland archipelago in Finland. The butterfly exists in small scattered populations in areas where its food plant grows, with each population having a relatively high probability of extinction. Previous studies had shown that extinction rates were correlated with population size, but Lei and Hanski [2] have demonstrated that parasitoid attack also influences extinction.

Melitaea cinxia is attacked by a parasitoid wasp, *Cotesia melitaearum*, which in this region appears to be specific to this host. Host populations of a particular size are more liable to become extinct if they are attacked by parasitoids. Cases of parasitoid, but not host, extinction, and parasitoid colonization of host patches have also been documented. The picture is complicated by the presence of other parasitoids in the system, including a further primary parasitoid that does best in patches where *C. melitaearum* is absent, and a hyperparasitoid with a very wide host range which shows a strong density-dependent response to *C. melitaearum* cocoons.

It is interesting that the father of parasitoid population biology, A. J. Nicholson [3], anticipated host-parasitoid metapopulations 50 years ago. The Nicholson-Bailey equations, the ur-model of parasitoid population dynamics, predict unstable local population dynamics and Nicholson suggested that persistence may occur through a dynamic process of colonization and extinction exactly as in a modern metapopulation.

A more curious example of spatial processes is provided by the work of Harrison and colleagues on the moth *Orgyia vetusta*, which forms persistent outbreaks on certain patches of its host plant (*Lupinus*) along the coastal strip of north California. Why do outbreaks persist in patches of contiguous bushes for many years without spreading? Experiments have excluded differences in host-plant quality, leading to the suggestion that parasitoids dispersing from the outbreak area might cause a halo of intense parasitism in the surrounding bushes that prevents the spread of an outbreak.

Earlier this year, Brodmann, Wilcox and Harrison [4] published the results of experiments showing that parasitism is highest in the areas surrounding the outbreak, as the parasitoid hypothesis predicts. But why should parasitoids dispense away from outbreaks and so presumably lower their reproductive fitness, and why do parasitoids not reduce population densities within the outbreak? Interference between searching parasitoids is a possible solution. Like the other two studies discussed here [1, 2], this work shows the importance of thinking spatially—but it raises as many questions as it answers.

REFERENCES

1. J. Roland and P. D. Taylor, *Nature* **386**, (1997), pp. 710-713.
2. G.-L. Lei and I. Hanski *Oikos* **75**, (1997), pp. 91-100.
3. A. J. Nicholson, *Rep. 26th Meet. Aust. NZ Assoc. Adv. Sci*, (1947), pp. 1-14.
4. P. A. Brodmann, C. V. Wilcox, and S. Harrison, *J. Anim. Ecol* **66**, (1997), pp. 65-72.

Insect Parasitoid Species Respond to Forest Structure at Different Spatial Scales

Nature, **386** (1997) 710–713.

J. Roland, P. D. Taylor

There is now a solid body of theoretical work [1–4] demonstrating that the spatial structure of the habitat combined with animal movement strongly influence host-parasitoid dynamics. The spatial pattern over which parasitoid search takes place can be affected by the distribution of the

Author affiliations: Jens Roland: Department of Biological Sciences, University of Alberta. Philip D. Taylor: Atlantic Cooperative Wildlife Ecology Research Network, Department of Biology, Acadia University.

hosts [5], by the spatial arrangement of the host's habitat [6] and by the spatial scale at which the parasitoid perceives variation in host abundance [7, 8]. Empirical work, however, has been largely restricted to small-scale field studies of less than one hectare [6, 9] with very few larger [10, 11]. Here we report initial results of a many-year, large-scale study that is among the first to examine the interaction between a population-level process (parasitism) and anthropogenic forest fragmentation at large and at multiple spatial scales. We demonstrate that parasitism by four species of parasitoids attacking the forest tent caterpillar, *Malacosoma disstria*, is significantly reduced or enhanced depending on the proportion of forested to unforested land. Each of the parasitoid species responds to this mosaic at four different spatial scales that correspond to their relative body sizes. Our data give empirical support to the argument that changes in landscape structure can alter the normal functioning of ecological processes such as parasitism, with large-scale population consequences [3, 4].

We determined the effect of forest structure (mosaic of cleared and forested land) on the rates of parasitism caused by four species of parasitic fly attacking an outbreak population of the forest tent caterpillar (*Malacosoma disstria*) in aspen forests near Edmonton, Alberta, Canada. Studies were done on a large-scale grid 420 km^2 in area (127 sample points) and a small-scale grid of 0.32 km^2 area (109 sample points) nested within the larger grid. At each sample point we estimated host population density, rates of parasitism caused by each species, and forest structure. Forest structure around each point was estimated at seven scales, from 53 m around each point up to 3,400 m.

Univoltine fly parasitoids dominate the parasitoid community that attacks forest tent caterpillar [12–14]. The four species we studied ranged in size from 34 mg for *Carcelia malacosomae* (Tachinidae) which attack host larvae directly, 41 mg for *Patelloa pachypyga* and 68 mg for *Leschenaultia exul* (Tachinidae) which lay eggs on foliage that are subsequently ingested by feeding host caterpillars, and 58 mg for *Arachnidomyia* (= *Sarcophaga*) *aldrichi* (Sarcophagidae) which larviposits on host cocoons. The three largest of these species are considered important in suppressing tent caterpillar outbreaks [12–14]. Because larvae of all of these flies develop in the late larval and pupal stage of

the host there is some competition among them. *A. aldrichi* attacks host pupae and is the last of the four to attack during the host's life cycle. Therefore, it competes with *C. malacosomae* and especially with *P. pachypyga* which remain within the host into the host's pupal stage [14, 15]. Larvae of *L. exul* exit before host pupation and are therefore affected little by competition with *A. aldrichi* [14].

Parasitism by all species varied markedly across the study area (Fig. 1) in response to both forest structure and the abundance of hosts (Table 1). Parasitism by the three largest fly species was greatest on hosts collected in contiguous forests, and was lower in those from fragmented forests (Table 1; positive coefficients reflect greater parasitism in contiguous forest). However, the spatial scale at which forest structure had its greatest effect differs among the parasite species. Parasitism by the largest species, *L. exul*, was most strongly correlated with forest structure measured at a scale of 850 m around each site. Parasitism by medium-sized flies *A. aldrichi* and *P. pachypyga* was most strongly related to forest structure within 425 m and 212 m, respectively. They caused less parasitism in the areas of greatest forest fragmentation (Fig. 1) and along large forest edges (Fig. 2). In contrast, the smallest fly, *C. malacosomae*, caused higher rates of parasitism in the fragmented forests (Fig. 1) and at forest edges (Fig. 2), and was affected by forest structure at the finest spatial scale (53 m). Parameter estimates for the effects of forest structure were not significantly different when estimated from data from the coarse-resolution and fine-resolution grids. The similarity of parameter estimates from the grids with different resolution suggests that there are no additional effects of landscape across the large-scale grid beyond those that can be estimated at a given sample point.

Rates of parasitism across the fine-resolution grid were spatially autocorrelated. Again, the distance to which they were autocorrelated differed among the four species. Parasitism by the two smallest species, *C. malacosomae* and *P. pachypyga*, were autocorrelated to distances of only 44 and 53 m, respectively, reflecting fine-grained variation in parasitism. Parasitism by the two larger fly species, *A. aldrichi* and *L. exul*, was autocorrelated to distances of 421 and 420 m, respectively, reflecting coarser-grained variation in parasitism. These patterns suggest greater movement by the larger fly species, reflected in the larger spa-

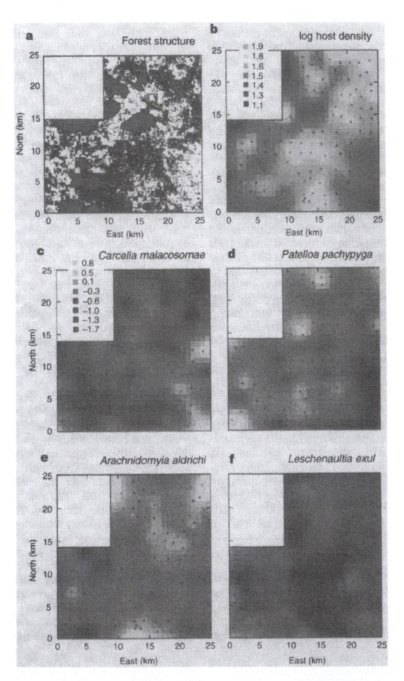

Fig. 1 a. Forest structure, **b.** log host density in 1995 (N_t) and **c–f.** parasitism by four fly parasitoid species attacking forest tent caterpillar at Ministik Hills, Alberta, Canada. Maps of parasitism are distance-weighted least-squares surface plots [21] for each of four parasitoid species attacking tent caterpillar across the coarse-resolution grid: **c.** *C. malacosomae*; **d.** *P. pachypyga*; **e.** *A. aldrichi*; and **f.** *L. exul*. Colours indicate log host density (b) and log odds-ratios of number parasitized by each species against number not parasitized (c–f). Dots are the 127 population sample sites. The red rectangle (in a) indicates the location of the fine-resolution grid (Fig. 2).

Table 1.

Parasitoid species	Pupal mass (mg)	53 m	106 m	212 m	425 m	850 m	1,700 m	3,400 m	N_{t-1}	N_t
Carcelia malacosomae	34	−0.77						0.10	−0.17	−0.26
Patelloa pachypyga	41			2.0				0.17		0.80
Arachnidomyia aldrichi	58				2.23				−0.12	0.52
Leschenauiltia exul	68					1.85			−0.24	0.28

Coefficients from logistic regression models estimating estimating the effects of forest structure (Fig 1a) and host density (Fig 1b) on the odds of being parasitized by each of four parasitoid species across the large, course-resolution grid. We show only the coefficients for the effects of forest structure at the spatial scale at which its effect was strongest, and for the effect of host density. Full models are contained in supplementary information.

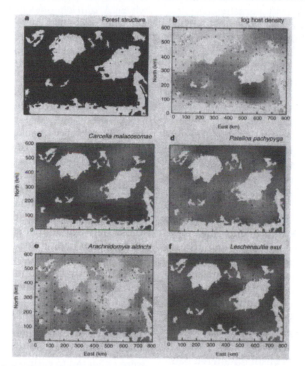

Fig. 2 a. Forest structure, **b.** log host density in 1995 (N_t), and **c–f.** parasitism across the 0.32 m^2 fine-resolution grid. Parasitism is estimated for **c.** *Carcelia malacosomae*; **d.** *Patelloe pachypyga*; **e.** *Arachnidomyia aldrichi*; and **f.** *Leschenaultia exul.* Methods and symbols as in Fig. 1.

tial scale at which they respond to forest structure.

In general, parasitism was higher in areas of high host density, a pattern most evident for *P. pachypyga* and *A. aldrichi*. Interestingly, the smallest parasitoid species again showed a pattern opposite to that of the other three species, causing higher rates of parasitism in areas of low host density (Table 1). For the two smallest parasitoid species (*C. malacosomae* and *P. pachypyga*), there was a significant interaction between host density

and forest structure on the rates of parasitism they cause. *C. malacosomae* responded positively to host density in partially cleared forests (at 53 m scale) but did not respond to host abundance in continuous forests. These patterns suggest that forest acts as a barrier to movement (aggregation) by *C. malacosomae*, or that their numbers increase little in continuous forest. *Patelloa pachypyga*, in contrast, responded strongly and positively to host abundance in continuous forests, and only weakly so in partially cleared forests, suggesting that clearings may inhibit movement by *P. pachypyga*. Theoretically, such limits on parasitoid redistribution undermine any stabilizing effect of parasitoid aggregation [15]. There were no such interactions for the two largest fly species, suggesting that any response to host abundance was consistent regardless of habitat structure.

Although our spatial analyses clearly show that the landscape patterning of forests influences the percentage of host attacked by parasites, the real question is whether this effect translates into altered host–parasitoid dynamics. Data from other studies suggest in fact that the effects of forest structure we document in this paper do indeed alter dynamics in a profound way. In particular, the three parasitoid species that are less effective in fragmented forests are those that normally dominate in declining populations of tent caterpillar in several parts of North America [12–14]. Outbreaks of tent caterpillar last longer in fragmented than in continuous forest [16]. On the basis of our analysis, a reduction in forest cover from continuous to 50 percent cover reduces the odds of parasitism by the three dominant parasite species to about half of that in intact forests. These species would be predicted to be most efficacious in only those areas with relatively large blocks of contiguous forests, 212 to 850 metres square in size (Fig. 3). The one parasitoid species that benefits from partial clearing of forests, *C. malacosomae* (Table 1), plays a minor role during tent caterpillar outbreak and collapse [12, 13]. Forest fragmentation, therefore, may exacerbate outbreaks of tent caterpillar by decoupling it from its natural enemies.

We suggest that the mechanism for altered parasitism is through the effects of habitat structure on movement. We have shown elsewhere [17] that the larger species (*A. aldrichi*) can "colonize" isolated forest stands up to 400 m from contiguous forest compared to only 125 m for the smaller species, *P. pachypyga*. Our analysis here also suggests that the ability of parasitoids to aggregate in response to host density is affected by landscape. It may be argued that forest fragmentation alters microclimate along forest edges, which different species may either favour or avoid [18]. However, we have controlled the effect of distance to edge at our large-grid sites by placing each sampling point 20 m into the forest; and we have removed the effect of distance to edge in the analysis of the fine-grid data.

Therefore, although there may be an effect of edge, the fragmentation effect detected here is not directly related to the edge. Differences in spatial patterns seen among the fly species may reflect a strong spatial pattern for one species, for example *A. aldrichi*, combined with its ability to out-compete the other species. However, such an effect on another species would be expected to result in a response to forest structure in the opposite direction (higher parasitism in the fragmented stands where *A. aldrichi* does poorly) but at the same spatial scale (425 m). The most likely competitors with *A. aldrichi* either show the same (positive) direction of their response to forest structure (*P. pachypyga*) or respond to forest structure at a much finer scale (*C. malacosomae*).

Changes in landscape structure such as forest fragmentation have been predicted to affect animals differentially depending on their size and the spatial grain of the habitat mosaic [19]. Populations of animals that evolved a specific "ambit" within, for example, a continuous forest system, may behave differently in areas of altered habitat structure. Successful predictions about the behaviour of complex systems and their response to disturbance may be possible using such simple relationships as those between animal size, spatial scale of search and grain of habitat structure. We reiterate the plea by others [4] that ecological studies be conducted at sufficiently large and multiple spatial scales to successfully extract such generalizations.

Methods

Estimates of parasitism and host abundance. At each sample point 50 to 100 late-instar caterpillars were collected to estimate parasitism by *C. malacosomae*, *P. pachypyga* and *L. exul*, and 50 to 100 host cocoons were collected 10 days later to estimate rates of parasitism by *A. aldrichi*. The collection of late-instar hosts to estimate parasitism by *C. malacosomae* and *Patelloa pachypyga* [12–14] was made before the attack by *A. aldrichi*, which is known to out-compete the other parasites in the host pupa. Hosts from each collection were reared all together and the number of emerging parasitoids of each species were counted. Density was estimated using a time-restricted search; the time taken to collect 50 cocoons (to a maximum of 15 mm) was recorded. If less than 15 mm were needed to collect 50, we estimated the number which would have been collected in 15.

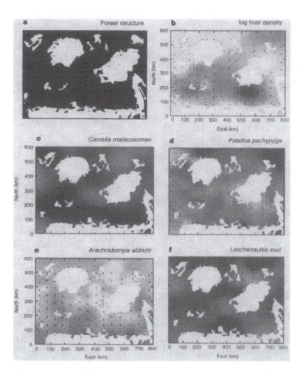

Fig. 3 The extent of forested habitat in which each parasitoid species would be most effective, disregarding effects of host density. Patterns are based on the effects of forest structure on parasitism (positive or negative) and the spatial scale at which each species responds to forest structure most strongly (from Table 1). Forests where the efficacy of each species is reduced from its maximum are deleted. Compare to Fig. 1a for the distribution of all forest.

Estimates of forest structure. Forest structure was estimated at each of the sample points across the 420 m² grid. Landscape structure was estimated at each point from a classified aerial photomosaic (1:20,000 scale) scanned at a resolution of 5.3 m per pixel. Using SPANS Geographic Information System software (Intera Tydac Technologies, Nepean, Ontario, Canada), we calculated the amount of forest in a series of seven squares, geometrically increasing in spatial scale from 53 m on a side to 3,400 m on a side (Table 1) and centred on each sampling point. The amount of forest within each square was scaled from 0 (virtually no forest) to 1 (complete coverage of continuous forest); it is therefore an index of forest continuity.

Model fitting. For each parasitoid species, models included as potential explanatory variables: the log-transformed density of forest tent caterpillar in the previous year (N_{t-1}, 1994), the log-transformed density of forest tent caterpillar in the current year (N_t, 1995), and terms for the location of each sample point on a cartesian grid (east, north, coefficients not shown) and the square of these locations. Location terms were included to account for any unknown historical patterns of spread of the outbreak across the coarse-resolution grid. Model-building proceeded in a stepwise fashion by separately testing the importance of landscape structure measured at each of the seven spatial scales in each of seven separate models (each including density and location terms). We selected the model with the landscape term that caused the maximum reduction in overall deviance of the respective model, and for which the coefficient was most significantly different from zero [20]. Once the best model was selected, additional terms were added to it to identify any additional effects of forest structure at scales smaller and larger than that which appeared to be most important. This strategy was used to identify the scale at which forest structure had its greatest effect, and then to examine whether additional terms contributed to the overall fit of the model. When testing for the effects of forest in the 3,400 m square surrounding a point, we used a subsample of points spaced by at least 4 km, thereby ensuring independence of forest structure estimates among points. Pupal mass was estimated from samples of 20 individuals of each fly species pooled from several sites. Full models are presented as supplementary information.

A second set of models (not shown) were fitted to data from the fine-resolution grid, but forest structure was only estimated within 53 m and 106 m around each sample point. Forest structure within 106 m of each point was again estimated using a subset of points to ensure independence of estimates. Distance of each sample point to the nearest forest edge was estimated and included in the model. Coefficients for the effect of forest structure (53 m and 106 m) on parasitism estimated from the fine-scale grid were not significantly different from those estimated from the coarse-resolution grid (53 and 106 m samples). All models were fit using S-PLUS software (MathSoft Inc., Seattle). Spatial autocorrelation of parasitism estimates for each species were based on the fit of a spherical model semivariogram with a minimum lag of 50 m, using GS$^+$ geostatistics software (Gamma Design Software, Plainwall, Michigan).

REFERENCES

1. M. P. Hassell, R. M. May, Spatial heterogeneity and the dynamics of parasitoid-host systems, *Ann. Zoo. Fennisi* **25**, (1998), pp. 55–61.
2. P. Kareiva, Population dynamics in spatially complex environments: theory and data, *Phil. Trans. R. Soc. B* **330** (1990), pp. 175–190.
3. R. M. May, in Large-scale Ecology and Conservssaon Biology (eds Edwards, P. I., May, R. M. and Webb. N. R.) (Blackwell Scientific, Oxford, 1994), pp. 1–17.
4. P. Kareiva, U. Wennergren, Connecting landscape pattern to ecosystem and population processes, *Nature* **373**, (1995), pp. 299–302.
5. I. K. Waage, Aggregation in field parasitoid populations: foraging time allocation by a population of *Diadegma* (Hymenoptera, Ichneumonidae) *Ecol. Entomol.* **8**, (1963), pp. 447–453.

6. A. Kruess & T. Tscharntke, Habitat fragmentation, species loss, and biological control, *Science* **264**, (1994), pp. 1581–1584.
7. P. A. Heads & J. H. Lawton, Studies on the natural enemy complex of the holly leaf-miner: the effects of scale on the detection of aggregative responses and the implications for biological control, *Oikos* **40**, (1983), pp. 267–276.
8. L. D. Rothman, D. C. Darling, Spatial density dependence: effects of scale, host spatial pattern and parasitoid reproductive strategy, *Oikos* **62**, (1991), pp. 221–230.
9. P. Kareiva, Habitat fragmentation and the stability of predator-prey interactions, *Nature* **326**, (1987), pp. 388–390.
10. D. J. Rogers & B. G. Williams, in *Large-scale Ecology and Conservation Biology* (eds Edwards, P. J., May, R. M. and Webb. N. R.), (Blackwell Scientific, Oxtord, 1994), pp. 247–271.
11. P. C. Marino, D. A. Landis, Effects at landscape structure on parasitoid diversity and parasitism in agroecosystems, *Ecol. Appl.* **6**, (1996), pp. 276–284.
12. W. L. Sippell, Outbreaks of the forest tent caterpillar, *Malacosoma disstria* Hbn., periodic defoliator of broad-leaved trees in Ontario, *Can. Entomol.* **94**, (1962), pp. 408–416.
13. J. A. Witter, H. M. Kulman, *The parasite complex of the forest tent caterpillar in northern Minnesota*, *Env. Entomol*, **8**, (1979), pp. 723–731.
14. D. Parry, Larval and pupal parasitism of the forest-tent caterpillar *Malacosoma disstria* Hübner (Lepidoptera, Lasiocampidae) in Alberta, Canada, *Can. Entomol.* **127**, (1995), pp. 877–893.
15. H. C. J. Godfray, *Parasitoids: Behavioral and Evolutionary Ecology* (Princeton Univ. Press, Princeton, 1994).
16. J. Roland, Large-scale forest fragmentation increase the duration of forest tent caterpillar outbreak, *Oecologia* **93**, (1993), pp. 25–301.
17. J. Roland, P. D. Taylor, in *Population Dynamics: New Approaches and Synthesis* (eds Cappuccino, N. and Price, P. W.) (Academic, San Diego, 1995, pp. 195–208).
18. R. M. Weseloh, Spatial distribution of gypsy moth (Lepidoptera: Lymantriidae) and some of its parisitoids within a forest environment, *Entomophaga* **17**, (1972), pp. 339–351.
19. C. S. Holling, Cross-scale morphology, ge-

ometry and dynamics of ecosystems, *Ecol. Monogr.*, (1992), pp. 447–502.
20. P. McCullagh, G. A. Nelder, *Generalized Linear Models* (Chapman and Hall, London, 1989).

21. L. Wilkinson, M. Hill, S. Miceli, G. Birkenbeuel, E. Vang, *SYSTAT for Windows: V.5* (Evanston, Illinois, 1992).

READING 13.4

Scope of the AIDS Epidemic in the United States

Commentary: This article was presented earlier in Reading 7.3. I return to it here because a form of the advection equation plays an important role in analyzing the time evolution of an epidemic. For example, consider the function $v(t, a)$, which gives the number of people at time $= t$ and age $= a$ who have AIDS. Figure 1A in the article gives some information about this function. Knowledge of the function is required if we are to have any hope of predicting the future course of the epidemic. In any event, I want to point out that this function v obeys a version of the advection equation. Indeed, since each person gets one year older after one year, the function v must obey the equation

$$\frac{\partial v}{\partial t} = -\frac{\partial v}{\partial a} - rv,$$

where $r = r(t, a)$ is a function of the variables t and a and equals the death rate minus the infectivity rate for people at time t and age a. In particular, this function r is the the key to understanding the function v. This article discusses what we know about r and v. (Note that when r is not a constant, then Chapter 13's solutions to the advection equation are not applicable.)

Exercises for Chapter 13

Exercises on Solutions to Equation (8) in Chapter 13

Here are some exercises to convince you that any function of the form given by Equation (8) of Chapter 13 satisfies Equation (7) of Chapter 13 in the case $r = 2$: Compute the $\frac{\partial}{\partial t}$ and $\frac{\partial}{\partial x}$ partial derivatives of the following functions and then verify that (7) is satisfied:

1. $u(t, x) = e^{-2t} \sin(x - 3t)$

2. $u(t, x) = e^{-2t}(x - 3t)^2$

3. $u(t, x) = e^{-2t} e^{-2(x-3t)/3} = e^{-2x}$

4. $u(t, x) = 3e^{-2t}$

5. $u(t, x) = e^{-2t}(1 + (x - 3t)^2)^{-1}$

Initial and Boundary Condition Problems

In the case $r = 2$, find the unique solution to Equation (7) of Chapter 13 with

6. $u(0, x) = \cos(x)$

7. $u(0, x) = e^{-4x}$

8. $u(0, x) = (1 + x^2)^{-1}$

9. $u(t, 0) = 1$

10. $u(t, 0) = \cos(t)$

11. $u(t, 0) = e^{-4t}$

12. Write the solution, $u(t, x)$, to the equation $\frac{\partial}{\partial t}u = -5\frac{\partial}{\partial x}u$ subject to the initial condition $u(0, x) = e^{-x^2}$. Do the same for the boundary condition $u(t, 0) = e^{-t^2}$.

13. Write the solution, $u(t, x)$, to the equation $\frac{\partial}{\partial t}u = -2\frac{\partial}{\partial x}u$ subject to the initial condition $u(0, x) = \sin(x)$. Do the same for the boundary condition $u(t, 0) = \sin(t)$.

14. Write the solution, $u(t, x)$, to the equation $\frac{\partial}{\partial t}u = -2\frac{\partial}{\partial x}u$ subject to the initial condition $u(0, x) = x^2$. Do the same for the boundary condition $u(t, 0) = t^2$.

Advection Solutions as Traveling Waves

To get a feeling for the solutions to Equation (7) of Chapter 13, consider a solution in the case that $r = 0$: $u(t, x) = f(x - 3t)$. To make matters concrete, take $f(s) = (1 + s^2)^{-1}$, a function that is peaked around the origin and that decays nicely to zero on both sides of the origin.

15. Sketch a graph of the function $y = u(0, x) = (1 + x^2)^{-1}$.

16. On the same graph paper, but with different colors, sketch a graph of the functions

    (a) $y = u(1, x)$
    (b) $y = u(2, x)$
    (c) $y = u(3, x)$

17. Can you see from your graphs in Exercise 16 that the function $u(t, x)$ describes a "bump" that is moving at a constant speed (3 meters per second) along the x-axis? Does this picture of a moving bump surprise you? In a sentence or two, state why this is a reasonable answer for the $r = 0$ case.

An Open-Ended Problem

18. Commentary 13.1 raises the question of how to predict the amount of seeding necessary to populate central Africa with malaria-resistant mosquitos in 50 years. I remarked there that the advection equation did not seem applicable to this problem. If you agree with me, explain why this equation is inappropriate. If you disagree, give a justification for its use.

14

Diffusion Equations

As in the previous chapter, the central issue here is that of predicting the behavior of a function that measures the density of some type of particle as a function of time and position. At the beginning of the previous chapter, I derived a completely tautological equation for this function. Let me remind you: I use t to denote the time coordinate and x to denote the space coordinate; and then I use u to denote the function. Thus, $u(t, x)$ measures the density of particles at time t and position x along some line. By keeping track of all possible fates of particles that start near x at time t, I argued that u should obey the equation

$$\frac{\partial}{\partial t}u(t, x) = -\frac{\partial}{\partial x}q(t, x) + k(t, x). \tag{1}$$

Here,

$q(t, x) =$ (the number of particles per second crossing x from left to right at time t)
 $-$ (the number of particles per second crossing x from right to left at time t), (2)

and

$k(t, x) =$ (the number of particles per second per unit length that appear at x at time t)
 $-$ (the number of particles per second per unit length that disappear at x at time t). (3)

With regard to (1), don't be misled by its stylish form, for (1) is only a fancy way of saying "What comes in either stays in or goes out again." However, this equation can become useful when the functions q and k are specified.

For example, in the last chapter, I considered the case where the particle motion is due entirely to the movement of the ambient fluid. Under this last assumption, I argued that the appropriate choice for q is $q(t, x) = c(t, x)u(t, x)$, where $c(t, x)$ is the velocity of the fluid at time t and position x. (In the last chapter, I took c to be a constant, but that extra simplification is often a poor approximation to reality.) In the previous chapter, I took $k(t, x)$ to equal $-ru(t, x)$, where r is a constant. The result of these choices yields the advection equation version of (1),

$$\frac{\partial}{\partial t}u(t, x) = -c\frac{\partial}{\partial x}u(t, x) - ru(t, x), \tag{4}$$

whose general solution has the form

$$u(t, x) = e^{-rt}f(x - ct) \tag{5}$$

with $f(s)$ any function of one variable, evaluated at $s = x - ct$.

I can't stress enough that (4) is appropriate only when the particle motion is, to a good approximation, due only to the motion of the ambient fluid. This may not be the case. For example, the fluid may be at rest, and the particles might invidually have random motions. In the latter case, a different choice for the function $q(t, x)$ is appropriate and the resulting version of (1) is called a ***diffusion equation***.

14.1   The Diffusion Equation

The archetypal diffusion equation has the form

$$\frac{\partial}{\partial t}u = \mu\frac{\partial^2}{\partial x^2}u + k(t, x), \tag{6}$$

where $k(t, x)$ is a function that can depend on $u(t, x)$, and where μ is a constant that is determined by the average speed of the particles. Here, $\frac{\partial^2 u}{\partial x^2}$ signifies the second derivative of the function u along the x-direction. (So keep t fixed as if it were a constant and pretend that you are taking the derivatives of a function of x only.)

More precisely, the constant μ is determined by the square root of the average of the square of the velocity of the particles. The derivation of the number μ from first principles requires some sophisticated probability theory. In practice, the constant μ can be measured in the laboratory using some fairly simple experiments. The constant μ is called the ***diffusion coefficient***.

14.2   The Derivation of the Diffusion Equation

Equation (6) is of the form of (1) with $q(t, x) = -\mu\frac{\partial}{\partial x}u(t, x)$. This general form for q can be justified in a heuristic sense as follows: When the motion of the particles is completely random, we expect that the number of particles that cross x per second at time t from left to right is proportional to the number of particles that are just to the left of x at time t. If the velocities of the particles are truly random, then roughly half of these particles will be moving to the left (and so not cross x), while half will be moving to the right, and so will cross x. Meanwhile, the number of particles per second that cross x from right to left at time t should be proportional to the number of particles that are just to the right of x at time t. Thus, we expect $q(t, x)$ to be positive if there are more particles just to the left of x at time t than just to the right, and we expect $q(t, x)$ to be negative if the converse is true. And we expect $q(t, x)$ to vanish if the number of particles just to the right is the same as the number that is just to the left. But this is essentially saying that $q(t, x)$ should be proportional to $-\frac{\partial}{\partial x}u(t, x)$, since $-\frac{\partial}{\partial x}u(t, x)$ is positive if there are more particles just to the left of x, and negative if there are more particles just to the right, and zero otherwise.

14.3   A Fundamental Solution

As with (4), the diffusion equation has a whole raft of solutions. The function

$$u(t, x) = \frac{a}{t^{1/2}}e^{-x^2/4\mu t} \tag{7}$$

is a fundamental solution to the particular version of (6) with $k(t, x) = 0$:

$$\frac{\partial}{\partial t}u = \mu\frac{\partial^2}{\partial x^2}u. \tag{8}$$

Here, a is any constant.

For any fixed t, the graph of $u(t, x)$ in (7) is the standard bell-shaped curve, but as t increases, the curve gets flatter and flatter. Its maximum height is at the origin, where $u(0, t) = a/t^{1/2}$. Thus, as t increases, the particle distribution does indeed diffuse out from a concentrated peak at small t to an almost uniformly small density at large t. See the sketches in Figure 14.1.

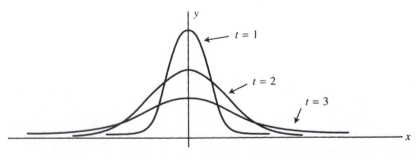

Figure 14.1

Note that at any fixed time t, the integral of $u(t, x)$ over the whole of the x-axis is constant. This is predicted by (7) since the term $k(t, x)$ that models the appearance and disappearance of particles is missing.

14.4 Lessons

Here are some key points from this chapter:

- Under the assumption that the particles are moving randomly, Equation (1) becomes Equation (6), the diffusion equation.

READINGS FOR CHAPTER 14

READING 14.1

Counting Polymers Moving Through a Single Ion Channel

Commentary: This article reports on an experiment that demonstrates a technique for counting polymers (and determining their diffusion coefficients) in a single molecular pore, in this case one of length $\sim$50 Angstroms and width $\sim$10 Angstroms. Here is the idea: The presence of a polymer molecule in the pore will change the electrical conductivity, and, conversely, by measuring the change in the conductivity, we can determine the number of polymers passing through the pore. For the purposes of this course, the point is that the polymer moves down the core by diffusion—its motion is essentially random. Thus, given an initial density of polymers, all on one side of the pore, the diffusion equation, $\partial_t u = \mu \, \partial_x^2 u$, can be used to predict how many polymers are in the pore at any given time if we know the diffusion coefficient μ. Conversely, if we know how many particles are in the pore, we can determine the diffusion coefficient. With this understood, direct your attention to the first equation on page 222 which is the very same diffusion equation, except that D is used to denote the diffusion constant instead of μ.

Counting Polymers Moving Through a Single Ion Channel

Nature, **370** (1994) 279–281.
Sergey M. Bezrukov, Igor Vodyanoy, V. Adrian Parsegian

The change in conductance of a small electrolyte-filled capillary owing to the passage of sub-micrometre-sized particles has long been used for particle counting and sizing. A commercial device for such measurements, the Coulter counter, is able to detect particles of sizes down to several tenths of a micrometre [1–3]. Nuclepore technology (in which pores are etched particle tracks) has extended the lower limit of the detection to 60-nm particles by using a capillary of diameter 0.45μm [4]. Here we show that natural channel-forming peptides incorporated into a bilayer lipid membrane can be used to detect the passage of single molecules with gyration radii as small as 5–15Å. From our experiments with alamethicin pores we infer both the average number and

---

Author affiliations: Sergey M. Bezrukov, Igor Vodyanoy, V. Adrian Parsegian: Division of Intramural Research, NIDDK. V. Adrian Parsegian: Laboratory of Structural Biology, DCRT, National Institutes of Health. Sergey M. Bezrukov: St. Petersburg Nuclear Physics Institute of the Russian Academy of Sciences. Igor Vodyanoy: Office of Naval Research.

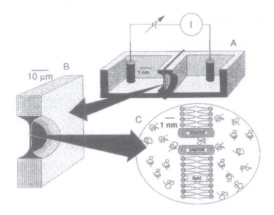

Fig. 1 Schematic representation of the cell (**A**), Teflon partition with a lipid bilayer (**B**), and a single ion pore exchanging polymers with bulk solution (**C**). Bilayer lipid membranes were prepared as described by Mantel and Mueller [18]. The membrane-forming solution was L-α-diphantonyl lecithin in n-pentane. Hexadecane in n-pentane (1:10) was used for aperture pre-treatment. Natural alamethicin purified as described elsewhere, was added as an ethanolic solution (after membrane formation) to the bathing solution from one side of the membrane only. Single channel spontaneous insertions were detected as current 'bursts' shown in Figure 2. PEGs of different M_r were added at 15%w/w concentration to aqueous 1 M NaCl solutions buffered at pH 6.2 by MES. All measurements were made at room temperature, at a transmembrane voltage of +150 mV on the side of peptide addition.

the diffusion coefficients of poly(ethylene glycol) molecules in the pore. Our approach provides a means of observing the statistics and mechanics of flexible polymers moving within the confines of precisely defined single-molecule structures.

The underlying idea of our experiment is very similar to the resistive pulse principle that has been used in Coulter counters since 1953. If a nonconducting particle suspended in a conducting medium moves into a small capillary, it decreases the conductance of the capillary relative to that of the capillary filled with the conducting medium alone. The magnitude of this decrease in conductance is related to particle size [3].

For a molecular pore of ~50Å length and ~10Å radius (similar to alamethicin pore dimensions [5, 6]), the brownian motion of particles dominates over directional flow at all reasonable hydrostatic pressure differences across the membrane. Estimates for flow transient time and diffusion relaxation time (see below) show that for 5Å-radius particles in water these times become equal only at hydrostatic pressure differences of ~10^8Pa, equivalent it to a 10-km-high water column.

To study diffusion-driven exchange of poly(ethylene glycol)s (PEGs) of different relative

molecular masses (M_r) in a molecular pore, we analysed polymer-induced conductance fluctuations of a single alamethicin channel [7] in the experiment schematically shown in Figure 1. Addition of polymers to the membrane-bathing solution changes the conductance of the pore-containing membrane: the amplitudes of conductance states decrease and the current noise of the open channel increases. Figure 2 illustrates these changes for three states of the alamethicin channel, comparing spontaneous current "bursts" in polymer-free solution with those recorded in solutions of 15 wt% PEG with M_r 600 (PEG 600).

Figure 3 shows that the magnitude of polymer-generated excess noise is different for different conductance levels and depends on polymer M_r.

Noise is not monotonic in polymer size. Rather, there is a clear maximum for M_r 600–1,000. Small polymers (M_r 200–400) produce relatively less noise, presumably because their effect is an average over a large number of small events. Large polymers ($M_r > 1,000$) are geometrically restricted from even entering the channel.

The average number of polymer molecules in the channel, $\langle N(w) \rangle$, can be determined from the degree of channel conductance reduction on addition

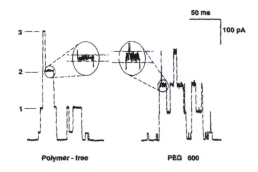

Fig. 2 Ion current of a single alamethicin channel as changed by polymer addition. (Left, polymer-free; right, after, addition of PEG 600). PEG 600 decreases the amplitude of different conductance levels (numbers 1, 2, and 3 at left) and induces additional current noise. Current records filtered by an 8-pole Bessel filter at 3 kHz are shown enlarged (2×) in insets for comparison. Different conductance levels of the channel correspond to different numbers of closely packed molecular pores (Fig. 1) of nearly uniform dimensions [5, 20]. The time/current scale in the right upper corner is common to both records.

of polymer [5, 8]. If the entry of a single polymer of M_r w lowers single-channel conductance by an amount $h(w)$, and if one is working at a level of occupancy $\langle N(w) \rangle$ such that $h(w) \langle (N(w) \rangle$ is much smaller than the mean conductance of polymer-free channel, one can write

$$h(w) \langle N(w) \rangle = h_{ch}(\infty) - h_{ch}(w)$$

Here h_{ch} is the mean conductance of the channel in the presence of polymers of M_r, w, and $h_{ch}(\infty)$ its conductance in the presence of totally excluded, large polymers. These average conductances are measured directly (see Figure 2).

We have verified previously [5] that the relative reduction of alamethicin channel conductance in the presence of small, easily penetrating PEG 200 is close to the relative reduction in the conductivity of bulk electrolyte solutions on polymer addition. This was also shown for another mesoscopic channel formed by staphylococcal α-toxin [9]. Taking into account the large size of these channels, this means that in the case of small polymers their incremental (negative) contribution to channel conductance is close to the contribution that one would expect for the conductance of an electrolyte layer whose thickness is equal to the channel length. This equivalence permits a straightforward derivation of $h(w)$ from the polymer-induced reduction in solution conductivity. As polymer size is increased, the channel starts to exclude polymers; the

proportional effect of polymer addition on channel conductance decreases. As long as the size of a polymer coil does not significantly exceed the size of the pore, and thus its shape and monomer density are not heavily distorted by the channel, $h(w)$ obtained in this way may serve as a good approximation.

An alternative way to determine $\langle N(w) \rangle$, and thus $h(w)$, is based on the observation that the reduction in solution conductivity is proportional only to the wt% concentration of polymer, that is, to monomer density, and does not depend on polymer M_r [5]. Supposing that the relative reduction in the pore conductance is also proportional to the density of monomers inside the pore, we calculate pore-bulk polymer partitioning as a function of polymer M_r, to obtain $\langle N(w) \rangle$.

Both methods give similar results; we use this second method as it employs less restrictive assumptions. It is also reassuring that polymer partitioning into the pore can be verified independently by measuring size-modulated osmotic response [8].

The mean value of square conductance fluctuations, $h(w)^2 \langle N(w) \rangle$, can also be calculated theoretically as a product of the low-frequency limit of conductance fluctuation spectral density $S_h(O)$ times the spectral width created by diffusion relaxation times. General considerations show that the bandwidth of these fluctuations should be proportional to the diffusion coefficient D of polymer in-

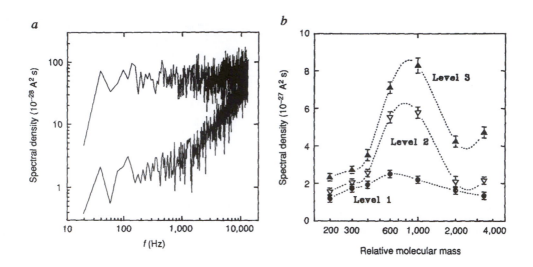

Fig. 3 Measurement of polymer-induced current fluctuations. **a**, Spectrum of open-channel noise in the presence of PEG 600 at level 2 (upper trace) is shown in comparison with the spectrum of the background noise obtained from the parts of the current recordings when the channel was closed (lower trace). **b**, Amplitude of the 'white' spectral part of the open-channel noise changes with polymer relative molecular mass M_r. The spectrum was averaged over a 200–2,000 Hz frequency range with the background subtracted. METHODS. In spectral measurements, an 8-pole Butterworth filter was employed, the corner frequency being chosen to be three-eighths of the sampling frequency (40kHz). To seperate open-channel noise from strong low-frequency contribution of channel switching between different conductance states, we first recorded the segments of the current corresponding to a particular conductance state into the computer memory; then, after cutting 0.2 ms off both ends of each segment to eliminate transient currents and after zeroing the mean value of each segment, we connected them into a 2,048-point vector to be used in Fourier transformation. This signal processing procedure is justified by the fact that characteristic times of current fluctuations within a particular state are orders of magnitude smaller than inter-state switching times. Nevertheless, to check for possible spectrum distortions due to a limited lifetime of the channel at a particlular state, we performed a specially designed test. A signal from a calibrated noise generator was admixed to the channel current recording and processed as described above. The spectra so obtained were then compared to those measured directly from the generator output. It turned out that the damping of the admixed noise spectra is significant only at frequencies lower than 50 Hz.

side the pore. To calculate the bandwidth for the case of a long pore we used a one-dimensional approximation. For the probability $P(x_0, x, \tau)$ to find a particle in position x at time τ if it had been in position x_0 at time $\tau = 0$

$$D(\partial^2 P(x_0, x, \tau)/\partial x^2) = \partial P(x_0, x, \tau)/\partial \tau$$

For a narrow channel, the probability that a particle that has left the channel will return diminishes rapidly as the particle moves from the mouth of the channel. Placing absorbing boundaries at both ends of the capillary, that is at $x = 0$ and $x = L$, we solve the problem using methods described elsewhere [10].

Formally, for the spectral density, $S_h(O)$, of conductance fluctuations at frequency $f = 0$, we obtain

$$S_h(O) = S_h(f)|_{f=0} = \langle (\delta h)^2 \rangle L^2/3D$$

where $\langle (\delta h)^2 \rangle$ is the time-averaged square fluctuation. The diffusion generated bandwidth of pore conductance fluctuations is $3D/L^2$. The corresponding time constant that describes averaged relaxation of polymer number fluctuation inside the pore is equal to $L^2/6\pi D$. This value is remarkably close to the intuitive result $L^2/12D$ reported by Feher and Weissman [11].

Experimentally, we obtain $S_h(0) = S_I(0)/V^2$ from the measured low-frequency 'white' part

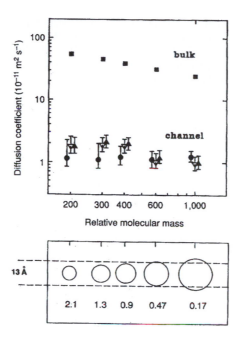

Fig. 4 Top, Diffusion coefficients of polymers inside the channel calculated from polymer-induced current noise (symbols for different levels correspond to those in Fig. 3b). An order of magnitude decrease in comparison with their values in bulk water is already expected from restricted-diffusion theory [13] for the smallest polymer, PEG 200. Bottom, PEG gyration diameters are compared with the size of the alamethicin channel opening (13Å). Numbers below the gyration diameters correspond to the average number of PEG molecules in the channel (level 1).

of the current noise spectrum $S_I(0)$ and transmembrane voltage V. Equating time-averaged and ensemble-averaged square fluctuations, $\langle(\delta h)^2\rangle$ and $h(w)^2\langle N(w)\rangle$, for the polymer diffusion coefficient inside the pore we have

$$D(w) = h(w)^2 L^2 \langle N(w)\rangle V^2 / 3 S_I(0)$$

The diffusion coefficients calculated for the polymers of different M_r are shown in Figure 4, together with their values in bulk aqueous solution. We assume channel length = 40Å and radius $R = 6.5$Å, based on results of measurements [5] and on molecular model considerations [6].

The smallest polymer used in our experiments, PEG 200, has a gyration radius of 4.0Å (ref. 12), already quite close to the pore radius. Formal application of restricted-diffusion theory [13] in this case gives the right order of magnitude for the reduction in diffusion rate (Figure 4) even though its

application is highly questionable here. Restricted-diffusion theory is developed for hard spheres, whereas PEG polymers in aqueous solutions exist as well-characterized random coils [14]. The observed order-of-magnitude-slower polymer diffusion in the pore can have many causes: size restriction, friction with the walls, higher viscosity of water in the pore, and so on. The absence of any significant dependence of the diffusion coefficient on the polymer M_r disagrees with the expected results of restricted-diffusion considerations for hard spheres, and might reflect the confinement of flexible coil in a pore whose dimensions are comparable to the size of the polymer. Although the viscosity of water inside the channel may be different from the bulk, the observed decrease in diffusion coefficient does not provide a quantitative measure of this.

It is important to note here that the diffusion coefficient obtained in our experiments describes transport of polymers only once they are inside the pore. To describe the overall diffusion of polymers across the whole membrane, one should take into account partitioning of polymers between the pore and bulk solution. Partitioning of polymers in the pore quickly decreases with increasing polymer M_r (Figure 4); so does diffusion across the membrane.

Fluctuation analysis of currents through ion channels, performed to obtain dwell times of charged permeant species, has been successfully used previously. Classic examples include studies on Ca^{2+} block of monovalent currents through calcium channels [15] and on Na^+ block of proton currents through the gramicidin A channels [16]. Now that strong evidence for the existence of protein-conducting channels is starting to emerge [17], the importance of studies on neutral-permeant transport is obvious. Using the approach outlined here, it should be possible to conduct such studies at a single-channel level.

REFERENCES

1. H. E. Kubitschek, *Nature* **182**, 234 (1958).
2. T. Allen, *Particle Size Analysis* 110–127 (Society for Anlytical Chemistry Publishers, London, 1967).
3. L. G. Bunville, *Modern Methods of Particle Size Analysis* (ed. H. G. Barth) 1–42 (Wiley, New York, 1984).
4. R. W. DeBlois *et al.*, *J. Colloid Interface Sci.* **61**, 323 (1977).
5. S. M. Bezrukov and I. Vodyanoy, *Biophys. J.* **64**, 16 (1993).
6. M. S. P. Sansom, *Eur. Biophys. J.* **22**, 105 (1993).
7. J. E. Hall *et al.*, *Biophys. J.* **45**, 233 (1984).
8. I. Vodyanoy *et al.*, *Biophys. J.* **65**, 2097 (1993).
9. O. V. Krasilnikov *et al.*, *FEMS Microbiol. Immunol.* **105**, 93 (1992).
10. S. M. Bezrukov and I. Vodyanoy, *Membrane Electrochemistry* (Advances in Chemistry Ser. No. 235, American Chemical Soc., Washington DC, in press).
11. G. Feher and M. Weissman, *Proc. Nat. Acad. Sci.* **70**, 870 (1973).
12. S. J. Kuga, *J. Chromatography* **206**, 449 (1981).
13. C. P. Bean, *Membranes* (ed. G. Eisenman) (Dekker, New York, 1972).
14. A. Couper and R. F. T. Stepto, *Trans. Faraday Soc.* **65**, 2486 (1969).
15. P. Hess and R. W. Tsien, *Nature* **309**, 453 (1984).
16. S. H. Heinemann and F. Sigworth, *J. Biochim. Biophys. Acta* **987**, 8 (1989).
17. S. M. Simon and G. Blobel, *Cell* **65**, 371 (1991).
18. M. Montal and P. Mueller, *Proc. Nat. Acad. Sci.* **69**, 3561 (1972).
19. T. M. Balasubramanian *et al.*, *J. Am Chem. Soc.* **103**, 6127 (1981).
20. L. G. M. Gordon and D. A. Haydon, *Phil. Trans. R. Soc. B* **270**, 433 (1975).

Exercises for Chapter 14

More Practice with Partial Derivatives

Compute the partial derivatives $\frac{\partial}{\partial t} f$ and $\frac{\partial}{\partial x} f$ in the cases when the function $f(t, x)$ is given:

1. $f(t, x) = \sin(x - 3t)$

2. $f(t, x) = e^{-t} \cos(x)$

3. $f(t, x) = t^{-1/2} e^{-x^2/t}$

4. $f(t, x) = (1 + (t - x)^2)^{-1}$

Visualizing x- and t-Dependent Functions

In Exercises 5–7, roughly sketch a graph of $f(1, x)$, $f(5, x)$, and $f(10, x)$ on the same piece of graph paper. Use different colors for $f(1, x)$, $f(5, x)$, and $f(10, x)$ so that you can tell them apart. Make the x-axis scale stretch from -20 to 20. (*Hint:* Use a calculator.)

5. $f(t, x) = e^{-(x-2t)^2}$

6. $f(t, x) = e^{-t}(1 + (t - x)^2)^{-1}$

7. $f(t, x) = t^{-1/2}e^{-x^2/t}$

Find the Equation Given the Function

8. The functions in Exercises 5 and 6 solve Equation (4) from Chapter 14 with the appropriate values in each case for the constants c and r. In each case, specify these values for c and r. Also, the function in Exercise 7 solves Equation (8) of Chapter 14 when the constant μ has a certain value. Specify this value for μ.

Diffusion Equation Problems

9. Verify that the function in Equation (7) from Chapter 14 gives a solution to Equation (8) of Chapter 14 by taking the appropriate derivatives.

10. Verify by taking the appropriate derivatives that the following are solutions to Equation (8) of Chapter 14:

   (a) $e^{\lambda t}e^{x(\lambda/\mu)^{1/2}}$ where λ is any constant

   (b) $e^{\lambda t}e^{-x(\lambda/\mu)^{1/2}}$ where λ is any constant

   (c) $a + bx$ where a and b are any constants

   (d) $e^{-\lambda t}\cos((\lambda/\mu)^{1/2}x)$ where λ is any constant

   (e) $e^{-\lambda t}\sin((\lambda/\mu)^{1/2}x)$ where λ is any constant.

11. Verify by taking the appropriate partial deriatives that $u(t, x) = \frac{a}{t^{1/2}}e^{rt}e^{-x^2/4\mu t}$ is a solution to

$$\frac{\partial}{\partial t}u = \mu\frac{\partial^2}{\partial x^2}u + ru \qquad (9)$$

when r is a constant.

12. Verify by taking the appropriate derivatives that the following are solutions to (9):

   (a) $e^{\lambda t}e^{x((\lambda-r)/\mu)^{1/2}}$ where λ is any constant with $\lambda > r$

   (b) $e^{\lambda t}e^{-x((\lambda-r)/\mu)^{1/2}}$ where λ is any constant with $\lambda > r$

   (c) $e^{rt}(a + bx)$ where a and b are any constants

   (d) $e^{\lambda t}\cos[((r - \lambda)/\mu)^{1/2}x]$ where λ is any constant with $\lambda < r$

   (e) $e^{\lambda t}\sin[((r - \lambda)/\mu)^{1/2}x]$ where λ is any constant with $\lambda < r$

15

Two Key Properties of the Advection and Diffusion Equations

We are studying differential equations for a function $u(t, x)$ of two variables, t and x. Two of the simplest such equations have the form

$$\frac{\partial}{\partial t} u = \mu \frac{\partial^2}{\partial x^2} u + ru \tag{1}$$

and

$$\frac{\partial}{\partial t} u = -c \frac{\partial}{\partial x} u + ru. \tag{2}$$

Here, $\mu > 0$ and c and r are constants. [The sign convention for r in Equation (2) and in Equation (3) on the next page is the opposite from that in the previous chapter; that used here is convenient for Chapter 16.]

15.1   These Equations Are Predictive

It is important to realize that both of these equations are *predictive* in the following sense:

- If you specify the value of u at $t = 0$ as a function of x, then both equations determine $u(t, x)$ for you at all times $t > 0$.

This property of being predictive is what makes these equations so useful. The predictive nature of these equations can be proved by making rigorous the following argument: If you know $u(0, x)$, then the equation tells you what $\frac{\partial}{\partial t} u$ is at $t = 0$ and at any x. This tells you (approximately) $u(\Delta t, x)$ for small Δt and all x. (But accuracy increases as Δt goes to zero.) Plug your newfound knowledge of $u(\Delta t, x)$ into the left side of Equation (1) or Equation (2) to learn what u is at $t = \Delta t$ and any x. This tells you (approximately) $u(2\Delta t, x)$ for small Δt and all x. Plug your newfound knowledge of $u(2\Delta t, x)$ into the left side of Equation (1) or Equation (2) to learn what u is at the next time step, $t = 2\Delta t$ and any $x \dots$, and so on.

15.2   Examples

We saw already that the solution to the second equation is given by

$$u(t, x) = e^{rt} u(0, x - ct). \tag{3}$$

Thus, knowing $u(0, x)$ for all x allows you to compute $u(t, x)$ for any t and x.

The answer for the first equation is more complicated:

$$u(t, x) = \frac{1}{(4\pi \mu t)^{\frac{1}{2}}} e^{rt} \int_{-\infty}^{\infty} u(0, s) e^{-(x-s)^2/4\mu t} ds. \tag{4}$$

[You should think of the function $e^{-(x-s)^2/4\mu t}$ as a function of three variables, x, s, and t. Then (4) integrates over the s-variable while keeping t and x fixed to obtain a function of x and t only.] The integral in (4) is often too complicated to be useful, so I won't be referring to it. However, the fact that (4) is so complicated means that we must develop some other techniques for finding solutions to (1).

15.3   Superposition

Both equations (1) and (2) obey the **superposition principle**. This principle is as follows: Suppose that $u_1(t, x)$ and $u_2(t, x)$ are both solutions to either (1) or (2) and that a_1 and a_2 are any real numbers. Then

$$u(t, x) = a_1 u_1(t, x) + a_2 u_2(t, x)$$

is a solution to (1) or (2), as the case may be. This superposition principle will be used over and over again, so please remember it. Equations for which the superposition principle holds are called **linear equations**. Generally, the superposition principle will fail if the equation involves powers of u greater than 1, or a function of u multiplying u or its derivatives, or functions of derivatives of u multiplying derivatives of u. For example, the equation $\frac{\partial}{\partial t} u = \mu \frac{\partial^2}{\partial x^2} u + ru^2$ does not obey the superposition principle.

Argue as follows to check the superposition principle for (2):

Step 1: Since a_1 and a_2 are constants, we compute

$$\frac{\partial}{\partial t}u = \frac{\partial}{\partial t}(a_1 u_1 + a_2 u_2) = a_1\left(\frac{\partial}{\partial t}u_1\right) + a_2\left(\frac{\partial}{\partial t}u_2\right). \tag{5}$$

This is because the derivative of a sum of functions is the sum of the derivatives, and the derivative of a real number times a function is the real number times the derivative of the function.

Step 2: Likewise, compute

$$\frac{\partial}{\partial x}u = \frac{\partial}{\partial x}(a_1 u_1 + a_2 u_2) = a_1\left(\frac{\partial}{\partial x}u_1\right) + a_2\left(\frac{\partial}{\partial x}u_2\right) \tag{6}$$

and multiply this last expression by $-c$ to get

$$-c\frac{\partial}{\partial x}u = -c\frac{\partial}{\partial x}(a_1 u_1 + a_2 u_2) = a_1\left(-c\frac{\partial}{\partial x}u_1\right) + a_2\left(-c\frac{\partial}{\partial x}u_2\right). \tag{7}$$

Step 3: Compute

$$ru = r(a_1 u_1 + a_2 u_2) = a_1(ru_1) + a_2(ru_2). \tag{8}$$

Step 4: Add the right sides of (7) and (8), and likewise add the left sides of these equations. Since the right side of each is equal to the left side of each, you will find that

$$-c\frac{\partial}{\partial x}u + ru = a_1\left(-c\frac{\partial}{\partial x}u_1 + ru_1\right) + -a_2\left(-c\frac{\partial}{\partial x}u_2 + ru_2\right). \tag{9}$$

Step 5: Finally, because u_1 and u_2 satisfy (2), the right-hand side of (9) is equal to

$$a_1\left(\frac{\partial}{\partial t}u_1\right) + a_2\left(\frac{\partial}{\partial t}u_2\right),$$

which is the right side of (5). Thus, the right side of (9) is equal to $\frac{\partial}{\partial t}u$ as it has to be if u is to satisfy Equation (2) as claimed.

Please convince yourself of the validity of the superposition principle for (1).

This superposition principle is *extremely important*; we shall use it repeatedly to construct complicated solutions to (1) and (2) from a basic collection of simple solutions. For example, a basic solution to (1) for $r = 0$ is

$$u_0(t, x) = \frac{1}{t^{1/2}}e^{-x^2/4\mu t} \tag{10}$$

or, more generally, if x_0 is any given point, then

$$u_{x_0}(t, x) = \frac{1}{t^{1/2}}e^{-(x-x_0)^2/4\mu t} \tag{11}$$

is a solution to the $r = 0$ version of Equation (1). Using the superposition principle, I can immediately conclude that I can select any number of points $\{x_0, x_1, \dots\}$ and a like number $\{a_0, a_1, \dots\}$ of numbers (i.e., $x_0 = 12, x_1 = -2.4, \dots$, and $a_0 = 5, a_1 = -.3$, $\dots$) and then

$$a_0 \frac{1}{t^{1/2}} e^{-\frac{(x-x_0)^2}{4\mu t}} + a_1 \frac{1}{t^{1/2}} e^{-\frac{(x-x_1)^2}{4\mu t}} + \cdots \tag{12}$$

is also a solution to (1).

15.4   Lessons

Here are some key points from this chapter:

- Both Equations (1) and (2) are predictive in the following sense: Pick any function of the variable x, call it $u_0(x)$, and then both equations have a unique solution $u(t, x)$ that equals your chosen function at $t = 0$. That is, $u(0, x) = u_0(x)$.

- Both equations obey the superposition principle. This is to say that if you have two solutions and you multiply each by a chosen number and add the resulting functions together, then the function of t and x so obtained also solves the equation.

READINGS FOR CHAPTER 15

READING 15.1

Diffusional Mobility of Golgi Proteins in Membranes of Living Cells

Commentary: This article reports on measurements of the diffusion coefficients (the constant μ in the equation $\partial_t u = \mu \partial_x^2 u$) for proteins that reside in the membranes of the Golgi complex. The reason for doing this is that the Golgi complex contains a number of resident components which help process and sort membrane proteins. These resident components are maintained in the Golgi complex despite a large flux of proteins and lipids through the complex. So, how are they maintained in the face of this flow? All models propose protein-protein or protein-lipid interactions which keep the resident components from flowing away. (That is, they stick to something.) To search for these interactions, the authors set up an experiment with fluorescently tagged membrane components to measure their diffusion coefficients within the Golgi and the endoplasmic reticulum. (The fluorescent tagging allows you to witness over time the motion of the tagged molecules.) Remarkably, all of the components under consideration had comparatively large diffusion coefficients—they move easily within the Golgi membrane and endoplasmic reticulum. On the face of it, this ease of movement is incompatible with the various explanations for the fact that these components are retained in the Golgi complex.

Diffusional Mobility of Golgi Proteins in Membranes of Living Cells

Science, **273** (1996) 797–801.

Nelson B. Cole, Carolyn L. Smith, Noah Sciaky, Mark Terasaki, Michael Edidin,
Jennifer Lippincott-Schwartz

The mechanism by which Golgi membrane proteins are retained within the Golgi complex in the midst of a continuous flow of protein and lipid is not yet understood. The diffusional mobilities of mammalian Golgi membrane proteins fused with green fluorescent protein from Aequorea victoria were measured in living HeLa cells with the fluorescence photobleaching recovery technique. The diffusion coefficients ranged from 3×10^{-9} square centimeters per second to 5×10^{-9} square centimeters per second, with greater than 90 percent of the chimeric proteins mobile. Extensive lateral diffusion of the chimeric proteins occurred between Golgi stacks. Thus, the chimeras diffuse rapidly and freely in Golgi membranes, which suggests that Golgi targeting and retention of these molecules does not depend on protein immobilization.

The Golgi complex contains a large number of resident components that play important roles in the processing and sorting of secretory and membrane proteins, but how these components are maintained in the Golgi despite a continuous flow of protein and lipid through the secretory pathway is currently a topic of debate [1]. Several mechanisms of Golgi protein retention have been suggested: oligomerization into structures too large to enter transport vesicles [2], lateral segregation into lipid microdomains [3], and recognition of retention or retrieval signals [4]. In these models specific protein-protein or protein-lipid interactions underlie Golgi protein retention.   Whether such interactions affect the dynamic properties of Golgi membrane proteins in vivo, including the diffusional mobility of Golgi proteins and trafficking of these proteins between Golgi stacks, has not been addressed.

To probe for interactions that might underlie the retention of Golgi membrane proteins, we examined the diffusional mobility of Golgi membrane components using fluorescence photobleaching recovery. (FPR). This technique is a powerful tool for investigating the environment of membrane proteins and has revealed several types of interactions that constrain the lateral diffusion of proteins in the plasma membrane [5]. The dynamic properties of intracellular membrane proteins have not been thoroughly explored by FPR because of a lack of appropriate fluorescent labels. Here, we used the Aequorea victoria green fluorescent protein (GFP) [6] as a tag to investigate the diffusional mobility of four Golgi membrane proteins [7], mannosidase II (Man II), β-1,4-galactosyltransferase (GalTase), and wild-type and mutant forms of KDEL receptor (KDELR), within the Golgi and the endoplasmic reticulum (ER). We also used fluorescence loss in photobleaching to examine the extent of lateral membrane continuity between Golgi stacks, and within the ER.

Man II and GalTase are "resident" enzymes of the Golgi complex, which function in carbohydrate processing. In contrast, KDELR is an itinerant Golgi component, which recycles to the ER when it binds KDEL ligand [8].   Substitution of asparagine for aspartic acid at position 193 in KDELR, denoted KDELR$_m$, prevents it from redistributing into the ER in the presence of high concentrations of ligand [9].   GFP was fused to full-length Man II at its lumenally oriented COOH-terminus (Man II-GFP), to GalTase truncated at position 60 in its lumenal domain (GalTase-GFP), and to the cytoplasmically oriented COOH-terminus of wild-type (KDELR-GFP) and mutated KDELR (KDELR$_m$-GFP) [10].

When each of the GFP-tagged proteins was expressed in HeLa cells, fluorescence was localized almost exclusively to juxtanuclear Golgi structures (Figure 1, A to D). Observations in cells treated with brefeldin A (BFA), which causes Golgi membranes

Author affiliations: N. B. Cole, N. Sciaky, J. Lippincott-Schwartz: Cell Biology and Metabolism Branch, National Institute of Child Health and Human Development. C. L. Smith: National Institute of Neurological Disorders and Stroke, National Institutes of Health. M. Terasaki: Department of Physiology, University of Connecticut. M. Edidin: Department of Biology, The Johns Hopkins University.

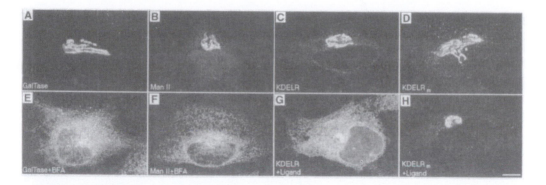

Fig. 1 Expression of Golgi-targeted GFP chimeras revealed by fluorescence microscopy of living HeLa cells transfected with GalTase-GFP, Man II-GFP, KDELR-GFP, or KDELR$_m$-GFP [26]. (**A** to **D**) Chimera distributions in untreated cells. Distributions of GalTase-GFP (**E**) and Man II-GFP (**F**) after a 30-min treatment with BFA (2 μg/ml), and KDELR (**G**) or KDELR$_m$ (**H**) after overexpression of lysozyme-KDEL (ligand). Each of the chimeras colocalized extensively with native protein labeled with antibodies in untreated, fixed cells. Scale bar = 10μm.

but not other membranes to redistribute into the ER [11], further supported a Golgi localization of the GFP chimeras (Figure 1, E and F). Overexpression of lysozyme-KDEL (ligand for KDELR) caused KDELR-GFP, but not KDELR$_m$-GFP, to redistribute into the ER (Figure 1, G and H), as observed in previous studies characterizing wild-type and defective KDEL receptors [9]. Thus, the GFP tag did not interfere with Golgi targeting of KDELR, Man II, and GalTase or with ligand-induced recycling of KDELR.

The diffusional mobilities of the GFP chimeras within Golgi membranes were examined in living cells at 37 degrees C by bleaching out fluorescence in a Golgi region with a high-energy laser scan of a confocal microscope and then using lower intensity illumination to image the recovery of fluorescence by means of diffusion of unbleached GFP chimeras into this region. Fluorescence recovery into the photobleached region was rapid for all of the GFP chimeras (Figure 2). The mobile fraction of GFP chimeras was determined in a series of images by calculating the ratio of fluorescence intensities for two regions of interest, one within the bleaching zone and the other outside of it, before photobleaching and after recovery (see boxed areas in Figure 2, D and E, for GFP-KDELR$_m$ and GFP-Man II, respectively). The ratio after recovery ranged between 85 to 98% of the ratio before photobleaching in multiple experiments performed with all of the chimeras. Thus, nearly all of the GFP chimeras were mobile in Golgi membranes. Recovery was due

to diffusion within a continuous membrane compartment rather than transfer of fluorescence by fusion of unbleached vesicles with the bleached region, because fluorescence recovered at the same rate in cells at reduced temperatures (22 degrees C) or upon energy depletion with a mixture of 2-deoxyglucose and sodium azide, conditions where vesicle transport has been found to be significantly slowed or blocked [12].

In contrast to cells in control medium, the Golgi of cells treated with aluminum fluoride (AlF) for 30 min showed essentially no recovery from photobleaching (Figure 2B). This effect of AlF appeared to be specific to Golgi because GFP chimeras localized within the ER after BFA treatment recovered normally after photobleaching in the presence of AlF (Figure 2C). AlF is known to cause extensive binding of peripheral protein complexes to Golgi membranes, similar to the effects of guanosine 5'-O-(3'-thiotriphosphate) (GTP-γ-S) [13], which could constrain the lateral movement of Golgi membrane proteins [14].

Quantitative studies of the mobility of the chimeras were carried out with a conventional (non-scanning) light microscope and laser system designed for FPR experiments (Table 1). Fluorescence loss and recovery was measured in a 2-μm stripe placed across the Golgi of individual cells at random orientations. The diffusion coefficients, D, calculated from these measurements ranged from 3×10^{-9} to 5×10^{-9} cm^{2}s^{-1}. These values are

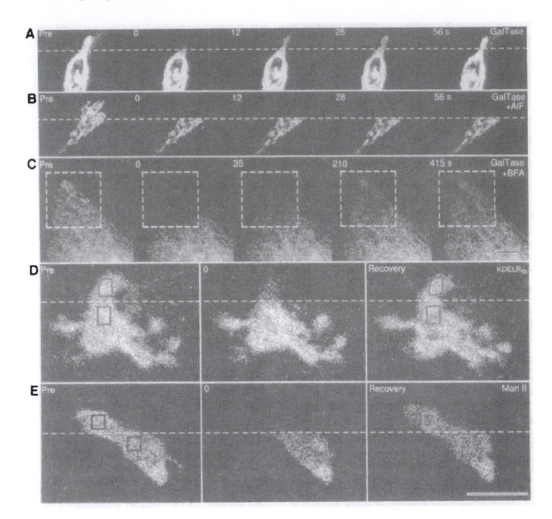

Fig. 2 Fluorescence photobleaching recovery of GFP chimeras in Golgi and ER membranes. Sequence of images of GalTase-GFP in Golgi of a control cell (**A**) and an AlF-treated cell (**B**) before and after photobleaching. The region of Golgi above the dotted line was photobleached immediately after the first image (Pre). In (**A**) the photobleached area rapidly recovered fluorescence, whereas in (**B**) there was essentially no recovery. (**C**) Sequence of images of GalTase-GFP in the ER of a BFA-treated cell before (Pre) and after photobleaching of the boxed region. Fluorescence returned to the photobleached region of the ER in BFA-treated cells whether or not AlF was present. The region of Golgi above the dotted line in (**A**) and (**B**) or of ER within the boxed area in (**C**) was photobleached by scanning it three times with the highest laser energy (100% power, 100% transmission). Recovery of fluorescence into the photobleached region was then observed by imaging the entire Golgi element at low energy (10% power, 3% transmission) at the times indicated after the bleach. Before photobleaching, the cells in (**B**) were treated with AlF (30 mM NaF and 50μM AlCl$_3$) for 30 min, and in (**C**) they were treated with BFA (2 μg/ml) for 1 hour. (**D** and **E**) Method used to calculate mobile fraction of GFP chimeras. Fluorescence intensities in the boxed regions of interest (ROIs) were measured before bleaching (Pre) and after recovery (Recovery). The ratio of bleached to unbleached ROIs at these two time points were used to calculate the mobile fraction (mobile fraction = ratio after recovery/ratio before bleaching). For KDELR$_m$-GFP, the mobile fraction was 98%, whereas for Man II-GFP it was 90%. The absolute fluorescence associated with Golgi elements was lower after photobleaching than before photobleaching because photobleaching removed a significant proportion of total fluorescence from the Golgi. The method described above for calculating mobile fraction is insensitive to this loss. Scale bar in (**C**) is equivalent to 10μm in (**C**) and 5μm in (**A**) and (**B**). Scale bar in (**E**) is 5μm for (**D**) and (**E**). A Quicktime movie sequence from the FPR experiment is available at http://www.uchc.edu/htterasaki/flip.html.

Table 1.

Chimera	Location	Condition	D value $(\times 10^{-9}\,\text{cm}^2\,\text{s}^{-1})$	Number measured
Man II-GFP	Golgi	Untreated	3.2 ± 0.3	18
GalTase-GFP	Golgi	Untreated	5.4 ± 0.4	22
GalTase-GFP	ER	BFA	2.1 ± 0.2	24
KDELR-GFP	Golgi	Untreated	4.6 ± 0.5	14
KDELR-GFP	ER	BFA	4.3 ± 0.5	15
KDELR-GFP	ER	+ Ligand	2.6 ± 0.3	15
KDELR$_m$-GFP	Golgi	Untreated	4.6 ± 0.5	15
KDELR$_m$-GFP	Golgi	+ Ligand	4.6 ± 0.4	24
KDELR$_m$-GFP	ER	BFA + Ligand	4.5 ± 0.5	38

Diffusion coefficients (D) of GFP chimeras in Golgi and ER membranes. HeLa cells were transfected with GalTase-GFP, Man II-GFP, KDELR-GFP, or KDELR$_m$-GFP cDNAs with or without lysozyme-KDEL (ligand) cDNA to allow expression of the chimeras. Cells were treated with or without BFA (2 µg/ml, for 30 min) before photobleaching [25]. D for Man II-GFP was obtained for cells at 37°C, whereas for GalTase-GFP, KDELR-GFP, and KDELR$_m$-GFP, D was obtained for cells at 22°C. D measured at 37°C for GalTase-GFP, KDELR-GFP, and KDELR$_m$-GFP closely resembled the values obtained at 22°C. The geometric assumption for calculating D was that recovery is due to one-dimensional diffusion [25]. Percentage recovery of fluorescence was 60 to 70% in Golgi and 80 to 90% in ER. This difference in recovery is likely because photobleaching removed a significant proportion of total fluorescence from the Golgi but not from the ER. The standard error of the mean is indicated for each chimera and condition.

comparable with the largest D known for any cell membrane protein, that for rhodopsin in rod outer segments [15]. They are three to five times those reported for antibody-labeled vesicular stomatitis virus G glycoprotein in the Golgi complex [16], and they are 10 to 30 times the D measured for many plasma membrane proteins, whose lateral mobility is typically constrained by interactions with each other, with components of the extracellular matrix, or with the cytoskeleton [15]. The high D values indicate that the lateral diffusion of the GFP chimeras in Golgi membranes was not hindered by interactions such as aggregation with other proteins.

The exceptionally high mobile fraction and diffusion coefficients of the GFP chimeras prompted us to investigate the extent to which Golgi membranes are continuous and whether all regions of the Golgi complex can contribute to recovery of fluorescence at a bleached site. A variation of FPR was used in which fluorescence loss outside the photobleached zone was monitored (called FLIP for fluorescence loss in photobleaching). A narrow box across the Golgi complex was bleached repeatedly and the fluorescence loss in Golgi membranes outside this box was followed over time. The GalTase-GFP fluorescence associated with Golgi elements extending outside the box was lost completely after 10 intense illuminations of the boxed region

(elapsed time was 360 s) (Figure 3B). This observation suggested that during this time period nearly all of the GalTase-GFP had diffused into the box to become photobleached. The Golgi complex at the upper right-hand corner of Figure 3B, which was contained within an adjacent cell, remained bright, indicating that the low-intensity laser illumination used for imaging between bleaching radiations led to minimal photobleaching. In some cells, fluorescence associated with particular Golgi sites near the bleached line was lost at different rates (Figure 3, A and D), suggesting differences in local densities or connectedness (or both) of adjacent Golgi stacks. Movement of GalTase-GFP between Golgi stacks appeared to be mediated by lateral diffusion rather than vesicular transport because FLIP was not slowed at 22°C or by energy depletion [17].

The FLIP experiments [18] were consistent with the exceptionally high mobility and diffusion rates of the GFP chimeras in Golgi membranes measured in the FPR studies, and the results imply that Golgi stacks normally are extensively interconnected, with Golgi membrane components moving rapidly from one part of the Golgi to another. Previous studies have demonstrated lateral movement of lipid, but not of protein, across the Golgi [19]. The tubulovesicular elements seen connecting membranes of adjacent Golgi stacks in ul-

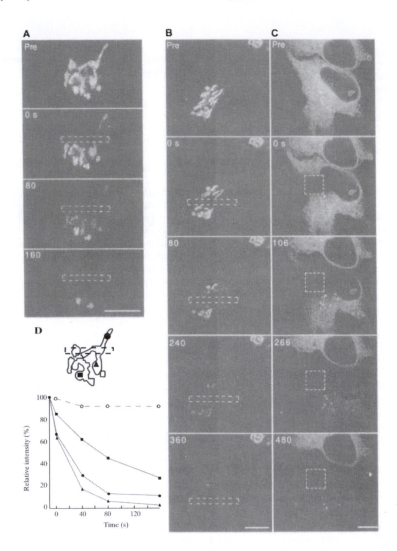

Fig. 3 FLIP of GalTase-GFP in Golgi and ER membranes. HeLa cells expressing GalTase-GFP were untreated (**A** and **B**) or treated with BFA (2 μg/ml) for 30 min (**C**) before the FLIP experiment, which was performed on a 37°C stage of a Zeiss LSM 410 microscope. A small rectangular region defined by the boxed area was repeatedly illuminated with the highest energy of the laser (100% power, 100% transmission, scan time of 30 s). Between each intense illumination, the entire field of view was imaged at low-power laser light (10% power, 3% transmission) to assess the extent that fluorescence outside the box was lost as a consequence of photobleaching within the box. In (**A**) and (**B**), areas of Golgi adjacent to the boxed area that was photobleached rapidly lost fluorescence, but a Golgi in an adjacent cell [(**B**), upper right hand corner] remained bright. Different zones within the Golgi in (**A**) lost fluorescence at different rates. The rate of fluorescence loss for three selected areas is plotted in (**D**). The open circles (**D**) show the intensity of fluorescence at the same time points in the Golgi of an adjacent cell, which remained bright. In (**C**), FLIP of a zone within the ER gradually depleted fluorescence from the entire ER of the affected cell, but not the adjacent cell above it. That Golgi and ER membranes inside the bleached zone were not damaged during exposure to the intense light was confirmed by fixing cells after photobleaching and staining with Golgi-specific antibodies, which revealed intact Golgi and ER structures. The possibility that regions on the edge of the illuminated zone are progressively bleached by light leakage during FLIP was ruled out by repeat of FLIP on fixed cells, which showed bleaching only in the area exposed to intense light illumination. Scale bars = 10μm. A Quicktime movie sequence from the FLIP experiment is available at http://www.uchc.edu/htterasaki/flip.html.

trastructural studies [20] could be the conduit by which these proteins and lipids diffuse between adjacent Golgi stacks. FLIP offers a simple method for providing insight into the nature and role of these membrane connections by revealing conditions of and rates of interstack diffusion. Because individual Golgi cisternae are not selectively photobleached in the FLIP experiments, the question remains as to whether intercisternal diffusion occurs within a single stack [21].

FLIP experiments were also used to investigate the mobility of GalTase-GFP redistributed into the ER by BFA (Figure 3C). Repeated photobleaching of a small rectangular region of the ER over 480 s resulted in the complete loss of ER-associated fluorescence from the photobleached cell (including its nuclear envelope). Thus, GalTase-GFP was free to diffuse throughout ER membranes, and all membranes of this compartment are completely interconnected, as expected [22].

To test whether the diffusibilities of the GFP chimeras in ER membranes were similar to those in Golgi membranes, we measured D by FPR of chimeras redistributed into the ER by either BFA treatment or overexpression of lysozyme-KDEL. Diffusion of the chimeras within ER membranes also was rapid with D ranging from 2.1×10^{-9} to 4.5×10^{-9} cm^{2}s^{-1} (Table 1), although D values for GalTase and KDELR (with overexpressed ligand) were only half as large as in Golgi membranes. Whether the lower D for these proteins reflects differences in their physical properties and interactions with luminal ER proteins, or reflects differences in the geometry of ER and Golgi membranes, remains to be determined.

These results have important implications for our understanding of Golgi membrane protein retention and mobility. The GFP-tagged proteins all diffused rapidly and freely within Golgi membranes, as well as within ER membranes. The high D and mobile fraction of these proteins appear inconsistent with mechanistic explanations of Golgi protein targeting and retention by means of the formation of large, relatively immobile protein complexes [2, 23]. Furthermore, rapid and widespread loss of fluorescence from the Golgi upon repeated bleaching of a small zone within the complex implies significant lateral diffusion rather than vesicular transport of the GFP-tagged proteins between Golgi stacks. Golgi models [24] thus need to account for how Golgi membranes maintain their

identity amidst rapid diffusion of resident components, and they need to explain the role played by these dynamic membrane properties in Golgi structure and function.

REFERENCES

1. I. Mellman and K. Simons, *Cell* **68**, 829 (1992); T. Nilsson and G. Warren, *Curr. Opin. Cell Biol.* **6**, 517 (1994); S. Munro, *EMBO J.* **14**, 4695 (1995).

2. T. Nilsson, P. Slusarewicz, M. H. Hoe, G. Warren, *FEBS Lett.* **330**, 1 (1993); C. E. Machamer, *Curr. Opin. Cell Biol.* **5**, 606 (1993).

3. M. S. Bretscher and S. Munro, *Science* **261**, 1280 (1993).

4. M. R. Jackson, T. Nilsson, P. A. Peterson, *J. Cell Biol.* **121**, 317 (1993); J. P. Luzio and G. Banting, *Trends Biochem. Sci.* **18**, 395 (1993).

5. M. Edidin, in *Mobility and Proximity in Biological Membranes*, S. Damjanovich, M. Edidin, J. Szollosi, L. Tron, Eds. (CRC Press, Boca Raton, FL, 1994), pp. 109–135.

6. D. C. Prasher, V. K. Eckonrode, W. W. Ward, F. G. Prendergast, M. J. Cormier, *Gene* **15**, 111 (1992); M. Chalfie, Y. Tu, G. Euskirchen, W. W. Ward, D. C. Prasher, *Science* **263**, 802 (1994).

7. Previous work suggested that Man II forms large oligomers within Golgi cisternae [2], which could constrain its mobility. In contrast, KDELR, in the absence of bound ligand, is not thought to oligomerize or be constrained in its membrane mobility [8, 9].

8. M. J. Lewis and H. R. B. Pelham, *Cell* **68**, 353 (1992).

9. F. M. Townsley, D. W. Wilson, H. R. B. Pelham, *EMBO J.* **12**, 2821 (1993).

10. The GFP coding region from cDNA pGFP10.1 [6], in which Ser^{65} was mutated to Thr [R. Heim, A. B. Cubitt, R. Y. Tsien, *Nature* **373**, 663 (1995)], was placed downstream of coding sequences containing murine Golgi α-mannosidase II [K. W. Moremen and P. W. Robbins, *J. Cell Biol.* **115**, 1521 (1991)], human galactosyltransferase [K. A. Masri, H. E. Appert, M. N. Fukuda, *Biochem. Biophys. Res. Commun.* **157**, 657 (1988)], or wild-type or a mutant form of the human homolog of the yeast ERD2 receptor, ELP1 [V. W. Hsu, N. Shah, R. D. Klausner, *Cell* **69**, 625 (1992)], also known as KDELR. All constructs were generated by standard one- or two-stage polymerase chain reaction methods. Man II-GFP contains full-length Man II, including the

NH$_2$-terminal cytoplasmic tail, uncleaved signal sequence, and complete lumenal domain, followed by the 13-amino acid MYC-derived epitope fused to full-length GFP. GalTase-GFP contains amino acids 1 through 60 of human galactosyltransferase, including the NH$_2$-terminal cytoplasmic tail, uncleaved signal sequence, and 17 amino acids of the luminal domain fused to full-length GFP. KDELR-GFP contains the full-length coding sequence of ELP1, followed by the 13-amino acid MYC-derived epitope, followed by full-length GFP. KDELR$_m$-GFP was generated by mutagenizing the aspartic acid to an asparagine residue at position 195 of ELP1, which is analogous to the mutation at position 193 in ERD2 described in [9]. All constructs were subcloned into the expression vectors pCDLSRα or pcDNA1 (Invitrogen) and transiently expressed in HeLa cells.

11. J. Lippincott-Schwartz *et al.*, *Cell* **60**, 821 (1990).

12. Energy depletion was performed as described [J. Donaldson *et al.*, *J. Cell. Biol.* **111**, 2295 (1990)]. This treatment dissociates coatomer complexes (which are believed to mediate vesicle traffic) from Golgi membranes of living cells, blocks export of proteins out of the ER, and prevents processing of newly synthesized proteins by Golgi enzymes. For further discussion of the inhibitory effects of energy depletion and reduced temperature on vesicle traffic, see C. J. Beckers *et al.*, *Cell* **50**, 523 (1987) and E. Kuismanen and J. Saraste, *Methods Cell Biol.* **32**, 257 (1989).

13. J. Donaldson *et al.*, *J. Cell Biol.* **112**, 579 (1991); P. Melancon *et al.*, *Cell* **51**, 1053 (1987).

14. AIF also transforms Golgi membranes into an array of coated buds and vesicles after 30 min of treatment, which could disrupt their continuity (P. Peters, L. Yuan, R. Klausner, J. Lippincott-Schwartz, unpublished electron microscope observations).

15. M. L. Wier and M. Edidin, *J. Cell Biol.* **103**, 215 (1986).

16. B. Storrie, R. Pepperkok, E. H. Stelzer, T. E. Kreis, *J. Cell Sci.* **107**, 1309 (1994).

17. An additional argument that vesicular traffic does not play a major role in the observed movement of the GFP chimeras during FLIP is that vesicles traveling through the cytoplasm should have an equal probability of fusing with acceptor membranes in stacks that are equidistant from a donor membrane. We found, however, that some stacks show very little loss of fluorescence during FLIP compared with others, even though they are equidistant from the zone of photobleaching. Moreover, there appears to be little or no interstack communication after microtubule depolymerization, when Golgi stacks reversibly scatter throughout the cytoplasm. These results are difficult to explain by vesicle traffic but are easily explained by differences in lateral continuities between Golgi stacks.

18. FLIP experiments at 37°C with cells expressing Man II-GFP and KDELR-GFP also showed loss of fluorescence throughout the Golgi complex, suggesting that these molecules diffuse rapidly between Golgi stacks. In cells expressing GFP chimeras in the Golgi, FLIP of a region of the cytoplasm that did not contain Golgi, but presumably did contain ER, did not result in significant loss of Golgi fluorescence over this time frame, suggesting that Golgi membranes are not in direct continuity with the ER.

19. M. S. Cooper, A. H. Cornell-Bell, A. Chernjavsky, J. W. Dani, S. J. Smith, *Cell* **61**, 135 (1990).

20. A. Rambourg and Y. Clermont, *Eur. J. Cell Biol.* **51**, 189 (1990); G. Griffiths *et al.*, *J. Cell Biol.* **108**, 277 (1989); P. Weidman, R. Roth, J. Heuser, *Cell* **75**, 123 (1993); K. Tanaka, A. Mitsushima, H. Fukudome, Y. Kashima, *J. Submicrosc. Cytol.* **18**, 1 (1986).

21. J. E. Rothman and G. Warren, *Curr. Biol.* **4**, 220 (1994); P. J. Weidman, *Trends Cell Biol.* **5**, 302 (1995).

22. L. A. Jaffe and M. Terasaki, *Dev. Biol.* **156**, 556 (1993).

23. The data, however, do not rule out the possibility that native Man II and GalTase ever oligomerize. They only indicate that such complex formation is not required for efficient Golgi targeting and retention of these proteins.

24. As one example, for models that assume the existence of functionally discrete Golgi cisternae or subcompartments [J. E. Rothman and L. Orci, *Nature* **355**, 409 (1992)], our findings would imply that mechanisms exist for ensuring that membrane continuities between adjacent stacks only form between homologous membranes (that is, cis to cis and trans

to trans). Otherwise, Golgi cisternae within a Golgi stack could not remain completely separate and distinct from each other.

25. The microscope system described in [15] was used in the quantitative FPR experiments. The FPR beam was imaged into the sample as a stripe 2 μm wide. Because the stripe extended across the entire width of the Golgi or ER and bleached through the whole depth, diffusion was into and out of a line bounded on its sides, and not on its end. Hence, recovery of fluorescence was due to one-dimensional diffusion. The imposition of one-dimensional geometry on a complicated membrane as well as the mathematics for this case are covered in C.-L. Wey, M. A. Edidin, R. A. Cone, *Biophys. J.* **33**, 225 (1981). Briefly, a tortuous diffusion path reduces the apparent D, so our measurements in that case would be an underestimation.

26. Cells were transfected with GFP chimera cD-NAs by $CaPO_4$ precipitation. Fluorescent cells were imaged at 37°C in buffered medium with a Zeiss LSM 410 confocal microscope system having a 100× Zeiss planapo objective (NA 1.4). The GFP molecule was excited with the 488 line of a krypton-argon laser and imaged with a 515–540 bandpass filter. Images were transferred to a Macintosh computer for editing and were printed with a Fujix Pictrography 3000 Digital Printer.

27. We thank R. Klausner, E. Siggia, J. Bonifacino, J. Zimmerberg, J. Donaldson, J. Presley, J. Ellenberg, and K. Zaal for valuable comments and suggestions, and M. Chalfie, K. Moremen, M. Fukuda, R. Poljak, and V. Hsu for generous gifts of reagents. M.E. is supported by grant R37 AI14584. Quicktime movies are available at http://www.uchc.edu/htterasaki/flip.html.

READING 15.2

Past Temperatures Directly from the Greenland Ice Sheet

Commentary: This article is included as a textbook example of the application of the diffusion/advection equation. It is not about biology, but I couldn't resist it.

The discussion concerns a strategy for determining past temperatures at the surface of the Greenland ice cap by measuring the temperature today of the Greenland ice cap as a function of depth. The idea is that temperature fluctuations at the surface propagate downward into the ice. For example, an unusually cold spell on the surface will chill the ice, with the cold propagating downward since ice chilled at the surface will cool ice below it and that ice will then cool ice further down. Likewise, warm temperatures at the surface will warm the ice near the surface, and the warmer ice will warm ice further down. So temperatures at the surface propagate down through the ice. Thus, by measuring the temperature T at time $t = 0$ (now) as a function of depth z from the surface, we can, in principle, reproduce T as a function of $t < 0$ at depth $z = 0$ if we know how the temperature changes propagate through the ice.

This simple picture leads to a diffusion equation for the temperature, $T(t, z)$, as a function of time $t \leq 0$ and depth z:

$$\rho c \frac{d}{dt} T = \frac{d}{dz} \left(K \frac{d}{dz} T \right) + f.$$

Here ρ is the density of ice, c and K are T-dependent functions that are determined by the physics of ice, and f is the surface heat production term. Unfortunately, this model is too simple because the ice is not stationary; it flows, too. Thus, measurements at depth under one spot at the surface do not sample temperatures that were determined by past temperatures at the surface at the measuring spot. To remedy this defect, we must make T a function of variables t = time, z = depth, and x = distance along the direction of flow. The equation to use is a mixed diffusion/advection equation, which is the equation on page 239. The symbol ∇ stands for the vector of derivatives, (∂_x, ∂_z). In long hand, the equation on page 239 reads

$$c\rho \frac{d}{dt}T = \frac{d}{dx}\left(K\frac{d}{dx}T\right) + \frac{d}{dz}\left(K\frac{d}{dz}T\right) - \rho c\left(v_x\frac{d}{dx}T + v_z\frac{d}{dz}T\right) + f.$$

Here, (v_x, v_z) is the velocity that controls the advection term. (Set $K = 0$ and you get a two-space dimensional version of our advection equation, set v_x and v_z to zero and you get a two-space dimensional version of our diffusion equation.)

Past Temperatures Directly from the Greenland Ice Sheet

Science, **282** (1998) 268-271.
D. Dahl-Jensen, K. Mosegaard, N. Gundestrup, G. D. Clow, S. J. Johnsen, A. W. Hansen, N. Balling

A Monte Carlo inverse method has been used on the temperature profiles measured down through the Greenland Ice Core Project (GRIP) borehole, at the summit of the Greenland Ice Sheet, and the Dye 3 borehole 865 kilometers farther south. The result is a 50,000-year-long temperature history at GRIP and a 7000-year history at Dye 3. The Last Glacial Maximum, the Climatic Optimum, the Medieval Warmth, the Little Ice Age, and a warm period at 1930 A.D. are resolved from the GRIP reconstruction with the amplitudes −23 kelvin, +2.5 kelvin, +1 kelvin, −1 kelvin, and +0.5 kelvin, respectively. The Dye 3 temperature is similar to the GRIP history but has an amplitude 1.5 times larger, indicating higher climatic variability there. The calculated terrestrial heat flow density from the GRIP inversion is 51.3 milliwatts per square meter. Measured temperatures down through an ice sheet relate directly to past surface temperature changes. Here, we use the measurements from two deep boreholes on the Greenland Ice Sheet to reconstruct past temperatures. The GRIP ice core (72.6°N, 37.6°W) was successfully recovered in 1992 [1, 2], and the 3028.6-m-deep liquid-filled borehole with a diameter of 13 cm was left undisturbed. Temperatures were then measured down through the borehole in 1993, 1994, and 1995 [3, 4]. We used the measurements from 1995 (Figure 1) [4], because there was no remaining evidence of disturbances from the drilling and the measurements were the most precise (±5 mK). Temperatures measured in a thermally equilibrated shallow borehole near the drill site are used for the top 40 m, because they are more reliable than the GRIP profile over this depth [5]. The present mean annual surface temperature at the site is −31.70°C. The 2037-m-deep ice core from Dye 3 (65.2°N, 43.8°W) was recovered in 1981. We used temperature data from 1983 measurements with a precession of 30 mK [6, 7]. The temperatures at the bedrock are −8.58°C at GRIP and −13.22°C at Dye 3. Calculations show that the basal temperatures have been well below the melting point throughout the past 100,000 years [8]. Because there are still climate-induced temperature changes near the bedrock, we included 3 km of bedrock in the heat flow calculation.

Past surface temperature changes are indicated from the shape of the temperature profiles (Figure 1). We used a coupled heat- and ice-flow model to extract the climatic information from the measured temperature profiles. The temperatures

---

Author affiliations: D. Dahl-Jensen, K. Mosegaard, N. Gundestrup, S. J. Johnsen, A. W. Hansen: Niels Bohr Institue for Astronomy, Physics and Geophysics, Department of Geophysics. G. D. Clow: USGS-Climate Program, Denver Federal Center. N. Balling: Department of Earth Sciences, Geophysical Laboratory, University of Aarhus.

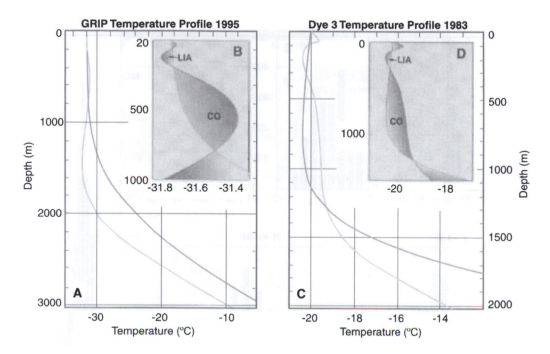

Fig. 1 The GRIP and Dye 3 temperature profiles [blue trace in (**A**) and (**C**)] are compared to temperature profiles [red trace in (**A**) and (**C**)] calculated under the condition that the present surface temperatures and accumulation rates have been unchanged back in time. (**A**) The GRIP temperature profile measured in 1995. The cold temperatures from the Glacial Period (115 to 11 ka) are seen as cold temperatures between 1200- to 2000-m depth. (**B**) The top 1000 m of the GRIP temperature profiles are enlarged so the Climatic Optimum (CO, 8 to 5 ka), the Little Ice Age (LIA, 1550 to 1850 A.D.), and the warmth around 1930 A.D. are indicated at the depths around 600, 140, and 60 m, respectively. (**C**) The Dye 3 temperature profile measured in 1983. Note the different shape of the temperature profiles when compared to GRIP and the different depth locations of the climate events. (**D**) The top 1500 m of the Dye 3 temperature profiles are enlarged so the CO, the LIA, and the warmth around 1930 A.D. are indicated at the depths around 800, 200, and 70 m, respectively.

down through the ice depend on the geothermal heat flow density (heat flux), the ice-flow pattern, and the past surface temperatures and accumulation rates. The past surface temperatures and the geothermal heat flow density are unknowns, whereas the past accumulation rates and ice-flow pattern are assumed to be coupled to the temperature history through relations found from ice-core studies [9–11]. The total ice thickness is assumed to vary 200 m as described in [9]. The coupled heat- and ice-flow equation is [7, 9, 12]

$$\rho c \frac{\partial T}{\partial t} = \nabla \cdot (K \nabla T) - \rho c \vec{v} \cdot \nabla T + f$$

where $T(x, z, t)$ is temperature, t is time, z is depth, x is horizontal distance along the flow line, $\rho(z)$

is ice density, $K(T, \rho)$ is the thermal conductivity, $c(T)$ is the specific heat capacity, and $f(z)$ is the heat production term. The ice velocities, $\vec{v}(x, z, t)$, are calculated by an ice-flow model [9,13]. Model calculations to reproduce a present-day temperature profile through the ice sheet are started 450,000 years ago (ka) at GRIP (100 ka at Dye 3), more than twice the time scale for thermal equilibrium of the ice-bedrock, so the unknown initial conditions are forgotten when generating the most recent 50,000-year temperature history (7000 years for Dye 3).

We developed a Monte Carlo method to fit the data and infer past climate. The Monte Carlo method tests randomly selected combinations of surface temperature histories and geothermal heat

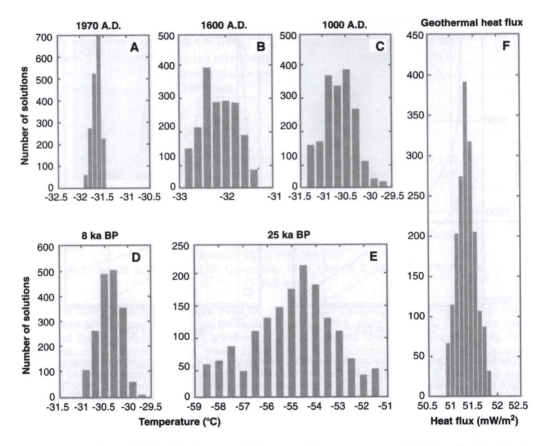

Fig. 2 (A through E) The probability distributions of the past surface temperatures at the Greenland Ice Sheet summit at selected times before present. They are constructed as histograms of the 2000 Monte Carlo sampled and accepted temperature histories [17]. All temperature distributions are seen to have a zone with maximum values, the most likely values, which are assumed to be the reconstructed surface temperature at these times [18]. (F) The probability distribution of the sampled geothermal heat flow densities. The most likely value is 51.3 mW/m^2.

flow densities by using them as input to the coupled heat- and ice-flow model and considering the resulting degrees of fit between the reproduced and measured temperature profiles [14–16]. Our results for each site are based on tests of 3.3×10^6 combinations of temperature histories and heat flow densities, of which 2000 solutions have been selected [17]. The 2000 temperature histories and heat flow densities are sampled with a frequency proportional to their likelihood [14, 15], and all accepted solutions fit the observations within their limits of uncertainty.

Histograms of the sampled geothermal heat flow densities and of the temperature histories at each

time before present can be made (for example, Figure 2). The distributions in general show that there is a most likely value, a maximum, at all times, which we refer to as the temperature history [18]. The distribution of accepted geothermal heat flow densities (Figure 2F) has a median of 51.3 ± 0.2 mW/m^2, which is slightly higher than the heat flow density from Archean continental crust across the Baffin Bay in Canada. A few heat flow measurements have been made from the coast of Greenland (36 and 43 mW/m^2), but these are not corrected for long-term climate variations and are minimum values [19]. The homogeneous thermal structure of ice is an advantage when the heat flow

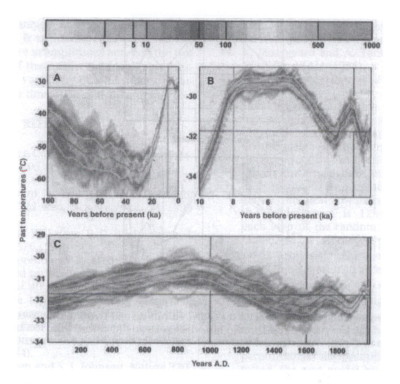

Fig. 3 The contour plots of all the GRIP temperature histograms as a function of time describes the reconstructed temperature history (red curve) and its uncertainty. The temperature history is the history at the present elevation (3240 m) of the summit of the Greenland Ice Sheet [21]. The white curves are the standard deviations of the reconstruction [18]. The present temperature is shown as a horizontal blue curve. The vertical colored bars mark the selected times for which the temperature histograms are shown in Figure 2. (**A**) The last 100 ky BP. The LGM (25 ka) is seen to have been 23 K colder than the present temperature, and the temperatures are seen to rise directly into the warm CO 8 to 5 ka. (**B**) The last 10 ky BP. The CO is 2.5 K warmer than the present temperature, and at 5 ka the temperature slowly cools toward the cold temperatures found around 2 ka. (**C**) The last 2000 years. The medieval warming (1000 A.D.) is 1 K warmer than the present temperature, and the LIA is seen to have two minimums at 1500 and 1850 A.D. The LIA is followed by a temperature rise culminating around 1930 A.D. Temperature cools between 1940 and 1995.

density and the temperature history are to be reconstructed [20].

Histograms from the GRIP reconstruction (Figure 3) show that temperatures at the Last Glacial Maximum (LGM) were 23 ± 2K colder than at present [21]. The temperatures at this time, 25 ka, reflect the cold temperatures seen on the measured temperature profile at a depth of 1200 to 2000 m. Alternative reconstructions of the ice thickness and accumulation rates all reproduce LGM temperatures within 2 K [9, 10, 22, 23]. The cold Younger Dryas and the warm Bolling/Allerod pe-

riods [24] are not resolved in the inverse reconstruction. The temperature signals of these periods have been obliterated by thermal diffusion because of their short duration [25]. After the termination of the glacial period, temperatures in our record increase steadily, reaching a period 2.5 K warmer than present during what is referred to as the Climatic Optimum (CO), at 8 to 5 ka. Following the CO, temperatures cool to a minimum of 0.5 K colder than the present at around 2 ka. The record implies that the medieval period around 1000 A.D. was 1 K warmer than present in Greenland. Two cold pe-

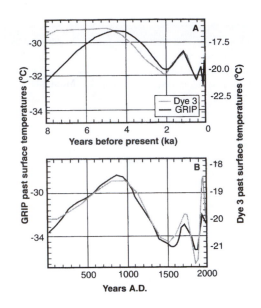

Fig. 4 The reconstructed temperature histories for GRIP (red curves) and Dye 3 (blue curves) are shown for the last 8 ky BP (**A**) and the last 2 ky BP (**B**). The two histories are nearly identical, with 50% larger amplitudes at Dye 3 than found at GRIP. The reconstructed climate must represent events that occur over Greenland, probably the high-latitude North Atlantic region.

riods, at 1550 and 1850 A.D., are observed during the Little Ice Age (LIA) with temperatures 0.5 and 0.7 K below the present. After the LIA, temperatures reach a maximum around 1930 A.D.; temperatures have decreased during the last decades [26]. The climate history for the most recent times is in agreement with direct measurements in the Arctic regions [27]. The climate history for the last 500 years agrees with the general understanding of the climate in the Arctic region [28] and can be used to verify the temperature amplitudes. The results show that the temperatures in general have decreased since the CO and that no warming in Greenland is observed in the most recent decades.

As seen in Figure 3, resolution decreases back in time [25, 29]. For the GRIP reconstruction, an event with a duration of 50 years and an amplitude of 1 K can be resolved 150 years back in time with a measurement accuracy of 5 mK; an event with a similar amplitude but a duration of 1000 years can be detected back to 5 ka. An event that occurred 50 ka will now be observed in the temperature profile at the bedrock. Climate events for times older than 50,000 years before present (ky BP) are not

well resolved [30]. At Dye 3, the reconstructed climate history extends only to 7 ka, because the ice is 1000 m thinner than at the summit and surface accumulation rate is 50% higher. The LGM is not well resolved in the Dye 3 record, and consequently the geothermal heat flow density is not uniquely determined [31]. On the other hand, the recent climate history has a higher resolution because of the increased accumulation (Figure 4).

The Dye 3 record is nearly identical with the GRIP record back to 7 ka, but the amplitudes are 50% higher. Thus, the resolved climate changes have taken place on a regional scale; many are seen throughout the Northern Hemisphere [27, 28, 32]. GRIP is located 865 km north of Dye 3 and is 730 m higher in elevation. Surface temperatures at the summit are influenced by maritime air coming in from the North Atlantic and air masses arriving from over northeastern Canada (associated with the Baffin trough) [28, 32, 33]. Temperatures at Dye 3 will be influenced to a greater degree by the North Atlantic maritime air masses. Dye 3 is located closer to the center of the highest atmospheric variability, which is associated with large interseasonal, interannual, and decadal temperature changes [32, 34]. It is therefore believed that the observed difference in amplitudes between the two sites is a result of their different geographic location in relation to variability of atmospheric circulation, even on the time scale of a millennium.

REFERENCES

1. Greenland Ice-Core Project (GRIP) members, *Nature* **364**, 203 (1993).
2. W. Dansgaard *et al.*, *ibid.*, p. 218.
3. N. S. Gundestup, D. Dahl-Jensen, S. J. Johnsen, A. Rossi, *Cold Reg. Sci. Technol.* **21**, 399 (1993).
4. G. D. Clow, R. W. Saltus, E. D. Waddington, *J. Glaciol.* **42**, 576 (1996).
5. The deep borehole is located in a building, and the liquid surface in the borehole is found at a depth of 40 m. The temperatures measured in the top 40 m are very disturbed, so we used measurements from an air-filled shallow borehole (100 m) near the borehole.
6. N. S. Gundestrup and B. L. Hansen, *J. Glaciol.* **30**, 282 (1984).
7. D. Dahl-Jensen and S. J. Johnsen, *Nature* **320**, 250 (1986).
8. D. Dahl-Jensen *et al.*, *J. Glaciol.* **43**, 300 (1997).
9. Between 50 and 20 ka, the ice thickness was 50 m less than at present, even though the ice sheet covered a larger area. The maximum ice thickness of 3230 m is found at 10 ka, after which the ice thickness gradually has decreased to the present 3028.6 m. The depression and uplift of the bedrock influences the elevation of the surface [S. J. Johnsen, D. Dahl-Jensen, W. Dansgaard, N. S. Gundestrup, *Tellus B* **47**, 624 (1995)].
10. K. M. Cuffey and G. D. Clow, *J. Geophys. Res.* **102**, 26383 (1997).
11. The past accumulation rates are determined by coupling them to the past (unknown) temperature through the relation $\lambda(T) = \lambda_0 \exp[0.0467(T - T_0) - 0.000227(T - T_0)^2]$, where $\lambda(T)$ is the accumulation rate at the surface temperature T, λ_0 is the present ice accumulation rate, which is 0.23 m/year at GRIP and 0.49 m/year at Dye 3, and T_0 is the present surface temperatures at the sites: $-31.7°C$ at GRIP and $-20.1°C$ at Dye 3, respectively [9].
12. S. J. Johnsen, *IAHS-AISH Publ.* **118**, 388 (1977).
13. S. J. Johnsen and W. Dansgaard, *NATO ASI Ser. I Global Environ. Change* **2**, 13 (1992).
14. K. Mosegaard and A. Tarantola, *J. Geophys. Res.* **100**, 12431 (1995).
15. K. Mosegaard, *Inverse Problems* **14**, 405 (1998).
16. Our Monte Carlo scheme is a random walk in the high dimensional space of all possible models, m (temperature histories and geothermal heat flow densities). The temperature history has been divided in 125 intervals (interval length is 25 ky at 450 ka and 10 years at present). Including the geothermal heat flow density as an unknown the model space is 126-dimensional. In each step of the random walk, a perturbed model, m^i_{pert} of the current model vector m^i is proposed. The next model becomes equal to m^i_{pert} with an acceptance probability $P_{\text{accept}} = \min\{1, \exp(-[S(m^i_{\text{pert}}) - S(m^i)])\}$, where $S(m) = \sum_j (g^j(m) - d^j_{\text{obs}})^2$, which is the misfit function measuring the difference between $g(m)$, the calculated borehole temperatures, and d_{obs}, the observed temperatures. If the perturbated model is rejected, the next model becomes equal to m^i and a new perturbed model is proposed. To ensure an efficient sampling of all possible models, we developed ways of choosing the temperature histories and geothermal heat flow densities to be tested. The main scheme to perturb the models is to randomly select one of the 126 temperature/heat flow density parameters and change its value to a new value chosen uniformly at random within a given interval. A singular value decomposition (SVD) of the matrix $G = \{\partial g_j / \partial m_i\}$, evaluated in a near-optimal model, yields a set of eigenvectors in the model space whose orientations reveal efficient directions of perturbation for the random walk. The SVD method is included as a possible method of perturbing models especially in the start of the process as it speeds the Monte Carlo scheme significantly.
17. Of the 3.3×10^6 models tested during the random walk 30% have been accepted by the Monte Carlo scheme [16]. Every 500 is chosen of those where the misfit function S [16] is less than the variance of the observations. The waiting time of 500 has been

chosen to exceed the maximum correlation length of the output model parameters. This is a necessary condition for the 2000 models to be uncorrelated. To further ensure that the output models were uncorrelated, the random walk was frequently restarted at several random selected points in the model space.

18. The probabilistic formulation of the inverse problem leads to definition of a probability distribution in the model space, describing the likelihood of possible temperature histories and geothermal heat flow densities. The Monte Carlo scheme is constructed to sample according to this probability distribution. The histograms in Figure 2 describe the probability distribution of the geothermal heat flow density and temperatures at times before present. The maxima in the histograms thus describe the most likely values. The method does not constrain the distributions to have a single maximum, indeed there could be histograms with several maxima, reflecting that more than one value of the temperature at this time would give a good fit to the observed temperature in the borehole. The histograms however, are all seen to have a well-defined zone with most likely past temperatures. A soft curve is fitted to the histograms and the maximum value is taken as the most likely value. The standard deviations shown in Figure 3 are derived as deviations from the maximum value.

19. J. H. Sass, B. L. Nielsen, H. A. Wollenberg, R. J. Munroe, *J. Geophys. Res.* **77**, 6435 (1972).

20. C. Clauser *et al.*, *ibid.*, **102**, 18417 (1997); L. Guillou-Frottier, J.-C. Marescal, J. Musset, *ibid.*, **103**, 7385 (1998); H. N. Pollack, S. J. Hurter, J. R. Johnson, *Rev. Geophys.* **31**, 267 (1993); W. G. Powell, D. S. Chapman, N. Balling, A. E. Beck, in *Handbook of Terrestrial Heat-Flow Density Determination* (Kluwer Academic, New York, 1988), pp. 167–222.

21. In order to produce a past temperature record from the calculated past surface temperatures, the temperatures have been corrected to the present elevation of the GRIP site (and Dye 3 site respectively) using the surface elevation changes described in [9] and a lapse rate of 0.006 K/m.

22. K. M. Cuffey *et al.*, *Science* **270**, 455 (1995).

23. D. Dahl-Jensen, in *Proceedings of the Interdisciplinary Inversion Workshop 2*, Copenhagen, 19 May 1993, K. Mosegaard, Ed. (The Niels Bohr Institute for Astronomy, Physics and Geophysics, University of Copenhagen, Copenhagen, 1993), pp. 11–14.

24. C. U. Hammer *et al.*, *Report on the stratigraphic dating of the GRIP Ice Core.* Special Report of the Geophysical Department (Niels Bohrs Institute for Astronomy, Physics and Geophysics, University of Copenhagen, Copenhagen, in press).

25. J. Firestone, *J. Glaciol.* **41**, 39 (1995).

26. The amplitude of the warming at 1930 A.D. must be considered to be more uncertain. The information leading to this result are the measured temperatures in an open shallow borehole, where air movements could influence the measurements.

27. D. Fisher *et al.*, *NATO ASI Ser. I Global Environ. Change* **41**, 297 (1996); J. W. C. White *et al.*, *J. Geophys. Res.* **102**, 26425 (1997); J. W. Hurrel, *Science* **269**, 676 (1995); P. Frich *et al.*, in *DMI Scientific Report 96-1* (Danish Meteorological Institute, Copenhagen, 1996).

28. R. G. Barry and R. J. Charley, *Atmosphere, Weather & Climate* (Routledge, London, ed. 6, 1992); J. Overpeck *et al.*, *Science* **278**, 1251 (1997); H. H. Lamb, *Climate History and the Modern World* (Routledge, London, ed. 2, 1995); N. W. T. Brink and A. Weidick, *Quat. Res.* **4**, 429 (1974).

29. G. D. Clow, *Palaeogeogr. Palaeoclimatol. Palaeoecol.* **98**, 81 (1992).

30. To comply with this resolution the time steps have been chosen with increasing length back in time. The increasing length of the time steps can be considered as an efficient way of calculating the mean temperatures in the intervals so full available resolution is kept but the calculations are rationalized.

31. In [7], it is argued that parameter combinations of mean glacial temperature, mean glacial accumulation, and geothermal heat flow density can be found that fit the Dye 3 measurements due to the reduced resolution of the climate history reaching further back than 7 ka. A combination with a geothermal heat flow density of 38.7 mW/m^2 was chosen corresponding to a mean glacial temperature 12 K colder than the present temperatures. If a value of 51 mW/m^2 is chosen as that found for our inversion, the mean glacial temperature is 19 K colder than the present, which is well in agreement with the results found for

the GRIP reconstruction. Comparison of the Dye 3 temperature history presented in [7] and that presented here shows a general good agreement for the last 7 ky. The history presented in [7] is more intuitive and less detailed, and the history has not been corrected for elevation changes. The ice thickness was assumed constant in this reconstruction.

32. L. K. Barlov, J. C. Rogers, M. C. Serreze, R. C. Barry, *J. Geophys. Res.* **102**, 26333 (1997).
33. R. A. Keen, *Occas. Pap.* **34** (Institute of Arctic and Alpine Research, University of Colorado, Boulder, 1980).
34. S. Shubert, W. Higgins, C. K. Park, S. Moorthi, M. Svarez, *An Atlas of ECMWF Analyses (1980-87). Part II: Second Moment Quantities*, NASA Tech. Memorandum 100762 (1990); F. Rex, *World Surv. Climatol.* **4**, 1 (1969).
35. This is a contribution to the Greenland Ice Core Project (GRIP), a European Science Foundation program with eight nations and the European Economic Commission collaborating to drill through the central part of the Greenland Ice Sheet. G.D.C. thanks the USGS Climate History Program and NSF for support.

Exercises for Chapter 15

1. Verify that the following functions obey the $r = 0$ version of Equation (1) in Chapter 15,

$$\frac{\partial}{\partial t}u = \mu \frac{\partial^2}{\partial x^2}u.$$

(Note that in doing so, you will also be verifying cases of the superposition principle.)

(a) $u(t, x) = 2\frac{1}{t^{1/2}}e^{-x^2/4\mu t} + \frac{1}{t^{1/2}}e^{-(x-1)^2/4\mu t}$

(b) $u(t, x) = 2\frac{1}{t^{1/2}}e^{-x^2/4\mu t} - \frac{1}{t^{1/2}}e^{-(x-1)^2/4\mu t}$

(c) $u(t, x) = 2\frac{1}{t^{1/2}}e^{-x^2/4\mu t} + 3\frac{1}{t^{1/2}}e^{-(x-1)^2/4\mu t}$

2. Verify that the following functions obey the $r = 0$ version of Equation (1) in Chapter 15. (This exercise is also a test of the superposition principle.)

(a) $u(t, x) = 2e^{\lambda t}e^{x(\lambda/\mu)^{1/2}} + 3e^{\lambda t}e^{-x(\lambda/\mu)^{1/2}}$ where λ is any constant

(b) $u(t, x) = 2e^{\lambda t}e^{x(\lambda/\mu)^{1/2}} - 3e^{\lambda t}e^{-x(\lambda/\mu)^{1/2}}$ where λ is any constant

(c) $u(t, x) = 2e^{\lambda t}e^{x(\lambda/\mu)^{1/2}} + 1 - 5x$ where λ is any constant

16

The No-Trawling Zone

As remarked in the previous chapter, various solutions to the diffusion equation

$$\frac{\partial}{\partial t}u = \mu\frac{\partial^2}{\partial x^2}u \tag{1}$$

spread out as time t increases as in the solution

$$u(t, x) = \frac{a}{t^{1/2}}e^{-x^2/4\mu t}. \tag{2}$$

Here, a is some constant. However, there are situations where spreading solutions as in (2) are not relevant. These occur when there are evident boundaries (in the x-direction) to the extent over which $u(t, x)$ can spread. The existence of such boundaries drastically diminishes the set of solutions to (1) that are relevant to the particular problem at hand. In fact, the use of "boundary" conditions to constrain the set of solutions to (1) is an important theme in the next few chapters.

This chapter presents the first example of this theme. The example is motivated by the discussion in the article presented in Reading 16.1 [*Fishing for Answers: Deep Trawls Leave Destruction in Their Wake–But for How Long?* by J. Raloff from *Science News*, **150** (1996) 268–271]. This article points out that bottom-dragging trawl-fishing nets have a profound effect on the environment of bottom-dwelling organisms due to the destruction and disturbance of the habitat. To protect bottom-dwelling creatures (lobsters, for instance), bottom-dragging nets can be banned in a strip of ocean of some given width. To balance economic demands of the fishing industry against environmental concerns, the relevant question is this: What is the minimum width of a long no-dragging zone that ensures the replenishment of lobster stocks in the zone? A lower bound for the width, R, of the no-dragging zone can be obtained with the help of a simple diffusion equation model for lobster behavior. Note that the analysis here is analogous to one in Edward Beltrami's book, *Mathematics for Dynamical Modeling*, in which a very similar problem is analyzed.

Before getting to the details, I need to warn you that my own students have found the material in this chapter and in the next few chapters to be the most difficult in the book. Part of the reason might be that the answers are not clean and simple and require many more intermediate steps. In particular, it is easy to lose sight of the goals on the long treks to them.

16.1 The Model

Model the strip in question by the part of the (x, y) plane where x is positive but less than the proposed width, R. The strip will be assumed to be much longer than its width, so I will pretend that the extent of the strip in the y-direction is much bigger than R and thus not relevant to the analysis. (Do you buy this approximation to reality?) Let $u(t, x)$ denote the number of lobsters per unit kilometer along the line of constant y. Suppose that lobsters are destroyed more or less immediately by effects of the net dragging trawlers if they venture outside the strip, where $x < 0$ or where $x > R$. Suppose that the lobster species in question will increase in population at a constant rate, r per lobster, per year, assuming adequate food supplies in the absence of habitat disturbance. (Note that the number r can be estimated by placing lobsters in a large aquarium and measuring their birth and death rates.)

The question is to estimate a size for R that will ensure that the number of lobsters in the strip does not decrease with time.

Since the lobsters are crawling randomly, we expect that the function $u(t, x)$ should obey a diffusion equation of the form

$$\frac{\partial}{\partial t} u = \mu \frac{\partial^2}{\partial x^2} u + k(t, x) \tag{3}$$

for some appropriate choice of constant μ and function k. The simplest choice for the function $k(t, x)$ is to take $k(t, x) = ru(t, x)$ with $r > 0$. This postulate is based on the assumption that the lobster population in an enclosed area will increase with constant growth rate per lobster. (This is realistic unless the lobster population gets too large.)

The diffusion coefficient μ is harder to measure than the constant r, since we would need to know the range of wandering of the species in question in the open ocean. In principle, we might put some radio transmitters on lobsters, release them, and then track their movements over some reasonable period of time. Assuming that the lobsters are not eaten by a fast-swimming predator during the course of the experiment, you would be able to derive an estimate for μ from the details of their wanderings. (If the radio transmitter gets lodged inside a fish that preys on your lobsters, then you would be unknowingly computing the diffusion coefficient of some other species.)

In any event, consider the model equation

$$\frac{\partial}{\partial t}u = \mu\frac{\partial^2}{\partial x^2}u + ru. \tag{4}$$

Here, we should be looking for solutions with $u(t, 0)$ and $u(t, R)$ suitably constrained to take into account that the strip ends where $x = 0$, and where $x = R$. One reasonable choice is for us to consider only solutions to (4) with

$$u(t, 0) = 0 \quad \text{and} \quad u(t, R) = 0 \tag{5}$$

for all times t. The extra conditions in (5) are called **boundary conditions**. The specification of boundary conditions is a crucial part of any application of the diffusion equation.

Using (4) and (5), the arguments that follow in this chapter find that the lobster population size in the strip will grow with time provided that

$$R > (\mu\pi^2/r)^{1/2}.$$

That is, according to the subsequent analysis, the minimum strip width for a sustainable lobster population in the strip is $(\mu\pi^2/r)^{1/2}$. In particular, the mathematics tells us the answer to the fisheries management question if we can determine (by observing lobsters) only two numbers, the diffusion coefficient μ and the ideal growth rate r.

The precise answer just given for the minimum strip width is not the important lesson in this chapter. The real point of this chapter is that I am introducing some new techniques for dealing with the diffusion equation. Thus, pay close attention to the techniques that I introduce.

There is something else to be learned from this chapter: Don't fail to notice that the diffusion equation is used to spotlight the relevant biological parameters. (In this case, they are the diffusion coefficient and the growth rate.)

16.2   Separating t and x Variables

As remarked earlier, there are myriad solutions to (4), and, as we shall soon see, there are less, but still many, to both (4) and (5). The issue is to find the relevant ones. Here is an important method for finding some particular solutions: Agree to search for a solution where $u(t, x)$ has the simple form

$$u(t, x) = A(t)B(x), \tag{6}$$

where A is a function of t and where B is a function of x.

It turns out that there are solutions to (4) and (5) that have the form in (6) and that they are even relevant to the biology. On the other hand, there are also lots of solutions that don't look like (6). Even so, I shall study solutions given by (6) because they are relatively easy to find. Of course, you might ask, "Shouldn't I consider other forms also so that I know precisely what is going on?" In principle, yes; but in practice, I don't know the general form of solution, so I must settle for less. [Although I won't say much about it in this chapter, it turns out that every solution to (4) and (5) can be written as a type of sum (via the superposition principle) of solutions that do obey the form in (6).]

In any event, with the assumption of (6), Equation (4) has the form

$$\left(\frac{d}{dt}A(t)\right)B(x) = \mu A(t)\left(\frac{d^2}{dx^2}B(x)\right) + rA(t)B(x). \tag{7}$$

Although it is by no means obvious, the next step is to divide both sides of this equation by $A(t)B(x)$ to find

$$\frac{1}{A(t)}\frac{d}{dt}A(t) = \mu\frac{1}{B(t)}\frac{d^2}{dx^2}B(x) + r. \tag{8}$$

Here is the key observation: Note that the left side of this last equation is a function only of time t, while the right side is a function only of x. A function *only* of time t that is also a function *only* of space x cannot depend on *either* time or space. That is, it must be constant. Therefore, we are forced to conclude that (8) can be true only if both of the following equations hold:

$$\frac{1}{A(t)}\frac{d}{dt}A(t) = \lambda \tag{9a}$$

$$\mu\frac{1}{B(t)}\frac{d^2}{dx^2}B(x) + r = \lambda. \tag{9b}$$

Here, λ is some constant.

Thus, under the assumption $u(t, x) = A(t)B(x)$, Equation (4) has become the two equations in (9). These equations are easier to solve than (4) because each involves only x or only t, not both. However, there is some expense, which is the introduction of the constant λ. [The constant λ is a manifestation of the aforementioned existence of lots of solutions to (4).] In principle, you are free to choose any value you wish for λ. Our choice for λ will be based on information from the trawl ban problem. We will see that some choices for λ result in a solution $u(t, x)$ that is unrealistic for our problem. [We will see that it results in $u(t, x)$ decreasing with time or being negative in places.] The constraint that the solution $u(t, x)$ represent a lobster number density that is growing with time will be seen to constrain further our choice for λ.

16.3   Solving (9)

The first equation in (9) is easy to solve; we solved it in Chapter 2. The answer is

$$A(t) = A(0)e^{\lambda t}. \tag{10}$$

Notice that $A(t)$ grows with time if λ is positive, and it decreases with time if λ is negative. Since $u(t, x) = A(t)B(x)$, and we are interested in sustaining our lobster population, we should only be interested in the case where the constant λ is zero or larger.

With the first equation in (9) understood, turn to the second equation. Given the constants r and μ in (9) and $\lambda > 0$, we must find a solution $B(x)$ to this equation that obeys

(a) $B(x) \geq 0$ for all x between 0 and R. This is so that $u(t, x) = A(t)B(x)$ can be rightly interpreted as a lobster density. (Negative density makes no sense.)

(b) $B(0) = 0$ so that $u(t, 0) = 0$ at left boundary of the trawl-ban strip.

(c) $B(R) = 0$ so that $u(t, R) = 0$ at the right boundary of the trawl-ban strip.   (11)

We shall see that the three constraints in (11) for the function are impossible to solve if R is too small. That is, there is some minimum value of R, R_0, such that when $R < R_0$, then there is no solution to (9b) that also obeys the constraints in (11). This R_0 will be the lower bound for our trawl-ban strip width.

The second equation in (9) can be rewritten as

$$\frac{d^2}{dx^2} B(x) = cB(x) \quad \text{with} \quad c = (\lambda - r)/\mu. \tag{12}$$

Here, I have introduced the new constant c to simplify notation. Its introduction is also meant to focus attention on the fact that solutions to (12) depend only on the specific combination of the parameters λ, μ, and r as c, and not otherwise on their individual properties.

In any event, with regard to (12), there are three cases to consider, depending on whether

$$c > 0 \tag{13a}$$

$$c = 0 \tag{13b}$$

or

$$c < 0. \tag{13c}$$

In each case, the general solution depends on two free parameters, α and β:

1. $c > 0$: $B(x) = \alpha e^{\sqrt{c}x} + \beta e^{-\sqrt{c}x}$, where $c = (\lambda - r)/\mu$

2. $c = 0$: $B(x) = \alpha + \beta x$

3. $c < 0$: $B(x) = \alpha \cos((-c)^{1/2}x) + \beta \sin((-c)^{1/2}x)$ where $c = (\lambda - r)/\mu$.

I will spend the last section of this chapter explaining where these solutions come from. Feel free to skip ahead to the final section to study the justification, but return here when you finish.

It is important that you internalize the following features of (16.3):

• The form of the solution depends on the sign of c and is thus determined by the choice of λ as well as r and μ.

- In each case ($c > 0$, $c = 0$, $c < 0$), the solution depends on two free parameters, which I have taken the liberty of calling α and β.

- In the $c > 0$ case, the solution is a sum of exponentials, one growing with x and the other shrinking. Also note that $\sqrt{c}$ multiplies x in the exponent in each of the exponentials.

- In the $c = 0$ case, the solution is a constant plus a multiple of x.

- In the $c < 0$ case, the solution is the sum of a cosine and a sine, where the square root of $-c$ multiplies x as the argument for both.

16.4   Constraining the Parameters

In principle, we are completely free to specify the constants α and β in (16.3). They are manifestations of the fact that (4) has lots of solutions. In practice, these constants (and also the constant λ) are constrained by the requirement that the solution B vanish at $x = 0$ and at $x = R$. Indeed, if $B(0) = 0 = B(R)$ and if

1. $\lambda - r > 0$, then $\alpha = 0$ and $\beta = 0$ are the only possibilities.

2. $\lambda - r = 0$, then $\alpha = 0$ and $\beta = 0$ are the only possibilities.

3. $\lambda - r < 0$, then $\alpha = 0$ and $\beta = 0$ or β is arbitrary and λ is constrained so

$$\sin(((r - \lambda)/\mu)^{1/2} R) = 0. \tag{14}$$

Do you take my word for the assertions in (14)? Are you sure you trust me? Since this is a very important point, let's check this in more detail. To start, the simplest case to consider is the case of (2). In this case, we have $c = 0$ and so the second case in (16.3) is the relevant one. With this understood, we see that the assertion that $B(0) = 0$ leads directly to the conclusion that $\alpha = 0$. With α now zero, the conclusion that $B(R) = 0$ leads to the conclusion that $\beta R = 0$ (i.e., that $\beta = 0$ too).

The next simplest case to consider is the case where $\lambda - r > 0$. Here, we have $c > 0$ so we are forced to use the first case in (16.3), the sum of exponentials. The constraint that $B(0) = 0$ translates to the condition that $\alpha + \beta = 0$ so β must equal $-\alpha$. Then the condition that $B(R) = 0$ reads $\alpha(e^{\sqrt{c}R} - e^{-\sqrt{c}R}) = 0$. Note that this is the difference between an exponential with a positive exponent and one with a negative exponent. As the former is greater than one and the latter less than one, the difference can't be zero. Thus, the only way to satisfy the boundary conditions is to take α and β both equal to zero.

Consider the final case where $\lambda - r < 0$. Here, $c < 0$ so we are forced to take the third case in (16.3) for the function B. Given this last point, consider that $B(0) = \alpha$, so if $B(0)$ is to vanish, we must take $\alpha = 0$. With $\alpha = 0$, then $B(R) = \beta \sin(\sqrt{-c}R)$. To make this zero, we can take $\beta = 0$, though this leads to a thoroughly boring function

u in (6), namely $u = 0$ everywhere. However, there is an alternate possibility, which occurs when λ, r, μ, and R collaborate so that $\sin((-c)^{1/2}R) = 0$. In this case, we don't need to take $\beta = 0$ and so we might have a chance at a reasonable $u(t, x)$.

To summarize, we see from (16.3) that the only case where the boundary conditions in (11b) can be satisfied without taking $u(t, x)$ to be identically zero occurs when $\lambda < r$. In this case, we require that (14) hold. Since the sine function vanishes only where its argument is a multiple of π, we see that (14) is the same as the requirement that

$$((r - \lambda)/\mu)^{1/2}R = n\pi \tag{15}$$

for some integer n. (With μ, r, and R fixed, this is a constraint on λ!) When (15) is satisfied, then

$$u(t, x) = \beta e^{\lambda t} \sin(n\pi x/R) \tag{16}$$

when $0 < x < R$. Here, $\beta > 0$ is a constant.

There is one last remark that constrains the possible integers n that can be used in (16). In fact, we shall see that only the case where $n = 1$ is relevant to the problem at hand. Indeed, the function $u(t, x)$ must be nonnegative where $0 < x < R$ to model lobster density realistically. Of all the integers n to take in (16), only the case where $n = 1$ has this property. See Figure 16.1.

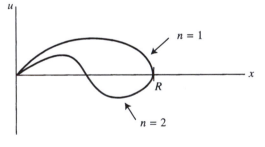

Figure 16.1

Thus, we are obliged to constrain further our consideration to the case $n = 1$ in (16) where

$$u(t, x) = \beta e^{\lambda t} \sin(\pi x/R) \tag{17}$$

and

$$\lambda = r - \mu \pi^2 / R^2. \tag{18}$$

16.5   Constraining R

With $u(t, x)$ in (17) understood, let us go back and see where we can get our claimed minimum estimate for the width R. For this purpose, it is crucial to remember that we want our lobster population to grow, and this requires $\lambda > 0$ in (17). Since λ is given by (18), we see from (18) that $\lambda > 0$ only if we constrain R so that $r - \mu\pi^2/R^2$ is positive. This requires

$$R > (\mu\pi^2/r)^{1/2}. \tag{19}$$

This last equation exhibits the asserted lower bound for the width of the no-fishing zone.

16.6   Lessons

Here are some key points from this chapter:

- Augmenting a differential equation such as (4) with boundary conditions such as in (5) severely constrains the set of solutions.

- Biologically relevant issues (such as positivity of a density or growth with time) can also be used to winnow out solutions to a differential equation.

- Solutions to a differential equation can sometimes be found to have the separation of variables form of (6). Even so, there may be many having this form.

- The general solution to (12) always has two free parameters, but the form of this solution changes depending on the sign of the constant c.

- The boundary conditions and conditions for biological relevance can both be used to eliminate from consideration many of the solutions to (12).

16.7   The Solutions to $\frac{d^2}{dx^2}B = cB$

This section constitutes an appendix of sorts whose purpose is to give some justification for the assertion that (16.3) describes the general solution to (12) in the various cases.

The simplest case to start with is the $c = 0$ case. Here, the equation asserts that the derivative of B's derivative is zero. Thus, B's first derivative is a constant, which I will call β. With this understood, you should recall from your first calculus course that a function whose derivative is the constant β is equal to $\beta x +$ constant. I have called the latter constant α.

In the general case, the justification requires two steps.

Step 1: This step verifies that $B(x)$ as given in each case of (16.3) and for any choice of α and β does indeed solve (12). To do this, you simply differentiate the given function twice and check that the result is equal to c times the original. This verification is left to you as an exercise.

Step 2: Having verified that the functions in (16.3) actually solve (12), this last step argues that they are the only solutions to (12). For this purpose, I am going to rewrite (12) so that it looks just like the two-component equations that we studied in previous chapters. Here is the trick for doing this: I will simply give the derivative of B a new name, P. Then I write two equations, where the first just tells me what P is,

$$\frac{d}{dx} B = P, \tag{20}$$

and the second is (12), but rewritten using P for the derivative of B:

$$\frac{d}{dx} P = cB. \tag{21}$$

Together, these last two equations carry the same information as (12).

The point now is that I have already given you an argument (in Section 8.1 of Chapter 8) for the fact that the solutions to the pair of equations given by (20) and (21) are completely determined by two parameters, the values of B and P at $x = 0$. In particular, in the $c > 0$ case, we have $B(0) = \alpha + \beta$ and $P(0) = \sqrt{c}(\alpha - \beta)$ and thus our α and β are the only free parameters in the general solution to (12). In the case where $c < 0$, we have $B(0) = \alpha$ and $P(0) = (-c)^{1/2}\beta$, so again the parameters α and β are the only free parameters in the general solution to (12).

READINGS FOR CHAPTER 16

READING 16.1

Fishing for Answers: Deep Trawls Leave Destruction in Their Wake—But for How Long?

Commentary: This article describes the habitat destruction that is caused by modern, bottom-hugging fishing trawls. The latter tend to ruin the ocean bottom habitat and thus seriously affect a basic part of the near-shore habitat. There is another modeling problem buried in here besides the one discussed in Chapter 16: Suppose that a sequence of bottom-hugging trawls depopulates the floor fauna along a very long swath of width R. Can we predict how long it will take for a certain species of bottom crawler to repopulate the swath? To obtain an answer, suppose that the width of the destruction zone lies along the x-axis from $x = 0$ to $x = R$. Let $u(t, x)$ denote the population density at time t and position x of the species in question. As remarked in Chapter 16, u can be modeled as a solution to the diffusion equation, $\frac{\partial}{\partial t} u = \mu \frac{\partial^2}{\partial x^2} u + ru$, where r is the birth-death rate and where μ is the diffusion coefficient. The initial conditions should be set so that $u(0, x) = 0$ for $0 \leq x \leq R$ and for x otherwise, $u(0, x) = 1$. The exact solution can be expressed as an integral:

$$u(t, x) = \int_{x < a < \infty} (4\pi \mu t)^{-1/2} \exp(-a^2/4\mu t) \, da$$

$$+ \int_{R-x < a < \infty} (4\pi \mu t)^{-1/2} \exp(-a^2/4\mu t) \, da.$$

These integrals are hard to evaluate, but even so, a bit of reflection leads to the conclusion that $u(t, R/2) \sim \exp(-R^2/16\mu t)$. Thus, repopulation at the center of the strip takes $t \sim R^2/\mu$.

Fishing for Answers: Deep Trawls Leave Destruction in Their Wake—But for How Long?

Science News, **150** (1996) 268-271.

Janet Raloff

In recent years, one after another of the world's major fisheries has collapsed or exhibited signs of severe stress. In hopes of saving their industry, fishers have been turning to species and fishing grounds they had formerly ignored.

For instance, some of those who had been plying their trade near shore now travel some 200 kilometers out. There, they harvest stocks along the continental shelves at depths of 300 meters. Others have even begun fishing the continental slopes—at such staggering depths as 1,200 meters.

Commercial fleets are increasingly investing in seabed equipment known as mobile gear. Dragged along the ocean floor at even the greatest of these depths, their trawls and dredges scoop up everything in their path, bringing to the surface whatever doesn't sift through their nets. Those nets inevitably snag some rocks, turning them over and destroying animals attached to them.

Lately, marine ecologists have begun showing up at these fishing grounds with their own, even higher tech gear. Their trawls, sleds, and dredges come equipped with video cameras, sidescan sonar, and computer-driven mechanical shovels that can sample the seabed at the touch of a button.

Their goal is not to catch fish but to haul in hard data documenting trawling's impact on tube worms, sponges, anemones, hydrozoans, urchins, and other denizens of the deep. Ten or even 20 pounds of these animals, which are generally smaller than the target fish, may be caught—and discarded as waste—for every pound of commercial catch. Caught in the roiling waters, some of the sea dwellers remain on the ocean floor, crushed, uprooted, or displaced after chains, bars, or metal doors have plowed through the sediment that had been their home.

Worthless by fishing standards, these critters provide food and habitat for some or all of the commercial fisheries under stress. In fact, argues Elliott A. Norse, director of the Marine Conservation Biology Institute in Redmond, Wash., trawling's toll on these largely ignored seafloor species may underlie the recent collapse of many commercial groundfish

stocks, which include cod, haddock, pollock, and flounder.

"What we've done is destroy the carrying capacity of the habitat to support those [fisheries] by removing the organisms that provide shelter for little fishes," he told *Science News*. "We're talking about destruction of marine habitat that is, if not equivalent, at least in the ballpark with clearcutting forests on land."

Not everyone concurs. "There's no question that certain habitats have taken a real pounding," says Andrew A. Rosenberg, northeast regional administrator of the National Marine Fisheries Service in Gloucester, Mass. Though he acknowledges that sharp declines in stocks of exploited fish, such as cod, have been "clearly associated with fishing," he adds, "I don't know that you'd conclude it's due to a clear-cutting type of effect on habitat."

Such a determination would require long-term monitoring of the nontargeted ocean floor communities—which, he notes, is not done today. So while he believes the clear-cutting issue "is a valid and important concern, [Norse's] conclusion may be a little premature."

Hoping to help resolve the issue, a number of research ventures have begun to identify vulnerabilities in the seafloor communities and to study how quickly damaged habitats bounce back. Their findings could influence whether and how fishing regulations might be modified to ensure that critical habitats receive a chance to recover.

A biological oceanographer who studies seafloor habitats, Les Watling of the University of Maine's Darling Marine Center in Walpole has become particularly concerned about fishing fleets moving into what had been inaccessible sites.

Rock-hopper gear, for example, introduced about 10 to 15 years ago, can roll over large seabed obstacles. Ropes that thread through a series of huge balls or rollers drag a net across the floor, often overturning rocks and, Watling says, "grinding to a pulp" any animals cemented to them.

In 1987, he videotaped Outer Falls, then a pristine community some 80 miles offshore in the Gulf of Maine. The boulder-strewn area teemed with ancient sponges, bushlike bryozoans, and other animals that form colonies and anchor themselves permanently onto solid footings.

Studies have shown that the fry of groundfish, such as cod, survive best in the shelter afforded by such structurally complex bottoms—seabeds strewn with cobbles or rocks and dense with organisms growing up from them. Areas like Outer Falls, some 100 meters below the surface, probably served as nurseries for vulnerable yearling fish, Watling says.

He could tell that Outer Falls hadn't yet been trawled 9 years ago, because "its stones were completely covered with animals." Marine fauna were even sandwiched into the crevices between rocks and the sediment. Fearing the area's rocky prominences, which would have ripped apart any nets dragged over them, trawlers had shunned this obviously old and stable community, Watling says.

When he returned to Outer Falls 3 years ago, "it looked like a hurricane had been through." Boulders had been overturned and the area's slow-growing colonial animals, which have no natural predators, had vanished. Judging by size, some of the lost sponges may have been at least 50 years old, Watling says. Because these slow-growing animals also take a long time to reestablish themselves, replacement of such mature communities could take a century.

He now suspects that "the biggest factor behind the decline of fish in the Gulf of Maine is the rock-hopper."

Canada's Department of Fisheries and Oceans is also concerned about rock-hoppers and associated gear tearing through seafloor communities. So for the past 3 years, its scientists have trawled and examined a small area that is closed to commercial fishing on the Grand Banks off Newfoundland.

They surveyed the local inhabitants before and after conducting a dozen trawling runs down each of three 13-kilometer-long corridors. This trawling simulates on a local scale a year's commercial fishing. A synthesis of the findings could be completed by early next year, says Donald C. Gordon Jr., an ecologist at the Bedford Institute of Oceanography in Dartmouth, Nova Scotia and a coleader of the project.

Preliminary findings from the first 2 years of the study indicate some early warning signs of ecological change, Gordon's team noted last June at

a small ecological conference on trawling held at the Darling Center. Sidescan sonar images revealed changes in the floor surface that persisted at least 1 year. Further acoustic studies detected millimeter-scale changes in the structure of the top 4.5 centimeters of sediment, where most animals live.

It looks as if trawling has homogenized the subsurface structure of the sandy sediment, a change that points to the removal or destruction of infauna-sediment-burrowing animals, says Peter Schwinghamer of the Bedford Institute. "Anything that's important to the infauna [here] is important to cod."

The net picked up fewer invertebrates—usually snow crabs, basket stars, and sea urchins—with each of the dozen passes of the trawl. However, what ended up in the trawl's net represents just a fraction of the damage to bottom dwellers. Many shell shards and other pieces of animals were visible on the seafloor.

In tropical waters half a world away, Ian Poiner and his colleagues at the Commonwealth Scientific and Industrial Research Organisation in Queensland are completing a 5-year study on the impact of prawn fishing between Australia's coast and the Great Barrier Reef. Like Gordon's team, they conducted a dozen research trawling runs down well-characterized corridors to simulate the intensity of local fishing. On average, Poiner notes, commercial trawls plow through most of these Australian waters at least once—and in many places up to eight times—annually.

With Japanese consumers willing to pay $25 to $35 per pound for tiger prawns, a single ship can earn $1 million in a couple of months of shrimping.

Poiner's findings, also reported at the Darling conference, showed that a single pass of the trawl removes some 5 to 20 percent of the seafloor animals. "So you get total depletion, certainly, by 10 or 12 trawls."

Here, mining the bottom does not appear to be hurting the short-term productivity of the exploited stock. One reason, Poiner suspects, is that across a given area, the trawls remove more predators than prawns.

Simon Thrush of New Zealand's National Institute of Water and Atmospheric Research in Hamilton has been working to estimate how quickly the disturbed seafloor communities recover. In one experiment, he kept patches of soft sediment covered with a concrete slab for 1 month. "We expected our plots to recover in 2 to 3 months," he says. However, 9 months later, the 0.2- to 3.2-meter-square test plots still exhibited less species diversity and lower abundances of each species than undisturbed tracts nearby.

In this environment, tube worms—on the menu of fish and birds the world over—normally make mats that cover the seafloor. Just 1 to 2 centimeters long, the worms glue fine sediment into fragile cylindrical homes that extend about 5 centimeters above the seabed. Thrush now suspects that their slow recolonization reflects his compaction of the sediment, which makes it difficult for would-be immigrants to remodel.

Because storms frequently reshape soft, sandy floors in waters less than 70 meters deep, Thrush notes, the conventional wisdom has held that trawling in these areas has minimal impact on sediment dwellers. "Our experiment illustrates that this is an oversimplification," he told *Science News.*

Major changes in sediment structure could also alter both the chemical form and the release of nutrients, thus affecting the habitability of the entire water column, according to Lawrence Mayer, a biogeochemist at the Darling Center.

Sediments supply about half of the nutrients in waters to depths of perhaps 200 meters, he notes. Studies have shown that a host of environmental factors can affect how bacteria manipulate chemicals in their vicinity. For instance, a sediment's geometry can influence whether bacteria release nitrogen in biologically useful forms that serve as natural fertilizers or in inactive compounds that most animals ignore.

"As we trawl," Mayer notes, "we convert the geometry of the ecosystem from one containing a small number of large burrows to one that contains a large number of small burrows." This reflects the replacement of larger animals by small opportunists.

Will trawling prompt sediments to act as a source or as a sink of fertilizer for continental shelf ecosystems? "I haven't the slightest idea," Mayer says. Too little research has been done on this "terribly complex system" to offer a useful gauge.

Today, fisheries are managed largely in terms of how many animals can be harvested without reducing the vitality of the population. The new trawling studies raise questions about the extent to which commercially fished stocks depend on habitats that are being degraded by seafloor trawling, Rosenberg says. He would like to see long-term monitoring of the ignored seabed communities to establish their role in the productivity of commercial fisheries.

So would Norse. Unfortunately, he says, this topic "has gotten very little attention" to date and even less research funding. Nor should the economic performance of commercial fisheries necessarily be the primary focus of such research, he argues. He would like to see the conservation of biodiversity accorded equal importance.

Toward that end, he advocates the development of marine reserves closed to fishing and other human disturbances.

Gordon, Schwinghamer, and some of their colleagues would also like to see the use of mobile gear in fisheries managed more conservatively, arguing that trawls and dredges should be permitted only in certain regions and be used only during specified periods, depending on the apparent vulnerability of the habitat and its role in the life cycle of other fishes.

Rosenberg would take more of a wait-and-see approach. He says that telling people not to trawl "is not a particularly viable strategy." He would like to see other management options explored through research that looks not only at biology but also at the sociology and economics of fishing.

John Williamson, a fisherman from Kennebunk, Maine, who does not use bottom trawls, worries that the answers to such questions may come too late.

Not long ago, he could motor out to where huge schools of fish congregated and reliably haul in the day's limit. Today, he says, "I'm not going to find a large concentration of fish anywhere"—and the situation is only getting worse. Already, he charges, it's as if fishers have been reduced to hunting down "small patches of fish in the middle of a barren desert."

Exercises for Chapter 16

1. Suppose that the density of a certain species of fish in a particular region of the ocean at time t and position x is given by the function

$$u(t, x) = R\frac{1}{\sqrt{t}}e^{-\frac{x^2}{4\mu t}},$$

where R and μ are positive constants. How would you determine R and μ if you had the graph of $u(1, x)$? How would you do it if, instead, you had the graph of $u(2, x)$?

2. In each of the cases $c > 0$, $c = 0$, and $c < 0$, verify, by differentiating the expression for $B(x)$ in (16.3) of Chapter 16, that the given function satisfies (12) in Chapter 16.

3. In each case, decide which of the given functions are not negative when the variable x is in the indicated range.

   (a) $\sin(\pi x)$, $\sin(3\pi x)$, $\cos(\pi x/2)$ for $0 \leq x \leq 1$

   (b) $\cos(\pi x/2)$, $\cos(3\pi x/2)$, $\cos(5\pi x/2)$ for $0 \leq x \leq 1/2$

   (c) $\sin(\pi x/4)$, $\sin(\pi x/2)$, $\sin(3\pi x/4)$, $\sin(\pi x)$ for $0 \leq x \leq 3/2$

4. Find all possible values of the constants α and β that make the function $B(x)$ given vanish at both $x = 0$ and $x = 1$.

   (a) $B(x) = \alpha e^{5x} + \beta e^{-5x}$

   (b) $B(x) = \alpha \cos(3\pi x/2) + \beta \sin(3\pi x/2)$

   (c) $B(x) = \alpha \cos(\pi x) + \beta \sin(\pi x)$

   (d) $B(x) = \alpha \cos(2\pi x) + \beta \sin(2\pi x)$

   (e) $B(x) = \alpha e^{3\pi x} + \beta e^{-3\pi x}$

17

Separation of Variables

The previous chapter introduced the technique of separation of variables for finding some special solutions to the diffusion equation. This technique works for other equations as well, and the discussion here is meant to provide another, slightly different example. In this example, we are going to model the distribution of a chemical (say a protein) in a developing embryo. (It is known that the concentration of some proteins at a given cell in an embryo determines whether or not a particular gene is expressed in that cell.)

This fairly schematic model is set up as follows: The embryo is modeled as a square of side length L (measured in appropriate units). Thus, the embryo encompasses those points in the (x, y) plane with $0 \leq x \leq L$ and $0 \leq y \leq L$. In this model, the thickness of the embryo is assumed to be unimportant. We let $u(t, x, y)$ denote the density of the protein in question at time t and position (x, y) in the embryo. Thus, we will restrict our attention to where both $0 \leq x \leq L$ and $0 \leq y \leq L$. I will suppose that the protein molecules move randomly through the embryo (sometimes this not a good assumption as proteins commonly bind to other molecules; indeed, the various molecules to which a given protein binds determines its effect in a cell or embryo) and that the protein is broken down in the cells of the embryo at some rate r. However, let me assume that the cells along the $x = 0$ edge of the embryo are rather special in that they actually produce the protein, and that this production is such that the protein concentration at a point $(0, y)$ is independent of time and is given by $u(t, 0, y) = \sin(\pi y/L)$. Note that this function is zero, where $y = 0$ and $y = L$, and peaks to 1 where $y = L/2$. See the graph in Figure 17.1.

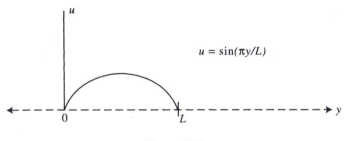

Figure 17.1

17.1   The Diffusion Equation

With the preceding model understood, we assume that the time evolution of the function $u(t, x, y)$ is such that it satisfies the diffusion equation,

$$\frac{\partial u}{\partial t} = \mu \left(\frac{\partial^2 u}{\partial x^2} + \frac{\partial^2 u}{\partial y^2} \right) - ru, \tag{1}$$

together with the boundary conditions (at all times t)

$$
\begin{aligned}
u(t, 0, y) &= \sin(\pi y/L) \\
u(t, L, y) &= 0 \\
u(t, x, 0) &= 0 \\
u(t, x, L) &= 0.
\end{aligned}
\tag{2}
$$

Here, μ is a constant that is determined by the protein in question and the intra-cellular medium.

17.2   Equilibrium Solutions

Let us now look for equilibrium solutions. This is a solution of (1) that obeys the boundary conditions in (2) and has no time dependence. That is, $\frac{\partial u}{\partial t} = 0$. In this case, the function u is a function only of x and y and obeys the equation

$$\mu \left(\frac{\partial^2}{\partial x^2} + \frac{\partial^2}{\partial y^2} \right) - ru = 0, \tag{3}$$

together with the boundary conditions in (2). In the case where $r = 0$, this last equation is called **Laplace's equation**, and in the case $r \neq 0$ it is called **Helmholtz's equation**.

17.3   Separation of Variables Revisited

We can try to find solutions to (3) by making the separation of variables assumption, namely that $u(x, y) = A(x)B(y)$. This last assumption is consistent with (3) if and only if the functions A and B obey the equation

$$\mu \left(B\frac{d^2 A}{dx^2} + A\frac{d^2 B}{dy^2} \right) - rAB = 0. \tag{4}$$

You should not take my word for this. Indeed, substitute the form $A(x)B(y)$ for u in (3) and differentiate to verify (4) yourself.

Now we employ the trick of dividing this last equation by AB, and so find that

$$\mu \left(A^{-1}\frac{d^2 A}{dx^2} + B^{-1}\frac{d^2 B}{dy^2} \right) = r. \tag{5}$$

At this point, it is crucial to note that $A^{-1}\frac{d^2 A}{dx^2}$ is a function only of x and $B^{-1}\frac{d^2 B}{dy^2}$ is a function only of y, and the only way these two functions can sum to a constant is if both are constant themselves. (No x-dependence in the former can be canceled by the latter as the latter has only y-dependence. Likewise, no y-dependence in the latter can be canceled by the former.) Thus, we learn that if $u = A(x)B(y)$ is going to satisfy (3), then A and B are constrained to obey

$$A^{-1}\frac{d^2 A}{dx^2} = r/\mu - \lambda$$

$$B^{-1}\frac{d^2 B}{dy^2} = \lambda. \tag{6}$$

Here, λ can (at this point) be any constant we like. In addition, we shouldn't forget the boundary conditions in (2). These translate to

$$A(0)B(y) = \sin(\pi y/L)$$
$$A(L)B(y) = 0$$
$$A(x)B(0) = 0$$
$$A(x)B(L) = 0. \tag{7}$$

17.4   Working Out the Answer

Now we can proceed to find solutions to these last two equations. To start, consider the second equation in (6) for B. Multiplying both sides by B, we see that this equation has the form $\frac{d^2 B}{dy^2} = \lambda B$. As you might recall from the previous chapter, the form of the general solution to this equation depends on whether $\lambda > 0$, $\lambda = 0$, or $\lambda < 0$:

- When $\lambda > 0$, then $B = \alpha e^{\sqrt{\lambda}y} + \beta e^{-\sqrt{\lambda}y}$.

- When $\lambda = 0$, then $B = \alpha + \beta y$.

- When $\lambda < 0$, then $B = \alpha \cos(|\lambda|^{1/2}y) + \beta \sin(|\lambda|^{1/2}y)$.   (8)

Here, α and β are arbitrary constants.

We can decide which case to take by using the first line of (7). This equation rules out the $\lambda > 0$ and $\lambda = 0$ cases of (8) and is consistent only with the $\lambda < 0$ case of (8) when $|\lambda|^{1/2} = \pi/L$. Then we can take $\alpha = 0$ and $\beta = 1$ so that $B(y) = \sin(\pi y/L)$. Note that this choice for $B(y)$ automatically fulfills the requirements in the third and fourth line of (7).

With B and λ determined, we can proceed to the first line of (6) and consider the equation for A: In this case, multiply both sides by A to find that A obeys the equation $\frac{d^2A}{dx^2} = (r/\mu + \pi^2/L^2)A$. Note that this has the schematic form $\frac{d^2A}{dx^2} = cA$, where $c > 0$ is the combination $r/\mu + \pi^2/L^2$. As we learned in the previous chapter, the $c > 0$ case of the equation $\frac{d^2A}{dx^2} = cA$ has the general solution

$$A(x) = \alpha e^{\sqrt{c}x} + \beta e^{\sqrt{-c}x},   \qquad (9)$$

where α and β are, at this point, arbitrary constants. Of course, if c here were to equal zero, then the general solution would have the form $A = \alpha x + \beta$; and were $c < 0$, then the general solution would be $A = \alpha \cos(|c|^{1/2}x) + \beta \sin(|c|^{1/2}x)$. In both cases, α and β are constants that we are free to choose. However, in this case, c is observedly positive.

With A given by (9) and $B(y) = \sin(\pi y/L)$, the first line of (7) requires that

$$\alpha + \beta = 1,   \qquad (10)$$

while the second line of (7) requires that

$$\alpha e^{\sqrt{c}L} + \beta e^{\sqrt{-c}L} = 0.   \qquad (11)$$

These last two equations can be solved for α and β. Indeed, use (11) to conclude that the constant $\beta = -\alpha e^{2\sqrt{c}L}$. Use this last result in (10) to write $\alpha = -1/(e^{2\sqrt{c}L} - 1)$. Then go back to (11) to find that $\beta = e^{2\sqrt{c}L}/(e^{2\sqrt{c}L} - 1)$.

Finally, with A and B determined, we have our solution u:

$$u(x, y) = (-e^{\sqrt{c}x} + e^{2\sqrt{c}L}e^{-\sqrt{c}x})(e^{2\sqrt{c}L} - 1)^{-1} \sin(\pi y/L),   \qquad (12)$$

where $c = r/\mu + \pi^2/L^2$.

17.5   What Should You Expect?

You should ask yourself the following question: Does the function in (12) conform to what you might expect for the equilibrium distribution? Indeed, you should think hard now about whether this $u(x, y)$ is reasonable from a biological point of view. For

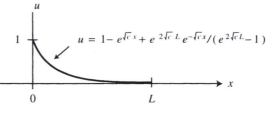

$$u = 1 - e^{\sqrt{c}\,x} + e^{2\sqrt{c}\,L}\,e^{-\sqrt{c}\,x}/(e^{2\sqrt{c}\,L} - 1)$$

Figure 17.2

example, look at u as x varies from 0 to L with y fixed at the midpoint, $L/2$. See the graph of $u(x, L/2)$ in Figure 17.2.

Notice that this function is largest at $x = 0$ and decreases steadily to zero as x increases to L. Do you think that this is reasonable? If the protein in question is made along the $x = 0$ line and nowhere else, doesn't it make sense that the protein's concentration is largest there? Since the protein takes time to diffuse to some point with positive x, and it is degraded at a rate r along the way, doesn't it make sense for the protein's concentration to decrease as x increases? Indeed, wouldn't you expect less protein at large x since protein created at $x = 0$ takes longer to get there and so has a greater chance of being degraded?

17.6   Lessons

Here are some key points from this chapter:

- Remember that an equilibrium solution to a diffusion equation is a solution with no time dependence.

- Study how the separation of variables procedure works in another example.

- When you have a solution in hand, you absolutely must ask yourself whether the solution conforms to your expectations as a biologist. A solution may not conform to your expectations for one or more of the following three reasons:

  (a) The solution is simply incorrect.

  (b) Your model is incorrect.

  (c) The model and solution are both correct; your intuition about the biology is incorrect.

  It is very important to find out which of these last three possibilities is in play.

Exercises for Chapter 17

In Exercise 1, consider a functions $h(x)$ for $0 \le x \le R$ that obeys

$$\frac{d^2}{dx^2}h = ch. \tag{13}$$

Here c is some fixed number. Remember from Chapter 16 that the general solution to (13) has two freely specifiable constants, α and β, and has the following form:

- When $c > 0$: $h(x) = \alpha e^{\sqrt{c}x} + \beta e^{-\sqrt{c}x}$

- When $c = 0$: $h(x) = \alpha + \beta x$

- When $c < 0$: $h(x) = \alpha \cos((-c)^{1/2}x) + \beta \sin((-c)^{1/2}x)$

In a typical application, the constants c and also α and β are specified by various requirements on h at $x = 0$ and at $x = R$.

1. In each case, find all constants c, α, and β for which a solution to (13) obeys

   (a) $h(0) = 0$ and $h(R) = 0$

   (b) $(\frac{d}{dx}h)(0) = 0$ and $(\frac{d}{dx}h)(R) = 0$

   (c) $(\frac{d}{dx}h)(0) = 0$ and $h(R) = 0$

   (d) $h(0) = 0$ and $(\frac{d}{dx}h)(R) = 0$

   (e) $h(0) = 0$ and $h(R) = 1$

   (f) $h(0) = -1$ and $h(R) = 1$

   (g) $h(0) = 1$ and $(\frac{d}{dx}h)(0) = 0$

   (h) $h(0) = -1$ and $(\frac{d}{dx}h)(R) = 3$

   (i) $(\frac{d}{dx}h)(0) = 1$ and $(\frac{d}{dx}h)(R) = 1$

2. Find a function $u(t, x)$ that is not everywhere zero and obeys

$$\frac{\partial}{\partial t}u(t, x) = 2\frac{\partial^2}{\partial x^2}u(t, x)$$

   when $0 \le x \le 10$, and that obeys, in each case,

   (a) $(\frac{\partial}{\partial x}u)(t, 0) = 0$ and $(\frac{\partial}{\partial x}u)(t, 10) = 0$ for all time t

   (b) $(\frac{\partial}{\partial x}u)(t, 0) = 0$ and $u(t, 10) = 0$ for all time t

   (c) $u(t, 0) = 0$ and $(\frac{\partial}{\partial x}u)(t, 10) = 0$ for all time t

   [*Hint:* Try finding a solution $u(t, x)$ of the form $A(t)B(x)$, where A is a function of t only and where B is a function of x only. Look in Chapters 16 and 17 for the strategy for dealing with such forms for u.]

3. The separation of variables strategy can also be applied to some forms of the advection equation. In particular, consider the equation

$$\frac{\partial}{\partial t} u = -c \frac{\partial}{\partial x} u + ru,$$

where r and c are constants. In terms of r and c (and possibly some freely specifiable constants), write all solutions $u(t, x)$ that obey the preceding equation and comply with the separation of variables assumption. In other words, write all solutions for which $u(t, x) = A(t)B(x)$.

18

The Diffusion Equation and Pattern Formation

This chapter begins the discussion of another important application of the diffusion equation, which is the modeling of pattern formation in biology. In this regard, you should know that the development of complicated patterns and structures in biological systems is currently under intense study. In particular, it is clear now that developmental patterns arise because

- Cells in an embryo react differently to different concentrations of certain molecules.

- Moreover, these concentrations are definitely not constant over the embryo.

Current investigations seek to determine answers to the following questions: First, which molecules are relevant to the formation of a particular structure? Next, how do the cells react to these molecules? Finally, how do the concentrations of the relevant molecules come to vary across the embryo?

The last question suggests a model of pattern formation that uses a diffusion equation. However, even if diffusion is the appropriate model, the story as uncovered so far seems anything but simple. (For example, often there are a multitude of molecules involved that affect each other's concentration through complicated feedback loops. When fully sorted out, the picture will be dazzling.)

In any event, here is a pertinent question: What is the simplest diffusion model that exhibits complicated spatial patterns? In particular, can a complex pattern arise from the varying concentration of a single chemical species? For example, can the varying concentration of a single chemical give a tiger stripes?

These are all hard questions—some not well understood at the time of this writing. In particular, I won't mislead you into thinking that answers will be provided here. Rather, this chapter and the next will study a toy model and some solutions to the toy model. Moreover, most of the solutions that I will consider don't provide interesting patterns.

With this last point understood, keep in mind that the models and their solutions are not really the point of the discussion that follows. Rather, the real purpose of Chapters 18 and 19 is to pinpoint some of the key issues that need to be addressed in any diffusion model approach to pattern formation. In particular, the crucial issue is that of the *stability* of a solution. In this regard, you should recall the discussion from Chapters 4, 9, and 10, where we studied stability in equations with only one variable, time. Here, we are going to consider the necessarily more complicated stability question for solutions to equations that involve time and space variables. Even so, you can be sure that stability is important here for the same reason as in the earlier chapters: An unstable solution is unlikely to arise in nature.

A second key issue is the manner in which the boundary conditions for the diffusion equation are derived.

As I said, the discussion here considers only the diffusion equation for a single chemical species. However, some of the commentaries for this chapter and the book *Mathematics for Dynamical Modeling* by Beltrami consider patterns that arise from varying concentrations of a pair of molecular species.

18.1   The Model

Consider the following simple model for the tiger's skin: Suppose that the yellow color of tiger hair is caused by high concentrations of a particular chemical in certain cells of a tiger embryo, and that the black color is caused by low concentrations of the same chemical in other cells. We shall suppose that the chemical concentration in the tiger embryo is described by a function, u, of time t and of one spatial coordinate only, say the x-coordinate. Thus, $u = u(t, x)$. In our model, suppose that the tiger embryo stretches from $x = 0$ to $x = L$.

18.2   The Equation

Next, assume that the chemical in question moves in a random way through the embryo. With this assumption given, then the distribution of u across the embryo is governed by a diffusion equation, which I will take to have the form

$$\frac{\partial}{\partial t}u = \mu \frac{\partial^2}{\partial x^2}u + f(u). \tag{1}$$

Here, f is a particular function of one variable. Note that $f(u)$ is a function on the (t, x) plane that depends (usually) on the function u. It is a function of (t, x) that depends on another function of (t, x). To be precise, $f(u)(t, x)$ is the value of the original function $f(u)$ at the point $u = u(t, x)$. Here are two simple examples:

(a) $f(u) = r_0 - r_1 u$, where $r_0 > 0$ and r_1 are constants

(b) $f(u) = r_1 u - r_2 u^2$, where r_1 and r_2 are positive constants.   (2)

In a realistic application, the function f in (1) is chosen so that the equation $\frac{d}{dt} v = f(v)$ for a function $v(t)$ describes the amount of the chemical as a function of time in a single, *isolated* cell; that is, in a context where no diffusion can occur. Thus, the form of the function f for (1) is a quantity that can be determined in principle by experiments on isolated cells. Indeed, the elegant thing about (1) is that it is a predictive equation for the dynamics of the chemical in the whole embryo that is obtained by adding just the diffusion term, $\mu \frac{\partial^2}{\partial x^2} u$, to the equation $\frac{d}{dt} v = f(v)$, which presumes to describe the dynamics of the chemical in a single cell.

Here is some related terminology: The f-term on the right side of (1) is sometimes called the **reaction** term, while the second derivative term is called the **diffusion** term. You will also see (1) called a **reaction-diffusion** equation.

Before going further, I warn you to be wary of the assumption that the chemical in question spreads by random motions. This is an assumption that must be verified by experiments. Other transport mechanisms are also feasible, and so the use of the diffusion equation constitutes a definite assumption about the underlying biochemistry that must be verified by experiments. [In this regard, see the article titled *Activin Signalling and Response to a Morphogen Gradient* by Gurdon, Harger, Mitchell, and Lemaire from Reading 18.4.]

18.3   The Boundary Conditions

It is important to think hard about the boundary conditions that nature imposes upon the function u at $x = 0$ and also at $x = L$. To decide upon these boundary conditions, it is necessary to return to the derivation of (1) in Chapter 14. Consider the total amount of chemical in the strip where $a \leq x \leq a + \Delta x$. This is given by

$$m(t, a) = \int_a^{a+\Delta x} u(t, x)\, dx.$$ (3)

When Δx is small, then u doesn't vary much over the integration interval and so $m(t, a)$ is approximately $u(t, a)\, \Delta x$. In any event, m satisfies the equation

$$\frac{\partial}{\partial t} m = q(t, a) - q(t, a + \Delta x) + \int_a^{a+\Delta x} f(u(t, s))\, ds.$$ (4)

Here, $q(t, a)$ is the rate at which the molecules pass $x = a$ from left to right minus the rate at which they pass $x = a$ from right to left. Since this is a process where the molecular motion is presumably random, we argued that $q(t, a) = -\mu \frac{\partial}{\partial x} u$.

Here again, you should question whether the relevant molecular motions here are random, and thus whether a diffusion model is appropriate. As I said, the random nature of the molecular motions constitutes a basic *assumption* that underlies our model. In any event, I'll keep this assumption.

Here is the key point: At $x = 0$, there should be *no* chemical passing from right to left, and also no chemical passing from left to right, since the embryo *ends* at $x = 0$. Thus, to be consistent, we are forced to require that $q(t, 0) = 0$. That is,

$$\left(\frac{\partial}{\partial x} u \right)(t, 0) = 0 \tag{5}$$

for all t. The identical argument at $x = L$ shows that $(\frac{\partial}{\partial x} u)(t, L) = 0$ for all t also.

In summary, we must augment to (1) the boundary conditions that

$$\left(\frac{\partial}{\partial x} u \right)(t, 0) = \left(\frac{\partial}{\partial x} u \right)(t, L) = 0. \tag{6}$$

18.4   The Equilibrium Solutions

We are now interested in an equilibrium (time independent) solution of (1) subject to the boundary conditions in (6). That is, we want to find a solution to (1) and (6) that has the property that $\frac{\partial}{\partial t} u = 0$. Our interest in an equilibrium solution implies that we are assuming that the stripe pattern, once set, does not change much with time. There are animals for which this is true, and there are others for which it evidently is not.

If u has no time derivative, then u is a function only of x [i.e., $u(t, x) = u_e(x)$]. And, because of (1) and (6), this $u_e(x)$ must satisfy the equation

$$\mu \frac{d^2}{dx^2} u_e + f(u_e) = 0 \tag{7}$$

with

$$\left(\frac{d}{dx} u_e \right)(0) = \left(\frac{d}{dx} u_e \right)(L) = 0. \tag{8}$$

(For now you should be thinking of u_e instead of u.)

18.5   Stability

Consider (7) and (8) for your favorite choice of function f [as in (2), or something bizarre: $f(u) = (1 + u^2)^{-1} \sin(\cos(u^3 e^u))$]. For the sake of argument, suppose that there is a solution $u_e(x)$ to (7) and (8) for this choice of function f. We should ask whether or not we will have any reasonable chance of seeing your solution in nature.

This is a stability question that we can formulate as follows: Suppose that $u_e(x)$ is a solution to (7) and (8), and suppose that we consider a tiny perturbation of $u_e(x)$ at time $t = 0$, namely $w(x)$. We shall now evolve $u(0, x) = u_e(x) + w(x)$ forward in time using (1) and (6) to obtain a solution $u(t, x)$ to (1) and (6) that at time 0 is equal to the sum of $u_e(x)$ and the perturbation, $w(x)$.

You might argue that your solution $u_e(x)$ to (7) and (8) is stable when the following condition holds:

- As long as the initial perturbation function $w(x)$ is small enough, then the resulting solution $u(t, x)$ to (1) and (6) that has $u(0, x) = u_e(x) + w(x)$ has the property that at *every* x, the values of $u(t, x)$ approach $u_e(x)$ as t gets large at each x.

Equivalently, we can say that $u_e(x)$ is stable if every solution to (1) and (6) of the form $u(t, x) = u_e(x) + w(t, x)$, with $w(0, x)$ everywhere small, has the property that $w(t, x)$ goes to zero as t gets large at each x.

Conversely, you might argue that the time-independent solution $u_e(x)$ is unstable when

- There are arbitrarily small (but not everywhere zero) initial perturbations [i.e., a choice of $w(x)$] so that the time $t > 0$ evolution $u(t, x)$ does not tend to $u_e(x)$ as t gets large for at least one x. Here, $u(t, x)$ obeys (1) and (6) and $u(0, x) = u_e(x) + w(x)$.

18.6   Linear Stability

In many cases, the preceding stablility criterion is essentially impossible to verify. Although this criterion might conform to our intuitive notion of stability, it is not always a practical criterion. In this section we provide the definition of an essentially equivalent and somewhat more practical criterion. By the way, if a solution is stable with respect to the following criterion, then it will be stable with respect to the definition given in the previous section. However, the converse is not necessarily true. On the other hand, the definition of stability given here also implies stability against slight changes in the *equation*, not just in the starting function $u(t, 0)$. Thus, the stability criterion given below is, in spite of its somewhat technical definition, more relevant to the real world.

In order to give the new stablility criterion, it is necessary first to digress with the introduction of a new function of x that is constructed out of the function $f(\cdot)$, which appears in (1) together with the equilibrium solution $u_e(x)$ to (7) and (8). So bear with me for a moment. This new function [which I call $z(x)$] is the function whose value at x is obtained by evaluating the u-derivative of f at the value $u = u_e(x)$. That is,

$$z(x) = \frac{df}{du}\bigg|_{u=u_e(x)} . \tag{9}$$

For example,

(a) When f is given by Equation (2a), then $z(x) = -r_1$.

(b) When f is given by Equation (2b), then $z(x) = r_1 - 2r_2 u_e(x)$.   (10)

Thus, $z(x)$ is a function of x that is determined jointly by the function f in (1) and by the solution $u_e(x)$ to (7) and (8).

With the function $z(x)$ understood, here is the new stability criterion:

STABILITY CRITERION: The solution $h(x)$ is a *stable* solution to (7) and (8) if and only if there is *no* pair (g, λ), where $g(x)$ is a not everywhere zero function of x for $0 \leq x \leq L$, where λ is a real number, and where the following additional constraints are satisfied:

- $\lambda \geq 0$

- $\lambda g = \mu \frac{d^2 g}{dx^2} + z(x)g$

- $\left. \frac{dg}{dx} \right|_{x=0} = \left. \frac{dg}{dx} \right|_{x=L} = 0.$ (11)

Conversely, if there is even one such pair (g, λ) that obeys the conditions in (11), then the equilibrium solution $h(x)$ is *unstable* in at least one of the following two ways:

1. There is a solution to (1) and (6) that at time $t = 0$ is very near to u_e at all x, and then moves away from $u_e(x)$ (for some x, anyway) as t gets large.

2. There is a slight perturbation of the equations [either (1) or (6)] that has a solution that also starts at time zero near to $u_e(x)$ at all x but then moves away from $u_e(x)$ (for some x, anyway) as t gets large.

18.7   Some Heuristic Justifications

This book will not give a rigorous argument to justify the stability criterion in the preceding section. However, the discussion that follows is meant to give some idea of the reasoning that leads to this criterion.

Step 1: The stability question involves solutions to (1) and (6) that are, at all points x, close to the equilibrium solution $u_e(x)$. These can be written as $u(t, x) = u_e(x) + w(t, x)$ where $|w|$ is supposed to be small at all x and at all t under consideration.

Step 2: As $|w|$ is small, why not use a Taylor's approximation to replace the potentially complicated $f(u_e(x) + w(t, x))$ with something simpler? For example, at each x, use the first-order Taylor's expansion (see Chapter 2) to replace $f(u_e(x) + w(t, x))$ by

$$f(u_e + w) \rightarrow f(u_e) + \left. \frac{df}{du} \right|_{u=u_e} w.$$ (12)

[Remember that $\left. \frac{df}{du} \right|_{u=u_e}$ is defined to equal $\lim_{w \to 0} \frac{f(u_e+w)-f(u_e)}{w}$ and so for small w, it is fair to approximate $\left. \frac{df}{du} \right|_{u=u_e}$ by $\frac{f(u_e+w)-f(u_e)}{w}$. Note that if we use this last approximation, then (12) becomes an equality.]

Use the replacement approximation in (12) to modify (1) so that the resulting equation reads

$$\frac{\partial}{\partial t}(u_e + w) = \mu \frac{\partial^2}{\partial x^2}(u_e + w) + f(u_e) + z(x)w. \tag{13}$$

[Remember the definition of $z(x)$ from (9).]

Note that this equation is *not* the same as (1) in general. However, we might hope (and this can be justified) that any solution to (13) with small $|w|$ gives a corresponding $u = u_e + w$ that looks very much like a solution to (1). Conversely, any solution to (1) of the form $u = u_e + w$ with $|w|$ small almost solves (13). [This is certainly the case when $|w|$ is identically zero as then (13) just restates (7) and u_e is independent of time.]

Step 3: Since u_e obeys (12) and is independent of time, (13) reads

$$\frac{\partial}{\partial t}w = \mu \frac{\partial^2}{\partial x^2}w + z(x)w. \tag{14}$$

This equation should be augmented with the boundary conditions for w that read

$$\left(\frac{\partial}{\partial x}w\right)(t, 0) = \left(\frac{\partial}{\partial x}w\right)(t, L) = 0. \tag{15}$$

Note that (15) is an equation for the function $w(t, x)$. In practice, to write (9), we usually have to know the function $u_e(x)$ that solves (7) and (8). Without $u_e(x)$, the function $z(x)$ that appears in (15) cannot be written. Equation (14) is simpler than (1) because the w appears in (14) only as a first power, not in some potentially complicated function like $f(u_e + w)$. For example, consider $f(u) = (1 + u^2)^{-1} \sin(\cos(u^3 e^u))$.

Step 4: The point of (14) and (15) is the following:

1. If w is a solution to (14) and (15) and if $|w|$ is very small at all points x, then the function of space and time $u_e(x) + w(t, x)$ is approximately a solution to (1) and (6). Indeed, the approximation gets better and better as $|w|$ gets smaller and smaller. Conversely, if (1) and (6) have a solution of the form $u(x, t) = u_e(x) + w(t, x)$ with $|w|$ small, then w will almost solve (14) and (15).

2. If w is a solution to (14) and (15) that has very small absolute value at all points x to start with, but that grows with time at some points x, then (1) and (6) will have a solution that is close to $u_e(x)$ to start with, but that departs from $u_e(x)$ as time increases. That solution will be approximately $u(t, x) = u_e(x) + w(t, x)$. Conversely, if (1) and (6) have a solution of the form $u(t, x) = u_e(x) + w(t, x)$ that starts at $t = 0$ with $|w|$ very small at all x, and that grows with time, then (14) and (15) will have also have a solution that starts small and grows with time. The latter will look very much like the original w, at least when t is small.

3. If all solutions w to (14) and (15) shrink in absolute value with increasing time, then all solutions to (1) and (6) that start near enough to the equilibrium solution $u_e(x)$ at $t = 0$ will approach $u_e(x)$ at all x as t increases. Thus, $u_e(x)$ will be stable.

Step 5: We can look for solutions of (14) and (15) using separation of variables. This is to say that we write $w(t, x) = A(t)g(x)$. Then our standard separation of variables argument (see Chapters 16 and 17) tells us that $A(t) = A(0)e^{\lambda t}$ and that $g(x)$ obeys the equation

$$\lambda g = \mu \frac{d^2 g}{dx^2} + z(x)g \tag{16}$$

with the boundary conditions $\frac{dg}{dx}|_{x=0} = \frac{dg}{dx}|_{x=L} = 0$. Thus, if there is a not everywhere zero solution $g(x)$ to this last equation with $\lambda \geq 0$ that obeys the boundary conditions, then $w(t, x) = e^{\lambda t}g(x)$ obeys (14) and (15) and doesn't shrink in absolute value as t increases. As remarked upon in Step 4, the existence of such a w implies that the equilbrium solution is unstable.

Step 6: The last step proves that when (14) and (15) have a solution that is not zero identically and that does not shrink in absolute value as t increases, then there is, in particular, a solution $w(t, x)$ with these properties that can be written as $w = e^{\lambda t}g(x)$, where $\lambda \geq 0$ and where g obeys (16) and the associated boundary conditions.

18.8   Lessons

Here are some key points from this chapter:

- Remember the notion of an equilibrium solution—there is no time dependence. So an equilibrium solution obeys a differential equation that involves only the space variables.

- Keep in mind that unstable solutions are unlikely to appear in nature.

- Learn the stability criterion in Section 18.6. This criterion is not simple, but even so, it is very important.

- Notice how (1) splits into a diffusion part, $\mu \frac{\partial^2}{\partial x^2} u$, and a reaction part, $f(u)$. The former models the spread of the chemical in the embryo while the latter models the behavior of the chemical in each cell.

- Don't forget that boundary conditions play a key role in the discussion of any diffusion equation. Moreover, the form of the boundary conditions constitute part of any model's assumptions.

- The boundary conditions used here were found by reconsidering the original derivation in Chapter 14 of the diffusion equation.

READINGS FOR CHAPTER 18

READING 18.1

Generic Modelling of Cooperative Growth Patterns in Bacterial Colonies

Commentary: This article reports on an experiment in which bacteria were grown in a dish under different growth conditions ranging from nutrient-poor to nutrient-rich environments. In some cases, the pattern of bacterial growth in the medium is striking, to say the least. The authors try to obtain the observed growth patterns in a model system that is governed in part by a diffusion equation. In particular, the Equation (1) in the article is the diffusion equation for the concentration $c(r, t)$ for the nutrients at position $r = (x, y)$ and time t. The first term on the left is the usual diffusion term, where D_c is the diffusion coefficient and ∇^2 is shorthand for $(\frac{\partial^2}{\partial x^2} + \frac{\partial^2}{\partial y^2})$. The second term on the left is meant to model the effect of bacteria eating the nutrients. In this regard, the symbol $\sum_{\text{active walkers}}$ is meant to indicate that you are summing up the effect of all of the active bacteria. Meanwhile, the symbol $\delta(\vec{r} - \vec{r}_i)$ signifies that there is no contribution from this term unless the position vector $\vec{r}$ coincides with the position vector $\vec{r}_i$ of the ith bacteria. Equation (3) in the article explains how the bacteria move from one time step to the next. (Note that vectors in the article have arrows over them and are not written in bold-face font.)

The equations are then evolved in time on a computer, with the result that lots of pretty patterns are produced, as seen later in the article.

Generic Modelling of Cooperative Growth Patterns in Bacterial Colonies

Nature, **368** (1994) 46–49.
Eshel Ben-Jacob, Ofer Schochet, Adam Tenenbaum, Inon Cohen, Andras Czirok, Tamas Vicsek

Bacterial colonies must often cope with unfavourable environmental conditions [1, 2]. To do so, they have developed sophisticated modes of cooperative behaviour [3–10]. It has been found that such behaviour can cause bacterial colonies to exhibit complex growth patterns similar to those observed during non-equilibrium growth processes in non-living systems [11]; some of the qualitative features of the latter may be invoked to account for the complex patterns of bacterial growth [12–18]. Here we show that a simple model of bacterial growth can reproduce the salient features of the observed growth patterns. The model incorporates random walkers, representing aggregates of bacteria, which move in response to gradients in nutrient concentration and communicate with each other by means of chemotactic 'feedback'. These simple features allow the colony to respond efficiently to adverse growth conditions, and generate self-organization over a wide range of length scales.

We have grown bacterial colonies, under different growth conditions [12, 13] ranging from a very low level of nutrient (0.1 g peptone per litre) to a very rich mixture (10 g peptone per litre), and from a soft substrate (~1% agar concentration) to a hard substrate (4% agar concentration). Growth was started with a droplet (5 μl containing ~10^5 bacteria) inoculation at the centre of Petri dishes incubated at 37°C and 30% humidity. The growth pat-

---

Author affiliations: Eshel Ben-Jacob, Ofer Schochet, Adam Tenenbaum, Inon Cohen: School of Physics and Astronomy, Raymond & Beverly Sackler Faculty of Exact Sciences, Tel Aviv University, Israel. Andras Czirok, Tamas Vicsek: Department of Atomic Physics, Eotvos University, Budapest, Hungary.

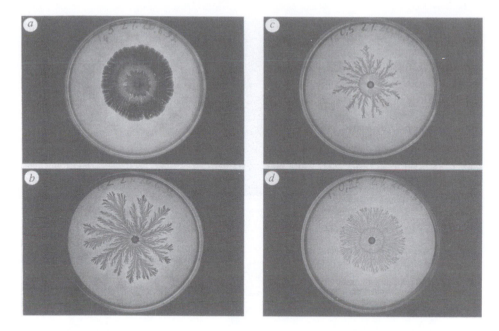

Fig. 1 Observed patterns of colonies grown on a substrate with 2% agar concentration. The peptone level is 5, 2, 0.5 and 0.25 g l^{-1} for **a**, **b**, **c** and **d** respectively. At high peptone levels the branches are wide, the pattern is very reminiscent of Hele-Shaw patterns [11] and the fractal dimension is close to two. As the peptone level is decreased, the patterns become more ramified (**b** and **c**), reminiscent of patterns observed during electro-chemical deposition [11] (**b**) and diffusion limited aggregation (DLA) [22] simulations (**c**). At even lower peptone levels the patterns become denser again (**d**). As explained in the text, we expect this phenomenon to result from chemotaxis signalling.

tern we describe are of bacteria derived from *Bacillus subtilis* strain 168 (refs 12, 13). The colonies adopt various shapes as growth conditions are varied (Fig. 1): patterns are compact at high peptone concentrations and become more ramified at low peptone concentrations (0.5 g l^{-1}), in agreement with results reported in refs 12–18. Surprisingly, at $<\sim0.25$ g peptone per litre, colonies adopt a more organized (well defined circular envelope), dense structure (Fig. 1d).

Optical microscopy reveals that the bacteria perform a random walk-like movement within a well defined envelope. The latter (Fig. 2) is formed presumably by chemicals that are excreted by the bacteria and/or by fluid drawn by the bacteria from the agar. The envelope propagates slowly as if by the action of effective internal pressure produced by the collective movement of the bacteria. At very low peptone concentrations the density of bacteria is low—the distance between bacteria is up to sev-

eral times the size of an individual bacterium. In this range, the bacteria are longer ($\sim5\mu$m) (Fig. 2e) and the movement seems to be more organized. At high agar concentration there is a boundary layer of high bacterial density at the leading tips of the growing branches (Fig. 2f). In this range, colonies display a pronounced structure in the perpendicular direction as well (Fig. 2c).

The growth of bacterial colonies presents an inherent additional level of complexity compared to non-living systems, as the building blocks themselves are living systems [15, 19, 20], each having its own autonomous (at times 'selfish') self-interest and internal degrees of freedom. To model the growth, we included the following generic features: (1) diffusion of nutrients; (2) movement of the bacteria; (3) reproduction and sporulation; (4) local communication. Diffusion of nutrients is handled by solving the diffusion equation for the nutrient

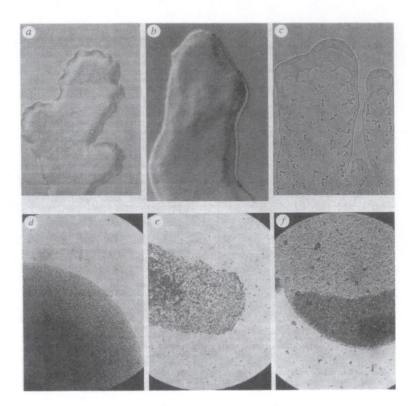

Fig. 2 Optical microscope observations of the colonies. **a** and **b**, Using a Numarski prism to indicate the sharpness of the envelopes. In **a** the envelope is rougher in the horizontal direction. **c**, Using transmitted light to show the complex three-dimensional structure of growth on substrates with a high agar concentration (2.5%). **d, e, f**, Micrographs of stained colonies. **d**, Intermediate values of peptone level and agar concentration. **e**, Low peptone level. **f**, High agar concentration: note the higher bacteria density at the tip in this case. Magnifications: **a, b, c**, ×28; **d, e, f**, ×280.

concentration c on a triangular lattice. The bacteria are represented by walkers, each of which should be viewed as a mesoscopic unit (coarse graining of the colony) and not as an individual bacterium.

Each walker is described by its location $\vec{r}_i$ and an internal degree of freedom ('internal energy' W_i), which affects its activity. The walker loses 'internal energy' at a rate e. To increase the internal energy it consumes nutrients at a fixed rate c_r, if sufficient food is available. Otherwise, it consumes the available amount. When there is not enough food for an interval of time (causing W_i to drop to zero), the walker becomes stationary (sporulation). When food is sufficient, W_i increases, and when it reaches some threshold t_r, the walker divides into two (reproduction).

The walkers perform off-lattice random walk within a well defined envelope (defined on the triangular lattice in Fig. 3a). Each segment of the envelope moves after it has been hit N_c times by the walkers. This requirement represents the local communication or cooperation in the behaviour of the bacteria. Note that, to a first approximation, the level of N_c represents the agar concentration, as more 'collisions' are needed to push the envelope on a harder substrate.

The model equations are:

$$\frac{\partial c(\vec{r},t)}{\partial t} = D_c \nabla^2 c(\vec{r},t)$$
$$- \sum_{\text{active walkers}} \delta(\vec{r} - \vec{r}_i) \min(c_r, c(\vec{r},t)). \quad (1)$$

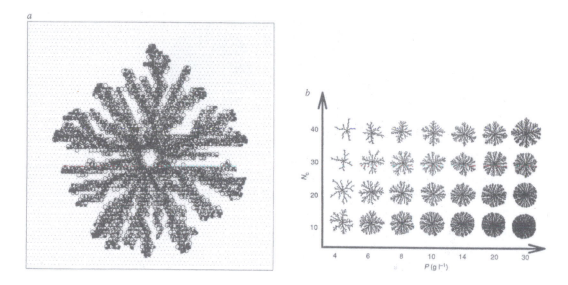

Fig. 3 The communicating walkers model. **a,** Close-up view of the model. The solid squares are the active walkers and the heavy dots are the stationary ones. Equation (1) is solved on the underlying triangular lattice. The hexagons mark the envelope. **b,** Results of numerical simulations of the communicating walkers model. The patterns are organized as a function of peptone level P (the initial value of c) and N_c (corresponds to the agar concentration). The typical system size is 600×600 and the typical number of the walkers is 10^5. Hence each walker represents $\sim 10^4$ bacteria. The observed patterns are compact for high peptone levels and become ramified for low peptone levels. For the same peptone level, the patterns are more branched for higher N_c. Both are consistent with experimental observations. Note that the fractal dimension becomes much smaller than the DLA fractal dimension [22]. It reflects the fact that the envelope propagation has non-trivial dependence on the gradient of the nutrients diffusion field.

This is the diffusion equation for the nutrients (D_c is the diffusion constant) which includes the consumption of food by the walkers (last term). The time evolution of W_i is given by:

$$\frac{dW_i}{dt} = \min(c_r, c(\vec{r}_i, t)) - e. \qquad (2)$$

At each time step, each of the active random walkers performs a random walk of step size d and angle Θ, uniformly chosen from the interval $[0, 2\pi]$. Thus the new location $\vec{r}_i$ is given by:

$$\vec{r}_i' = \vec{r}_i + d(\cos\Theta, \sin\Theta). \qquad (3)$$

If the step $\vec{r}_i \rightarrow \vec{r}_i'$ crosses the envelope, the step is not performed and a counter on the appropriate segment of the envelope is increased by one. When a segment counter reaches N_c, the envelope segment is shifted by one lattice step. Results of numerical simulations of the model are shown in

Fig. 3b. As in the growth of bacterial colonies, the patterns are compact at high peptone levels and become fractal with decreasing food level. For a given peptone level, the patterns are more ramified as the agar concentration increases. Clearly the results shown in Fig. 3 are very encouraging and do capture features of the observed patterns. However, there are some crucial qualitative differences. The most dramatic one is the ability of the bacteria to develop organized patterns at very low peptone levels (Fig. 1d), a feature which is not captured by this version of the model.

As the environmental conditions become more hostile (low peptone concentration or hard surface), a higher level of cooperation is required for a more efficient response of the colony. Non-local communication and transfer of information between each of the individuals and the colony might be needed. Can chemotactic signalling provide the colony with the means to do so? Generally, chemotaxis means

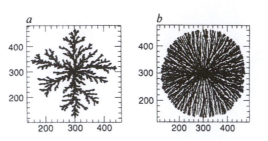

Fig. 4 Effect of chemotaxis signalling on the communicating walkers model. **a,** In the absence of the chemotaxis for $P = 10g \ l^{-1}$ and $N_C = 40$. **b,** In the presence of chemotaxis (for the same values of P and N_C). The pattern becomes denser with radial thin branches and well defined circular envelope, in agreement with experimental observations. The numbers on the axes represent the trigonal lattice sites to indicate the system size.

movement of the microorganisms in response to a concentration gradient of certain chemicals [6–10]. Ordinarily, the involvement is along the gradient either in the direction of the gradient or the opposite direction. The chemotactic response may be to an external chemical field or to one produced by the microorganisms; the latter may be called chemotactic signalling or communication. Moreover, it is well known [4] that excretion of the signal can be triggered by an external stress. Such non-local communication enables each bacterium to obtain information about the state of the colony as a whole and to respond to it. For example, the migration of *Dictyostelium* under low nutrient conditions depends on chemotactic signalling via cyclic adenosine monophosphate (cAMP) [22]. In this case, each of the microorganisms may excrete and consume cAMP and move according to its concentration gradient.

Here we include a simple version of chemotactic communication in the hope of identifying the generic features that it induces. Each of the stationary walkers (or alternatively, walkers which have been exposed to a low level of food) produces a communication chemical at a fixed rate s_r, (in an attempt to drive other bacteria away), and each of the active walkers consumes the chemical at a fixed rate c_c. As we show below, this simplified version is sufficient to capture the qualitative features of the growth. A more realistic model would include a dependence of these rates on the concentrations of nutrients and chemotactic signalling compounds. The equation of the communication field in the present model is given by:

$$\frac{\partial s(\vec{r},t)}{\partial t} = D_s \nabla^2 s(\vec{r},t)$$

$$+ \sum_{\text{stationary walkers}} \delta(\vec{r} - \vec{r}_i)s_r$$

$$- \sum_{\text{active walkers}} \delta(\vec{r} - \vec{r}_i)\min(c_c, s(\vec{r},t)).$$

$$(4)$$

The movement of the active bacteria changes from pure random walk (equal probability to move along any direction) to a random walk with a bias along the gradient of the communication field (high probability to move in the direction of the signalling material).

In Fig. 4 we show that the inclusion of such chemotaxis signalling indeed produces the desired phenomena. The pattern changes from fractal to a dense structure with thin branches and a well defined circular envelope. Moreover, the number of walkers (bacteria density) in the colony is much lower than in the absence of chemotaxis. We find that a chemotactic response to gradients of nutrient concentration does not reproduce this effect. Here we have simply introduced chemotactic signalling when the peptone level becomes low; in reality, there might be additional control mechanisms which change the intensity of the chemotaxis communication as the bacteria go through phenotypic transformations. We suspect that the observed perpendicular growth at high agar concentration also results from chemotactic signalling.

Our goal in this work is to demonstrate that apparently complex behaviour in biological systems can be elucidated by relatively simple, generic modelling in conjunction with a close comparison to experimental observations.

REFERENCES

1. R. Y. Steiner, M. Doudoroff & E. A. Adelberg, *The Microbial World* (Prentice-Hall, New Jersey, 1957).
2. J. A. Shapiro, *Scientific American* **253** (6), 62–69 (1988).
3. J. A. Shapiro & D. Trubatch, *Physica* **D49**, 214 (1991).
4. E. O. Budrene & H.C. Berg, *Nature* **349**, 630 (1991).
5. J. O. Kessler, *Contemp. Phys.* **26**, 147 (1985).
6. J. Adler, *Science* **166**, 1588 (1969).
7. J. M. Lacklie (ed.), *Biology of the Chemotactic Response* (Cambridge Univ. Press, 1981).
8. P. Devreotes, *Science,* **245**, 1054 (1989).
9. H. O. Berg & E. N. Purcell, *Biophys. J.* **20**, 193 (1977).
10. R. Nossal, *Expt. Cell Res.* **75**, 136 (1972).
11. E. Ben-Jacob & P. Garik, *Nature* **343**, 523 (1990).
12. E. Ben-Jacob, H. Shmueli, O. Schochet & A. Tenenbaum, *Physica* **A187**, 378 (1992).
13. E. Ben-Jacob, A. Tenenbaum, O. Schochet & O. Avidan, *Physica* **A202**, 1 (1994).
14. T. H. Henrici, *The Biology of Bacteria* 3rd edn (Heath, Boston, 1948).
15. A. L. Cooper, *Proc. R. Soc. B* **171**, 175 (1968).
16. H. Fujikawa & M. Matsushita, *Phys. Soc. Japan* **53**, 3875, (1989).
17. T. Matsuyama & M. Matsushita, *Crit. Rev. Bact.* **19**, 117 (1993).
18. J. Schindler & T. Rataj, *Binary* **4**, 66 (1992).
19. C. Allison & C. Hughes, *Sci. Prog.* **75**, 403 (1991).
20. J. Henrichsen, *Bacter. Rev.* **36**, 478 (1972).
21. D. Kessler & H. Levine, *Phys. Rev. E* (in press).
22. L. M. Sander, *Nature* **322**, 789 (1986).

READING 18.2

Dynamics of Stripe Formation;
A Reaction-Diffusion Wave on the Skin of the Marine
Angelfish Pomacanthus;
Letters to Nature

Commentary: The first article is a commentary on the second. The second article investigates the mechanism of stripe formation on the angelfish Pomacanthus. The article claims to find good evidence that the stripe pattern is governed by a reaction-diffusion equation—that is, an equation that has the schematic form $\frac{\partial}{\partial t} u = \mu \frac{\partial^2}{\partial x^2} + f(u)$, where u is the concentration of some pigment and where $f(\cdot)$ is a function that is evaluated at u. The authors use a particular reaction-diffusion equation to try to obtain model stripes that evolve in time like those on the fish. The point is that as the fish grows from a fry, the stripe pattern necessarily changes—new stripes are added. The way new stripes appear is characteristic of certain kinds of stripe forming mechanisms. In any event, consider the equation in Figure 1 of the second article for the variable A, which is the concentration as a function of x and t of a hypothetical "activator" molecule. This equation also involves the variable I, which is the concentration as a function of x and t of a hypothetical "inhibiting" molecule. Note that this equation is a diffusion equation for A with a "reaction term": $c_1 A + c_2 I + c_3 - g_4 A$, which depends on both A and I. Meanwhile, the next equation is a diffusion equation for I with a reaction term that depends on A. The minus sign in front of D_A in the article's diffusion equation for A is probably a misprint.

The final reading here consists of two letters to the editor of *Nature*. The first asserts that the angelfish pattern cannot be caused by a reaction-diffusion equation. The authors argue that the conclusions in the original article (the second article here) are erroneous. The authors of the original article then respond by saying that there are many mechanisms that can explain the fish patterns, and theirs is just one possible way to do it.

Dynamics of Stripe Formation

Nature, **376** (1995) 722-723.
Hans Meinhardt

Stripes are a common feature in developing organisms, but how are pattern elements generated that have a long extension in one dimension and a short extension in the other? Kondo and Asai [1] investigate the formation of stripes on tropical fish, inviting a fresh look at this process in embryogenesis. Considered in the light of zebrafish genetics, their results open up a possible route to a molecular explanation of stripe formation.

Stripes have been a focus of interest ever since the discovery that the pair-rule gene *fushi tarazu* is expressed in seven stripes in the fruitfly *Drosophila*. Stripe-like expression is now known to be associated with most other pair-rule and segment-polarity genes [2] and is also pronounced, in the ocular dominance columns for example, during neuronal development. Various models have been put forward to explain stripe patterning, but these have been undermined by the finding that transcription of pair-rule genes is initiated stripe by stripe by specific promoter regions under the influence of particular concentrations of gap gene products (for example, see an early News and Views on the topic entitled "Making stripes inelegantly" [3]).

The important finding of Kondo and Asai [1] is that during growth of a tropical fish, new stripes become intercalated or existing stripes can be split in two. The point of splitting can move in a zipper-like fashion over the field, enlarging the area with an increased stripe number. This dynamic regulation is quite unlike the control of stripe formation by the rigid coordinate system that operates in *Drosophila*. Surprisingly, this stripe-forming mechanism still functions at later stages in many cells.

Some of the regulatory features of stripe formation in fishes have close parallels in the *Drosophila* embryo. The dorsoventral organization is achieved by a high nuclear concentration of the *Dorsal* protein along the ventral side. Thus, the region of high *Dorsal* concentration is a stripe with a long antero-posterior but a small dorsoventral extension. It is brought into position by a repulsive action of the *gurken-torpedo* system, with their high points at the dorsal site [4]. An impairment of the *gurken-torpedo* system can lead to a split of the high *Dorsal* stripe in much the same way as is observed on the growing fishes.

Observations of pair-rule gene activity in the more primitive short-germ band insect *Tribolium* also suggest a dynamic regulation. Rapid cell proliferation takes place close to the posterior pole. New stripes of the proteins encoded by *hairy* [5] or *even-skipped* [6] appear there whenever the distance to an existing stripe becomes too large. In *Drosophila*, with progressing subdivision into more and more cells, a second set of seven *even-skipped* stripes becomes inserted between the primary seven stripes [7]. The new stripes are thinner, as was observed on the fishes.

In his pioneering paper [8], Turing showed that two interacting substances with different diffusion rates can generate stable patterns in space, the necessary condition for pattern formation being local autocatalysis and a long-range inhibition [9]. Since then, patterns consisting of patches and stripes have been observed in defined chemical systems [10, 11].

What causes stripes instead of patches *in vivo*? Different models [5, 12, 13] have in common a limitation of autocatalysis at high activator concentrations that is independent of the amount of diffusible antagonist: if the self-enhancing reaction has an upper limit, then the fast-spreading antagonistic reaction must be limited too. Activated cells must therefore tolerate other activated cells in the neighbourhood, leading to an enlargement of the patches until the antagonistic reaction is balanced. On the other hand, an immediate neighbourhood

---

Author affiliation: Hans Meinhardt: Max-Planck-Institut für Entwicklungsbiologie, Tübingen, Germany.

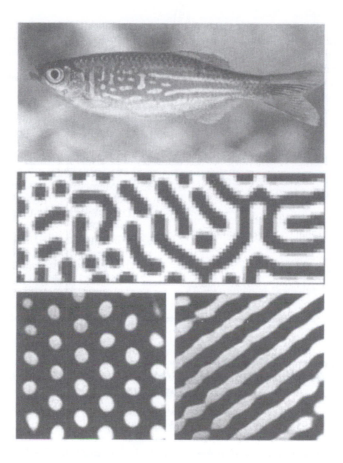

Fig. 1 Stripes and patches derived from three different systems, neatly illustrating the convergence of results obtained by molecular genetics, computer modelling and chemistry. Stripes and patches in: top, *Danio malabaricus*, a relative of the zebrafish; centre, a computer simulation made with increasing saturation of autocatalysis [13] (going from left to right), enforcing the transition from patches to stripes; bottom, a chemically defined reaction [11]. Photograph of *D. malabaricus* courtesy of Rohan Pethiyagoda.

of non-activated cells is essential to maintain the activity of activated cells, either to get rid of the inhibitor or to obtain the necessary substrate.

These requirements of large activated patches and an environment of only non-activated cells seem to contradict each other, but this is not the case: in a stripe-like activation pattern, each activated cell has an activated neighbour and non-activated cells are close by. A variant of this scheme is important for insect segmentation. Stripe formation also occurs if two cell states exclude each other locally but activate each other over a long

range. A long common border between two adjacent stripes allows an efficient mutual activation. This prediction [14] has been confirmed for the segment-polarity genes [2].

It has long been known that local self-enhancement occurs in *even-skipped* and *fushi tarazu*, as expected from the general theory. Other components involved in cell-cell communication are predicted, and two recently discovered pair-rule genes are good candidates for this function [15, 16]. So, the rigid mechanism of stripe formation in *Drosophila* may be a late evolutionary modification of a genuine patterning process that can be observed in very different developmental situations, including those found in *Drosophila* itself.

But the stripes on the fish still call for more explanation: those shown in Figure 1 of the paper by Kondo and Asai [1] are very narrow with respect to the spaces in between. All the models I know of can only produce stripes and interstripes of the same size.

REFERENCES

1. S. Kondo & R. Asai, *Nature* **376**, 765 (1995).
2. P. W. Ingham & A. Martinez-Arias, *Cell* **68**, 221 (1992).
3. M. Akam, *Nature* **341**, 282 (1989).
4. S. Roth & T. Schupbach, *Development* **120**, 2245 (1994).
5. R. J. Sommer & D. Tautz, *Nature* **361**, 448 (1993).
6. N. H. Patel, B. G. Condron & K. Zinn, *Nature* **367**, 429 (1994).
7. M. Frasch, T. Hoey, C. Rushlow, H. Doyle & M. Levine, *EMBO J.* **6**, 749 (1987).
8. A. M. Turing, *Phil. Trans. R. Soc.* **B237**, 37 (1952).
9. A. Gierer & H. Meinhardt, *Kybernetik* **12**, 30 (1972).
10. V. Castets, E. Dulos, J. Boissonade & P. De Kepper, *Phys. Rev. Lett.* **64**, 2953 (1990).
11. Q. Ouyang & H. L. Swinney, *Nature* **352**, 610 (1991).
12. M. J. Lyons & L. G. Harrison, *Dev. Dynamics* **195**, 201 (1992).
13. H. Meinhardt, *Development* (suppl.) 169 (1989).
14. H. Meinhardt & A. Gierer, *J. Theor. Biol.* **85**, 429 (1980).
15. S. Baumgartner, D. Martin, C. Hagios & R. Chiquet-Ehrismann, *EMBO J.* **13**, 3728 (1994).
16. J. F. Colas, J. M. Launay, O. Kellermann, P. Rosay & L. Maroteaux, *Proc. Natn. Acad. Sci. U.S.A.* **92**, 5441 (1995).

A Reaction-Diffusion Wave on the Skin of the Marine Angelfish Pomacanthus

Nature, **376** (1995) 765–768.
Shigeru Kondo, Rihito Asai

In 1952, Turing proposed a hypothetical molecular mechanism, called the reaction-diffusion system [1], which can develop periodic patterns from an initially homogeneous state. Many theoretical models based on reaction-diffusion have been proposed to account for patterning phenomena in morphogenesis [2–4], but, as yet, there is no conclusive experimental evidence for the existence of such a system in the field of biology [5–8]. The marine angelfish, **Pomacanthus**, has stripe patterns which are not fixed in their skin. Unlike mammal skin patterns, which simply enlarge proportionally during their body growth, the stripes of **Pomacanthus** maintain the spaces between the lines by the continuous rearrangement of the patterns. Although the pattern alteration varies depending on the conformation of the stripes, a simulation program based on a Turing system can correctly predict future patterns. The striking similarity between the actual and simulated pattern rearrangement strongly suggests that a reaction-diffusion wave is a viable mechanism for the stripe pattern of **Pomacanthus**.

When juveniles of *Pomacanthus semicirculatus* are smaller than 2 cm long, they have only three dorsoventral stripes (Figure 1(a)). As they grow, the intervals of the stripes get wider proportionally until the body length reaches 4 cm. At that stage, new stripes emerge between the original stripes (Figure 1(b)). As a result, all the spaces between the stripes revert to that of the 2-cm juvenile. New lines are thin at first, but gradually get broader. When the body length reaches 8-9 cm, an identical process is repeated (Figure 1(c)).

Author affiliations: S. Kondo: Kyoto University Centre for Molecular Biology and Genetics, Kyoto, Japan. R. Asai: Kyoto University Seto Marine Biological Laboratory, Wakayama, Japan.

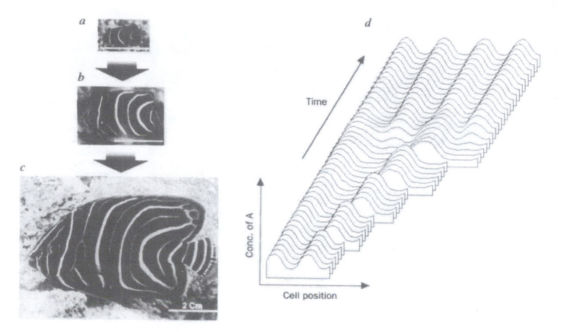

Fig. 1 Rearrangement of the stripe pattern of *Pomacanthus semicirculatus* and its computer simulation. **a-c**, Photographs of the juvenile of *P. semicirculatus*. Ages are approximately 2 months (**a**), approximately 6 months (**b**) and approximately 12 months (**c**). Scale bars, 2 cm. **d**, Computer simulation of the reaction-diffusion wave on the growing one-dimensional array of cells. One of the five cells is forced to duplicate periodically (once in 100 iterations). Concentration of activator is represented as the vertical height. The equations for calculation are as follows:

$$\frac{dA}{dt} = c_1 A + c_2 I + c_3 - D_A \frac{d^2 A}{dx^2} - g_A A, \quad \frac{dI}{dt} = c_4 A + c_5 - D_I \frac{d^2 I}{dx^2} - g_I I$$

where A and I are the concentration of the activator molecule and the inhibitor molecule, respectively, D_A and D_I are the diffusion constants, g_A and g_I are the decay constants, and $D_A = 0.007$, $D_I = 0.1$, $g_A = 0.03$, $g_I = 0.06$, $c_1 = 0.08$, $c_2 = 0.08$, $c_3 = 0.05$, $c_4 = 0.1$, $c_5 = 0.15$. Upper and lower limits for the synthesis rates of the activator ($c_1 A + c_2 I + c_3$) and inhibitor ($c_4 A + c_5$) are set as $0 < c_1 A + c_2 I + c_3 < 0.18$ and $0 < c_4 A + c_5 < 0.5$. These upper and lower limits are set to avoid unrealistic situations. A moderate upper-limit value of the activator synthesis rate is required to get a pattern of stripes rather than spots [15] (spots are obtained if this value is exceeded). We used the kinetics of Turing [1]. Other stripe-forming interactions [12, 15], in which the upper and lower limit is a natural outcome of the kinetics, can simulate the fish pattern rearrangement reported here.

The reaction-diffusion system used here consists of two hypothetical molecules (activator and inhibitor) which control the synthesis rate of each other. Figure 1(d) shows a computer simulation of a reaction-diffusion wave on a growing array of cells. At time 0, the field width is adjusted to be twice the intrinsic wavelength, calculated from the equations used in this simulation. One of the five cells is forced to duplicate periodically. As the field enlarges, all waves widen evenly. When the field length reaches about twice the original length, new peaks appear in the middle of the original peaks, as observed in *P. semicirculatus*, and the wavelength reverts to that of the original.

The juvenile of *P. imperator* has concentric stripes, which increase in number in a manner similar to that of *P. semicirculatus*. But when the *P. imperator* becomes an adult, the stripes become

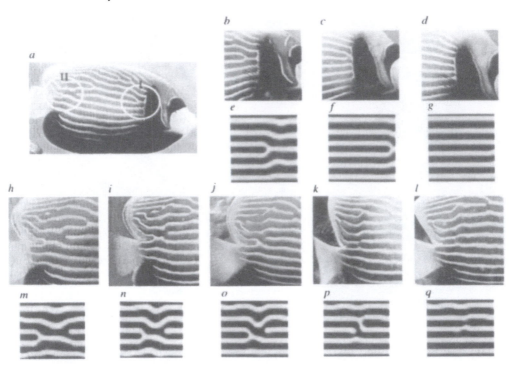

Fig. 2 Rearrangement of the stripe pattern of *Pomacanthus imperator* (horizontal movement of branching points) and its computer simulation. **a**, An adult *P. imperator* (approximately 10 months old). **b**, Close-up of region I in **a**. **c, d**, Photographs of region I of the same fish taken two (**c**) and three (**d**) months later. **e**, Starting stripe conformation for the simulation (region I). **f, g**, Results of the calculation after 30,000 (**f**) and 50,000 (**g**) iterations. **h**, Close-up of region II in **a**. **i-l**, Photographs of region II of the same fish taken 30 (**i**), 50 (**j**), 75 (**k**) and 90 (**l**) days later, respectively. **m**, Starting stripe conformation for the simulation (region II). **n-q**, Results of the calculation after 20,000 (**n**), 30,000 (**o**), 40,000 (**p**) and 50,000 (**q**) iterations, respectively. Fish (Fish World Co. Ltd (Osaka)) were maintained in artificial sea water (Martin Art, Senju). Skin patterns were recorded with a Canon video camera and printed by a Polaroid Slide Printer. In the simulated patterns, darker colour represents higher concentrations of the activator molecule. Equations and the values of the constants used, as Figure 1.

parallel to the anteroposterior axis by a process of continuous cutting and joining of the lines (data not shown). As they grow, the number of lines increases proportionally to body size, and the spaces between the lines are kept at an even width. The stripe pattern of *P. imperator* usually contains several branching points (Figure 2(a)). During growth, the branching points move horizontally like a zip, resulting in addition of new lines. Figure 2(b-d) shows a branching point moving in the anterior direction until it fuses with the border of the stripe region. In Figure 2(h-l), two branching points meet and disappear leaving a new line. This type of re-

arrangement also happens in the simulation of the reaction-diffusion system, by setting a homologous conformation as a starting pattern (Figure 2(e-g, m-q)). In Figure 2(e), the field height is adjusted to be six times the intrinsic wavelength. The waves in the right half are slightly extended, which causes loss of stability in this region. The rightward movement of the branch restores the stability of the righthand region. It is notable that not only the final conformation, but also each intermediate stage (Figure 2(n-p)), look quite similar to the actual pattern change that occurs in the fish (Figure 2(i-k)).

Branching points located on more dorsal or ven-

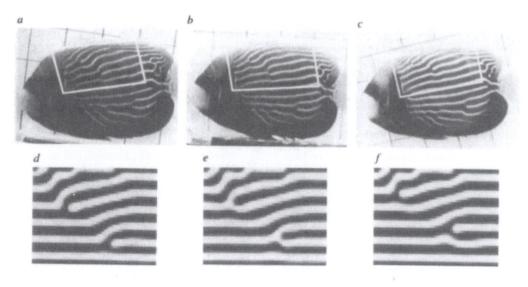

Fig. 3 Rearrangement of the stripe pattern of *P. imperator* (switch of joint) and its computer simulation. **a,** Photograph of a young *P. imperator* (approximately 7 months old). **b, c,** Photographs of the same fish taken 6 (**b**) and 12 (**c**) days later. **d,** Starting stripe conformation for the simulation. Pattern changing in the region surrounded by the white box was simulated. **e, f,** Results of the calculation after 2,000 (**e**) and 5,000 (**f**) iterations, respectively.

tral regions behave differently. As shown in Figure 3(a-c), they move vertically by switching at a joint. This phenomenon can be simulated by the program used in Figure 1 and Figure 2 only by setting a different starting pattern (Figure 3(d-f)). In the simulation, a local region that contains a branching point is less stable than a region without a branch point, and joint switching tends to occur. The direction of joint switching is determined by the conformation of neighbouring lines. In our simulation, the line under the branching point is straighter than the line above. Because the curving line is less stable than the straight line, the joint switches in the upper direction. If both upper and lower lines are symmetrical to the branched line, horizontal movement of the branch point occurs (Figure 2). In the case of actual young fish, the lines in the middle region are usually straight, but in the dorsal and ventral regions the lines are curved. Branching points always move farther away from the middle region which consists of straight lines.

The times required for these pattern changes also suggest a mechanistic homology between actual fish and the simulations. In the simulation of joint switching, one change of joint can take place very quickly (in less than 1,000 iterations of calculation), because the change in pattern is quite local. For the horizontal movement of the branching point (from Figure 2(e to g)), more than 50,000 iterations are required because it is necessary for the upper lines and lower lines to 'slide' in order to evolve to a new pattern of stripes that are evenly spaced. In the case of real fish, the joint changes also occur quickly. In the fastest case we have observed, it took place in two days (data not shown), whereas the change from Figure 2(b to d) took more than three months.

Although we do not have any information about the molecules which are involved in the pattern-forming reaction, it is possible to estimate roughly the diffusion coefficients of the molecules by comparing the simulation and the actual pattern changing of fish stripe. The stripe spacing is approximately 0.5 cm in *P. imperator*, and approximately 10 grids in the simulated patterns (Figure 2); a grid in the simulation therefore represents 0.05 cm. The pattern change from Figure 2(h to l) took 90 days (7,776,000 seconds) in reality, and 50,000 iterations in the simulation. The time step for the simulation therefore corresponds to 155.5 seconds. These values give diffusion coefficients of 1.125×10^7 cm^{2}s^{-1} and 1.608×10^{-6}cm^{2}s^{-1} for the activator and the inhibitor, respectively. Both values are in the range of the diffusion coefficients of proteins in aqueous media [9]. However, the diffusive molecules may be smaller than proteins, because the diffusion rate of molecules is usually much smaller in real biological systems than in aqueous media.

In some other biological systems, the insertion of new structures during growth have been observed and simulated [3, 10–14]. The novel features of the work reported here are that the inserted structure is a stripe and that the underlying mechanism is operative for a long period. The reaction-diffusion wave is a kind of standing wave. Therefore, to determine that a given pattern is consistent with a reaction-diffusion wave, it is necessary to impose some disturbance on the field and to see how the pattern responds. The pattern alteration of the *Pomacanthus*, accompanied by skin growth, can be taken as a natural experiment to help elucidate the underlying mechanisms which govern pattern formation. From the striking similarity between the actual and the simulated pattern alteration, it is highly probable that the mechanism is a reaction-diffusion system. Because the pattern-forming mechanism is maintained in adult skin, it should be possible to identify the molecules involved.

REFERENCES

1. A. M. Turing, *Phil. Trans. R. Soc.* **B237**, 37 (1952).
2. S. A. Kauffman, *Pattern Formation* (eds G. M. Malacinsky & S. Bryant.) 73–102 (Macmillan, New York, 1984).
3. H. Meinhardt, *Models of Biological Pattern Formation* (Academic, London, 1982).
4. J. D. Murray, *Scient. Am.* **258**, 80 (1988).
5. A. T. Winfree, *Nature* **352**, 568 (1991).
6. I. Lengyel & I. R. Epstein, *Science* **251**, 650 (1991).
7. Q. Ouyang & H. L. Swinney, *Nature* **352**, 610 (1991).
8. R. Pool, *Science* **251**, 627 (1991).
9. E. A. Dawes, *Quantitative Problems in Biology* (Longman, London, 1956).
10. E. Bunning & H. Z. Sagromsky, *Z. Naturf.* **B3**, 203 (1948).
11. T. C. Lacalli, *Phil. Trans. R. Soc.* **B294**, 547 (1981).
12. H. Meinhardt, *Rep. Progr. Phys.* **55**, 797 (1992).
13. L. A. Segel & J. L. Jackson, *J. theor. Biol* **37**, 545 (1972).
14. V. B. Wigglesworth, *J. exp. Biol.* **17**, 180 (1940).
15. H. Meinhardt, *Development* (suppl.) **107**, 169 (1989).

Letters to Nature

Nature, **380** (1996) 678.
T. Hofer and P. K. Maini
S. Kondo and R. Asai

SIR—Kondo and Asai [1] interpret observations on the time evolution of skin patterns of the angelfish (*Pomacanthus*) as the first instance of a Turing (reaction-diffusion) pattern in biology. But we believe that reaction-diffusion systems per se cannot provide a mechanistic basis for one of the main patterns reported in [1].

Reaction-diffusion systems are characterized by

an intrinsic spatial wavelength of the self-organized concentration pattern, that is, the distance between adjacent peaks of chemical concentrations is determined solely by the system parameters (kinetic constants and diffusion coefficients). Although on a two-dimensional domain such as the fish skin, several equidistant geometrical arrangements of the concentration peaks are possible, the nonlin-

Author affiliations: Thomas Hofer and Philip K. Maini: Centre for Mathematical Biology, Mathematical Institute, University of Oxford. Shigeru Kondo: Kyoto University Centre for Molecular Biology and Genetics, Kyoto, Japan. Rihito Asai: Kyoto University Seto Marine Biological Laboratory, Wakayama, Japan.

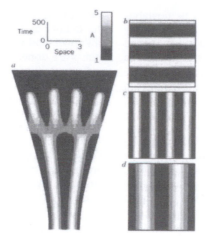

Fig. 1 Behaviour of the Turing system proposed in ref. 1 on a growing square domain (with the signs of the diffusion terms corrected). **a,** Concentration plot of A in a horizontal cross-section of the domain; time increases from bottom to top (see separate scales). The 4-stripe pattern produced by the first period-doubling is unstable, rearranges into a 3-stripe pattern perpendicular to the original pattern, and the stripe contours terminate. **b–d,** Snapshots of the stripe patterns corresponding to a (domains scaled to same size): **b,** initial 2-stripe pattern ($t = 500$); **c,** after period-doubling ($t = 3,000$); and **d,** after rearrangement into 3 stripes ($2 + 2$ half-stripes, $t = 4,000$, corresponding to the dark region in a). Simulations: equations scaled to the form $\partial u / \partial t = s^2 f(u) + D \nabla^2 u$, and solved with a standard ADI scheme on a fixed domain (mesh size 0.2, time step 0.05) with zero flux boundary conditions; increase in s is equivalent to increase in (domain length)^2 [2], here $s(t) = \sqrt{0.15 + 10^{-7} t^2}$. Patterning sequence is sensitive to the speed of domain growth and for faster growth rates the transitions become less controlled; we found transitions from 2 stripes to higher modes (5 stripes and more) with subsequent rearrangements.

ear terms of the reaction dynamics usually select only one of these possibilities—for the system chosen by Kondo and Asai, a regular array of stripes. These two features, an intrinsic wavelength and a strong tendency to form stripes, are the essential ingredients of the simulations they presented in [1]. Many pattern-forming systems other than reaction-diffusion are known which select an intrinsic spatial wavelength and pattern geometry [2], among them biologically relevant mechanisms involving chemotactic or haptotactic cell movement and mechanical forces [3]. Therefore, there is no justification for equating observed patterns with a particular mechanism, as suggested in [1].

Although our point does not exclude the possibility that a Turing system underlies the *Pomacanthus* skin patterns, we demonstrate here that its properties are not sufficient to explain perhaps the most striking observation of the paper, the regular insertion of new stripes between older ones dur-

ing the growth of *Pomacanthus semicirculatus*. We have solved the authors' reaction-diffusion equations on a growing, two-dimensional domain—a more realistic representation of the fish skin than the one-dimensional domain used in [1].

Our results show that regular stripe-doubling sensitively depends on the artificial geometrical constraints of the one-dimensional domain (see Figure 1). As the restriction of one-dimensionality is removed, complete spatial rearrangement of the pattern occurs on the growing domain, which clearly is not seen in the fish. This behaviour is readily explained by the two properties of Turing systems emphasized above. As the domain grows bigger, new stripes should be added, one at a time, approximately conserving the spatial wavelength. Initially, the preexisting pattern appears to force a different sequence of stripe additions to occur, corresponding to the 'period-doubling' behaviour sometimes seen in one-dimensional systems [3].

However, this situation turns out to be unstable, and the whole pattern rearranges perpendicularly to the old one to form a new stripe pattern enlarged by one stripe. This behaviour does not depend on the aspect ratio of the domain; we have found complete perpendicular rearrangement of pattern even on very narrow (quasi-one-dimensional) domains. Thus, the patterning dynamics must involve an interplay of the mechanism that sets the distance between adjacent stripes and some form of 'memory' that conserves the location of old stripes. The 'memory' could be provided by pigment cells forming stable aggregations [4]. More specific quantitative models based on experimentally implicated mechanisms are needed to formulate testable predictions on the origin of the dynamic *Pomacanthus* skin patterns.

KONDO AND ASAI REPLY—With respect to Hofer and Maini's first criticism, we agree that many pattern-forming systems can explain the phenomenon we observed. These models have in common a set of interactions involving local activation/lateral inhibition coupled with the appropriate nonlinearities [5]. The most important message of our report [1] is that a dynamical mechanism like Turing's is viable for the fish patterns. It should therefore be possible to identify the real molecular

mechanism by experiments. Of course, at present the details of the fish-patterning mechanism are unknown, and will not be understood until experiments are done.

Second, Hofer and Maini claim that a two-dimensional simulation of the *P. semicirculatus* pattern is more realistic than the one-dimensional simulation in our paper. This is by no means clear. All the stripe lines of *P. semicirculatus* are perpendicular to the body axis and there are no branch points. These features suggest the presence of a directional preference forcing the stripes to run in the same direction. A one-dimensional simulation captures some of the character of this system better than does an isotropic two-dimensional simulation.

REFERENCES

1. Kondo, S. & Asai, R. *Nature* **376**, 765–768 (1995).
2. Cross, M. C. & Hohenberg, P. C. *Rev. Mod. Phys.* **65**, 851–1112 (1993).
3. Murray, J. D. *Mathematical Biology* (Springer, Berlin, 1993).
4. Le Douarin, N. M. Curr. *Topics Dev. Biol.* **16**, 31–85 (1980).
5. Meinhardt, H. *Nature* **376**, 722–723 (1995).

READING 18.3

Complex Patterns in a Simple System

Commentary: For the purposes of this book, the point of this article is to illustrate the sorts of bizarre patterns that can arise with a pair of diffusion equations. The article considers functions $U(t, x, y)$ and $V(t, x, y)$, which are meant to represent the concentrations as functions of time t and space coordinates x and y of two different but interacting chemical species. Their time evolution is modeled by a pair of diffusion equations:

$$\frac{\partial}{\partial t}U = D_u\left(\frac{\partial^2}{\partial x^2}U + \frac{\partial^2}{\partial y^2}U\right) - UV^2 + F(1 - U)$$

$$\frac{\partial}{\partial t}V = D_v\left(\frac{\partial^2}{\partial x^2}V + \frac{\partial^2}{\partial y^2}V\right) + UV^2 - (F + k)V.$$

[This is Eq. (2) in the article.] Here, D_u is the diffusion coefficient for U, and D_v the same for V. Meanwhile, F and k are constants. The author runs these equations on a computer and learns that his computer approximation to the true solutions produces bizarre patterns when D_u and D_v, F and k are chosen appropriately.

In any event, the equations above are t, x, y generalizations for two unknowns of our simpler, t, x diffusion equation $\frac{\partial}{\partial t}u = \mu\frac{\partial^2}{\partial x^2}u + ru$ for the one unknown u. (In the case of the paper, the term ru is replaced by a more complicated expression in U and V.)

Complex Patterns in a Simple System

Science, **261** (1993) 189–192.
John E. Pearson

Numerical simulations of a simple reaction-diffusion model reveal a surprising variety of irregular spatiotemporal patterns. These patterns arise in response to finite-amplitude perturbations. Some of them resemble the steady irregular patterns recently observed in thin gel reactor experiments. Others consist of spots that grow until they reach a critical size, at which time they divide in two. If in some region the spots become overcrowded, all of the spots in that region decay into the uniform background.

Patterns occur in nature at scales ranging from the developing *Drosophila* embryo to the large-scale structure of the universe. At the familiar mundane scales we see snow-flakes, cloud streets, and sand ripples. We see convective roll patterns in hydrodynamic experiments. We see regular and almost regular patterns in the concentrations of chemically reacting and diffusing systems [1]. As a consequence of the enormous range of scales over which pattern formation occurs, new pattern formation phenomenon is potentially of great scientific interest. In this report, I describe patterns recently observed in numerical experiments on a simple reaction-diffusion model. These patterns are unlike any that have been previously observed in theoretical or numerical studies.

The system is a variant of the autocatalytic Selkov model of glycolysis [2] and is due to Gray and Scott [3]. A variety of spatio-temporal patterns form in response to finite-amplitude perturbations. The response of this model to such perturbations was previously studied in one space dimension by Vastano *et al.* [4], who showed that steady spatial patterns could form even when the diffusion coefficients were equal. The response of the system in one space dimension is nontrivial and depends

both on the control parameters and on the initial perturbation. It will be shown that the patterns that occur in two dimensions range from the well-known regular hexagons to irregular steady patterns similar to those recently observed by Lee *et al.* [5] to chaotic spatio-temporal patterns. For the ratio of diffusion coefficients used, there are no stable Turing patterns.

Most work in this field has focused on pattern formation from a spatially uniform state that is near the transition from linear stability to linear instability. With this restriction, standard bifurcation-theoretic tools such as amplitude equations have been developed and used with considerable success [6]. It is unclear whether the patterns presented in this report will yield to these now-standard technologies.

The Gray-Scott model corresponds to the following two reactions:

$$U + 2V \rightarrow 3V \tag{1}$$
$$V \rightarrow P$$

Both reactions are irreversible, so P is an inert product. A nonequilibrium constraint is represented by a feed term for U. Both U and V are removed by the feed process. The resulting reaction-diffusion equations in dimensionless units are:

$$\frac{\partial U}{\partial t} = D_u\nabla^2 U - UV^2 + F(1-U)$$
$$\frac{\partial V}{\partial t} = D_u\nabla^2 V + UV^2 - (F+k)V \tag{2}$$

where k is the dimensionless rate constant of the second reaction and F is the dimensionless feed rate. The system size is 2.5 by 2.5, and the diffusion coefficients are $D_u = 2 \times 10^{-5}$ and $D_v = 10^{-5}$. The

Author affiliation: John E. Pearson: Center for Nonlinear Studies, Los Alamos National Laboratory.

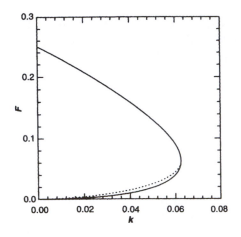

Fig. 1 Phase diagram of the reaction kinetics. Outside the region bounded by the solid line, there is a single spatially uniform state (called the trivial state) ($U = 1$, $V = 0$) that is stable for all (F, k). Inside the region bounded by the solid line, there are three spatially uniform steady states. Above the dotted line and below the solid line, the system is bistable: There are two linearly stable steady states in this region. As F is decreased through the dotted line, the nontrivial stable steady state loses stability through Hopf bifurcation. The bifurcating periodic orbit is stable for k less than 0.035 and unstable for k more than 0.035. No periodic orbits exist for parameter values outside the region bounded by the solid line.

boundary conditions are periodic. Before the numerical results are presented, consider the behavior of the reaction kinetics which are described by the ordinary differential equations that result upon dropping the diffusion terms in Eq. 2.

In the phase diagram shown in Figure 1, a trivial steady-state solution $U = 1$, $V = 0$ exists and is linearly stable for all positive F and k. In the region bounded above by the solid line and below by the dotted line, the system has two stable steady states. For fixed k, the nontrivial stable uniform solution loses stability through saddle-node bifurcation as F is increased through the upper solid line or by Hopf bifurcation to a periodic orbit as F is decreased through the dotted line. (For a discussion of bifurcation theory, see chapter 3 of [7].) In the case at hand, the bifurcating periodic solution is stable for k less than 0.035 and unstable for k more than 0.035. There are no periodic orbits for parameter values outside the region enclosed by the solid line. Outside this region the system is excitable. The trivial state is linearly stable and globally attracting. Small perturbations decay exponentially but larger perturbations result in a long excursion through phase space before the system returns to the trivial state.

The simulations are forward Euler integrations of the finite-difference equations resulting from discretization of the diffusion operator. The spatial mesh consists of 256 by 256 grid points. The time step used is 1. Spot checks made with meshes as large as 1024 by 1024 and time steps as small as 0.01 produced no qualitative difference in the results.

Initially, the entire system was placed in the trivial state ($U = 1$, $V = 0$). The 20 by 20 mesh point area located symmetrically about the center of the grid was then perturbed to ($U = 1/2$, $V = 1/4$). These conditions were then perturbed with $\pm 1\%$ random noise in order to break the square symmetry. The system was then integrated for 200,000 time steps and an image was saved. In all cases, the initial disturbance propagated outward from the central square, leaving patterns in its wake, until the entire grid was affected by the initial square perturbation. The propagation was wave-like, with the leading edge of the perturbation moving with an approximately constant velocity. Depending on the parameter values, it took on the order of 10,000 to 20,000 time steps for the initial perturbation to spread over the entire grid. The propagation velocity of the initial perturbation is thus on the order of

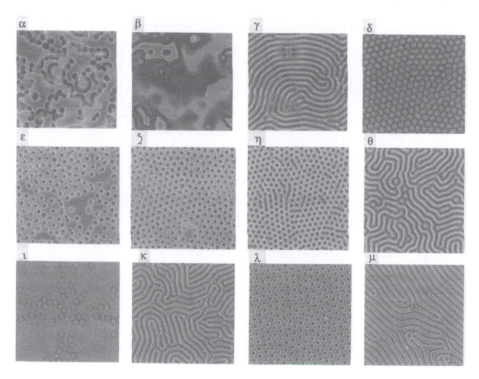

Fig. 2 The key to the map. The patterns shown in the figure are designated by Greek letters, which are used in Figure 3 to indicate the pattern found at a given point in parameter space.

1×10^{-4} space units per time unit. After the initial period during which the perturbation spread, the system went into an asymptotic state that was either time-independent or time-dependent, depending on the parameter values.

Figures 2 and 3 are phase diagrams; one can view Figure 3 as a map and Figure 2 as the key to the map. The 12 patterns illustrated in Figure 2 are designated by Greek letters. The color indicates the concentration of U with red representing $U = 1$ and blue representing $U \approx 0.2$; yellow is intermediate to red and blue. In Figure 3, the Greek characters indicate the pattern found at that point in parameter space. There are two additional symbols in Figure 3, R and B, indicating spatially uniform red and blue states, respectively. The red state corresponds to $(U = 1, V = 0)$ and the blue state depends on the exact parameter values but corresponds roughly to $(U = 0.3, V = 0.25)$.

Pattern α is time-dependent and consists of fledgling spirals that are constantly colliding and annihilating each other: full spirals never form. Pattern β is time-dependent and consists of what is generally called phase turbulence [8], which occurs in the vicinity of a Hopf bifurcation to a stable periodic orbit. The medium is unable to synchronize so the phase of the oscillators varies as a function of position. In the present case, the small-amplitude periodic orbit that bifurcates is unstable. Pattern γ is time-dependent. It consists primarily of stripes but there are small localized regions that oscillate with a relatively high frequency ($\approx 10^{-3}$). The active regions disappear, but new ones always appear elsewhere. In Figure 2 there is an active region near the top center of pattern γ. Pattern δ consists of regular hexagons except for apparently stable defects. Pattern η is time-dependent: a few of the stripes oscillate without apparent decay, but the remainder of the pattern remains time-independent. Pattern ι is time-dependent and was observed for only a single parameter value.

Patterns θ, κ, and μ resemble those observed by

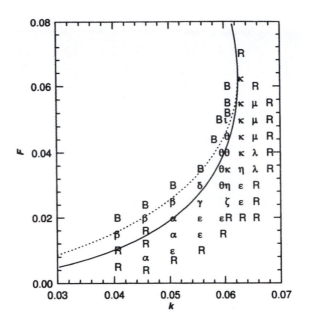

Fig. 3 The map. The Greek letters indicate the location in parameter space where the patterns in Figure 2 were found; B and R indicate that the system evolved to uniform blue and red states, respectively.

Lee *et al.* [5]. When blue waves collide, they stop, as do those observed by Lee *et al.* In pattern μ, long stripes grow in length. The growth is parallel to the stripes and takes place at the tips. If two distinct stripes that are both growing are pointed directly at each other, it is always observed that when the growing tips reach some critical separation distance, they alter their course so as not to collide. In patterns θ and κ, the perturbations grow radially outward with a velocity normal to the stripes. In these cases if two stripes collide, they simply stop, as do those observed by Lee *et al.* I have also observed, in one space dimension, fronts propagating toward each other that stop when they reach a critical separation. This is fundamentally new behavior for nonlinear waves that has recently been observed in other models as well [9].

Patterns ε, ζ, and λ share similarities. They consist of blue spots on a red or yellow background. Pattern λ is time-independent and patterns ε and ζ are time-dependent. Note that spots occur only in regions of parameter space where the system is excitable and the sole uniform steady state is the red state ($U = 1$, $V = 0$). Thus, the blue spots cannot persist for extended time unless there is a gradient present. Because the gradient is required for the existence of the spots, they must have a maximum size or there would be blue regions that were essentially gradient-free. Such regions would necessarily decay to the red state. Note that these gradients are self-sustaining and are not imposed externally. After the initial perturbation, the spots increase in number until they fill the system. This process is visually similar to cell division. After a spot has divided to form two spots, they move away from each other. During this period, each spot grows radially outward. The growth is a consequence of excitability. As the spots get further apart, they begin to elongate in the direction perpendicular to their motion. When a critical size is achieved, the gradient is no longer sufficient to maintain the center in the blue state, so the center decays to red, leaving two blue spots. This process is illustrated in Figure 4. Figure 4A was made just after the initial square perturbation had decayed to leave the four spots. In Figure 4B, the spots have moved away from each other and are beginning to elongate. In Figure 4C, the new spots are clearly visible. In Figure 4D, the replication process is complete. The subsequent evolution depends on the

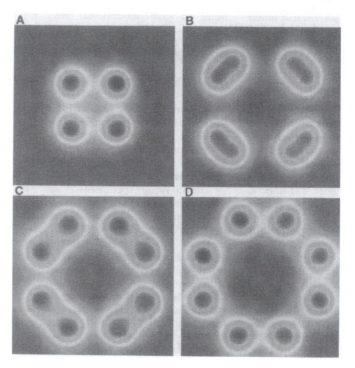

Fig. 4 Time evolution of spot multiplication. This figure was produced in a 256 by 256 simulation with physical dimensions of 0.5 by 0.5 and a time step of 0.01. The times t at which the figures were taken are as follows: (**A**) $t = 0$; (**B**) $t = 350$; (**C**) $t = 510$; and (**D**) $t = 650$.

control parameters. Pattern λ remains in a steady state. Pattern ζ remains time-dependent but with long-range spatial order except for local regions of activity. The active regions are not stationary. At any one instant, they do not appear qualitatively different from pattern ζ Fig. 2 but the location of the red disturbances changes with time. Pattern ε appears to have no long-range order either in time or space. Once the system is filled with blue spots, they can die due to overcrowding. This occurs when many spots are crowded together and the gradient over an extended region becomes too weak to support them. The spots in such a region will collapse nearly simultaneously to leave an irregular red hole. There are always more spots on the boundary of any hole, and after a few thousand time steps no sign of the hole will remain. The spots on its border will have filled it. Figure 5 illustrates this process.

Pattern ε is chaotic. The Liapunov exponent (which determines the rate of separation of nearby trajectories) is positive. The Liapunov time (the in-

verse of the Liapunov exponent) is 660 time steps, roughly equal to the time it takes for a spot to replicate, as shown in Figure 4. This time period is also about how long it takes for a molecule to diffuse across one of the spots. The time average of pattern epsilon is constant in space.

All of the patterns presented here arose in response to finite-amplitude perturbations. The ratio of diffusion coefficients used was 2. It is now well known that Turing instabilities that lead to spontaneous pattern formation cannot occur in systems in which all diffusion coefficients are equal. (For a comprehensive discussion of these issues, see Pearson and co-workers [10, 11]; for a discussion of Turing instabilities in the model at hand, see Vastano *et al.* [12].) The only Turing patterns that can occur bifurcate off the nontrivial steady uniform state (the blue state). Most of the patterns discussed in this report occur for parameter values such that the nontrivial steady state does not exist. With the ratio of diffusion coefficients used here, Turing pat-

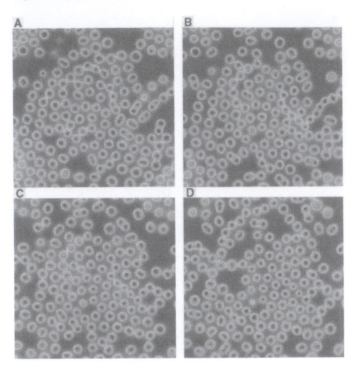

Fig. 5 Time evolution of pattern ε. The images are 250 time units apart. In the corners (which map to the same point in physical space), one can see a yellow region in (**A**) to (**C**). It has decayed to red in (**D**). In (**A**) and (**B**), the center of the left border has a red region that is nearly filled in (**D**).

terns occur only in a narrow parameter region in the vicinity of $F = k = 0.0625$, where the line of saddle-node bifurcations coalesces with the line of Hopf bifurcations. In the vicinity of this point, the branch of small-amplitude Turing patterns is unstable [12].

With equal diffusion coefficients, no patterns formed in which small asymmetries in the initial conditions were amplified by the dynamics. This observation can probably be understood in terms of the following fact: Nonlinear plane waves in two dimensions cannot be destabilized by diffusion in the case that all diffusion coefficients are equal [13]. During the initial stages of the evolution, the corners of the square perturbation are rounded off. The perturbation then evolves as a radial wave, either inward or outward depending on the parameter values. Such a wave cannot undergo spontaneous symmetry breaking unless the diffusion coefficients are unequal. However, I found symmetry breaking over a wide range of parameter values for

a ratio of diffusion coefficients of 2. Such a ratio is physically reasonable even for small molecules in aqueous solution. Given this diffusion ratio and the wide range of parameters over which the replicating spot patterns exist, it is likely that they will soon be observed experimentally.

Recently Hasslacher et al. have demonstrated the plausibility of subcellular chemical patterns through lattice-gas simulations of the Selkov model [14]. The patterns discussed in the present article can also be found in lattice-gas simulations of the Selkov model and in simulations carried out in three space dimensions. Perhaps they are related to dynamical processes in the cell such as centrosome replication.

REFERENCES

1. G. Nicolis & I. Prigogine, *Self-Organization in Non-Equilibrium Systems* (Wiley, New York, 1977).

2. E. E. Selkov, *Eur. J. Biochem.* **4**, 79 (1968).

3. P. Gray & S. K. Scott, *Chem. Eng. Sci.* **38**, 29 (1983); *ibid.* **39**, 1087 (1984); *J. Phys. Chem.* **89**, 22 (1985).

4. J. A. Vastano, J. E. Pearson, W. Horsthemke & H. L. Swinney, *Phys. Lett. A* **124**, 6 (1987). *ibid.*, p. 7; *ibid.*, p. 320.

5. K. J. Lee, W. D. McCormick, Q. Ouyang & H. L. Swinney, *Science* **261**, 192 (1993).

6. P. Hohenberg & M. Cross, *Rev. Mod. Phys.* **65**, 3 (1993).

7. J. Guckenheimer & P. Holmes, *Nonlinear Oscillations, Dynamical Systems, and Bifurcations of Vector Fields* (Springer-Verlag, Berlin, 1983), chap. 3.

8. Y. Kuramoto, *Chemical Oscillations, Waves, and Turbulence* (Springer-Verlag, Berlin, 1984).

9. A. Kawczynski, W. Comstock & R. Field, *Physica D* **54**, 220 (1992); A. Hagberg & E. Meron, University of Arizona preprint.

10. J. E. Pearson & W. Horsthemke, *J. Chem. Phys.* **90**, 1588 (1989).

11. J. E. Pearson & W. J. Bruno, *Chaos* **2**, 4 (1992); *ibid.*, p. 513.

12. J. A. Vastano, J. E. Pearson, W. Horsthemke & H. L. Swinney, *J. Chem. Phys.* **88**, 6175 (1988).

13. J. E. Pearson, *Los Alamos Publ.* LAUR 93-1758 (Los Alamos National Laboratory, Los Alamos, NM, 1993).

14. B. Hasslacher, R. Kapral & A. Lawniczak, *Chaos* **3**, 1 (1993).

15. I am happy to acknowledge useful conversations with S. Ponce-Dawson, W. Horsthemke, K. Lee, L. Segel, H. Swinney, B. Reynolds, and J. Theiler. I also thank the Los Alamos Advanced Computing Laboratory for the use of the Connection Machine and A. Chapman, C. Hansen, and P. Hinker for their ever-cheerful assistance with the figures.

READING 18.4

Direct and Continuous Assessment by Cells of Their Position in a Morphogen Gradient;

Activin Signalling and Response to a Morphogen Gradient

Commentary: A central question in biology is what determines the fate of cells in a developing embryo. For example, how do cells that become skin "know" they are to be skin, while those that become bone know that they are to be bone? Current belief has it that signaling molecules (called morphogens) are produced by certain cells and diffuse through the embryo. The morphogen concentration (or, more probably, concentrations of some number of morphogens) determines the cell fate by determining which genes become active. Of course, the signaling molecules are produced by cells which are 'told to do so' by the concentrations of other signaling molecules. Thus, there is a cascade of diffusable, chemical signals in the embryo that initiate gene activity at different times and places depending on their relative concentrations and timing of their appearance.

This article describes an experiment that determines that certain amphibian cells respond directly to changing morphogen concentrations in a "ratchet-like" manner. That is, the response of the cells is not a linear function of the morphogen concentration; rather it is a step function where a given response occurs when the morphogen concentration lies between two values, and when increased or decreased above or below these values, a distinctly different response occurs. The morphogen used here is a protein called Activin.

The term "gradient" in the title of this article and the next has the following interpretation: First, the term refers to the gradient of the function which measures the morphogen concentration. Second, the gradient of any function of some coordinates [say, x or (x, y) or (x, y, z)] is just the coordinate-dependent vector whose components are the partial derivatives of the function in question.

The second article reports on another experiment whose purpose was to determine whether a certain morphogen in an embryo occurs via diffusion or via some other mechanism. (We can imagine a more direct method whereby the signal is passed directly between neighboring cells with little or no help from intercellular diffusion. This might work much like a relay hand-off.)

In this experiment, the authors found that the selection of genes expressed by a cell was determined by the cell's distance from a source of activin (the morphogen). They report that the activin signal spread over at least 10 cell diameters, and they give evidence that this spread is by passive diffusion. In particular, passive diffusion is inferred because the response can bypass cells that neither respond to the morphogen nor synthesize activin. (Of course, it may be that the signal is transmitted via some other molecule and that the activin synthesis is due to this other signaling molecule's interaction with a cofactor that may or may not exist in any given cell.)

Direct and Continuous Assessment by Cells of Their Position in a Morphogen Gradient

Nature, **376** (1995) 520–521.
J. B. Gurdon, A. Mitchell, D. Mahony

According to the morphogen gradient concept [1–5], cells in one part of an embryo secrete diffusible molecules (morphogens) that spread to other nearby cells and activate genes at different threshold concentrations. Strong support for the operation of a morphogen gradient mechanism in vertebrate development has come from the biochemical experiments of Green and Smith [6, 7], who induced different kinds of gene expression in amphibian blastula cells exposed to small changes in activin concentration. But the interpretation of these experiments has been complicated by recent reports [8–10] that cells tested for gene expression 3 hours after exposure to activin fail to show the graded response previously reported at 15 hours [6, 7], a result suggesting that cells recognize their position in a gradient by an indirect mechanism. Here we conclude from the *in situ* analysis of blastula tissue containing activin-loaded beads [11] that cells respond directly to changing morphogen concentrations, in a way that resembles a ratchet-like process.

To analyse early responses to an activin gradient, we had first to determine how soon gene activation can be seen following exposure to the morphogen. Figure 1 shows that the time of *Xbra* [12] RNA synthesis is related to the developmental age of responding cells, and not to the time of first exposure to activin, a conclusion reached previously for certain other kinds of gene expression [13] or cell behaviour [14] in amphibian embryos. Figure 1 also shows that, in animal caps first exposed to activin at stage 9.5 [15], *Xbra* expression can be detected by RNase protection within 1 hour. However, by *in situ* hybridization, a longer time is needed for *Xbra* expression to be seen at a distance from the activin source. When beads containing a high concentration of activin (strong beads; see legend to Fig. 1) are enclosed in stage-9 animal caps and sectioned for *in situ* hybridization as described before

---

Author affiliations: J. B. Gurdon, A. Mitchell, D. Mahony: Wellcome CRC Institute, Cambridge, UK, and Department of Zoology, University of Cambridge, UK.

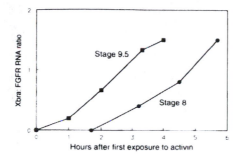

Hours after first exposure to activin

Fig. 1 The time course of *Xbra* expression is related to developmental stage and not to the time elapsed since activin exposure. Animal caps from stage 8 or stage 9.5 blastulae (separated by 1.5 h; ref. 15) were used to enclose activin-containing beads in sandwiches [11]. These were cultured for the times shown when they were frozen for RNase protection analysis. Each sample was probed for *Xbra* and FGF receptor RNA, and analysed on the same gel. FGF receptor was used solely for quantitative comparison of samples. Affigel blue beads (Biorad) were incubated in a solution of 0.1% BSA containing human recombinant activin (17 nM, strong beads) for 30 min at 37°C. They were then kept for up to 2 weeks at +4°C until used. Each sandwich contained 5 beads.

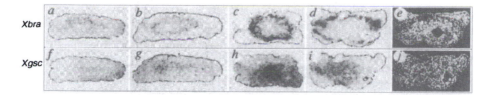

Fig. 2 Early spatial changes of expression of *Xbra* and *Xgsc* in activin-bead sandwiches. Strong beads (incubated in 15 nM activin) were implanted into stage 8.5 animal caps. Beads were removed after 25 min (a, f), 50 min (b, g) or 2 h (c, h). Disbeaded sandwiches were resealed for further culture. d, e, i, j, Samples fixed at 3 h. The total culture time for all samples was 3-4 h, so that they had reached control stage 10.5 when fixed. Histological sections were processed by *in situ* hybridization, as described [11], for expression of *Xbra* (a-e) or *Xgsc* (f-i). e, j, Same sections as in d and i, but photographed to show nuclear staining, and therefore an even distribution of cells throughout the section.

[11], *Xbra* RNA can be seen well separated from the beads by *Xbra*-negative cells 2 hours after bead implantation (Fig. 2c). Therefore within 2 hours a gradient of activin seems to have been established, and cells have been able to respond to their appropriate activin concentration.

We can now ask how cells respond to the forming morphogen gradient during the first 2 hours after activin exposure, while the gradient is still being established. This provides a test of whether cells respond to a morphogen gradient directly or only after complex interactions among themselves. We might expect there to be an early time when cells close to a strong signalling source would experience a low concentration of activin, such as would

induce *Xbra*. After the time needed for activin concentration to build up, these same cells should have received too high a concentration of the morphogen to express *Xbra*. They should therefore discontinue *Xbra* expression and initiate a higher-grade response, such as *Xgsc* [16], if they respond to the gradient in a simple and direct way. The time between activin bead implantation and the accumulation of enough *Xbra* messenger RNA to be seen by *in situ* methods is 1-2 hours. We have therefore tested the above proposition by removing beads at early times after implantation, and culturing disbeaded sandwiches until the normal time of *Xbra* expression about 2 hours later. We find that a 25-minute exposure to strong activin beads induces

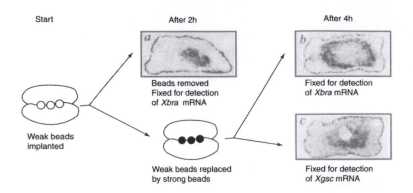

Fig. 3 Cells can switch their gene response to a changing morphogen gradient. Weak beads, incubated in 1.5 nM activin, were implanted into animal cap sandwiches at stage 9. Two hours later (at control stage 10), some were fixed and *in situ* hybridized for *Xbra* expression (a). At the same time, others in the same series had their weak beads removed and replaced by strong beads. After a further 2 h (to control stage 10.75), they were fixed and *in situ* hybridized for expression of *Xbra* (b) or *Xgsc* (c), in nearby sections from the same conjugate.

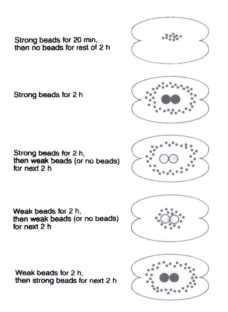

Fig. 4 Diagram to show the results of activin bead implant and exchange experiments. In each figure the outer line represents an animal cap sandwich. The small spots indicate *Xbra*-positive cells. The lower four figures show 2 activin beads in each sandwich. Strong (15 nM) beads are indicated by dark filled circles, and weak (1.5 nM) beads by light filled circles. Within the timescale of these experiments, cells can increase but not decrease their level of response, suggesting a ratchet mechanism. Our conclusions are not affected by whether the morphogen gradient is created by diffusion or by a relay process.

Xbra expression in cells close to where the beads were located (Fig. 2a). After a 50-minute exposure, the strongest *Xbra* expression has been extended away from the bead location (Fig. 2b), and an exposure of 2 hours or more results, as just described, in a ring of *Xbra*-positive cells separated from the bead location by several diameters of *Xbra*-negative cells (Fig. 2c–e). Activin induced *Xgsc* expression is first seen 1–2 hours later than that of *Xbra*, and only in cells that are close to the bead location and that expressed *Xbra* at an earlier time (Fig. 2f–j). As previously observed [11], distal *Xgsc* expression overlaps with proximal *Xbra* expression (see also Fig. 3b, c). These are the results that we would expect of a simple and direct response of cells to a forming morphogen gradient.

We are now in a position to approach the problem of how cells recognize their position in a morphogen gradient, and specifically to ask whether they do so continuously or only at a particular time. To distinguish these ideas, we carried out a bead transfer experiment in two parts. First, weak beads were implanted into stage-9 sandwiches that were cultured for 2 hours when (at the equivalent of stage 10) they were fixed; *Xbra* expression was seen, as expected, in only those cells located close to the beads (Fig. 3a). The second part of the experiment was initiated as in the first part, but now the weak beads were replaced by strong beads after 2 hours, and the resulting sandwiches cultured for a further 2 hours (until stage 10.75). *In situ* analysis showed *Xbra* expression in a ring distant from the beads (Fig. 3b); notably, the cells nearest the beads were now *Xbra*-negative; they were also *Xgsc*-positive (Fig. 3c), as expected for response to a high activin concentration. It has previously been established that cells do not move significantly under these conditions [11]. Therefore cells that have already made an early response to their position in a morphogen gradient can subsequently change their response to a higher level. We conclude that cells can assess morphogen concentration continuously, and not only at one particular time.

Lastly we have asked whether cells change their response in a downward as well as upward direction, and have carried out the series of bead replacement experiments summarized in Fig. 4. Conjugates exposed to weak or strong beads show the same response at 4 hours, whether or not beads are removed after 2 hours. But cells exposed first to weak beads and then to strong beads change their

Xbra expression to the distal (strong) type, as described. In contrast, strong beads replaced by weak beads result in a continued strong-bead response. From the combination of our results (Fig. 4), we conclude that cells express genes characteristic of the highest concentration of morphogen to which they have been exposed within their period of competence [17], as if by a ratchet-like process. Whether activin or some other related molecule such as Veg1 (refs 18–20) is a natural mesoderm inducer, our results, which differ in design from those previously reported [8–10,21,22], strongly suggest that cells interpret their position in a morphogen gradient in a remarkably simple and direct way.

REFERENCES

1. L. Wolpert, *Development* (suppl.) **107**, 3 (1989).
2. P. A. Lawrence, *Cell* **54**, 1 (1988).
3. P. Biickell & S. Tickle, *BioEssays* **11**, 145 (1989).
4. J. M. W. Slack, *From Egg to Embryo* 2nd edn (Cambridge Univ. Press, 1991).
5. J. Cooke, *BioEssays* **11**, 93 (1995).
6. J. B. A. Green & J. C. Smith, *Nature* **247**, 391 (1990).
7. J. B. A. Green, H. V. New & J. C. Smith, *Cell* **71**, 731 (1992).
8. J. B. A. Green, J. C. Smith & J. C. Gerhart, *Development* **120**, 2271 (1994).
9. N. Symes, C. Yordan & M. Mercola, *Development* **120**, 2339 (1994).
10. P. A. Wilson & O. A. Metton, *Curr. Biol.* **4**, 676 (1994).
11. J. B. Gurdon, P. Harger, A. Mitchell & P. Lemaire, *Nature* **271**, 487 (1994).
12. J. C. Smith, B. M. J. Price, J. B. A. Green, O. Weigel & B. G. Hermann, *Cell* **67**, 79 (1991).
13. J. B. Gurdon, S. Fairman, T. J. Mohun & S. Brennan, *Cell* **41**, 913 (1985).
14. K. Symes & J. C. Smith, *Development* **101**, 339 (1987).
15. P. O. Nieuwkoop & I. Faber, *Normal Table of Xenopus laevis* (Daudin)(North-Holland, Amsterdam, 1956).
16. K. W. Y. Cho, B. Blumberg, H. Steinbeisser & E. M. De Robertis, *Cell* **67**, 1111 (1991).
17. E. A. Jones & H. R. Woodland, *Development* **101**, 557 (1987).
18. G. H. Thomsen & O. A. Mellon, *Cell* **74**, 433 (1993).

19. P. O. Vize & O. H. Thomsen, *Trends Genet.* **10**, 371 (1994).
20. S. Schulte-Merker, J. C. Smith & L. Dale, *EMBO J.* **12**, 3533 (1994).
21. C. L. Ferguson & K. V. Anderson, *Cell* **71**, 451 (1992).
22. J. Jiang & M. Levine, *Cell* **72**, 741 (1993).

Activin Signalling and Response to a Morphogen Gradient

Nature, **371** (1994) 487–492.

J. B. Gurdon, P. Harger, A. Mitchell, P. Lemaire

Using combinations of amphibian embryo tissues, it is shown that the selection of genes expressed by a cell is determined by its distance from a source of activin, a peptide growth factor contained in vegetal cells and able to induce other cells to form mesoderm. This long-range signal spreads over at least 10 cell diameters in a few hours. It does so by passive diffusion, because it can by-pass cells that do not themselves respond to the signal nor synthesize protein. These results provide direct support for the operation of a morphogen concentration gradient in vertebrate development.

Much current work is directed towards identifying molecules involved in embryonic induction and to describing their developmental effects. It is thought that some kinds of inducer molecules, acting as morphogens are synthesized and emitted by cells in one region of an embryo and become diluted in concentration as they spread to cells distant from this point. A favoured hypothesis is that the type of gene activation and cell differentiation undergone by responding cells is related to the concentration of inducer that they receive [1, 2].

The best evidence that this mechanism operates in vertebrate embryos comes from the important work of Green and Smith [3, 4]. They exposed dissociated *Xenopus* blastula cells to different concentrations of an inducer, namely activin, and found that the type of genes expressed in these cells is related to the concentration of inducer used. It has yet to be shown, however, that a similar effect can take place in normal solid tissue rather than in dissociated cells. Furthermore there is a clear need to test the morphogen concept directly by asking whether the kind of response that cells of a normal tissue make to a signal is determined by inducer concentration and therefore by their position with respect to the source of the morphogen.

Here we describe experiments in which we show (1) that an embryonic signalling process can spread for at least 300 μm through solid tissue; (2) that the selection of genes activated is related to the distance of cells from a source of signal, and presumably therefore to the concentration of a signalling molecule; and (3) that the signal is transmitted by passive diffusion rather than by active cell-by-cell amplification. Our results therefore provide direct support for the fundamental concept that selective gene activation takes place in response to morphogen concentration in normal embryonic tissue of a vertebrate.

Inducer concentration and response location

To ask whether there is a relationship between the strength of an inducer and the distance from its source at which particular genes are activated, we have used the following experimental system. A Nieuwkoop conjugate is constructed in which the vegetal hemisphere of a blastula is placed in contact with a blastula animal cap containing the fluorescent intracellular marker rhodamine lysinated dextran (RDLx). The vegetal hemisphere is taken from an embryo which had been injected at its two-cell stage with varying amounts of activin messenger RNA synthesized *in vitro* (Fig. 1a). The conjugates are cultured until the early gastrula stage, when they are fixed, sectioned and hybridized *in situ* for expression of the pan-

Author affiliations: J. B. Gurdon, P. Harger, A. Mitchell & P. Lemaire: Wellcome CRC Institute, Cambridge, UK; and Department of Zoology, University of Cambridge, UK.

mesodermal marker *Xbrachyury (Xbra)* (ref. 5). In normal embryos at the mid-gastrula stage, *Xbra* mRNA is seen to be localized in an internal ring in the marginal equatorial zone (Fig. 2e). Figure 2a–c shows that increasing amounts of activin mRNA in vegetal cells cause *Xbra* RNA to appear in animal cells located progressively further away from the activin source. The results of 25 analyses are summarized in Table 1. Many of the *Xbra*-expressing cells seen in Fig. 2a–c subsequently progress to muscle, as shown by their acquisition of strong nuclear staining for MyoD, a marker of muscle progenitor cells (Fig. 2d), and when cultured further express terminal muscle marker genes (not shown). We conclude that there is a direct relationship between the amount of inducer mRNA in vegetal cells and the distance from them at which mesoderm genes are expressed.

We have asked whether this distance effect of activin mRNA can be generated only by those cells that are natural sources of mesoderm-inducing signals. The injected activin mRNA might stimulate the production of some other endogenous inducer in these cells. If, on the other hand, the effect is related directly to activin itself, we might expect it to be generated by cells that are not themselves normal mesoderm inducers, for example by animal pole cells injected with activin mRNA (Fig. 1b). This is indeed the result obtained when activin-containing animal caps are conjugated to other non-injected animal caps (Table 1; see also Figs 3e and 4h). We conclude that the amount of activin mRNA required to induce *Xbra* expression at a given distance from the source of the signal is very similar whether it is supplied by the naturally inducing (vegetal) cells or by other (animal) cells which never normally release this signal.

It has recently been reported [6, 7] that the treatment of dissociated cells with activin protein fails to reveal a sharp concentration dependence of gene expression when this is assayed at an early time (after $2\frac{1}{2}$ h), in the way that was previously described for assays at 15 h [3, 4]. Our results described above were obtained with conjugates cultured for 5–6 h analysed at the same early stage ($10\frac{1}{2}$) as used by Green *et al.* [6]. When we tested *Xbra* expression at 1-h intervals from the time of conjugation, we saw no expression at $3\frac{1}{2}$ h, but at $4\frac{1}{2}$ h the light expression seen already revealed the same distance effect seen in Fig. 2a–c.

A minimum estimate of the rate at which this signalling process takes place comes from our observation that *Xbra* RNA can be seen 5 h after conjugation at 23 °C at a distance of at least 280 μm from the source. The rate of spread is therefore not less than 55 μm h^{-1}, and is almost certainly faster, because animal caps are not sufficiently large to determine the maximum distance that can be achieved, and because other genes are probably activated further from the source than *Xbra*.

Cell movement and cell division

Our results so far described establish the spread of an *Xbra*-inducing effect, but they do not establish its cellular basis.

One possibility is the cells containing activin mRNA signal to the responding cells with which they are in direct contact, and these cells then move away to a distant position within the responding tissue. A variant of this idea is that a responding cell in contact with an inducing cell divides to generate daughters and granddaughters, which are now separated from the inducing cells. Both these explanations suppose that the stronger the *Xbra* signal received, the further the induced cells or their daughters move away from the source of the signal.

The important principle behind these ideas is that the signalling process might be transmissible only by direct contact, and that the subsequent spatial relationship of induced cells would be established by cell movement and/or cell division. Alternatively, the signalling process could involve the spread of a signalling molecule itself through cells that do not themselves move.

To distinguish these ideas, we have constructed conjugates with fluorescein (FLDx)-containing animal cap cells as a thin reaggregated layer between inducing vegetal cells and a rhodamine (RLDx)-containing animal cap (Fig. 1c). If cell movement is involved in generating the distance effect, we would expect the FLDx animal cap cells closest to a strong activin mRNA-vegetal cell source to move into the more distant RLDx-animal cell region. In fact, no such cell movement took place in conjugates fixed at the early gastrula stage (Fig. 3a, b), and *Xbra*-synthesizing cells are clearly located in the distal RLDx cells (Fig. 3c, d). Cell movement is not therefore responsible for the distance effect.

To determine the possible role of cell division, we have cultured conjugates in medium containing 5 μg ml^{-1} cytochalasin, a procedure previously found to inhibit cell division and cell movement

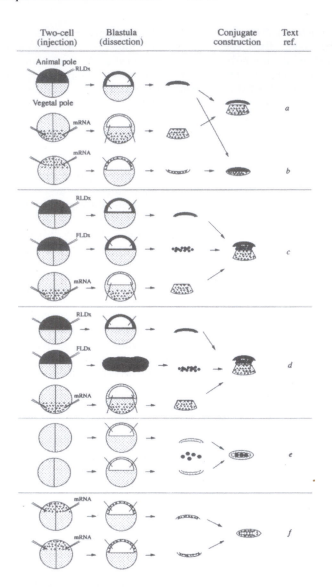

Fig. 1 a–f, Diagrams illustrating the design of experiments. See text for further details. Red, cells containing the lineage marker rhodamine lysinated dextran (RLDx); green, cells containing fluorescein lysinated dextran (FLDx). Black spots represent injected mRNA. Blue circles (in e) represent beads containing activin protein.

[3]. Its effect is seen by the presence of multinucleated cells (not shown). Although cytochalasin reduces the distance over which signalling occurs, *Xbra* RNA is seen in cells separated by several cell diameters from the activin source (Fig. 3e, f).

We conclude that cell movement and cell divi-sion do not account for the separation of *Xbra*-expressing cells from the source of the inducer. Therefore, this process must involve signal trans-mission through or past cells which do not them-selves express *Xbra*. Cells that transmit, but do not seem to respond to a signal can extend in our ex-

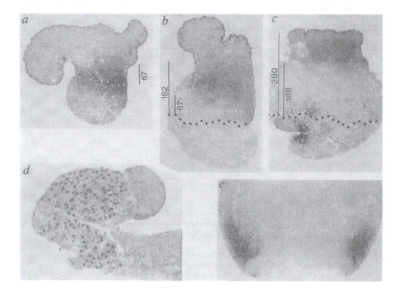

Fig. 2 Mesoderm and muscle gene expression in animal-vegetal conjugates of *Xenopus* blastulae. Conjugates were prepared at stage 8 by standard procedures [17], and cultured until fixation at stage 11 (**a–c, e**) or stage 18 (**d**). After histological sectioning, slides were hybridized to a digoxigenin-labelled RNA probe specific for *Xbra* [13] (**a–c, e**, purple), or stained with an anti-XMyoD antibody [47] (**d**, black nuclei). In **a–c**, the animal part of each conjugation was lineage-labelled with RLDx: *in situ* hybridization to sections reduces fluorescence to too low a level for photography, but fluorescence was present in all tissue above the dotted lines in **a–c**. The vegetal part of each conjugate was injected with activin mRNA: **a**, 2 pg: **b**, 7 pg: **c**, 20 pg. In **a–c**, the vertical lines indicate distances in μm of *Xbra* expression from mRNA-containing vegetal tissue. **e**, Section of a normal stage-11 embryo.

periments for about 200 μm or six cell diameters from its source (Fig. 2c).

Signalling by passive diffusion

A fundamental question concerning intercellular signalling is whether transmission takes place by passive diffusion through or past cells, rather than by active amplification of the signal. In the experimental system under discussion here, those responding cells that are nearest to the activin mRNA-containing vegetal cells and which do not themselves express *Xbra* might nevertheless synthesize activin, and so pass the signal along to their neighbours by a relay process.

To distinguish passive diffusion from relay amplification we have inserted non-responding (that is, non-competent) cells between inducing vegetal and responding animal tissues (Fig. 1d). As non-competent cells, we have used endoderm (progenitor intestine) cells from a stage-40 tadpole, labelled with FLDx; these cells cannot induce *Xbra* in an

animal cap sandwich (Fig. 3g), nor can they express *Xbra* in response to mesoderm-inducing signals. Nevertheless the *Xbra*-inducing signal from activin mRNA vegetal cells is transmitted successfully through the non-competent cells (Fig. 3h).

This leaves open the possibility that the non-competent cells might respond to their adjacent activin mRNA-containing vegetal cells by synthesizing some other (non-activin) inducer, which causes *Xbra* expression in distant cells but not in themselves. This possibility has been tested by inhibiting protein synthesis in the layer of non-competent cells. Previous work has shown that a 1-h treatment with cycloheximide at 5 μg ml^{-1} inhibits protein synthesis for a period of 4 h, after which it slowly recovers. [9–11]. Conjugates were therefore constructed with a layer stage-40 endoderm cells which had been incubated in 10 μg ml^{-1} cycloheximide for 2 h, and these were placed in a layer between inducing and responding tissues (Fig. 1d). *Xbra* was

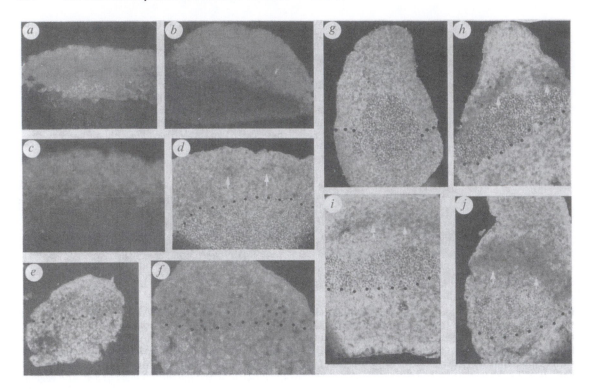

Fig. 3 *Xbra* expression in conjugates fixed at the early gastrula stage ($10\frac{1}{4}$). **a–d,** A layer of reaggregated FLDx animal caps (red). No significant cell movement has taken place (**a, b**), and *Xbra* staining, which is mainly nuclear at this stage, is seen in the red RLDx cell region (white arrows, **d**): **c** and **d** are the same sections photographed respectively before and after *in situ* hybridization. **e, f,** Conjugates composed of two animal caps, of which the one below the black dots contained mRNA but no lineage label; that above the dotted line contained RLDx but no mRNA. Conjugates were cultured in $5\mu g$ ml^{-1} cytochalasin. *Xbra* nuclear staining is seen in a band (dark purple spots). **g, h,** Activin signalling induces *Xbra* expression after passing through non-competent cells. Stage-40 reaggregated endoderm cells, recognized by the refractile yolk platelets, do not induce *Xbra* in an animal cap sandwich (**g**), but permit signal transmission from activin mRNA-containing animal tissue not containing mRNA (above the refractile endoderm tissue, white arrows). **i, j,** Activin signalling can be transmitted by cycloheximide-inhibited stage-40 endoderm cells (same design as **g, h**). White arrows indicate *Xbra* expression (purple) in RLDx-labelled responding tissue.

expressed in the responding animal cells (RLDx-labelled), but not by the cycloheximide-inhibited endoderm cells which are readily recognized by their content of yolk platelets (Fig. 3i–j).

We conclude that the long-range signalling capable of activating *Xbra* can be transmitted passively through non-responding cells, or more likely past these cells via the extracellular matrix.

It has not been possible to distinguish passive diffusion from relay amplification in other experimental systems.

Activin protein can mediate spatial control

Although distant *Xbra* expression in our experiments is dependent on overexpressing activin mRNA in the inducing cells, it is not clear whether activin protein is directly responsible for the distance effect, or whether this might be achieved through some other component released by the message-injected cells. To resolve this uncertainty, we have inserted agarose beads loaded with activin protein into animal cap sandwiches (Fig. 1e). At

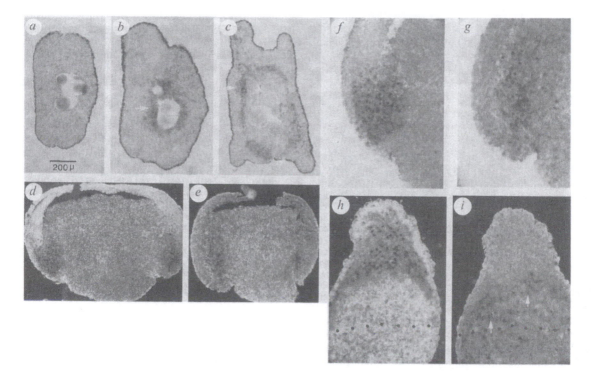

Fig. 4 *Xbra* and *Xgscd* expression in embryos or conjugates fixed at stage $10\frac{1}{2}$. **a–c**, Activin protein shows the same distance effect as mRNA-containing vegetal cells. Agarose beads incubated in nil, 1nM or 4 nM activin protein were included in animal cap sandwiches and processed for expression of *Xbra* RNA (white arrows). In subsequent experiments, smaller amounts of a more active preparation of activin gave similar results. **f–g**, Comparison of *Xbra* (**d, f**) and *Xgscd* (**e,g**) expression in normal stage $10\frac{1}{2}$ embryos, analysed by *in situ* hybridization of digoxigenin probes to sections. **h, i**, Adjacent sections of the same animal-animal conjugate stained for *Xbra* (**h**) or *Xgscd* (**i**). The mRNA-injected animal tissue extends downwards from the dotted lines, the upper, responding animal cap (above the dotted lines) having been RLDx-lineage labelled. In **h**, the strong nuclear and diffuse purple cytoplasmic stain for *Xbra* is seen to be above and well separated from the signalling cells below the dotted line. In **i**, the region above the dotted line, unstained in **h**, contains *Xgscd*-stained nuclei (white arrows). There is a partial overlap of *Xbra* and *Xgscd* staining cells.

a low activin dose, we see *Xbra* gene expression in a ring of cells close to the activin-containing beads (Fig. 4a, b). When beads are loaded with activin, at a much higher concentration, *Xbra* expression is moved to 100–200 μm or up to six cell diameters away from the source (Fig. 4c). Therefore we can reproduce, with purified activin protein, the same relationship between the concentration of inducer and the location of gene activation as we have described for cells injected with activin mRNA. This supports our view that the biological activity of these mRNA-containing cells results from the release and spread of activin protein itself. We should however emphasize that signalling induced by mRNA-injected cells rather than by protein-loaded beads approximates more closely to the normal mesoderm induction process.

Activation of *Xgscd* compared with *Xbra*

We next investigated whether genes other than *Xbra* can be induced by the signalling process we have described. In particular, we wish to know, which, if any, genes are expressed in considerable space between a strong activin mRNA source and the nearest *Xbra* RNA, as seen in Fig. 2c. *Xgoosecoid* (*Xgscd*) is one of several genes expressed in only the

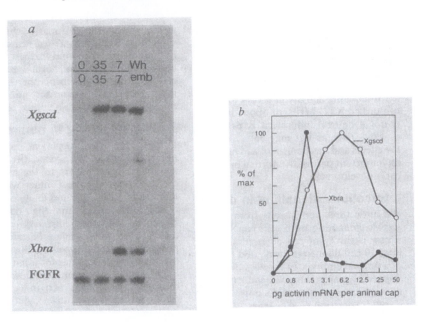

Fig. 5 Quantitation of gene activation by activin mRNA-containing animal caps, **a**, Activin mRNA (0, 7, or 35 pg) was injected into the animal pole of eggs, and animal cap doubles prepared, each cap of the conjugate having the same amount of mRNA. Wh emb, Whole stage $10\frac{1}{4}$ embryos, for comparison; FGFR, FGF receptor. **b**, Graphic representation of another experimental series, in which animal cap doubles were frozen at stage 10. Analyses in **a** and **b** were by RNase protection. Samples were frozen at stage $10\frac{1}{2}$. For methods see ref. 13.

dorsal lip region of an early gastrula [12]. Using *in situ* hybridization to sectioned material [13], a procedure we prefer to whole-mount hybridization on account of the more even access to deep tissues, we see *Xgscd* expression, like that of *Xbra*, as a diffuse cytoplasmic and strong nuclear stain (Fig. 4e). In normal early gastrulae, *Xbra* is precisely localized to an equatorial ring (Fig. 4d, f), whereas *Xgscd* is expressed in different but partly overlapping region on the dorsal but not ventral side (Fig. 4e, g, dorsal side on the left).

Conjugates which include animal or vegetal tissue with large amounts (20 pg or more) of activin mRNA were sectioned alternately onto two slides, which were reacted with *Xbra* or *Xgscd* antisense RNA probes. As expected *Xbra* was expressed at a distance from the signalling tissue; however, *Xgscd* was expressed in the intervening but partially overlapping space between the activin source and cells expressing *Xbra* (Fig. 4h, i), a pattern expected in view of the normal expression of these genes. This result supports the proposition that activin,

or other morphogens, activates the most dorsal responding genes nearest its source where its concentration is presumed to be highest.

To determine the relative expression of *Xbra* and *Xgscd* in a more quantitative way than can be achieved by *in situ* hybridization, we have used Rnase protection. To simplify interpretation, we have analysed conjugates in which both component tissues were prepared in the same way, for example by injection of high, low or nil levels of activin mRNA (Fig. 1f). Thus we might predict that *Xgscd* would be expressed most strongly in animal caps with the greatest amount of activin mRNA, whereas *Xbra* should be most strongly represented in middle-dose activin mRNA samples. This is in fact the result obtained (Fig. 5a), because relative to normal embryos, the expression of *Xgscd* is stronger after activin than after no activin mRNA injection, wheras *Xbra* is 15 times stronger after a middle (7ng) than a high (35 ng) mRNA injection. In two more extensive experiments (not shown), activin mRNA, was injected in small increments from 0 to 50 pg per animal cap, and conjugates were analysed after 5 and 7 h of incubation. *Xgscd* expression was greatest after injection of 3-12 pg activin mRNA, whereas at both time points *Xbra* was expressed most strongly at 1.5 pg of activin mRNA, and was virtually eliminated by higher doses (Fig. 5b).

In summary, our *in situ* results show that the expression of *Xbra* and *Xgscd* is spatially related to the source of an inducing signal. In all, we have discussed three levels of response to activin signalling. These are *Xgscd* nearest the signal source, *Xbra* in the middle distance, and furthest away a nil response probably corresponding to the normal ectodermal fate of uninduced ectoderm. Our experiments therefore provide direct support for the hypothesis that gene activation is determined by the spatial relationship of cells to a source morphogen.

Activin signalling in normal development

Does activin signalling play an important part in *Xenopus* development? Cells of the second tier of a 32-cell *Xenopus* embryo are major contributors to the mesoderm and its derived cell types [14, 15], but require contact with third-tier cells to become mesoderm [16]. For this and many other reasons [17–20], it is believed that second-tier cells and their daughters are normally induced by vegetal cells to be redirected from an epidermal to a mesodermal fate, by the so-called Nieuwkoop induction. Several gene products are able to mediate this process, and activin is one of the strongest candidates for a natural inducer [21–23]. Although its mRNA is not synthesized until well after the Nieuwkoop induction has started [24], an activin-like protein is present in eggs [25] (its distribution not known). Activin A was the first mesoderm inducer to be purified from normal cell extracts [26], and has mesoderm-inducing activity at a very low concentration (0.2 ng ml^{-1} of purified XTC-MIF, or about 10 pM activin). Even if activin is not a natural inducer, it is probable that a closely related molecule such as Vg1 may be [27, 28]. It is therefore very likely that the effects and mechanism of activin signalling described here are an important part of the normal mesoderm-forming processes in early development.

Intercellular signalling in other systems

We draw three conclusions from our results. The first is that signals emanating from cells can initiate new gene activity and a new pathway of cell differentiation in other cells located at least 10 cell diameters away, and this takes place in solid tissue of a normal embryo. In the vertebrates, this conclusion has been implied by many embryonic induction experiments since the time of Spemann and Mangold's dorsal lip grafts [29], as well as by the ectopic expression of *Wnt* genes [30, 31]. In *Drosophila*, *wingless (wg)* has effects on cells, and its proteins can spread a few cell diameters from its source [32, 33].

A second more important result of our experiments is to strongly support the operation of a morphogen gradient, an idea suggested by other work. In *Drosophila*, it has been shown that a change in the concentration of *bicoid* protein determines the position of nuclei expressing *hunchback* [34, 35]. In this case, however, signalling takes place in a syncytium by the spread of RNA or protein through cytoplasm. In contrast, *decapentaplegic (dpp)* encodes an intercellular signalling molecule belonging, like activin, to the TGF-β family of growth factors, and is thought to be distributed in a dorso-ventral gradient during the post-syncytial stage of development [36]. When *dpp* RNA is injected into syncytial embryos, 2–4 fold increases in concentration lead to the formation of progressively more dorsal structures [37]. It has been suggested that *hedgehog (hh)* may also act as a morphogen in the patterning of the *Drosophila* epidermis [38], though its mode of action may be indirect [39]. Within the vertebrates, evidence has accumulated in favour of the idea that the zone of polarizing activity in the chick limb bud is a source of morphogen whose concentration determines the type of digit to be formed [40]. The implantation of preloaded beads into a limb bud generates a gradient of retinoic acid that is related to the sequence of digits formed [41], though not in any simple way to the early expression of genes [42]. In amphibia, increasing amounts of *Xgscd* can dorsalize ventral tissue in a dose-dependent way [43]; however, as a transcription factor, *Xgscd* (like *bicoid*) is a response to, and not an agent of, intercellular signalling, and might be a downstream consequence of the activin effects we report here. Our own results on activin concentration agree with those of Green and Smith [3, 4] and extend them most significantly by showing that the concentration of activin released by signalling cells determines which cells in a responding field of normal solid tissue will express a particular type of gene. Our results therefore provide direct evidence for the importance of morphogen gradient in normal development.

The third conclusion from our work concerns the mechanism by which signalling molecules exert their effect through a mass of cells. In previous work it has not been possible to determine whether the long-range effects of *wg* or *hh* result from diffusion of a signalling molecule or from a cascade or relay mechanism [44]. In an entirely different system, Placzek *et al.* [45] have shown that a chemoattractant emanating from the floor plate of a neural tube can spread through neural epithelium over a distance of 250 μm and can cause a change in the orientation of growing axons towards the source of the influence. Our results reported here with non-competent cells in which protein synthesis was inhibited seem to provide the strongest evidence so far that a peptide growth factor can progress past many cells, activate genes by a diffusion rather than a relay or cascade mechanism, generating a concentration gradient as it does so.

Overall, our results add direct support to the operation of a morphogenetic gradient in vertebrate development, an idea which has been prominent in the minds of embryologists for many years [46].

REFERENCES

1. L. Wolpert, *Development* (suppl.) 3–12 (1989).
2. J. B. A. Green & J. C. Smith, *Trends Genet.* **7**, 245 (1991).
3. J. B. A. Green & J. C. Smith, *Nature* **347**, 391 (1990).
4. J. B. A. Green & J. C. Smith, *Cell* **71**, 731 (1992).
5. J. C. Smith, B. M. J. Price, J. B. A. Green, D. Weigel & B. G. Hermann, *Cell* **67**, 79 (1991).
6. J. B. A. Green, J. C. Smith & J. C. Gerhart, *Development* **120**, 2271 (1994).
7. P. A. Wilson & D. A. Melton, *Curr. Biol.* **4**, 676 (1994).
8. J. B. Gurdon, *Development* **105**, 27 (1989).
9. J. B. Gurdon, K. Kao, K. Kato & N.D. Hopwood, *Development* (suppl.) 137–142 (1992).
10. S. Cascio & J.B. Gurdon, *Development* **100**, 297 (1987).
11. F. M. Rosa, *Cell* **57**, 965 (1989).
12. K. W. Y. Cho, B. Blumberg, H. Steinbeisser & E. M. De Robertis, *Cell* **67**, 1111 (1991).
13. P. Lemaire & J. B. Gurdon, *Development* **120**, 1191 (1994).
14. L. Dale & J. M. W. Slack, *Development* **99**, 527 (1987).
15. S. A. Moody, *Devl. Biol.* **122**, 300 (1987).
16. J. B. Gurdon, T.J. Mohun, S. Fairman & S. Brennan, *Proc. Natn. Acad. Sci. U.S.A.* **82**, 139 (1985).
17. P. D. Nieuwkoop, *Wilhelm Roux Arch. Dev. Biol.* **162**, 341 (1985).
18. O. Nakamura, H. Takasaki & M. Ishihara, *Proc. Jap. Acad.* **47**, 313 (1971).
19. J. C. Smith, *EMBO J.* **12**, 4463 (1993).
20. J. M. W. Slack, *Mech. Dev.* **41**, 91 (1993).
21. A. Hemmati-Brivanlou & D. A. Melton, *Nature* **359**, 609 (1992).
22. G. Thomsen *et al.*, *Cell* **63**, 485 (1990).
23. R. A. Cornell & D. Kimmelman, *Development* **120**, 453 (1994).
24. E. A. Jones & H.R. Woodland, *Development* **101**, 557 (1987).
25. M. Asashima *et al.*, *Proc. Natn. Acad. Sci. U.S.A.* **88**, 6511 (1991).
26. J. C. Smith, B. M. J. Price, K. Van Nimmen & D. Huylebroeck, *Nature* **345**, 729 (1990).
27. G. H. Thomsen & D. A. Melton, *Cell* **74**, 433 (1993).
28. L. Dale, G. Matthews & A. Colman, *EMBO J.* **12**, 4471 (1993).
29. H. Spemann & H. Mangold, *Roux's Arch. Entw. Mech.* **100**, 599 (1924).
30. A. P. McMahon, *Trends Genet.* **8**, 236 (1992).
31. L.W. Burrus, *BioEssays* **16**, 155 (1994).
32. F. Gonzalez, L. Swales, A. Bejsovic, H. Skaer & A. Martinez-Aria, *Mech. Dev.* **35**, 43 (1991).
33. J. P. Couso, S.A. Bishop & A. Martinez-Arias, *Development* **120**, 621 (1994).
34. W. Driever & C. Nusslein-Volhard, *Cell* **54**, 83 (1988).
35. G. Struhl, K. Struhl & P. M. Macdonald, *Cell* **57**, 1259 (1989).
36. D. St. Johnston & C. Nusslein-Volhard, *Cell* **68**, 201 (1992).
37. E. L. Ferguson & K. V. Anderson, *Cell* **71**, 451 (1992).
38. J. Heemskerk & S. DiNardo, *Cell* **76**, 449 (1994).
39. K. Basler & G. Struhl, *Nature* **368**, 208 (1994).
40. C. Tickle, B. Alberts, L. Wolpert & J. Lee, *Nature* **296**, 564 (1982).
41. Tickle, J. Lee & G. Eichele, *Devl Biol.* **109**, 82 (1985).
42. J. Brockes, *Neuron* **2**, 1285 (1989).
43. C. Neihrs, H. Steinbeisser & E. M. De Robertis, *Science* **263**, 817 (1994).
44. J-P. Vincent & P. A. Lawrence, *Cell* **77**, 909 (1994).

45. M. Placzek, M. Tessier-Lavigne, T. M. Jessell & J. Dodd, *Development* **110**, 19 (1990).
46. P. A. Lawrence, *Cell* **54**, 1 (1988).

47. N. D. Hopwood, A. Pluck, J. B. Gurdon & S. M. Dilworth, *Development* **114**, 31 (1992).

Exercises for Chapter 18

1. Suppose that $f(u_e) = au_e$ with $a > 0$. Show that Equations (7) and (8) from this chapter can be solved with u_e not identically zero if a can be written in terms of L and μ as $a = \mu n^2 \pi^2 / L^2$ for some integer n.

   In solving this exercise, remember from the discussion in Chapter 16 that solutions to the equation

   $$\frac{d^2 h}{dx^2} = ch$$

   (with c a constant) have the following form:

   1. If $c > 0$, then $h(x) = \alpha e^{\sqrt{c}\, x} + \beta e^{-\sqrt{c}\, x}$.

   2. If $c = 0$, then $h(x) = \alpha + \beta x$.

   3. If $c < 0$, then $h(x) = \alpha \cos((-c)^{1/2} x) + \beta \sin((-c)^{1/2} x)$.

   Here, you get to choose the constants α and β as you wish. [Note, however, that when working through this exercise, the boundary conditions in Equation (8) of this chapter will restrict the possible choices for these constants.]

2. Suppose that $f(u_e) = -au_e$ with $a > 0$. Show that there are no solutions to both Equations (7) and (8) of this chapter except for $u_e = 0$ everywhere.

3. If nature only allows the function $f(h)$ to equal ah for some number a, give an argument (based on your answers to Exercises 1 and 2) why you should never expect to see striped tigers. (*Hint:* How likely is it for a and μ and L to satisfy some special relationship?)

4. Answer the following:

   (a) Show that $u_e(x) = r_0/r_1$ is an equilibrium solution to Equations (1) and (6) of this chapter with f given as in Equation (2a) of this chapter. Then write the version of Equation (11) of this chapter [i.e., find the correct $z(x)$] for this f and u_e.

   (b) Show that $u_e(x) = r_1/r_2$ is an equilibrium solution to Equations (1) and (6) of this chapter with f given as in Equation (2b) of this chapter. Then write the version of Equation (11) of this chapter for this choice of f and $u_e(x)$.

(c)   Show that $u_e(x) = 0$ is an equilibrium solution to Equations (1) and (6) of this chapter with f given as in Equation (2b) of this chapter. Then write the version of Equation (11) of this chapter for this choice of f and $u_e(x)$.

5. Find $f(u_e(x))$ and $(\frac{d}{du}f)(u_e(x))$ in the case where $f(u)$ and $u_e(x)$ are given by

(a) $f(u) = \sin(u)$ and $u_e(x) = \cos(2x)$

(b) $f(u) = \cos(2u)$ and $u_e(x) = \sin(x)$

(c) $f(u) = (1 + u^2)$ and $u_e(x) = (1 + x^2)$

19
Stability Criterion

This chapter continues the discussion about the stability criterion for equilibrium solutions to a simple diffusion equation and presents some examples.

Recall that we are considering solutions $u(t, x)$ to the equation

$$\frac{\partial}{\partial t}u = \mu\frac{\partial^2}{\partial x^2}u + f(u) \tag{1}$$

that obey, for all time t, the boundary conditions

$$\frac{\partial}{\partial t}u = 0 \text{ at } x = 0 \quad \text{and} \quad \frac{\partial}{\partial x}u = 0 \text{ at } x = L. \tag{2}$$

Here, $f(u)$ is shorthand for $f(u(t, x))$, where f is a function of a single variable. Two simple examples are

$$f(u) = r_0 - r_1 u, \text{ where } r_0 > 0 \text{ and } r_1 \text{ are constants} \tag{3a}$$

$$f(u) = r_1 u - r_2 u^2, \text{ where } r_1 \text{ and } r_2 \text{ are positive constants.} \tag{3b}$$

Suppose that a time-independent solution to (1) and (2) has been found; thus the function $u(t, x)$ has no t-dependence. In this case, I will write $u(t, x) = u_e(x)$ to stress the lack of t-dependence. By definition, u_e is an equilibrium solution to (1) and (2) and so satisfies

$$\mu\frac{d^2}{dx^2}u_e + f(u_e) = 0, \tag{4}$$

311

with the boundary conditions

$$\frac{d}{dx}u_e = 0 \text{ at } x = 0 \text{ and } x = L. \tag{5}$$

As stressed in the preceding chapter, equilibrium solutions that are not stable are not generally seen in nature.

19.1   A Stability Definition

Here is the stability criterion we are using (it is equivalent to the one given in Section 18.5 of Chapter 18):

- The solution $u_e(x)$ is *stable* if there is *no* pair of number $\lambda \geq 0$ and function $g(x)$ (with g not zero at every x) that obey

$$\lambda g = \mu \frac{d^2}{dx^2}g + (\frac{d}{ds}f)(u_e(x))g \tag{6a}$$

$$\frac{d}{dx}g = 0 \text{ at } x = 0 \text{ and } x = L. \tag{6b}$$

Meanwhile, instability is characterized by the following:

- The solution $u_e(x)$ is *unstable* if there exists even one pair consisting of a number $\lambda \geq 0$ and a function $g(x)$ (not identically zero) that solves the two preceding equations, (6a), (6a).

[Remember that in (6), $(\frac{d}{ds}f)(u_e)$ is typically a function of x if u_e itself has x-dependence; its value at the point x is the value of the function $\frac{d}{ds}f$ at the point $s = u_e(x)$. In Chapter 18, I denoted this new function of x by $z(x)$. Here, I shall use $\left(\frac{df}{ds}\right)(u_e)$ so that we don't forget that it is determined both by $f(s)$ and the particular equilibrium solution $u_e(x)$.]

I want to stress that the preceding is absolutely equivalent to the definition in Section 18.5 of Chapter 18.

19.2   Justification for This Definition

In the previous chapter I indicated the rationale behind the definition of stability given in Section 19.1. Bear with me here as I say a bit more on this point.

In particular, although it is not obvious, this definition of stabililty ensures the following: If u_e is stable in the sense just given, then every time-dependent solution $u(t, x)$ to (1) and (2) that has $u(0, x)$ near enough to $u_e(x)$ for every x will actually approach $u_e(x)$ as t gets large at all values of x. Conversely, if u_e is unstable as defined in Section 19.1, then one of the following two possibilities will occur:

- If $\lambda > 0$, then there is a solution $u(t, x)$ to (1) and (2) that has $u(0, x)$ close to $u_e(x)$ for all values of x, but that gets very far from $u_e(x)$ at some values of x as t gets large.

- If $\lambda = 0$, there are some arbitrarily small perturbations of the equations [either (1) or (2)] that have solutions that start near $u_e(x)$ at time 0 and then move away as time gets large. (7)

To see the case for the first point, suppose that $\lambda > 0$ and g solve (6). Introduce $w(t, x) = \alpha e^{\lambda t} g(x)$, with α very small. Notice that w satisfies

$$\frac{\partial}{\partial t} w = \mu \frac{\partial^2}{\partial x^2} w + \left(\frac{df}{ds}(u_e(x)) \right) w \tag{8}$$

and $\frac{d}{dx} w = 0$ at $x = 0$ and $x = L$. Unless g is identically zero, then $|w(t, x)|$ will grow with increasing time t. This is because of the factor $e^{\lambda t}$ with $\lambda > 0$ that appears in w. Moreover, for small α, the function $u(x, t) = u_e(x) + w(x, t)$ is almost a solution to (1) that obeys the boundary conditions in (2). Indeed, this function $u(t, x) = u_e(x) + w(t, x)$ solves the equation

$$\frac{\partial}{\partial t} u = \mu \frac{\partial^2}{\partial x^2} u + f(u_e) + \left(\frac{d}{ds} f \right)(u_e) w. \tag{9}$$

[Here, I have used the fact that u_e satisfies (4).] Check for yourself that the boundary conditions in (2) are satisfied.

Now, (9) is almost (1) as it is within $|w(x, t)|^2$ of satisfying (1) at any value of x and t. Yet since $|w|$ does not shrink with increasing time (assuming $\lambda \geq 0$), so (9) has a solution that starts very near u_e at time zero [if $|w(0, x)|$ is chosen small] and that does not move toward u_e as time increases. The fact that (9) is almost the same as (1) when $|w|$ is small allows us to prove that at least one of the possibilities in (7) will occur.

Conversely, we can show that any $u(t, x)$ that solves (1) and (2) and that starts very close to $u_e(x)$ at $t = 0$ (for all x) and does not approach $u_e(x)$ as t gets large (for all x) can be written as a sum of functions of the form $\alpha e^{\lambda t} g(x)$, where λ is a number and where g almost solves (6). And at least one of the functions in this sum for w must have nonnegative λ. This can be used to prove that there is a pair (λ, g) that actually solves all of (6). (This converse is not supposed to be obvious. It is quite hard to prove—but it is true nonetheless.)

19.3 An Example

Consider, for example, the case where f is given by (3a) and where $u_e(x)$ is the constant solution r_0/r_1. Is this constant $u_e(x)$ stable or not? According to (6), the solution $u_e(x) = r_0/r_1$ is stable if there is *no* pair of number $\lambda \geq 0$ and function $g(x)$ that solves (6). If there *is* a pair $\lambda \geq 0$ and $g(x)$ that solves (6), then $u_e(x)$ is unstable.

With the preceding understood, note that in the case where f is given by (3a), then $(\frac{d}{ds} f)(u_e(x)) = -r_1$ so (6) is

$$\lambda g = \mu \frac{d^2}{dx^2} g - r_1 g \tag{10a}$$

$$\frac{d}{dx} g = 0 \text{ at } x = 0 \text{ and at } x = L. \tag{10b}$$

Note as well that (10a) is the same as requiring that

$$\frac{d^2}{dx^2} g = \mu^{-1}(\lambda + r_1)g, \tag{11}$$

which is of the form $\frac{d^2}{dx^2} g = cg$, with $c = \mu^{-1}(\lambda + r_1)$.

Recall from Chapter 16 that solutions to the equation $\frac{d^2}{dx^2} g = cg$ have the following general form:

1. When $c > 0$, then $g(x) = \alpha e^{\sqrt{c}x} + \beta e^{-\sqrt{c}x}$.

2. When $c = 0$, then $g(x) = \alpha + \beta x$.

3. When $c < 0$, then $g(x) = \alpha \cos((-c)^{1/2}x) + \beta \sin((-c)^{1/2}x)$. $\tag{12}$

Here, α and β are constants that will now be constrained by the requirement in (10b). In this regard, there are two possibilities. The first has $r_1 > 0$, in which case $c > 0$ and so we are forced to use the first case in (12). [Remember that to test for stability, we need only worry about solutions to (10) with $\lambda \geq 0$. All solutions with $\lambda < 0$ give perturbations that evolve back to the time-independent solution as t gets large.] Thus, in the case $r_1 > 0$, all solutions to (10a) are given by (12.1). The constants α and β are determined by the boundary conditions in (10b). The condition at $x = 0$ requires

$$\sqrt{c}(\alpha - \beta) = 0, \tag{13}$$

which is solved only by taking $\alpha = \beta$. The condition at $x = L$ requires

$$\alpha \sqrt{c}(e^{\sqrt{c}L} - e^{-\sqrt{c}L}) = 0, \tag{14}$$

which is solved only by taking $\alpha = 0$. Thus, we conclude that there are no nontrivial pairs (g, λ) of function g solving (10) with $\lambda \geq 0$. This implies that the constant solution is stable in the case where $r_1 > 0$.

The other possibility has $r_1 < 0$. In this case $c = \mu^{-1}(\lambda - |r_1|)$. The case where $c > 0$ goes as before. There are no solutions to (10) that have λ bigger than $|r_1|$. However, consider the case where $\lambda = |r_1|$. In this case, $c = 0$, so (12.2) is relevant and $g(x) = \alpha$ solves both (10a) and (10b). So we conclude that the $r_1 < 0$ case does not have a stable constant solution.

19.4 Boundary Conditions

As seen in many examples in previous chapters, the behavior of solutions to (1) depends a good deal on the nature of the boundary conditions that we impose at $x = 0$ and $x = L$. Most of the examples over the past few chapters used the boundary conditions in (2). This is because these chapters considered a specific morphogenesis problem. However, there are other situations where different boundary conditions are appropriate. For example, in the no-trawling zone problem from Chapter 16, we used the boundary conditions $u(t, 0) = 0 = u(t, L)$.

This brings up an important point:

The stability criterion for a given equilibrium solution depends in part on the choice of boundary conditions.

The point is that the boundary conditions in the stability criterion must match those that are imposed on the solution u. For example,

- Suppose $u_e(x)$ is an equilibrium solution to (1) with the boundary conditions $u(t, 0) = 0$ and $u(t, L) = 0$. In this case, the stability criterion reads as follows:

  "The solution u_e is unstable when there exists a pair (λ, g) where $\lambda \geq 0$ and where $g(x)$ is a function of x that is not zero everywhere and that obeys

  $$\lambda g = \mu \frac{d^2}{dx^2} g + \left(\frac{df}{ds}\right)(u_e)g \text{ with the conditions } g(0) = g(L) = 0.$$

  If no such pair (λ, g) exists, then u_e is stable."

- By way of contrast, suppose instead that $u_e(x)$ is an equilibrium solution to (1) with the boundary conditions given by (2). In this case, the stability criterion reads as follows: "The solution u_e is unstable when there exists a pair (λ, g) where $\lambda \geq 0$ and where $g(x)$ is a function of x that is not zero everywhere and that obeys

  $$\lambda g = \mu \frac{d^2}{dx^2} g + \left(\frac{df}{ds}\right)(u_e)g \text{ with the conditions } \frac{d}{dx} g(0) = \frac{d}{dx} g(L) = 0.$$

  If no such pair (λ, g) exists, then u_e is stable."

Notice the difference in the two cases. The point is that the boundary conditions for g must match those for u.

19.5 Lessons

Here are some key points from this chapter:

- Learn the stability criterion.

- Note that (6) is an equation whose form depends on the equilibrium solution $u_e(x)$ through its appearance in $(\frac{d}{ds} f)(u_e(x))$. Take a different equilibrium solution and you will get a different version of (6a). Indeed, this is how the stability criterion accounts for the fact that some solutions will be stable and others unstable.

- Note as well that the boundary conditions in (6b) are determined by the boundary conditions that we chose in (2). If we had chosen different boundary conditions in (2), then the boundary conditions for g in (6b) would change accordingly.

Exercises for Chapter 19

In the following eight exercises, take the function f as in Equation (3b) of Chapter 19, where r_1 and r_2 are specified as follows:

1. With $r_1 = r_2 = 1$, is the solution $u_e = 1$ to Equations (4) and (5) of Chapter 19 stable or not?

2. With $r_1 = r_2 = 1$, is the solution $u_e = 0$ to Equations (4) and (5) of Chapter 19 stable or not?

3. With $r_1 = 1$ and $r_2 = -1$, is the solution $u_e = -1$ to Equations (4) and (5) of Chapter 19 stable or not?

4. With $r_1 = 1$ and $r_2 = -1$, is the solution $u_e = 0$ to Equations (4) and (5) of Chapter 19 stable or not?

5. With $r_1 = -1$ and $r_2 = 1$, is the solution $u_e = -1$ to Equations (4) and (5) of Chapter 19 stable or not?

6. With $r_1 = -1$ and $r_2 = 1$, is the solution $u_e = 0$ to Equations (4) and (5) of Chapter 19 stable or not?

7. With $r_1 = -1$ and $r_2 = -1$, is the solution $u_e = 1$ to Equations (4) and (5) of Chapter 19 stable or not?

8. With $r_1 = -1$ and $r_2 = -1$, is the solution $u_e = 0$ to Equations (4) and (5) of Chapter 19 stable or not?

9. Consider a function $u(t, x)$ that solves the equation

$$\frac{\partial}{\partial t} u = 2\frac{\partial^2}{\partial x^2} u + ru.$$

(a) For what values of r is there a nontrivial (i.e., not everywhere zero), time-independent solution to this equation with the property that $u = 0$ at $x = 0$ and $u = 0$ at $x = 1$?

(b) For what values of r is there a nontrivial, time-independent solution to this equation with the property that $u = 0$ at $x = 0$ and at $x = 10$?

(c) For what values of r is there a nontrivial, time-independent solution to this equation with the property that $u = 0$ at $x = 0$ and at $x = 20$?

Remember from Chapter 16 that solutions to the equation $\frac{\partial^2}{\partial x^2} h = ch$ have the form

$$
\begin{aligned}
&c > 0 : h = e^{\sqrt{c}x} + \beta e^{-\sqrt{c}x} \\
&c = 0 : h = \alpha + \beta x \\
&c < 0 : h = \alpha \cos((-c)^{1/2}x) + \beta \sin((-c)^{1/2}x).
\end{aligned}
\tag{15}
$$

10. Consider the same equation as in Exercise 9. We are concerned in this problem with solutions that grow in absolute value as t increases.

(a) Consider u having form $u(t, x) = e^{\sigma t} h(x)$, where σ is some constant to be determined. The equation for u determines an equation for $h(x)$. Show that the equation for h has the form

$$
\frac{d^2}{dx^2} h = ch
$$

for a suitable constant c. Determine the constant c in terms of σ and r.

(b) What are the correct boundary conditions for h if the boundary conditions for u are as follows: $\frac{\partial}{\partial x} u = 0$ at $x = 0$ and at $x = 10$?

(c) Remember that the solutions to the equation for h have the form given in Equation (15). Use the boundary conditions in Exercise 10b to constrain the allowed values of the constants σ and r. In this regard, remember that we want a solution $u(t, x)$ that satisfies the boundary conditions in Exercise 10b and does increases in absolute value at some x with increasing time.

11. Redo Exercises 10b–10c using the boundary conditions $u = 0$ at $x = 0$ and at $x = 10$.

20

Summary of Advection/Diffusion

We have been considering two basic types of equations to predict the behavior of concentrations that depend on space as well as time. The concentration at a point x at time t is the value of the function $u(t, x)$. Our two basic equations are

$$\text{The advection equation: } \frac{\partial}{\partial t}u = -c\frac{\partial}{\partial x}u + f(u) \tag{1a}$$

$$\text{The diffusion equation: } \frac{\partial}{\partial t}u = \mu\frac{\partial^2}{\partial x^2}u + f(u). \tag{1b}$$

In these equations, the constants c and μ must be determined experimentally. Also, f is a function of one variable that must also must be determined experimentally.

20.1 When to Use Advection and When to Use Diffusion

The advection equation is relevant when the particles of "stuff" move along the x-axis at a constant speed (this speed is c). For example, the spread of pollution in a flowing river can be modeled by an advection equation.

The diffusion equation is relevant when each particle of "stuff" moves in an essentially random fashion. For example, the spread of pollution in stationary water can be modeled by a diffusion equation.

318

Often, there is advection and diffusion, in which case the relevant equation is

$$\frac{\partial}{\partial t}u = \mu\frac{\partial^2}{\partial x^2}u - c\frac{\partial}{\partial x}u + f(u). \qquad (2)$$

I'll say little more about this last equation. [For an application of (2), see the article in Reading 15.3, *Past Temperatures Directly from the Greenland Ice Sheet*, by Dahl-Jensen, Mosegaard, Gundestrup, Clow, Johnsen, Hansen, and Balling, which appears in *Science*, **282** (1998) 268–271.]

20.2   Solutions and Boundary Conditions

For the advection equation, we have seen that there are many solutions. However, once the behavior of u at $t = 0$ (for all x) *or* the behavior of u at $x = 0$ (for all t) has been specified, then there is a unique solution. Roughly speaking, this solution looks like a lump of concentration that is traveling along the x-axis with speed c.

For the diffusion equation, we have seen that there are also many solutions. However, the number of solutions is limited by imposing boundary conditions. For example, if we are interested in the behavior of u for x constrained between two values, say $a \le x \le b$, and if we know that the stuff cannot pass beyond $x = a$ or $x = b$, then we impose the boundary conditions $\frac{d}{dx}u = 0$ at $x = a$ and at $x = b$ (for all t).

20.3   Equilibrium Solutions

An equilibrium solution to either the advection or the diffusion equation is a solution $u(t, x)$ that is independent of time t. Thus, $u = u_e(x)$, where u_e obeys the relevant boundary conditions and, in the advection case, u_e also obeys the equation

$$0 = -c\frac{d}{dx}u_e + f(u_e). \qquad (3)$$

In the diffusion case, u_e will obey the equation

$$0 = \mu\frac{d^2}{dx^2}u_e + f(u_e). \qquad (4)$$

20.4   Stability

In most cases, an equilibrium solution $u_e(x)$ will not be observed in nature unless it is stable. We didn't discuss stability for the advection equation since its solutions are determined uniquely by specifing the boundary conditions at $t = 0$ for all x, or at $x = 0$ for all t. In the case of the advection problem, we should consider stability against some small change in the boundary conditions. (If the boundary conditions are slightly changed, does the new solution differ by a lot or by a little from the old one? If the new solution always differs marginally from the old, then the solution to the advection equation is stable.)

For the diffusion problem, the solution u_e is stable when *every* $u(t, x)$ that starts near $u_e(x)$ for all x and that solves (1b) with the relevant boundary conditions tends to $u_e(x)$ as t gets large for all x. Stability also requires that every solution $u(t, x)$ to any slight perturbation of (1b) stays near $u_e(x)$ for all x if it starts near $u_e(x)$ for all x.

In the diffusion problem, this definition of stability is equivalent to the assertion that the solution u_e is stable if there is no $\lambda \geq 0$ and function g (not identically zero) that obeys

$$\lambda g = \frac{d^2}{dx^2} g + \left(\frac{d}{ds} f \right) (u_e(x)) g. \tag{5}$$

Also, $u_e(x) + e^{\lambda t} g(x)$ must obey the given boundary conditions for solutions $u(t, x)$ to Equation (1b). [Note that the original equilibrium solution u_e appears in the last term of (5). As for $(\frac{d}{ds} f)(u_e(x))$, here is an example: If $f(s) = s(1 - s)$, then $(\frac{d}{ds} f)(u_e(x)) = 1 - 2u_e(x)$.]

20.5 Example for Advection

The following example will consider equilibrium for the advection equation

$$\frac{\partial}{\partial t} u = -\frac{\partial}{\partial x} u - (100)^{-1} u(1 - u). \tag{6}$$

Here, we are interested in $u(t, x)$ for $x \geq 0$. Impose the boundary condition that $u(t, 0) = 1/2$. This is a model for the following scenario: We are concerned with bacteria concentrations downstream on a river from a sewage treatment plant. The river flows toward positive x at a speed of 1 meter per second. The treatment plant is at $x = 0$, where the concentration of bacteria is constantly monitored. The concentration at $x = 0$ is independent of time and equal to $1/2$. We would like to know the concentration downstream.

Consider that $u(t, x) = u_e(x) = (e^{x/100} + 1)^{-1}$ is a time independent solution to (6) where $x \geq 0$. It also satisfies the required boundary conditions at $x = 0$. Notice that $u_e(x)$ tends rapidly to zero as x gets large, and so this is a reasonable density of bacteria for along the positive x-axis. For example, $u_e(1000) \sim e^{-10}$.

20.6 Example for Diffusion

Consider the diffusion equation

$$\frac{\partial}{\partial t} u = \frac{\partial^2}{\partial x^2} u + u - u^2. \tag{7}$$

Supppose that we are interested in solutions for $-\pi/2 \leq x \leq \pi/2$ and that for boundary conditions, we are told to require that $u(t, -\pi/2) = u(t, \pi/2) = 3$.

I leave it to you to verify that $u(t, x) = u_e(x) = 3(1+\cos(x))^{-1}$ is an equilibrium solution to (7) that satisfies the boundary conditions. If you roughly graph this solution, you will see that it has its largest value (which is 3) at the boundaries, $x = \pm\pi/2$; and its smallest value (which is 3/2) in the middle, at $x = 0$.

A rectangular lake (used by anglers) of constant depth and width is stocked with trout by releasing fish at two opposite ends. The concentration of fish is monitored at these ends, and fish are released to maintain the concentration at these two ends to equal 3 fish per 100 cubic meters. You are asked to predict the concentration of fish as a function of distance from one of the ends of the lake. Do you think that (7) (with its boundary condition) is a reasonable model for this problem? Is the solution $u_e(x)$ believable?

If you believe that $u_e(x)$ might be a reasonable equilibrium distribution of fish, then you should verify that this $u_e(x)$ is stable. Remember that the solution $u_e(x)$ is stable if there are absolutely no pairs (λ, g) where $\lambda \geq 0$ is a number and where $g(x)$ is a function, not everywhere zero, that satisfies

$$\lambda g = \frac{d^2}{dx^2} g + [(\cos(x) - 5)/(1 + \cos(x))]g \qquad (8)$$

with $g(-\pi/2) = g(\pi/2) = 0$. [If g satisfies these boundary conditions, then $u(t, x) = e^{\lambda t} g(x) + u_e(x)$ will satisfy the boundary conditions for Equation (7).]

20.7   The Maximum Principle

There is a neat trick for analyzing (8) that is called the maximum principle. It doesn't work all of the time, but it is useful nonetheless. In the example at hand, I use the maximum principle to prove that the solution $u_e(x)$ is stable.

To use the trick, suppose that you have found a pair (λ, g) that solves (8) and the boundary conditions. Moreover, g can't be zero everywhere. I want to convince you that λ must be negative. To see that such is the case, I remark first that g can't be zero everywhere, so it must take a maximum and a minimum and these can't be the same. (Remember that g vanishes at $x = \pm\pi/2$ so g can't be constant without violating the condition of not vanishing everywhere.) Anyway, either the maximum of g is positive or the minimum of g is negative or both.

For the sake of argument, suppose that $g > 0$ at its maximum. At a maximum, the graph of g is concave down, so the first term on the right side of (8) is not positive. That is, $\frac{d^2}{dx^2} g \leq 0$. Also, since the cosine function is no larger than 1, and g is positive at its maximum, the second term on the right side of (8) is definitely negative—that is, the number $[(\cos(x) - 5)/(1 + \cos x)]g < 0$ at g's maximum. Thus, the right side of (8) is negative at any point x where g has a positive maximum. Thus, the left side of (8) has to be negative at any point where g has a positive maximum and this requires $\lambda < 0$.

Of course, it may be that g has no positive maxima, that is, $g \leq 0$ everywhere. Then, since g is not zero identically, g must have a negative minimum. At the point x where this happens, the graph of g is turning up and so $\frac{d^2}{dx^2} g \geq 0$ at a negative minimum. At the same point, $[(\cos(x) - 5)/(1 + \cos(x))]$ is negative and g is also negative, so their product is positive. Thus, the right side of (8) is positive where g has a negative minimum. With this last point understood, look at the left side of (8). It can be positive where g is negative only if $\lambda < 0$. Thus, any solution to (8) with g not zero identically has $\lambda < 0$ and we have proved stability.

In general, similar maximum principle arguments can be used to analyze the stability equation (5) for other choices of the function f and the corresponding equilibrium solutions. However, I should warn you that the application of the argument that I just gave has to be modified if the boundary conditions do not say for g itself to vanish on the boundaries. Also, the maximum prinicple argument is going to be much harder to make if there are points where the function $(\frac{d}{ds} f)(u_e(x))$ in (5) is positive. Indeed, the nonpositivity of the function $(\cos(x) - 5)/(1 + \cos(x))$ was a pivotal fact in the preceding example.

In any event, the key point in all applications of the maximum principle is this: If g is *any* function of x, then $\frac{d^2 g}{dx^2} \leq 0$ at g's local maxima, while $\frac{d^2 g}{dx^2} \geq 0$ at g's local minima.

20.8   Lessons

Here are some key points from this chapter:

- The advection equation is used when the particle motion is due to the motion of the ambient fluid. The diffusion equation is used when the particles move by random drift.

- If both fluid motion and random drift are important, then there is a mixed advection/diffusion equation to use.

- Unstable equilibrium solutions are generally not useful, so we should verify stability.

- Study the application of the maximum principle to the problem of verifying the stability of an equilibrium point. In this regard, it can prove to be a very efficient (though not foolproof) tool.

READINGS FOR CHAPTER 20

READING 20.1

Helping Neurons Find Their Way

Commentary: This is an article about recent research toward understanding how nerve cell axons (the nerve cell projections that make connections to other nerve cells) find the cells they should connect to. The question up for grabs here is this: How are nerve cell axons guided to their connections? The article describes the recent discovery of proteins made by nerve cells that diffuse into surrounding tissue and either attract or repel nerve cell axons. In fact, some proteins do both, depending on their concentration. There appear to be a number of short-range attracting and repelling proteins plus a number of long-range ones.

The notion of short-range and long-range can be modeled in a simple diffusion equation of the form $\frac{\partial}{\partial t}u = \mu\frac{\partial^2}{\partial x^2}u - ru$, where r is positive. The larger r is, the shorter the range. To see this, consider an equilibrium solution of this equation in an embryo where x runs from 0 to L. Suppose that the protein whose concentration is given by u is kept at a fixed concentration (say 1) at $x = 0$, and that the concentration at $x = L$ is zero. Thus, we are considering our equilibrium diffusion equation $\mu\frac{\partial^2}{\partial x^2}u - ru = 0$ for a function $u(x)$ with the boundary conditions $u(0) = 1$ and $u(L) = 0$. This equation can be rewritten to read $\frac{\partial^2}{\partial x^2}u = cu$, where $c = r/\mu$. Note that $c > 0$, so the general solution has the form $u(x) = \alpha e^{\sqrt{c}x} + \beta e^{-\sqrt{c}x}$. The boundary condition $u(0) = 1$ requires the sum $\alpha + \beta = 1$, and the boundary condition $u(L) = 0$ requires $\beta = -\alpha e^{2\sqrt{c}L}$. Plugging into this the $u(0) = 1$ condition that $\beta = 1 - \alpha$, we find that $1 = \alpha(1 - e^{2\sqrt{c}L})$; thus $\alpha = -1/(e^{2\sqrt{c}L} - 1)$. This then gives $\beta = e^{2\sqrt{c}L}/(e^{2\sqrt{c}L} - 1)$ and $u(x) = (-e^{\sqrt{c}x} + e^{\sqrt{c}(2L-x)})/(e^{2\sqrt{c}L} - 1)$. Figure 20.1 is a graph of this function using $c = r/\mu$ when r is big and when r is small.

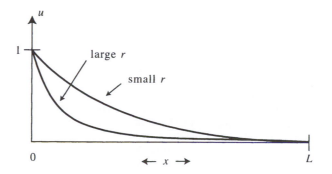

Figure 20.1

Helping Neurons Find Their Way

Science, **268** (1995) 971–973.
J. Marx

A flurry of recent papers suggests that the growing axons of neurons are guided to their targets by diffusible chemorepellents as well as by attractants.

If you've ever been reduced to despair by a last-minute Christmas Eve effort to assemble a bicycle as a gift for one of your children, take heart. Your assignment could be worse. Much worse. Try assembling the nervous system of a developing embryo. While you would have more than one frantic evening to accomplish the job, in higher organisms the task requires making billions of precise connections between nerve cells, as well as between nerve cells and muscles and other body tissues. Often the axons, the nerve cell projections that make the connections, must travel distances of up to several feet—cosmic distances for cells—through a maze of many tissues to find their targets. And, as with bicycle assembly, the consequences of failure aren't trivial. As neurobiologist Lou Reichardt, a Howard Hughes Medical Institute (HHMI) investigator at the University of California, San Francisco (UCSF), puts it, "The nervous system doesn't work unless it's wired properly."

The obvious significance of axon guidance has made understanding the system a very high priority for neurobiologists. But its complexity has made it a difficult problem to crack. Indeed, until the past few years, the maze was almost impenetrable. But a flood of recent publications from labs in Europe, Japan, and the United States shows that researchers are finally beginning to find their way through the axonal guidance labyrinth. The latest developments are reported in a remarkable flurry of eight papers appearing in the May and June issues of *Neuron* and in the 19 May issue of *Cell*.

This recent growth spurt started last year, when axonal guidance research got a big boost from the discovery by a group led by Marc Tessier-Lavigne, also an HHMI investigator at UCSF, of netrin-1. This protein is secreted by neuronal target cells into the surrounding tissue, where it attracts the correct axons. Developmental neurobiologists had been looking for such a diffusible chemoattractant for decades, but this was the first molecule that seemed to fit the bill. The new flood of papers fills out the axonal guidance story by showing that secreted chemorepellents—proteins that tell growing axons to "stay away"—are every bit as important as those that say "come here."

Researchers believe that there may be many such diffusible proteins that repel the searching axons, complementing the many secreted attractive factors that they believe exist. So far they have found three proteins with chemorepellent activity. These include, ironically, netrin-1, which apparently serves a dual purpose in axonal guidance, along with proteins known as semaphorin II and semaphorin III (or collapsin), which belong to a different family. And they have found that chemorepellent molecules appear to be secreted throughout the developing nervous system, occurring in the brain as well as the spinal cord and influencing pathfinding by several different types of neurons.

Adrian Pini of University College London, whose research 2 years ago provided the first evidence for diffusible chemorepellents, says that these latest discoveries "round off the possibilities for axonal guidance." He was referring to the fact that neurobiologists believe that there are four possible types of signals that could guide axons toward their targets: short-range repulsive and attractive cues provided by molecules on nerve cell surfaces, plus the diffusible chemoattractants and chemorepellents, which can act over longer distances.

The cell adhesion molecules involved in attraction were the first to come on the axonal guidance scene. In the early 1980s, researchers found that nerve cells carry molecules on their surfaces that can cause them to adhere to some other cells when they come into contact. The first evidence for short-range repulsion came later, toward the end of the 1980s, when three groups provided evidence that embryonic tissues carry membrane proteins that cause nerve cells to back away on contact. At the time, this finding was something of a surprise, according to axonal guidance researcher Corey Goodman, an HHMI investigator at the University of California, Berkeley. The reason: neurobiologists were "looking for 'turn-ons,' not 'turn-offs,'" he says.

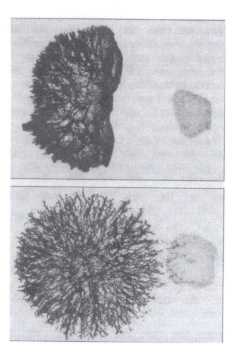

Fig. 1 Selective. Cells secreting semaphorin III repel one type of sensory neuron (top), while a second type is unaffected and can extend all the way to the cells.

Researchers soon succeeded in identifying some of the cell surface molecules involved in both short-range attraction and repulsion, and these discoveries focused attention on contact-mediated cues for guiding axons to their targets. "Molecules involved in guidance were thought to be acting through adhesion," says Jonathan Raper of the University of Pennsylvania School of Medicine, whose team was one of those that discovered that neurons may be repelled when they contact another cell.

Despite this emphasis on the short-range, contact-mediated cues, however, there was also reason to believe that axons might be guided by signals that act over much longer distances as well. Indeed, the suggestion that neuronal targets might release such guidance molecules dates all the way back to work performed a century ago by Spanish neurobiologist Santiago Ramón y Cajal. Finding these molecules was another matter, however, and neurobiologists didn't get their hands on a good candidate until last year, when Tessier-Lavigne's group came up with netrin-1 (*Science*, 28 October 1994, p. 568).

Their work showed that netrin-1 is made by the floor plate, a strip of tissue that runs along the lower (ventral) surface of the developing spinal cord, where it helps guide the axons of a particular type of spinal cord neuron: the commissural neurons, which relay information about pain and temperature to the brain. While these neurons are born in the dorsal (upper) surface of the spinal cord, their axons don't grow directly to the brain. They first project down to the ventral surface of the cord and cross the floor plate. Tessier-Lavigne's group showed that netrin-1 may help guide them along their way by acting as a diffusible positive signal to draw the commissural nerve axons to the floor plate.

One more to go

With this discovery, researchers had bagged examples of three of the four possible types of axonal guidance cues. Still missing were diffusible chemorepellents. But those wouldn't be long in coming. Pini had provided evidence for their existence in 1993, in experiments Tessier-Lavigne describes as "seminal work." In that research Pini exploited a culture system devised by Andrew Lumsden of Guy's Hospital in London and Alun Davies of St. Andrews University in Scotland that had already shown its merits by demonstrating the existence of diffusible attractants. The system involves culturing two samples of nerve tissue some distance apart within a collagen gel. Because the tissue samples aren't in contact, the only way they can influence each other is by releasing a substance that diffuses through the gel. In the experiments of Pini and his collaborators, axons of some neurons actually turned and grew away from other neurons, indicating that the others were secreting repellents.

At the time, the specific identities of the repellents weren't known. But even as the Tessier-Lavigne group was showing that netrin-1 is a chemoattractant for commissural neurons, there were also hints that this molecule might be a chemorepellent for other nerve cells. The first hint came from analysis of netrin-1's sequence, which revealed that it is the chicken equivalent of a protein from the nematode *Caenorhabditis elegans*, UNC-6, whose gene had been cloned by Ed Hedgecock of Johns Hopkins University, Joseph Culotti of Mount Sinai Hospital in Toronto, and their colleagues.

Studies of neuronal growth during development in *unc-6* mutants indicate that the protein is involved in guiding axons of two classes of neurons that migrate in opposite directions. This suggested that UNC-6 attracts the one while repelling the other. The DNA sequence similarity suggested that the same might be true for netrin-1, and Tessier-Lavigne and UCSF colleague Sophia Colamarino set out to find out if that was in fact the case.

In the 19 May issue of *Cell*, they provide the answer—which is a resounding yes. For these experiments, the researchers chose embryonic trochlear motor neurons, which help control eye movements. Unlike the axons of commissural neurons, trochlear neuron axons grow away from the floor plate, not toward it, a growth pattern that suggests the floor plate might repel the axons of the trochlear neurons. And that is exactly what Colamarino and Tessier-Lavigne found when they put trochlear neurons into the gel culture assay with floor plate tissue. To prove that the effect was due to netrin-1, the researchers transferred the netrin-1 gene into a type of nonneuronal cell that otherwise has no effect on axon guidance. The altered

cells also repelled trochlear neuron axons. "The bi-functionality [of netrin-1] mirrors the dual action of UNC-6," Tessier-Lavigne says. "In one fell swoop, we've identified a diffusible attractant and a diffusible repellent."

Netrin-1's dual guidance effects may not be limited to commissural and trochlear neurons. In the May and June issues of *Neuron*, a team led by Fujio Murakami of Osaka University in Japan describes results showing that axons of brain neurons respond to floor plate tissue much as spinal neurons do: Those that normally cross the floor plate during development are attracted by it, while those that don't cross it are repelled. "My guess is that many neurons share the [same guidance] mechanism," Murakami says. Further support for that idea comes from Sarah Guthrie of Guy's Hospital and Pini, who also report in the June issue of *Neuron* that the floor plate repels axons from certain motor neurons in the brain and spinal cord.

The Osaka team found that netrin-1 attracts the same neurons attracted by the floor plate in their studies, but haven't yet examined the protein's effects on the neurons that are repulsed. And while the London groups haven't done molecular studies to try to find out what was causing the effect they saw, the investigators predict that netrin-1 will be involved. "We haven't yet tested netrin-1 on the neurons, but we would be pretty surprised if it doesn't work," Guthrie says.

And netrin-1 isn't the only diffusible molecule that has a repelling effect on other axons. Semaphorin III/collapsin also plays that role, at least in culture. This protein has two names because it was discovered independently by two groups. Berkeley's Goodman and his colleagues originally discovered the first member of this family, semaphorin I, in the grasshopper as a cell surface protein that influences axon steering and inhibits axon branching. They then went on to identify genes for related proteins in the fruit fly and human. Two of these, semaphorins II and III, turned out to be secreted molecules, suggesting that they might be diffusible guidance molecules, although how they might function was something of a mystery.

Meanwhile, Pennsylvania's Raper and his colleagues had noticed that the growing tips of some axons from chickens collapse when they contact certain others in culture. Their assumption: The collapse of the growth cones, as the axonal tips are called, could well have been caused by some inhibitory substance on the axon they encountered. (This in fact was one of the early experiments indicating the existence of chemorepellent substances for neurons.) The Raper group set out to identify the molecule that causes the growth cone collapse, which they called "collapsin." They ultimately cloned and sequenced the gene in 1993, and its sequence revealed that collapsin is the chicken equivalent of semaphorin III, whose gene had just been cloned by the Goodman group.

In search of a role

While this work suggested that semaphorin III is a chemorepellent, the protein's exact role has been unclear. But another paper in the May issue of *Neuron* provides some answers. This work, a joint effort by the groups of Berkeley's Goodman, UCSF's Tessier-Lavigne, and Carla Shatz, also an HHMI investigator at Berkeley, shows that the protein specifically repels growing axons of one group of sensory neurons in the rat without affecting those of another. What's interesting, says Lumsden, is that the results are "thoroughly consistent" with the behavior of those neurons in the living animal.

The axons of both sets of neurons studied by the California team enter the spinal cord through the dorsal surface. After entering the cord, the axons of one set (which transmit sensory information about muscle stretch and position) proceed to the ventral region. The axons of the other group (which transmit pain and temperature information) stop in the upper dorsal region. Using the standard collagen gel culture assay, the three-way collaboration showed that semaphorin III repels the axons of this latter group but not those that normally penetrate the ventral cord. They also found that semaphorin III is made in the ventral half of the cord, putting it in just the right place to act as a "sieve" that can keep out pain and temperature neurons while letting the others in.

There is still a caveat about the roles proposed for both semaphorin III and netrin-1, however. Most of the work so far has been done with cultured neurons, and the ultimate proof of what the proteins do awaits the creation of animals in which the genes have been knocked out to see whether that produces the expected neuronal guidance defects.

In addition to the collaborative work on semaphorin III, evidence is piling up for other diffusible chemorepellents. The Goodman group, in a paper in the 19 May issue of *Cell*, presents data showing that semaphorin II has inhibitory effects on growing axons, although it apparently acts by a mechanism different from that of semaphorin III.

In this case, the evidence does come from studies in a living organism. Goodman and his colleagues genetically engineered fruit flies so that semaphorin II would be made in a muscle that does not normally express the protein. They found that this change blocked the ability of certain neurons to form synapses, the specialized connections between neurons and their targets, with the muscle. This suggests that semaphorin II acts much closer in than semaphorin III, which repels neurons before they get near the wrong place.

But these differences may only scratch the surface of the possible activities of semaphorin family members. For one thing, the family is growing by leaps and bounds. In the May issue of *Neuron*, Heinrich Betz and his colleagues at the Max-Planck-Institut für Hirnforschung in Frankfurt, Germany, report cloning the genes for four new mouse semaphorins, while Raper's team has cloned four new genes from the chicken. (The paper will be in the June issue of *Neuron*.) A search of the databases indicates that humans also have several semaphorins, Goodman says.

Although the researchers do not yet know what the new semaphorins do, they have some clues. The Betz and Raper teams have found that each semaphorin gene has a specific expression pattern, with different genes being turned on in different tissues. This suggests that each repellent helps guide a different set of neurons. "As soon as you see there is a big family [of semaphorins], you can build in a great deal of specificity with just these repellent molecules," Goodman says. Indeed, agrees Lumsden, discovery of this family of molecules "suggests this is going to be a major factor in governing the projection patterns of neurons."

What's more, because some of the semaphorin genes are expressed in tissues, such as the lung, and others may function in the immune system, both places where they might be expected not to function in neuronal guidance, the importance of the proteins may extend beyond the nervous system. "We are all focusing on growth cone guidance, but they could be functioning in a variety of other things," Goodman speculates.

As the work progresses, researchers will try

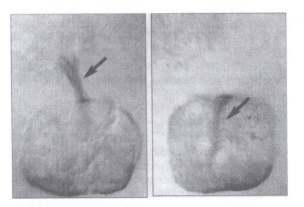

Fig. 2 Hostile territory. Trochlear motor neurons (arrows) emerging from a neural explant grow into control cells (left) but are repelled by cells secreting netrin-1 (right).

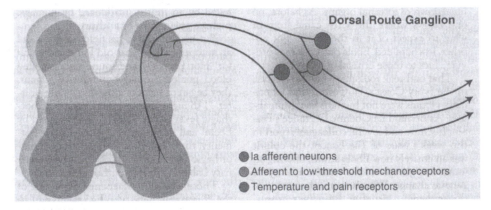

Dorsal Route Ganglion

Ia afferent neurons
Afferent to low-threshold mechanoreceptors
Temperature and pain receptors

Fig. 3 A molecular sieve? Semaphorin III produced in the ventral half of the embryonic spinal cord (dark tan) may repel the axons of temperature- and pain-sensory neurons while allowing in those of Ia afferent neurons that respond to muscle stretch.

to decipher which neurons respond to which semaphorins, netrins, and other diffusible neuronal guidance molecules. They also want to know whether any of the semaphorins do double duty as netrin-1 does. Having such a system of diffusible guidance cues that repel, as well as attract, could be very useful to the developing nervous system, Pini suggests. If guidance works on such a "push-pull system," he says, "it would be a very efficient way of getting axons from one place to another over long distances."

An equally important question concerns the nature of the machinery by which neurons respond to the guidance molecules. This is especially true,

Goodman notes, for molecules such as netrin-1 that can have opposite effects on different types of neurons. The reason for the differences must lie in the receptors and other components of the response machinery, as studies by the Culorti and Hedgecock groups already suggest. They have evidence that a protein called UNC-5 may be the receptor through which UNC-6 exerts its repellent—but not its attractive—effects on neurons.

And then there's the issue of what this growing repertoire of axonal guidance cues might mean clinically. Although the researchers are currently interested in the basic biology of the system, there has always been hope that understanding the molecules that guide neurons to their destination might lead to better treatments for people with spinal cord injuries and other nerve damage. Attractive molecules might be used, for example, to help direct regenerating neurons to their destinations. But researchers now know they will have to take the chemorepellents into consideration, too.

But despite the many mysteries remaining, the work has at least put diffusible guidance cues on a firm footing. And as Raper points out, the large number of unresolved issues has its own appeal. The work is "very exciting," he says, "because it raises a whole plethora of questions it will probably take years to answer." And, like diffusible molecules, those questions are drawing researchers over long distances toward the answers that are their targets.

READING 20.2

New Protein Appears to be Long-Sought Neural Inducer;
Neural Induction by the Secreted Polypeptide Noggin

Commentary: The first article is a nontechnical summary of the second. The latter reports on the discovery of a secreted protein (called "noggin") that appears to induce the neuronal development in frog embryos. Presumably, noggin is produced in certain embryonic cells and diffuses from them. The concentration of noggin at a given cell is conjectured to help determine the cell's fate. A high concentration of noggin appears to induce the cell to become a neural cell.

The first article illustrates the state of developmental biology circa 1993, whereby the goal was simply to determine the signaling proteins. There was little concern for diffusion equations and concentration gradients for the simple reason that the relevant diffusing proteins were as yet unidentified. The situation has not changed a great deal in the intervening years. In this regard, applications of sophisticated mathematics to embryo development problems may well be premature. There is just too much basic biochemistry to do first.

New Protein Appears to Be Long-Sought Neural Inducer

Science, **262** (1993) 653–654.
Marcia Barinaga

Some of the most critical events in our lives happen before we are even born. Take for example gastrulation, when the ball-shaped embryo buckles, and cells stream inward to form three layers which respond to a myriad of chemical signals that tell them which mature tissues to become. Most of our tissues and organs—including the brain and nervous system—have their origins in this key event. In the case of the nervous system, part of the top cell layer, called the ectoderm, follows orders to become a plate of neural tissue that later folds inward to form the spinal cord and brain. This "neural induction" was discovered more than half a century ago, and although biologists have repeatedly searched for the molecular signal that triggers it, they have always come up empty-handed.

Now, in the subsequent article, developmental biologist Richard Harland and his co-workers at the University of California, Berkeley, report finding that an embryonic protein they discovered last year, called "noggin," acts as a neural-inducing signal in frog embryos. The noggin gene has been found in rodents as well, suggesting it may perform similar functions in mammals. "It's an exciting result," says Salk Institute developmental biologist Chris Kintner. "It brings a phenomenon [neural induction] that has been known for many years to a better molecular understanding." Adds developmental biologist Jonathan Cooke of the National Institute for Medical Research in London: "Lots of people are interested in noggin. It is the only protein that has been found that is a bona fide direct neural inducer."

It's not only developmental biologists who are excited. Noggin has also been found in the brains of adult rats, raising some neuroscientists' hopes that the molecule may be a new type of chemical signal, with multiple roles in nervous system development and function. If that is the case, noggin might someday prove useful as a treatment for nerves damaged by disease or trauma, a possibility that has at least one biotech company interested. But in spite of the manifest enthusiasm, researchers caution that much more work is needed before they accept definitively that noggin is a player in either neural development or the adult brain.

Still, Harland's results already represent the solution to a mystery that began with experiments—some dating to the early decades of this century—in which researchers transplanted pieces of tissue from one developing embryo into new locations in another and watched to see what happened. These transplants revealed that the yolky bottom half of the embryo, known as the vegetal hemisphere, produces a chemical signal that tells a band of cells at the embryo's equator to become mesoderm, the tissue that streams inward at gastrulation, and later forms muscle, blood, and bones.

The transplantation work also showed that the mesoderm releases a welter of chemical signals. Even before gastrulation begins, one patch of mesoderm—called the "Spemann organizer" after its discoverer, German embryologist Hans Spemann—begins sending signals that tell part of the mesoderm to make dorsal, or back-of-the-body, structures such as the vertebral column, as well as signals that tell the adjacent part of the ectoderm to form the neural plate.

Researchers lost no time in starting to look for the molecules that determine mesoderm formation and the subsequent events, but for many years their efforts were unsuccessful. The hunt for the neural inducing factor proved particularly confusing—not because there were no results, but because the work produced an embarrassment of riches. Indeed, a long list of compounds, most of them biologically irrelevant, could trigger ectoderm to become neural

tissue in the newt and salamander embryos commonly studied at that time. Even though those experiments were done in the 1930s and '40s, they had a lasting—and chilling—effect on the field, says developmental biologist Jonathan Slack of the Imperial Cancer Research Fund at the University of Oxford. "Even when I came in, in the mid-1970s, people said neural induction is completely nonspecific, you can't do anything with it."

It wasn't until the late 1980s that researchers began finding a way out of this morass, when they began to discover molecules that figure in mesoderm induction in a different amphibian, the African clawed frog *Xenopus laevis*. But even in *Xenopus*, the sought-after neural-inducing molecule was nowhere to be found. It was that absence that drew the attention of both Harland and postdoc Bill Smith, who joined the lab in 1991. "There had been a lot of attention on mesoderm induction," says Harland, "and so when Bill came he decided to make a concentrated effort to find a neural inducer."

Smith began by searching through a collection of messenger RNAs (mRNAs) from gastrulation-stage frog embryos, to see whether any of their protein products would cause the formation of neural tissue in embryos that had been experimentally treated to prevent them from making any dorsal parts on their own. "We thought if we were lucky," says Harland, "we would get little patches of disorganized neural tissue." Instead, they found that a protein made from one pool of mRNAs would "rescue" the treated embryos, enabling them to form apparently normal dorsal structures. Smith went on to clone the gene for the protein that was responsible for the rescue, which the researchers named noggin because high doses produce embryos with exceptionally large heads.

The fact that noggin caused dorsal tissue to form in frog embryos didn't necessarily mean it was the long-sought neural inducer. Other dorsalizing factors had already been found, and some of them caused formation of neural tissue indirectly, by triggering the production of dorsal mesoderm, including the Spemann organizer; signals from the organizer then caused neural induction. To prove that noggin was different from these factors, the Harland team had to show it would induce neural tissue formation in pure ectoderm that had been removed from an embryo.

At first noggin seemed to fail this test, but graduate student Anne Knecht kept the team's hopes alive when she showed that injecting noggin mRNA into early embryos causes neural tissue to form in pure ectoderm that was removed from the developing embryo before it could receive mesodermal neural-induction signals. Subsequently the group found the reason for noggin's original strike-out: There wasn't enough noggin in the preparation used. A more concentrated preparation made by graduate student Teresa Lamb, as well as pure noggin protein provided by collaborators at Regeneron Pharmaceuticals in Tarrytown, New York, both passed the direct neural-inducing test.

Noggin passed another key test when the Harland group showed that the protein is made in the Spemann organizer, which is known to be the site of production of the neural inducer. All in all, those results make noggin a solid candidate for the neural inducing factor. "It has two things going for it," says Salk's Kintner. "It has the right biological activity and it is in the right place at the right time."

But though the body of evidence points in the direction of noggin, the proof isn't airtight. The most worrisome problem is that high concentrations of the protein—20 times higher than those required for inducing dorsal mesoderm—are needed to trigger formation of neural tissue from ectoderm. Harland suggests that such high concentrations may be necessary because of the artificial nature of the experiments: When ectoderm tissue is removed from the embryo, it tends to curl up into a ball, making it hard for noggin to penetrate. But some skeptics suggest that there's another, less pleasing explanation. Noggin "may not be the endogenous [neural-inducing] factor" after all, says developmental neurobiologist Jane Dodd of Columbia College of Physicians and Surgeons. This concern was echoed by several other researchers who suggested that at high enough concentration noggin might simply mimic the effects of the actual, and as yet undiscovered, inducer.

To rule out that possibility, noggin needs to meet yet another test. "In order to prove that a substance is really doing a particular job in vivo," offers Oxford's Slack, "if you take it away, the process has to fail." One way to do this would be to create "knockout" animals, in which the noggin gene is inactivated, to see what effect gene loss has on embryonic development. Harland is working with Andrew McMahon at Harvard University to knock out the noggin gene in mice and see how their nervous systems develop in its absence.

Even if noggin proves necessary for nervous system development, Harland points out, that doesn't mean the puzzle of early nervous system formation has been fully solved. The Berkeley group has found that neural tissue induced by noggin has characteristics suggesting it is destined to become the forward part of the brain, implying that other molecules are needed to trigger the formation of hind brain and spinal cord. And even for forebrain, noggin gets the differentiation process going, but isn't sufficient to make the tissue to go on to become full-fledged neurons.

Despite those caveats, noggin's neural inducing ability remains an exciting finding, and researchers would like to know how it acts. The protein's amino acid sequence suggests it is secreted into the extracellular fluid, and it presumably exerts its effects by binding to a receptor on target cells. But noggin bears no more than a minor resemblance to other known proteins, and, as a result, the Berkeley team still has few clues to noggin's function.

One intriguing suggestion about its mode of action comes from developmental biologist Doug Mehon of Harvard University, and is based on a discovery made recently in his lab by then-postdoc Ali Hemmati-Brivanlou, who is now at Rockefeller University. Hemmati-Brivanlou found that blocking the action of activin, a signaling protein that acts early in mesoderm induction, caused ectodermal cells to differentiate into neural tissue. "It's as if neural [development] is a default pathway," says Melton. Noggin's role, he says, may be to block the actions of activin in ectodermal cells, allowing them to express their basic "neural nature." That might be a surprise to neurobiologists, adds Melton, because traditional thinking has often assumed that the neural state is "the highest to which a cell can aspire," rather than being merely a default state.

That's unlikely to be the last surprise noggin has in store for the neuroscience community. Another could be noggin's as yet unknown role in the adult nervous system. Noggin is "expressed in discrete patterns in the adult nervous system," says George Yancopolous of Regeneron, who has been studying noggin in collaboration with Harland. "I think it's a very good bet that it's serving a variety of roles." That finding, along with the fact that noggin doesn't resemble any known families of molecules that act as signals in the nervous system, has the Regeneron team particularly excited. "The major reason we're interested in noggin is certainly the fact that noggin could be the first member of a new family [of signaling molecules]," says Yancopolous.

One thing that sets noggin apart from many other neural signaling molecules is that while many of the other signals are also found in tissues outside the nervous system, noggin is not. "Whatever its role is," Yancopolous says, "it might be quite specific to the nervous system." If that role proves to have therapeutic applications, he adds, noggin's specificity may mean it will have few side effects on non-neural tissue. In that event, noggin might not only be the solution to the 50-year embryological mystery of the brain's beginnings, but it would also be poised to make some contributions to the study and therapy of the mature nervous system as well.

Neural Induction by the Secreted Polypeptide Noggin

Science, **262** (1993) 713–718.
Teresa M. Lamb, Anne K. Knecht, William C. Smith, Scott E. Stachel, Aris N. Economides, Neil Stahl, George D. Yancopolous, Richard M. Harland

The Spemann organizer induces neural tissue from dorsal ectoderm and dorsalizes lateral and ventral mesoderm in *Xenopus*. The secreted factor noggin, which is expressed in the organizer, can mimic the dorsalizing signal of the organizer. Data are presented showing that noggin directly induces neural tissue, that it induces neural tissue in the absence of dorsal mesoderm, and that it acts at the appropriate stage to be an endogenous neural inducing signal. Noggin induces cement glands and anterior brain markers, but not hindbrain or spinal cord markers. Thus, noggin has the expression pattern and activity expected of an endogenous neural inducer.

Induction of the vertebrate nervous system is best understood in amphibians. Early experiments showed that transplants of Spemann's organizer (dorsal mesoderm) to the ventral side of a host gastrula result in twinned embryos. Although it was expected that the secondary embryo would be derived exclusively from the transplant, the organizer recruited the secondary nervous system from host tissues that would usually form skin [1, 2]. The induction of this patterned nervous system has been investigated intensively, but little is known about the molecular nature of the factors responsible for the induction [3].

In contrast to neural induction, much progress has been made in understanding how mesoderm is induced. The mesoderm (which forms notochord, muscle, heart, mesenchyme, and blood) is induced in the equatorial region of the embryo [4]. Candidates for the endogenous inducers include members of the fibroblast growth factor (FOF) family and activin [5, 6]. Members of the wnt family (a family of cyteine-rich secreted proteins originally defined by the segment polarity gene *wingless* in *Drosophila* and the murine protooncogene *int-1*) and noggin modify the kind of mesoderm induced by activin and FOF, and may be important in the formation of the dorsal mesoderm [6]. The use of dominant negative receptors for both FOF [7] and activin [8] in *Xenopus* embryos suggests that the signaling pathways activated by these molecules are essential for proper mesoderm formation.

Until recently, there were no candidate molecules for the organizer signal that instructs lateral mesoderm to become muscle [1, 9–11]. Noggin, a secreted protein lacking similarity to other known inducing factors, is expressed in the organizer, and noggin protein can dorsalize ventral mesoderm [12]. Thus, noggin appeared to be a good candidate for this signal.

Early attempts to identify neural inducing factors were not productive because the newt and salamander ectoderm was poised to become neural, and therefore many heterologous chemicals (such

Author affiliations: T. M. Lamb, A. K. Knecht, W. C. Smith, S. E. Stachel, and R. M. Harland: Department of Molecular and Cell Biology, University of California, Berkeley. A. N. Economides, N. Stahl, and G. D. Yancopolous: Regeneron Pharmaceuticals, Inc.

as methylene blue) could elicit neuronal differentiation [3]. Recently, neural induction has been studied in *Xenopus* embryos, which do not develop neural tissue so readily. Apart from isolated instances where a phorbol ester was used to induce neural tissue [3], no substances have been purified on the basis of their neural inducing activity. Activin, a mesoderm inducer, can promote formation of neural tissue in the blastula stage, but this is due to a secondary induction by the dorsal mesoderm that activin induces [13-15]. However, in the gastrula, activin is ineffective at promoting the formation of neural tissue, since the gastrula ectoderm loses competence to form mesoderm in response to activin. In contrast, the endogenous neural inducer, dorsal mesoderm, can induce neural tissue until the end of gastrulation [16].

Studies in which the inducing effects of activin and dorsal mesoderm have been compared provide two criteria for the activities of an authentic neural inducer [15]. First, the molecule should be able to induce neural tissue from animal cap ectoderm in the absence of dorsal mesoderm. If neural induction proceeds in the absence of dorsal mesoderm, it is considered a direct induction. Second, competent ectoderm should be responsive to the neural inducer at the gastrula stage, when dorsal mesoderm can still induce neural tissue [15, 16]. In addition to displaying these activities, an endogenous neural inducer must be present at the right time and place to account for normal neural development. Finally, if a factor is required for neural induction, elimination of the activity should block normal neural development.

The *noggin* gene is expressed at the right time and place to be a neural inducer. The *noggin* cDNA was cloned because its RNA is able to rescue ventralized embryos [17]. Ectopic *noggin* expression in the gastrula partially rescues ventralized embryos [12], an indication that noggin can mimic organizer signals. Zygotic *noggin* expression begins at the late blastula stage in the dorsal mesoderm and continues in the gastrula stage organizer [17]. Later, *noggin* is expressed in the organizer derivatives, the head mesoderm, and notochord; the notochord directly underlies the neural plate and has been shown to be a potent neural inducer [18, 19]. We now show that noggin activity satisfies the two criteria expected of an authentic neural inducer.

Direct neural induction by noggin. To determine whether noggin induces neural tissue directly,

we added medium containing *Xenopus* noggin [20] to blastula animal caps and assayed the expression of neural and mesoderm specific transcripts. The markers used in a ribonuclease (RNase) protection assay [21] were NCAM [22, 23], which is a cell adhesion molecule expressed throughout the nervous system [24]; an isoform of β-tubulin [25-27] expressed in the hind brain and spinal cord; a neurally expressed intermediate filament gene, XIF3 [28]; and muscle actin [29]. *Xenopus* noggin induces high levels of NCAM and XIF3 expression (Fig. 1B, lane 8), without inducing muscle actin (lane 13), while control medium fails to induce either muscle or neural tissues (lanes 7 and 12). In contrast, purified activin [30] induces muscle actin (lane 11) and all three neural markers (lane 6), demonstrating its ability to generate neural tissue indirectly. Noggin induces very little β-tubulin expression, while inducing high levels of NCAM, but activin induction has the converse effect.

Although noggin does not induce muscle in late blastula animal caps, noggin might induce other types of dorsal mesoderm. To address this, we asked whether noggin induces the expression of the early mesoderm markers goosecoid or brachyury (*Xbra*) [31-33]. Goosecoid marks organizer tissue and subsequently head mesoderm, while *Xbra* appears to be expressed in all mesodermal precursors early, and subsequently is expressed in posterior mesoderm and notochord. Animal caps were treated at stage 9 (st9) and collected at stage 11 (st 11), when expression of goosecoid and *Xbra* in the whole embryo is high [32, 33]. Noggin does not induce expression of these genes (Fig. 1C, lane 5), while the mesoderm inducer activin induces both goosecoid and *Xbra* expression (lane 4) [32, 33]. Untreated animal caps show no expression of these mesodermal markers, and the amounts of RNA in the collected animal caps are comparable, as assessed by the ubiquitously expressed marker EF-1α [34]. Thus, noggin induces neural tissue in the apparent absence of mesoderm as expected for a direct neural inducer.

Neural induction by purified noggin. To determine whether noggin protein is sufficient to induce neural tissue, COS cells were transfected with a human noggin expression plasmid, and noggin was purified to apparent homogeneity from the conditioned medium (Fig. 2) [35]. This purified noggin is capable of neural induction (Fig. 3A, see below), therefore additional factors that may have been

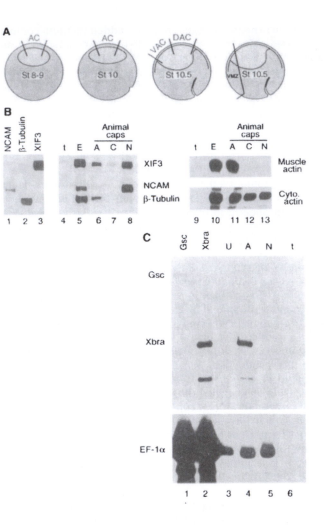

Fig. 1 (A) Experimental design competent animal cap (AC) ectoderm was dissected from staged embryos. The st10.5 ventral AC (VAC) and ventral marginal zones (VMZ) were also dissected. Explants were washed once in low Ca^{2+}, low Mg^{2+} Ringer (LCMR) solution [9] and then placed in treatment medium containing factor diluted in LCMR + 0.5 percent bovine serum albumin (BSA). Explants cultured to late stages (st20+) were removed from treatment medium 6 to 16 hours after the start of treatment and placed in LCMR. When explants reached the desired stage they were either harvested for RNA, or they were fixed for whole mount, in situ hybridization or antibody Gsc staining. (B) Neural induction by noggin in the absence of muscle. Lanes 1 to 3 show specific fragments protected by NCAM, β-tubulin, and XIF3 probes, respectively, in whole st24 embryo RNA [21]. Lanes 4 to 8 show protection by the mixture of these three probes, while lanes 9 to 13 show protection by an actin probe on tRNA(t), st24 embryo ANA Xbra (E), and RNA collected from st9 AC treated with 50 pM activin (A), 25 percent of concentrated (1:20) control CHO cell medium (C), or 25 percent of concentrated (1:20) noggin conditioned CHO cell medium (N). Ubiquitously expressed cytoskeletal actin (29) used as a loading control shows that the amounts of RNA in all treatments were comparable (lanes 11 to 13). (C) Expression of early mesoderm markers in activin but not noggin induced animal caps. Animal caps were dissected from st8 embryos, treated as described (legend to Fig 3A) and harvested at st11. Lanes 1 and 2 show goosecoid and Xboa, respectively, probe protection by st10.5 whole embryo RNA Lanes 3 to 6 show protection by a mix of these two probes. Relative amounts of RNA are demonstrated by separate EF-1a probe protection (U. untreated).

present in the crude medium are not required. Consistent with their 80 percent amino acid identity, both *Xenopus* and human noggin can induce neural tissue in *Xenopus*.

Purified noggin directly induces the expression of neural specific transcripts; however, it is possible that this is a transient induction. To address this, we treated animal caps with noggin and cultured to the late tailbud stage (st30) for antibody staining with the 6F11 antibody to NCAM, which marks the entire neural tube of a normal embryo [36, 37] (Fig. 4). Noggin-treated animal caps express this antigen, whereas untreated animal caps do not.

Neural induction at the gastrula stage. The organizer signal induces neural tissue from gastrula ectoderm. To assess the ability of noggin to induce neural tissue at different stages, we treated animal caps taken from blastula (st8), late blastula (st9), early gastrula (st10), and ventral animal caps from mid-gastrula (st10.5) embryos (Fig. 1A) with purified human noggin. Animal caps from similar stages were treated with activin medium [30] to contrast its effects with those of noggin (Fig. 3A). Noggin can induce neural tissue in animal caps taken from all of these stages without inducing either the notochord and somite marker, collagen type II [38, 39], or muscle actin (Fig. 3A). In this experiment responsiveness to noggin appeared to decline at the later stages, since there was a reduction of NCAM transcripts induced in animal caps. Upon repeating a similar experiment (40) twice, we found responsiveness to noggin at st8 and at st10.5 to be similar, indicating that there was not a loss of competence to noggin at the gastrula stage. Activin, however, promotes neural tissue formation only in conjunction with the induction of dorsal mesoderm, such as muscle and notochord. We have confirmed that the ability of activin to induce dorsal mesoderm, and consequently neural tissue [41], declines rapidly at the gastrula stage (Fig. 3A, lane 12) [13, 15]. Thus, noggin induces neural tissue in animal caps at the time of normal neural induction, a time when mesoderm inducers are inactive.

Noggin can induce neural tissue in the absence of muscle; however, in some experiments noggin when added to gastrula (but not blastula) animal caps induced neural tissue and muscle. While the animal caps come from a region of the embryo that does not normally form mesoderm, there is evidence that gastrula animal caps receive a weak mesoderm-inducing signal. By itself, the signal that

spreads into the gastrula animal cap is insufficient to induce mesoderm, but in the presence of either Xwnt-8 [42] or noggin, muscle can form. Since noggin can induce muscle from ventral mesoderm, it is not surprising that noggin added to ectoderm that has received a weak mesoderm-inducing signal also induces muscle. One interesting corollary of the induction of muscle is that the kinds of neural tissue seen in the explant are modified. Induction in explants that contain no muscle usually yields NCAM expression, but if muscle is present, expression of both NCAM and β-tubulin is seen. This phenomenon is demonstrated (i) in the secondary neural induction by activin in st9 animal caps (Fig. 1B) and (ii) in the comparison of neural tissue induced by noggin in ventral marginal zones and animal caps (see Fig. 5). In the ventral marginal zones and animal caps in which muscle is present, both NCAM and β-tubulin are expressed, whereas induced animal caps without muscle show only NCAM expression. This result suggests that the presence of mesoderm influences the type of neural tissue induced by noggin.

Neural induction after injection of DNA coding for *noggin*. To strengthen the arguments that noggin alone is the inducing activity and that noggin can induce neural tissue in gastrula animal caps, we have used an alternative experimental approach. Noggin expression was directed to gastrula stage animal caps by injecting the plasmid pCSKA-*noggin* into the animal pole of embryos at the one cell stage. In this plasmid, *noggin* is under the control of the cytoskeletal actin promoter, which turns on mRNA expression at the onset of gastrulation [12]. The animal caps were dissected at the blastula stage and then matured to tailbud stages for molecular analysis. Animal caps injected with the *noggin* plasmid expressed NCAM without expressing either muscle or notochord markers (Fig. 3B). A control plasmid directing expression of *lacZ* induced no neural or mesodermal tissue as expected. This experiment shows that ectopic *noggin* expression is sufficient to induce neural tissue and refutes the possibility that a minor contaminant in the purified preparation was the active factor. Furthermore, *noggin* expressed in this manner is active at the gastrula stage, the time of neural induction in embryos.

Dose dependence. To determine how much noggin protein is required for neural inducing activity, we did a dose response experiment. In

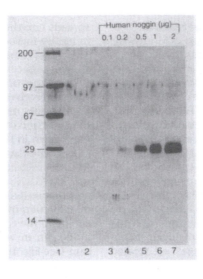

Fig. 2 Human noggin run on a 12 percent SDS-PAGE under reducing conditions. Proteins were visualized by silver staining. Lane 1 shows molecular size standards. Lanes 2 to 7 show 0. 0.1, 0.2, 0.5, 1, and 2 μg of purified human noggin.

addition to determining the doses required for neural induction in animal caps, we examined the effect of noggin dose on the dorsalization of ventral marginal zones (VMZ) [12] in order to compare the doses required for these two types of inductions. Stage 9 animal caps or st10.5 VMZs were treated with purified human noggin, and NCAM and β-tubulin were used to assay neural induction, while muscle actin was used as a marker of dorsal mesoderm. Neural but not muscle induction by noggin occurs in animal caps only at a dose of 1 μg/ml ($\sim$ 10 nM). Since activin can induce muscle at picomolar doses, the noggin dose requirement for neural induction seems quite high. Ventral marginal ones are reproducibly induced to form muscle at doses of 50 ng/ml and above. This experiment shows that neural induction in animal caps requires a dose (1 μg/ml) that is 20 times higher than that required for dorsalization of VMZ (Fig. 5).

Several observations may account for the apparently high dose requirement. First, for maximal neural induction by dorsal mesoderm, the tissues must be left in contact through most of neurulation [16]. Animal caps close up rapidly, inhibiting factor access [43] and consequently reducing the effective dose. The VMZs are much slower to close up, resulting in a longer exposure. This might account for both the large difference in dose required for the

two kinds of induction and for the high absolute dose requirement for neural induction in animal caps. Second, it is likely that noggin is not the only neural inducer active in the embryo. The somites [18, 19] and the neural plate [2, 44] have neural inducing activity and noggin transcripts are not detected there. Thus, it is plausible that noggin is only one of several neural-inducing activities. Noggin induces neural tissue in ventral marginal zones at the same doses that dorsalize them to generate muscle, whereas other experiments show that induction of a similar amount of muscle at this stage by activin does not result in neural induction. Therefore, the mesoderm present may be producing an additional factor that reduces the noggin dose requirement for neural induction, yet by itself cannot induce neural tissue. Third, it may be that only a small fraction of the purified protein is active, and the experiment results in an overestimation of the amount of protein needed for neural induction. Finally, it is possible that the accessibility of exogenously added soluble noggin is lower than that of noggin protein secreted endogenously.

Patterning. Embryonic neural tissue initially forms as a tube with no obvious anterioposterior (A-P) pattern. Subsequently, brain structures, eyes, and the spinal cord form. Formation of A-P neural pattern requires the presence of dorsal meso-

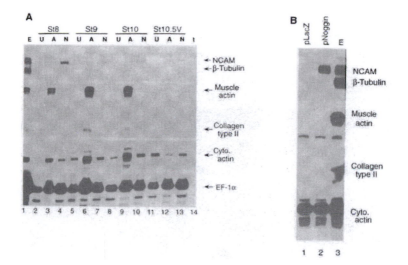

Fig. 3 (A) Staged treatments of animal caps with purified noggin and with activin; direct compared to indirect neural induction. Animal caps were dissected as shown in Fig. 1A and treated with LCMR plus 0.5 percent BSA (U), a 20 percent dilution of activin-conditioned medium [30] (A), or purified human noggin 1 μg/ml (N). RNA isolated from treated animal caps (lanes 2 to 13) along with sf22 whole embryo RNA (lane 1) and tRNA (lane 14) was probed for NCAM, β-tubulin, muscle, cytoskeletal actins, collagen type II, and EF-1α. (B) Noggin expression directed to gastrula stages by plasmid injection induced neural tissue directly. One cell stage embryos were injected with 20 pg of pCSKA*lacZ* or pCSKA*noggin* [12] into the animal pole. Animal caps from injected embryos were dissected at st8 to st9 and cultured until st20 when they were harvested for analysis by RNase protection.

derm, whether it be adjacent to the responding ectoderm in a planar configuration [23, 45–47], or directly beneath it in a vertical interaction [16, 19, 45, 48]. Both of these types of interactions occur normally, and both probably contribute to the resulting pattern [49]. Noggin produced by the dorsal mesoderm could be responsible for inducing general neural tissue, or it may also be active in patterning. Initially we observed that noggin induces cement glands. In situ hybridization [50] confirms the expression of a cement gland specific transcript, *XAG-1* [43] in noggin treated, but not control treated animal caps (Fig. 4). Since cement glands are induced organs of ectodermal origin found anterior to the neural plate, noggin may induce anterior neural structures also. To determine whether noggin induces patterned neural tissue, we used ocxA from *Xenopus* [51] as a marker of anterior brain, *En-2* [52] as a marker of the mid brain-hind brain boundary, and *Krox-20* [53] as a marker of the third and fifth rhombomeres of the hind brain. Antibodies to X1Hbox6 mark posterior hind brain and spinal cord structures [54]. Nog-

gin induces *otxA* (Fig. 4); however, we have not detected *En-2*, *Krox20*, or X1Hbox6, suggesting that these more posterior markers are not induced by noggin. Noggin does not appear to induce expression of three antigens that are characteristic of various subclasses of differentiated neural cells. These include 209 [18], which stains most neural tissue, including peripheral neurons; Tor 25.44 [55], which stains sensory neurons; and Tor 23 [55], which stains a variety of neurons, including motor neurons. Furthermore, noggin treated animal caps cells failed to grow neuronal processes when plated on an appropriate growth matrix. Thus, noggin can induce neural tissue, but it fails to cause differentiation of mature neurons, a process that presumably requires additional factors.

To conclude, we have presented two kinds of evidence that noggin protein can induce neural tissue directly. First, neural tissue is induced in the absence of induced mesoderm. Second, neural tissue is induced in gastrula stage ectoderm that has lost competence to form mesoderm, but retains competence to be neuralized. Such ectoderm, when

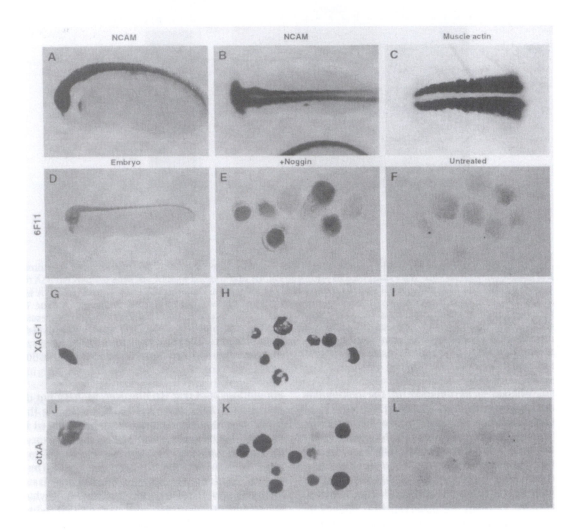

Fig. 4 In situ hybridization and antibody staining. Tailbud embryos stained for NCAM showing side and dorsal views (A and B); NCAM RNA was detected only in the neural tube and not the somites. For comparison, somites of a tailbud embryo stain with muscle actin, dorsal view (C). Neural specific 6F11 antibody staining [62] at st30 (D to F). Some cement gland pigment remained in these embryos after bleaching as seen in (D); however, this pigment is distinct from antibody staining. The inner mass of staining in the noggin-treated animal caps was due to the 6F11 antibody. Cement gland specific XAG-1 transcripts detected at st23 in whole embryos (G), human noggin treated (1 μg/ml) animal caps (17 of 30 explants were XAG positive) (H), untreated animal caps (1 of 31 explants were XAG positive) (I). Anterior brain *otxA* transcripts detected at st35 in whole embryos (J), human noggin treated (1 μg/ml) animal caps (14 of 20 explants were *otxA* positive) (K), untreated animal caps (0 of 16 explants were *otxA* positive) (L). No *en-2* ($n = 28$), *Krox20* ($n = 25$), or X1hBox6 ($n = 10$) expression was detected in noggin treated animal caps. Whole embryos have anterior to the left.

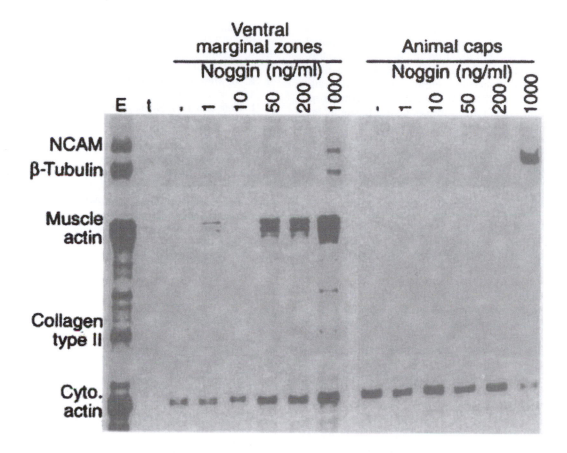

Fig. 5 Dose response of ventral marginal zones and animal caps to human noggin protein. st10.5 VMZs and st9 animal caps were dissected as shown in Fig. 1A, and treated with human noggin at 0, 1, 10. 50, 200, and 1000 μg/ml (lanes 3 to 8 and 10 to 15, respectively). RNA collected from treated explants and from control whole embryos, both aged to st26, was analyzed by RNase protection, with the probes NOAM, β-tubulin, actin, and collagen type II. In this experiment, muscle induction at the dose of 1 ng/ml was stronger than at 10 ng/ml, and there was a low level of muscle actin expression in the uninduced VMZs. In repeated experiments, muscle induction was observed only at the doses of 50 ng/ml and above.

treated with activin, can no longer form neural tissue by an indirect induction. Since noggin is a secreted protein that is expressed in the Spemann organizer and its derivatives, noggin appears to be the only factor yet described that satisfies these criteria to be an endogenous neural inducer.

The type of neural tissue induced by noggin in the absence of mesoderm appears to be of an anterior nature because we detected *otxA*, but not *En-2*, *Krox20*, X1Hbox6, or β-tubulin expression. In the presence of mesoderm, however, the nature of neural tissue induced by noggin appears to be more caudal in that β-tubulin, a marker of hind brain and spinal cord, is expressed in addition to NCAM. These results may support the idea that neural pattern arises from an initial activation or neuralization that results in specification of forebrain, followed by transformation or caudalization to produce more posterior structures [2, 49, 56]. Noggin's activity fits with a role in the initial neuralization.

Mechanism of noggin action. Dissociation of ectodermal cells for an extended period results in formation of neural tissue [57, 58]. This suggests that in normal ectoderm a signal may be distributed that prevents neuralization and promotes development of skin. Since inhibition of the activin receptor also promotes neuralization [8], this signal may be mediated in part by the activin receptor. That noggin also induces neural tissue by antagonizing the activin receptor seems unlikely because at the blastula stage activin synergizes with noggin to form dorsal mesoderm [59]. Cloning of the noggin receptor should clarify the signal transduction pathway by which noggin mediates neural induction.

Noggin has been reported to contain a conserved spacing of seven cysteines characteristic of a motif found in Kuniti class protease inhibitors [60]. The possibility that noggin acts as a protease inhibitor rather than interacting directly with a receptor is worth consideration, since dorsalventral pattern formation in *Drosophila* requires a protease cascade to initiate ventralspecific signaling [61].

We have shown that noggin has direct neural inducing activity. Noggin is made at the correct place and time to be an endogenous neural inducer. It can induce neural tissue at the gastrula stage, which is the time of endogenous neural induction. It is not yet clear that the physiological concentration of noggin is sufficient to be active in neural induction. To prove that noggin is a physiological neural inducer, it will be necessary to inhibit the noggin signal, or signaling pathway. However, noggin is found in the embryo and has the activities expected of an endogenous neural inducer. The noggin protein provides a valuable reagent to study the signal transduction pathway and early events in neural tissue formation.

REFERENCES

1. H. Spemann. *Embryonic Development and Induction* (Yale Univ. Press, New Haven, CT. 1938).
2. V. Hamburger, *The Heritage of Experimental Embryology: Hans Spemann and the Organizer* (Oxford Univ. Press, New York, 1988).
3. S. F. Gilbert and L Saxén, *Mech. Devel.* **41**, 73 (1993).
4. P. D. Nieuwkoop, *Wilhelm Roux Arch. Entwickslungmech. Org.* **162**, 341 (1969).
5. T. M. Jessell and D. A. Melton, *Cell* **68**, 257 (1992).
6. H. L. Sive, *Genes Dev.* **7**, 1 (1993).
7. E. Amaya. T. J. Musci, M. W. Kirschner. *Cell* **66**, 257 (1991).
8. A. Hemmati-Brivanlou and D. A. Melton, *Nature* **359**, 609 (1992).
9. R. M. Stewart and J. C. Gerhail, *Development* **109**, 363 (1990).
10. L. Dale and J. M. Slack, *ibid.* **100**, 279 (1987).
11. L. A. Leltice and J. M. W. Slack, *ibid.* **117**, 263 (1993).
12. W. C. Smith. A. K. Knecht, M. Wu, R. M. Harland, *Nature* **361**, 547 (1993).
13. J B Green, G Howes, K. Symes, J. Cooke, J. C. Smith, *Development* **106**, 173 (1990).
14. J. B. A. Green and J. C. Smith, *Nature* **347**, 391 (1990).
15. C. R. Kintner and J. Dodd, *Development* **113**, 1495 (1991).
16. C. R. Sharpe and J. B. Gurdon, *ibid.* **109**, 765 (1990).
17. W. C. Smith and R. M. Harland, *Cell* **70**, 829 (1992).
18. E. A. Jones and H. R. Woodland. *Development* **107**, 785 (1989).
19. A. Hemmati-Brivanlou, R. M. Stewart, R. M. Harland, *Science* **250**, 800 (1990).
20. Conditioned medium was made from Chinese hamster ovary (CHO) cells after selecting for CHO cells stably transfected with *Xenopus* noggin. Dihydrofolate reductase-deficient (dhfr^{-1}) CHO parental cells (J. Papkoff, Syntex Research) were transfected with a *Xenopus* noggin expression plasmid. Selection and amplification were carried out as described [63]. The presence of noggin transcripts was tested by RNA Northern analysis. Clone B3 secreted noggin protein, since B3 conditioned medium was capable of dorsalizing ventral marginal zones. Furthermore, labeling B3 proteins with [^{35}S]methionine revealed noggin protein as a band of ~30 kD on reducing SDS-PAGE, and a bend of 60 kD on nonreducing SDS-PAGE (indicating that it forms the expected dimer). These properties matched those of the noggin protein previously produced in *Xenopus* oocytes [12]. B3-conditioned medium was collected in a mixture of one part of α-modified Eagle medium and nine parts of serum-free medium (CHO-S-SFMII) (Gibco-BRL). The cells conditioned the medium for 3 days. Control medium from parental cells (CHO dhfr$^-$) was collected in the same way. The medium was concentrated (Centriprep-1O; Amicon) to 5 percent of its

original volume.

21. RNase protection data were obtained as described [64], except that the preparation was digested at room temperature (22°C) with RNase T1 (Calbiochem 556785) alone at 10 U/ml. Animal caps (20 to 30) were harvested for each lane, and 80 percent of this material was used for neural markers and 10 percent for muscle actin and collagen type II. For goosecoid and brachyury, 20 caps were used. Exposures ranged from 12 hours to 5 days. In all cases, films were sensitized by pre-flashing. To confirm the absence of expression of mesoderm markers, phosphor imaging, a more sensitive detection method, was used.

22. K. Balak, M. Jacobson, J. Sunshine, U. Rutishauser, *Dev. Biol.* **119**, 540 (1987).

23. C. R. Kintner and D. A. Melton, *Development* **99**, 311 (1987).

24. Although polyclonal antibodies to NCAM have been reported to show NCAM in somites and chordamesoderm [22]. NCAM RNA has not been detected outside the nervous system from gastrulation until late tailbud stages [23] (Fig. 5). NCAM RNA and protein are detected in myoblasts from about stage 36 (A. K. Knecht and R. U. Harland, unpublished data).

25. K. Richter. H. Grunz, 1.8. Dawid, *Proc. Natl. Acad. Sci. USA.* **85**, 8086 (1988).

26. P. J. Good, K. Richter, I. B. Dawid, *Dev. Biol.* **137**, 414 (1990).

27. ____, *Nucleic Acids Res.* **17**, 8000 (1989).

28. C. R. Sharpe, A. Pluck, J. B. Gurdon, *Development* **107**, 701 (1989).

29. J. B. Gurdon, S. Fairman, T. J. Mohun, S. Brennan, *Cell* **41**, 913 (1985).

30. Purified activin was obtained from J. Vaughn end W. Vale. Activin conditioned medium was made by transfection of COS cells with the plasmid pKB590 (a gift of T. Jessel, Columbia University).

31. K. W. Cho, B. Blumberg, H. Steinbeisser, E. M. De Robertis, *Cell* **67**, 1111 (1991).

32. B. Blumberg, C. V. E. Wright, E. M. De Robertis, K. W. Y. Cho, *Science* **253**, 194 (1991).

33. J. C. Smith, B. M. J. Price, J. B. A. Green, D. Weigel, B. G. Herrmann, *Cell* **67**, 79 (1991).

34. P. A. Kriag. S. M. Vamum, W. M. Wormington, D. A. Mellon, *Dev. Biol.* **133**, 93 (1989).

35. COSm5 cells were transfected with a human noggin expression plasmid (D. M. Valenzuela, E. Rojas, L. Nuñez, A. N. Economides, T. Lamb.

R. Harland, N. Stahl, G. D. Yancopoulos, unpublished data). Cells were allowed to condition DMEM (Specialty Media) for 2 to 3 days. The medium was centrifuged and passed through a 0.2-μm cellulose acetate filter. The cleared medium was pumped on to a MonoS (Pharmacia) column, which was washed with 40 mM phosphate buffer (pH 7.3), 150 mM NaCl, and 1 mM EDTA. Proteins were eluted in a linear gradient with 40 mM phosphate buffer (pH 8.5), 1.8 M NaCl, and 1 mM EDTA. Noggin elutes at 0.8 M NaCl and is ≥90 percent pure by reducing SDS-PAGE. Conditioned medium from cells transfected with human or *Xenopus* noggin expression plasmids have dorsalizing activity in VMZs at similar dilutions, suggesting that there is little difference in the specific activity of these two proteins.

36. A. Hemmati-Brivanlou and R. M. Harland, *Development* **106**, 611 (1989).

37. W. A. Harris and V. Hartenstein, *Neuron* **6**, 499 (1991).

38. J. J. Bieker and M. Yazdani-Buicky, *J. Histochem. Cytochem.* **40**, 1117 (1992).

39. E. Amaya, P. A. Stein, T. J. Musci, M. W. Kirschner, *Development* **118**, 477 (1993).

40. Animal caps at st10.5 were obtained by cutting animal caps at st8 or st9 and aging them to st10.5 in a 1:1 mixture of LCMR and Ca^{2+}- and Mg^{2+}- free medium (CMFM) [65]. This medium kept the animal caps from ciosing up, without dissociation of cells, so that factors could be applied at later stages (st10.5). Activin treatment of these animal caps confirmed the loss of competence to induce mesoderm at st10.5, indicating that animal cap aging is parallel to whole embryo aging. Noggin treatment of these aged animal caps resulted in NCAM induction as strong as that of animal caps treated at st8, demonstrating that there was no loss of responsiveness to noggin in gastrula animal caps. Control medium did not induce NCAM.

41. In the experiment shown, a high dose of activin was given; under these conditions, only a small amount of neural tissue was made, perhaps because the induced mesoderm leaves little ectoderm remaining to be neuralized.

42. S. Y. Sokol, *Development* **118**, 1335 (1993).

43. H. L. Sive, K. Hattori, H. Weintraub, *Cell* **58**, 171 (1989).

44. J. Cooke, J. C. Smith, E. J. Smith, M. Yaqoob, *Development* **101**, 893 (1987).

45. M. Servetnick and R. M. Grainger, *Dev. Biol.* **147**, 73 (1991).

46. J. E. Dixon and C. R. Kintner, *Development* **106**, 749 (1989).

47. A. Ruiz i Altaba, *ibid.* **108**, 595 (1990).

48. T. Doniach. C. R. Phillips, J. C. Gerhart, *Science* **257**, 542 (1992).

49. T. Doniach, *J. Neurobiol,* **24** (10), 1256 (1993).

50. R. M. Harland, *Methods Cell Biol.* **36**, 675 (1990). Recent batches of antibody to digoxigenin gave high backgrounds, but this was alleviated by substitution of phosphate buffer with 2 percent Boehringer Mannheim blocking reagent in 100 mM maleic acid, 150 mM NaCl, pH 7.5 for the antibody blocking and incubation steps (T. Doniach, personal communication).

51. To isolate *Xenopus otx* clones, a tadpole head cDNA library [36] was screened with a mouse *otx* cDNA (S.-L An and J. Rossant, Toronto) at low stringency. The clone used to make the probe, pXOT21.2, represents a class designated *otxA*. By in situ hybridization, transcripts are first detected before gastrulation throughout the marginal zone, but quickly become restricted to the superficial layer on the dorsal side. During neurulation a large anterior domain including both neural and nonneural tissues expresses the gene. After a decline in expression in the tailbud tadpole, the gene is again expressed specifically in the brain and eyes. At this stage there is no *otxA* expression in the mesoderm.

52. A. Hemmati-Brivanlou, J. R. de la Torre, C. Holt, R. M. Harland, *Development* **111**, 715 (1991).

53. D. G. Wilkinson, S. Bhatt, P. Chavrier, R. Bravo, P. Chamay, *Nature* **337**, 401 (1989).

54. C. V. Wright, E. A. Morita, D. J. Wilkin, E. M. De Robertis, *Development* **109**, 225 (1990).

55. P. D. Kushner, A. Hemmati-Brivanlou, R. M. Harland, unpublished data.

56. J. M. Slack and D. Tannahill, *Development* **114**, 285 (1992).

57. S. F. Godsave and J. M. Slack, *ibid.* **111**, 523 (1991).

58. H. Grunz and L. Tacke, *Cell Differ. Devel.* **28**, 211 (1989).

59. W. C. Smith, unpublished data.

60. N. Q. McDonald and P. D. Kwong, *Trends Biochem. Sci.* **18**, 208 (1993).

61. D. St. Johnston and C. Nusslein-Vollhard, *Cell* **68**, 201 (1992).

62. W. A. Harris and V. Hartenstein, *Neuron* **6**, 499 (1991).

63. F. M. Ausubel *et al.*, Eds., *Current Protocols in Molecular Biology* (Wiley-lnterscience, New York, 1989), vol. 2. p. 16.14.1.

64. D. A. Melton *et al.*, *Nucleic Acids Res.* **12**, 7035 (1984).

65. T. D. Sargent, M. Jamrich, I. B. Dawid, *Dev. Biol.* **114**, 238 (1986).

66. We thank J. Rossant and T. Jessel for sending their (unpublished) plasmids; J. Papkoff for CHO cells; W. Harris for the 6F11/XAN3 antibody; J. Vaughn and W. Vale for purified activin; J. Green for RNase protection probes and advice; T. Doniash and T. Musci's laboratory for help with in situ hybridizations and antibody to X1HBox6; R. Grainger and C. Kintner for useful discussions; and J. Green, C. Schatz. B. Meyer, and J. Erickson for suggestions about the manuscript. Supported by NIH grants (R.M.H.), an NSF predoctoral fellowship (T.M.L), Howard Hughes predoctoral fellowship (A.K.K.), and a postdoctoral fellowship from the American Cancer Society (W.C.S.).

READING 20.3

Mobility of Photosynthetic Complexes in Thylakoid Membranes

Commentary: The recent focus in developmental biology has been to sort out the relevant signaling molecules. Without a complete list of the players, it is premature to study the diffusional dynamics. Meanwhile, other areas of biology are somewhat more advanced, and the dynamic aspects are now the focus. This article studies the diffusional properties of protein complexes that are involved in cyanobacteria photosynthesis. To quote the first paragraph, "The structure of many photosynthetic pigment-protein complexes have now been determined, but a real understanding of the photosynthetic membrane at the molecular level will also require knowledge of the organization and dynamics of these complexes in the intact membrane." The article makes in-vivo measurements of the diffusion constants for certain light-harvesting complexes in thylakoid membranes of a type of cyanobacteria. The reader is specifically directed to Box 1 in the article, that exhibits an equation that should, by now, be familiar.

Mobility of Photosynthetic Complexes in Thylakoid Membranes

Nature, **390** (1997) 421–424.

Conrad W. Mullineaux, Mark J. Tobin, and Gareth R. Jones

The structures of many photosynthetic pigment-protein complexes have now been determined [1–6], but a real understanding of the photosynthetic membrane at the molecular level will also require knowledge of the organization and dynamics of these complexes in the intact membrane. Using fluorescence recovery after photobleaching (FRAP) [7] and a scanning confocal microscope, we have made direct measurements *in vivo* of the lateral diffusion of light-harvesting complexes and reaction centres in the thylakoid membranes of the cyanobacterium *Dactylococcopsis salina* [8]. We find that the phycobilisomes (the accessory light-harvesting complexes of cyanobacteria) diffuse quite rapidly, but that photosystem II is immobile on the timescale of the measurement, indicating that the linkage between phycobilisomes and photosystem II is unstable. We propose that the lateral diffusion of phycobilisomes is involved in regulation of photosynthetic light-harvesting (state 1–state 2 transitions). The mobility of the phycobilisomes may also be essential to allow the synthesis and repair of thylakoid membrane components.

FRAP has been widely used for measuring the lateral diffusion of components of eukaryotic cell membranes [9]. Typically, mammalian cells with large, flat surfaces are used. The components to be studied are labelled with a fluorescent tag, and a focused laser beam is used to bleach an area of the labelled membrane. The recovery of fluorescence in the bleached area is monitored with the same laser. This indicates the rate at which unbleached chromophores diffuse into the bleached area. A refinement of the technique uses a scanning confocal microscope: the beam is scanned over a small area of the cell surface to bleach the chromophore and then over a larger area to record a sequence of images [10].

The long-range diffusion coefficient for LHCII, the integral membrane light-harvesting complex of green plants, has been estimated from rapid membrane-fractionation studies [11]. However, there has been no direct measurement of the mobility of photosynthetic complexes in a native membrane. Photosynthetic pigment protein complexes contain fluorescent chromophores, so there is no need for an added fluorescent label in a FRAP measurement. We do not know of any photosynthetic membrane with the extended flat surface required for conventional FRAP measurements. However, the cyanobacterium *Dactylococcopsis salina* has unusually large cells (typically about 4–8 μm across and 35–80 μm long) [8]. As in many cyanobacteria, the thylakoid membranes are concentric cylinders aligned along the long axis of the cell (A. E. Walsby, personal communication). We have exploited the rotational symmetry of the cells by bleaching a plane across the short axis of the cell. This generates a one-dimensional bleaching profile along the long axis of the cell (Fig. 1). The evolution of the bleaching profile can be interpreted in terms of the one-dimensional diffusion equation (Box 1).

Cyanobacteria are prokaryotes that perform ocygenic photosynthesis. Like green plants, cyanobacteria posess the chlorophyll-containing photosys-

---

Author affiliations: Conrad W. Mullineaux: Department of Biology, University College London. Mark J. Tobin and Gareth R. Jones: CLRC Daresbury Laboratory.

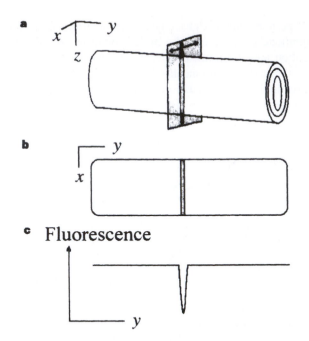

Fig. 1 The geometry of our one-dimensional FRAP measurements. a, The objective of the confocal microscope produces an illumination spot elongated in the z-direction. Scanning the spot in the x direction along the short axis of the cell causes photobleaching in the x-z plane. b, The fluorescence from the cell is imaged by scanning the confocal spot in the x-y plane. The reduced optical resolution in the direction means that the measured signal is the fluorescence from the full depth of the cell in the z direction. The image shows a photobleached line across the cell in the x direction. c, The image is integrated across the full width of the cell in the x direction to give a one-dimensional bleaching profile fluorescence versus position on the long axis of the cell.

tem I and photosystem II reaction centers [12]. The accessory light-harvesting complexes of cyanobacteria are the phycobilisomes which are aggregates of phycobiliproteins anchored to the cytoplasmic surface of the thykaloid membrane [2]. Phycobilisomes are typically hemidiscoidal and ~70 nm in diameter, with a relative molecular mass of 4,500K–15,000K (ref. 2). Phycobilisomes are not present in green plants, which use integral membrane chlorophyll-binding proteins instead [4].

FRAP was measured using a scanning confocal microscope with two lasers as alternative light sources (see Methods). The 442 nm laser excites chlorophyll a of photosystem I and photosystem II. Only photosystem II fluoresces significantly at room temperature [13], so essentially all the fluorescence detected is from photosystem II. The 633 nm laser excites the phycocyanin and allophycocyanin chromophores in phycobilisomes [2].

Fluorescence excitation and emission spectra for *Dactylococcopsis* cells (not shown) are typical for cyanobacteria. In the excitation spectrum, there is a small peak at 440 nm (chlorophyll a) and larger peaks at 580–625 nm (phycocyanin [2]). There is no indication of any unusual, highly fluorescent chromophores which might be detected in our measurements in addition to the phycobilisome components and the chlorophylls of photosystem II.

Figure 2 shows typical sequences of fluorescence images. The derived bleaching profiles are shown in Fig. 3. With phycobilisome excitation (Figs 2a and 3a), the initial bleaching profile is gaussian, with a half-width ($1/e^2$) of 2.5 μm. The bleaching is much broader than the confocal spot (about 0.3 μm): this must be due to scattering of the bleaching light by the multiple layers of membrane. The initial width and depth of the bleach varied in different measurements (Table 1). At the centre of the bleach, 70–

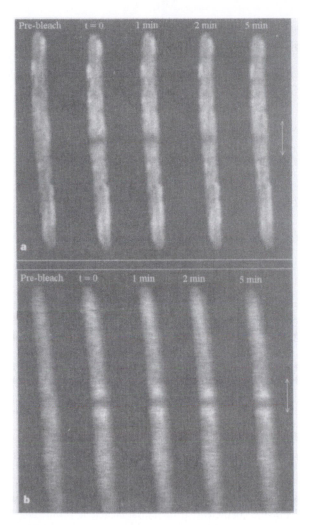

Fig. 2 Selected fluorescence images from typical sequences recorded before bleaching, immediately after bleaching, and at various times thereafter. a, Excitation at 633 nm. showing fluorescence from the phycobilisomes. b, Excitation at 442 nm, showing fluorescence from photosystem II. Scale bars. 10 μm.

80% of fluorescence was normally lost (Table 1). We are therefore measuring the behaviour of the bulk population of phycobilisomes, not a minor subpopulation or a chromophore not associated with the phycobilisomes. Within a few minutes, the bleaching profile spreads, becoming broader and shallower. This shows that phycobilisomes are diffusing, with unbleached phycobilisomes diffusing into the bleached area (Figs 2a and 3a).

With photosystem II excitation, the initial bleaching profile is complex: a central bleached zone is flanked by areas of increased fluorescence (Figs 2b and 3b). The fluorescence increase must be the result of mild exposure to the scattered bleaching laser. Light exposure sufficient to disrupt photochemistry without bleaching a significant proportion of the chlorophyll would cause a fluorescence increase. The bleaching profile evolves very little on the timescale of the measurement. At longer times (10–20 min), the bleach becomes shallower. There

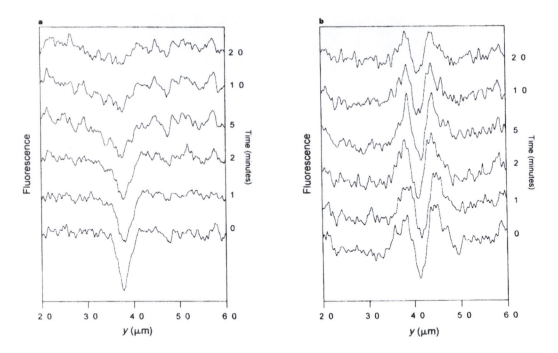

Fig. 3 Selected one-dimensional bleaching profiles derived from the sequences of fluorescence images shown in Fig. 2. a, Excitation at 633 nm (phycobilisome fluorescence). b, Excitation at 442 nm (photosystem II fluorescence).

Box 1.

The one-dimensional diffusion equation is

$$\frac{\partial C}{\partial t} = D\frac{\partial^2 C}{\partial y^2} \tag{1}$$

where t is time, y is distance along the long axis of the cell, $C_{(y,t)}$ is the concentration of of the bleached chromophore and D is the diffusion coefficient. If the initial bleaching profile is gaussian, so that: $C_{y,t=0} = C_{y=0,t=0}\exp[-2y^2/R_0^2]$, where R_0 is the half-width $(1/e^2)$ of the bleach, which is centred at $y = O$, then the solution of equation (1) (assuming that the cell is very long in comparison to the width of the bleach) is: $C_{y,t} = C_{y=0,t=0}R_0(R_0^2 + 8Dt)^{-1/2}\exp[-2y^2/(R_0^2 + 8Dt)]$. Note that the bleaching profile remains gaussian, but becomes broader and shallower with time.

At time t, the half-width $(1/e^2)$ of the bleach is given by: $R_{(t)} = [R_0^2 + 8Dt]^{1/2}$, and the peak concentration of bleached chromophore is given by $C_{y=0,t} = C_{(y=0,t=0)}R_0[R_0^2 + 8Dt]^{-1/2}$.

---

is no corresponding spread of the bleach, suggesting that in situ processes are occurring rather than diffusion. In eukaryotic cells, immobility of membrane proteins is generally ascribed to attachment to the cytoskeleton [9]. There is no evidence for a cytoskeleton capable of performing a similar role in cyanobacteria. However, photosystem II complexes in cyanobacteria are frequently arranged in long parallel rows [6, 14, 15]. Tight coupling to such structures may prevent diffusion.

The diffusion coefficient can be obtained from the evolution of the depth of the bleach (Box 1). This calculation for phycobilisomes is shown in Fig. 4. The independent calculation from the spread

Table 1. Diffusion coefficents for phycobilisomes in different cells.

Experiment	Maximum initial bleaching*	$R_0(\mu m)$	$D \ (cm^2s^{-1})$
1	0.78	2.51 ± 0.06	$(6.3 \pm 0.5) \times 10^{-11}$
2	0.80	3.94 ± 0.10	$(17.3 \pm 0.6) \times 10^{-11}$
3	0.63	3.02 ± 0.17	$(22.1 \pm 3.0) \times 10^{-11}$
4	0.75	3.49 ± 0.09	$(18.8 \pm 1.0) \times 10^{-11}$
5	0.77	3.48 ± 0.13	$(5.6 \pm 0.9) \times 10^{-11}$

In each case, the diffusion coefficient was calculated as for Fig. 4.
*Fraction of fluorescence bleached.

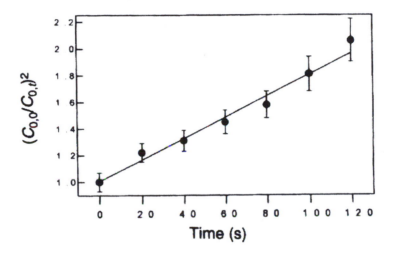

Fig. 4 Calculation of the diffusion coefficient D for phycobilisomes from the time-dependence of the maximum bleach depth, $C_{y=0,t}$ in the measurement series shown in Figs 2a and 3a. A plot of $(C_{y=0,t=0}/C_{y=0,t})^2$ against time should give a straight line with gradient $8D/R_0^2$ where R_0 is the initial half-width $(1/e)^2$ of the bleach (see Box 1). In this case, R_0^2 was $6.3 \pm 0.3 \ \mu m$; weighted linear regression gives a plot gradient of $0.0080 \pm 0.0005s^{-1}$, so the diffusion coefficient D is calculated as $(6.3 \pm 0.5) \times 10^{-11} cm^2 s^{-1}$.

of the bleach gives a similar result of $(5.4 \times 0.7) \times 10^{-11} cm^2 s^{-1}$ (not shown). Together with the linearity of the plot of Fig. 4, this indicates that our measurement does not greatly perturb the system. Phycobilisomes diffuse at similar rates in the originally bleached area and in the adjacent unbleached areas of the membrane. Experiments on different cells gave a range of values for D (Table 1). The diffusion coefficients are within the range typical for 'slow' membrane proteins, with diffusion slowed by transient binding to immobile components [9]. The rate of diffusion of phycobilisomes could be limited either by transient binding to immobile reaction

centres in the membrane, or by steric hindrance in the cytoplasm. The diffusion coefficient for phycobilisomes is considerably greater than that estimated for the integral membrane light-harvesting complex of green plants [11].

The association of phycobilisomes with photosystem II must be transient and unstable, because phycobilisomes diffuse whereas photosystem II is immobile. Efficient energy transfer from phycobilisomes to photosystem II could occur if the steady-state proportion of photosystem II-coupled phycobilisomes were sufficiently high. This is likely, given the dense packing of phycobilisomes and photosystem II on the membrane surface [16]. The phycobilisome rod elements obscure much of the cytoplasmic surface of the membrane [15, 16]. The diffusion of phycobilisomes may therefore be necessary to allow access to ribosomes and other cytoplasmic components involved in the synthesis, regulation and repair of the reaction centres. The turnover of photosystem II is unusually rapid, because the complex must be continuously repaired after photodamage [17, 18]. The diffusion of the phycobilisomes may also be essential for state 1–state 2 transitions, which regulate energy transfer from phycobilisomes to photosystem II and photosystem I over a timescale of seconds to minutes [14, 19]. State transitions have been proposed to involve the decoupling of phycobilisomes from photosystem II and binding to photosystem I [14, 19–21]. Our results suggest that the phycobilisomes are the mobile element. A calculation based on our measured diffusion coefficients indicates that the diffusion of phycobilisomes from photosystem II to photosystem I could be complete within 100 m. The rate of state transitions could be limited by the signal transduction pathway rather than the diffusion of the phycobilisomes.

Methods

Growth and preparation of cells. Dacrylococcopsis salina [8] was grown at 30°C and with moderate illumination in liquid growth medium [8]. Cells were spread on 0.6% agar containing the same growth medium, covered with a 0.2-mm quartz coverslip and placed on a stage heated to 30°C under the microscope objective. The objective was immersed in a drop of glycerol on the coverslip.

FRAP measurement. We used the scanning confocal microscope Syclops [32] at the Daresbury Laboratory. The lasers used were He-Ne (633 nm, 150 mW) or He-Cd (442 nm, 60 mW). The laser light was focused, passed through a 15-μm pinhole and focused onto the sample with a 10× objective lens. The vertical (Z) resolution was 2.6 μm (FWHM) and the resolution in the xy-plane was about 0.32 μm with 633-nm light, and 0.23 μm with 442 nm light (FWHM). The laser was scanned over the sample with oscillating mirrors. Fluorescence from the sample was separated from the excitation light with a polarizing beamsplitter and long-pass filters (Schott RG665), passed through a 40-μm pinhole and detected with a cooled photomultiplier (Hamamatsu R3896). The sample was bleached by scanning the laser repeatedly in one dimension across the cell for 20 s. The laser power was then reduced tenfold with a neutral density filter and the cell was imaged by scanning over a square of 75 × 75 μm in the xy-plane. There was no detectable photobleaching during the recording of successive image scans.

Data analysis

Fluorescence images were aligned and an image recorded before the bleach was subtracted. A one-dimensional bleaching profile was extracted (Fig. 1) using Optimas image analysis software (Optimas Corporation). To extract the width and depth of the bleach, the bleaching profile was fitted as a gaussian curve using Sigmaplot (Jandel Scientific).

REFERENCES

1. Krauss, N. *et al.* Photosystem I at 4-Å resolution represents the first structural model of a joint photosynthetic reaction centre and core antenna system. *Nature Struct. Biol.* **3**, 965-973 (1996).
2. Sidler, W. A. in *The Molecular Biology of Cyanobacteria* (ed. Bryant, D. A.) 139–216 (Kluwer Academic, Dordrecht, 1994).
3. McDermott. G. *et al.* Crystal structure of an integral membrane light-harvesting complex from photosynthetic bacteria. *Nature* **374**, 517–521 (1995).
4. Kühlbrandt, W., Wang, D. N. & Filiyoshi, Y. Atomic model of plant light-harvesting complex by electron crystallography. *Nature* **367**, 614–621 (1994).
5. Holzenburg, A., Bewley, M. C., Wilson. F. H., Nicholson. W. V. & Ford, K. C. Three-dimensional structure of photosystem II, *Nature* **363**, 470–473 (1993).
6. Boekema, E. J. *et al.* Supramolecular structure of the photosystem II complex from green plants and cyanobacteria. *Proc. Natl. Acad. Sci. USA* **92**, 175–179 (1995).
7. Thomas, J. & Webb, W. W. in *Non-invasive Techniques in Cell Biology* 129–152 (Wiley-Liss, New York, 1990).
8. Walsby, A. E., van Rijn, J. & Cohen, Y. The biology of a new gas-vacuolate cyanobacterium, *Dactylococcopsis salina* sp. nov. in Solar Lake. *Proc. R. Soc. Lond. B*, **217**, 417–447 (1983).
9. Zhang, F., Lee, G. M. & Jacobson, K. Protein lateral mobility as a reflection of membrane microstructure. *BioEssays* **15**, 579–588 (1993).

10. Kubitscheck, U., Wedekind, F. & Peters, R. Lateral diffusion measurement at high spatial resolution by scanning microphotolysis in a confocal microscope. *Biophys. J.* **67**, 948–956 (1994).

11. Drepper, F., Carlberg, I., Andersson, B. & Haehnel, W. Lateral diffusion of an integral membrane protein: Monte Carlo analysis of the migration of phosphorylated light-harvesting complex II in the thylakoid membrane. *Biochemistry* **32**, 11915–11922 (1993).

12. Barber, J. & Andersson, B. Revealing the blueprint of photosynthesis. *Nature* **370**, 31–34 (1994).

13. Holzwarth, A. R. Fluorescence lifetimes in photosynthetic systems. *Photochem. Photobiol.* **43**, 707–725 (1986).

14. Bald, D., Kruip, J. & Rogner, M. Supramolecular architecture of cyanobacterial thylakoid membranes: how is the phycobilisome connected with the photosystems? *Photosynth. Res.* **49**, 103–118 (1996).

15. Giddings, T. H., Wasmann, C. & Staehelin, L. A. Structure of the thylakoids and envelope membranes of the cyanelles of *Cyanophora paradoxa. Plant Physiol.* **71**, 409–419 (1983).

16. Mustardy, L., Cunningham. F. X. & Gantt, E. Photosynthetic membrane topography: quantitative *in situ* localisation of photosystems I and II. *Proc. Natl. Acad. Sci. USA* **89**, 10021–10025 (1992).

17. Adir, N., Shochat, S. & Ohad, I. Light-dependent D1 protein synthesis and translocation is regulated by reaction centre II. *J. Biol. Chem.* **265**, 12563–12368 (1990).

18. Allen, J. F. Protein phosphorylation in regulation of photosynthesis. *Biochem. Biophys. Acta* **1098**, 275–335 (1992).

19. Soitamo, A. J. *et al.* Overproduction of the D1:2 protein makes *Synechococcus* cells more tolerant to photoinhibition of photosystem II. *Plant Mol. Biol.* **30**, 467–478 (1996).

20. Mullineaux, C. W., Bittersmann, E., Allen, J. F. & Holzwarth, A. R. Picosecond time-resolved fluorescence emission spectra indicate decreased excitation energy transfer from the phycobilisome to photosystems II in the cyanobacterium *Synexhococcus* 6301. *Biochim. Biophys. Acta* **1015**, 231–242 (1990).

21. Mallineaux, C. W. Excitation energy transfer from phycobilisomes to photosystem I in a cyanobacterial mutant lacking photosystem II. *Biochim. Biophys. Acta* **1184**, 71–77 (1994).

22. van der Oord, C. J. R. *et al.* High-resolution confocal microscopy using synchrotron radiation. *J. Microsc.* **182**, 217–224 (1996).

Exercises for Chapter 20

1. For each of the following cases, indicate whether modeling with an advection equation or a diffusion equation is more appropriate.

(a) The concentration of snake venom in the blood after a snake bite

(b) The spread of genetically engineered mosquitos released from a research station

(c) The spread of an oil slick in the Gulf Stream

(d) The spread of an oil slick in Lake Michigan

2. In each of the following problems, you are going to use the maximum principle to try to answer the following question: Is there a pair (λ, g) where λ is a positive number and g is a function on the interval $0 \leq x \leq 1$ that vanishes at $x = 0$ and $x = 1$, is nonzero somewhere, and solves the indicated equation? In particular, either write "no" and justify your answer using the maximum principle, or write "I can't tell with the maximum principle" and explain what goes wrong.

(a) $\lambda g = \frac{\partial^2}{\partial x^2} g - (x^2 + 1)g$

(b) $\lambda g = \frac{\partial^2}{\partial x^2} g + (x^2 + 1)g$

(c) $\lambda g = \frac{\partial^2}{\partial x^2} g + (x^2 - 1)g$

(d) $\lambda g = \frac{\partial^2}{\partial x^2} g + (2x^2 - 1)g$

21

Traveling Waves

The purpose of this chapter and the next is to present an example of some surprising behavior of nonequilibrium solutions of reaction-diffusion equations. In particular, reaction-diffusion equations admit time-dependent solutions that behave just like traveling waves, and this chapter describes just such a solution.

To begin the discussion, here is a (hypothetical) biological context for the diffusion problem: Suppose that mice can be infected by a virus that is harmless to them yet virulent in people. (In this regard, you should look at the articles referenced in Commentary 21.1.) I will suppose that once a mouse is infected, it stays infected, and infected mice pass the infection to some percentage of the uninfected mice that they encounter. Furthermore, each mouse roams over some small territory more or less at random, and this is how it can meet uninfected mice.

With the preceding understood, let us imagine that at time $t = 0$, mice in coastal California are 100% infected, but that mice in eastern California and in the rest of the United States are uninfected. We expect that the viral infestation will slowly move eastward due to random interactions between infected mice and uninfected mice. The question here is the following: Can we say anything about the speed at which the infestation moves eastward? In particular, can we predict when the infestation will arrive on the east coast?

In an attempt to solve this problem, we shall adapt the simplifying model of the United States as an infinitely long strip whose width and topography are irrelevant to this discussion. See Figure 21.1.

Here, the coordinate x is assumed to be very negative in California and very

351

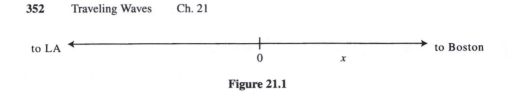

to LA

0
x

to Boston

Figure 21.1

positive on the east coast. My demotion of topography to irrelevance is open to serious debate. For example, I should first verify that mice can easily swim the Mississippi River.

Introduce a function $u(t, x)$ of time t and position x to denote the proportion of mice at time t and position x that are infected with the virus. At $t = 0$, we are told that $u(0, x)$ has the form in Figure 21.2.

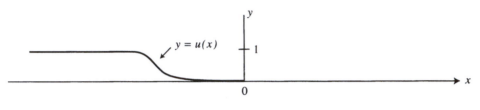

Figure 21.2

We might guess that $u(t, x)$ is controlled by a diffusion equation of the form

$$\frac{\partial}{\partial t}u = \frac{\partial^2}{\partial x^2}u + ru(1 - u), \tag{1}$$

where $r > 0$ is a number whose value can be determined by studying the rate at which laboratory mice are infected by the virus. Indeed, here is my argument for using (1): With regard to the terms on the right side of (1), I put in the diffusion term, $\frac{\partial^2}{\partial x^2}u$, to account for the random movement of infected mice. (I took the diffusion constant to equal to 1 only to simplify the notation.) I put in $ru(1 - u)$ for the reaction term because in the absence of diffusion (such as in a laboratory), I would expect that the percentage of infected mice should be governed by a standard logistics equation, $\frac{d}{dt}u = ru(1 - u)$. Indeed, remember from Chapter 3 that this last equation has the following property: Start with any nonzero fraction of infected mice and the fraction increases toward its limit, 1. Thus, as in any reaction-diffusion equation, the right side of (1) consists of the diffusion term, which takes spatial dependence into account, and the reaction term, which takes into account the underlying biology of the infection. Do you think my model is reasonable?

Equation (1) is called Fisher's equation, and there is a nice discussion of it in Beltrami's book, *Mathematics for Dynamical Modeling*. [In fact, I learned of the traveling wave solutions to (1) from Beltrami's book.]

21.1   The Traveling Wave Assumption

We might look for a solution to (1) that has the form of a traveling wave, this being a solution $u(t, x)$ that has the form $u(t, x) = f(x - ct)$ with $f(\cdot)$ a function of one variable. Here $c > 0$ is a number that is, in part, at our disposal. [Thus, $y = f(x)$ is the graph in Figure 21.2.] Notice that a function $u(t, x)$ of this form maintains the same shape over time, but the boundaries where u differs from 1 and from 0 are translated eastward (toward positive x) at a constant speed that is given by the number c. Thus, the number c is of crucial importance if we wish to know when the viral infestation reaches the east coast. (The fact that the form of u stays the same except for translation eastward at a constant speed is the justification for calling the solution a traveling wave.)

This chapter demonstrates that there are traveling wave solutions to (1). In this regard, the discussion in this section follows closely an analogous discussion in Beltrami's book, *Mathematics for Dynamical Modeling*.

If we assume that $u(t, x) = f(x - ct)$, then (1) will constrain the function f. The constraints on f can be found by using the chain rule:

$$\frac{\partial}{\partial t} f = \left(\frac{d}{ds} f\right) \frac{\partial}{\partial t} s = -c \frac{d}{ds} f$$

$$\frac{\partial}{\partial x} f = \left(\frac{d}{ds} f\right) \frac{\partial}{\partial x} s = \frac{d}{ds} f$$

$$\frac{\partial}{\partial x} f = \left(\frac{d}{ds}(\frac{\partial}{\partial x} f)\right) \frac{\partial}{\partial x} s = \frac{d^2}{ds^2} f. \tag{2}$$

Here, I have introduced a "dummy variable" s so I am thinking of f as a function of this dummy variable s, and when writing $u(t, x) = f(x - ct)$, I mean $u(t, x) = f(s)$, where s is set equal to $x - ct$. Substituting (2) into (1) shows that f must obey

$$-c \frac{d}{ds} f = \frac{d^2}{ds^2} f + rf(1 - f). \tag{3}$$

Notice that the assumed traveling wave form for $u(t, x)$ has turned the two-variable equation for u into a one-variable equation for the function f. This is a key simplification.

With regard to solutions to (3), it is important to note that we are looking for a solution that obeys the following auxiliary conditions:

(a) $0 \le f \le 1$.

(b) $f(s) \to 1$ as s gets very negative.

(c) $f(s) \to 0$ as s gets very positive. $\tag{4}$

The first condition is necessary to interpret $u(t, x) = f(x - ct)$ as a fraction of mice with the virus. The second condition models the known initial constraint that the mice in coastal California are already infected. The third constraint models the known initial constraint that the mice on the east coast are uninfected at $t = 0$. Ideally, we want to find a solution that has the form of Figure 21.2 at $t = 0$.

21.2   The First-Order Form

The first question we will ask is whether or not (3) has a solution that obeys (4). If the answer is no, then we have learned that our traveling wave assumption is unrealistic.

It proves useful to rewrite (3) as a pair of equations, each with one derivative. In fact, (3) is equivalent to

$$\frac{d}{ds}f = p$$

$$\frac{d}{ds}p = -cp - rf(1 - f). \qquad (5)$$

Indeed, the first equation in (5) simply renames $\frac{d}{ds}f$, and with this new name, the second equation in (5) is the same as (3). I rewrite (3) in this way so that I can use material from Chapters 5–12 in the subsequent analysis.

In particular, we learned in Chapter 6 that phase plane analysis provides a powerful tool for studying equations such as (5). Figure 21.3 is a phase plane diagram for the equations (5).

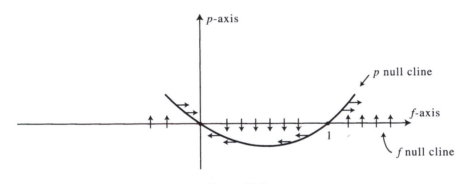

Figure 21.3

Notice that there are two equilibrium points,

$$f = 0, p = 0 \quad \text{and} f = 1, p = 0. \qquad (6)$$

The stability matrix at the first equilibrium point is

$$\mathbf{A} = \begin{pmatrix} 0 & 1 \\ -r & -c \end{pmatrix}. \qquad (7)$$

The determinant of this matrix is equal to r and the trace is equal to $-c$, so the first equilibrium point in Figure 21.2 is stable.

The stability matrix at the second equilibrium point is given by (7) except that r replaces $-r$ in the lower left-hand entry. Thus, this one has determinant $-r$, which means that the second equilibrium point is unstable.

21.3   Our Solution in the Phase Plane

We are interested in a solution to (5) whose trajectory in the phase plane, depicted in Figure 21.3, tends to the origin as s gets large and positive, and tends to the unstable equilibrium point $\left(\begin{smallmatrix}1\\0\end{smallmatrix}\right)$ as s gets very negative. I claim that when $c^2 > 4r$, there is such a trajectory in the lower half of the phase plane (that is, where $p < 0$).

The argument has two parts to it. The first part argues that there is a trajectory in the $p < 0$ part of the plane that tends to $\left(\begin{smallmatrix}1\\0\end{smallmatrix}\right)$ as s gets very negative and that, for moderately negative values of s, has $p < 0$ and $f < 1$. Part 2 of the argument shows that when $c^2 \geq 4r$, then the triangle such as that pictured in Figure 21.4 is a basin of attraction for (5), and that any trajectory in this basin must tend to $\left(\begin{smallmatrix}0\\0\end{smallmatrix}\right)$ as s gets large. (A ***basin of attraction***, also called a ***trapping region***, is a region in the plain with the following property: Once a trajectory enters this region, then it can never escape. Think of a basin of attraction as a region where no trajectories ever leave.)

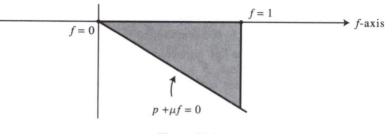

Figure 21.4

Both Parts 1 and 2 make use of the following observations:

Observation 1: On the line segment where $p = 0$ and $0 < f < 1$, $\frac{d}{ds}p < 0$, so every trajectory that crosses this line segment moves from above to below.

Observation 2: On the half-line where $p < 0$ and $f = 1$, $\frac{d}{ds}f < 0$, so every trajectory that crosses this half-line moves from right to left.

21.4   Trajectories that Leave $\left(\begin{smallmatrix}1\\0\end{smallmatrix}\right)$

Draw a small quarter-circle around the equilibrium point $\left(\begin{smallmatrix}1\\0\end{smallmatrix}\right)$ as pictured in Figure 21.5.

Using Observation 1, we find that trajectories that hit the upper end of the quarter-circle must pass above the equilibrium point $\left(\begin{smallmatrix}1\\0\end{smallmatrix}\right)$ as s gets very negative. Using Observation 2, we find that trajectories that hit the lower end of the quarter-circle must pass below the equilibrium $\left(\begin{smallmatrix}1\\0\end{smallmatrix}\right)$ point as s gets very negative. Thus, as we move from the upper end to the lower end of the quarter-circle, we must pass through at least one point that lies on a trajectory that tends precisely toward $\left(\begin{smallmatrix}1\\0\end{smallmatrix}\right)$ as s gets very negative.

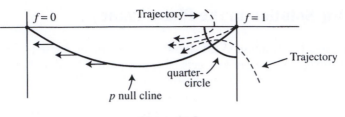

Figure 21.5

21.5   The Triangle as a Basin of Attraction

We have seen in Observation 1 that trajectories cannot exit the triangle in Figure 21.4 through its top. We have seen in Observation 2 that trajectories cannot exit the triangle in Figure 21.5 from its right side. If a trajectory is to exit the triangle in Figure 21.4, then this must happen through its hypotenuse. I will now demonstrate that when $c^2 > 4r$, no trajectory exits through the hypotenuse either.

Indeed, the hypotenuse is the line segment where $p + cf/2 = 0$ with $0 < f \leq 1$. To the right of this line segment, $p + cf/2 > 0$. To the left, $p + cf/2 < 0$. Thus, if a trajectory crosses this line segment from left to right at some $s = s_0$, then $p(s) + cf(s)/2$ must change from being negative (for $s < s_0$) to being positive (for $s > s_0$). That is,

$$\frac{d}{ds}(p(s) + cf(s)/2) > 0 \tag{8a}$$

at $s = s_0$ such that

$$p(s_0) + cf(s_0)/2 = 0 \quad \text{and} \quad 0 < f(s_0) \leq 1. \tag{8b}$$

If (8a,b) always holds, then we have demonstrated that no trajectory can exit the triangle through its hypotenuse side.

The derivative in (8a) can be computed (in a sense) using (5). Substitute (5) into (8a) to conclude that a trajectory on the hypotenuse of the triangle at $s = s_0$ has

$$\frac{d}{ds}(p + cf/2) = -cp(s_0) - rf(s_0)(1 - f(s_0)) + cp(s_0)/2. \tag{9}$$

Now use (8b) to replace $p(s_0)$ with $-cf(s_0)/2$ in (9) and thus rewrite (9) as

$$\frac{d}{ds}(p + cf/2) = c^2 f(s_0)/4 - rf(s_0) + rf(s_0)^2 = f(s_0)(c^2/4 - r + rf(s_0)). \tag{10}$$

Since $f(s_0)$ is assumed here to be positive, the far right-hand expression in (10) is evidently positive if $c^2/4 > r$. This is precisely what we wished to demonstrate.

21.6   Trajectories that Tend Toward $\left(\begin{smallmatrix} 0 \\ 0 \end{smallmatrix}\right)$

We have seen that when $c^2 > 4r$, then the triangle in Figure 21.4 is a basin of attraction. We have also seen that there is a trajectory in this triangle that tends to the equilibrium point $\left(\begin{smallmatrix} 1 \\ 0 \end{smallmatrix}\right)$ as s gets very negative. As s gets very positive, this trajectory cannot leave the triangle, and it is not hard to see that this trajectory tends to $\left(\begin{smallmatrix} 0 \\ 0 \end{smallmatrix}\right)$ as s gets very positive. (*Hint:* To prove that this must be the case, note that every triangle of the form in Figure 21.6 that is congruent to the triangle in Figure 21.4 is also a basin of attraction when $c^2 > 4r$.)

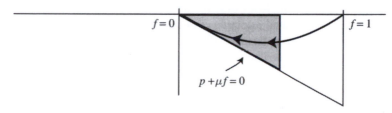

Figure 21.6

Thus, we have shown that (5) has a trajectory that tends to the unstable equilibrium point (where $f = 1$) when s gets very negative and that tends to the stable equilibrium point (where $f = 0$) as s gets very large. The function $f(s)$ for this trajectory gives our traveling wave solution to (1) by setting $u(t, x)$ to equal $f(x - ct)$.

Question: Does it concern you that we have found solutions for any value of the speed c as long as $c^2 > 4r$? That is, there are traveling wave solutions with speeds that are arbitrarily large. What are the implications of this discovery for the mice/virus problem? Do you believe the resulting prediction for the mice/virus problem?

21.7   Lessons

Here are some key points from this chapter:

- The traveling wave form for a solution u to a reaction-diffusion equation has $u(t, x) = f(x - ct)$, where c is a constant and where f is a function of just one variable.

- Reaction-diffusion equations can have time dependent solutions that look like traveling waves.

- The assumption of the traveling wave form for u turns the differential equation with two variables into one with just a single variable.

- Phase plane analysis can be used to analyze the behavior of solutions to a differential equation with one variable, one unknown, and terms with two or fewer derivatives.

- A basin of attraction in the phase plane is a region where trajectories enter only; they cannot leave.

READINGS FOR CHAPTER 21

READING 21.1

Hantavirus Outbreak Yields to PCR;
US braces for Hantavirus Outbreak

Commentary: The first article tells the remarkable story of the discovery of the disease agent for a cluster of fatal respiratory infections that struck the Four Corners area of the American Southwest in the early 1990s. The cause was a virus (named the hantavirus) that is endemic in the local mice population. Evidently, the virus spread from mice to people.

The second article is a short report that predicts that 1998 should have been a troublesome year for the hantavirus infection inasmuch as the mouse population in the Southwest had recently boomed and as much as 30% of the local mice carried the virus.

Chapter 21 used the hantavirus outbreak as background for an introduction to the traveling wave aspects of the diffusion equation. Of course, the real world is not as simple as the model in Chapter 21. First, as the article notes, hantavirus has been found in mice as far east as Maryland. Presumably, the virus is ubiquitous in mice in the United States. Moreover, were there a virus that infected mice living only west of the Continental Divide and were you to ask for the speed with which the infection spreads east, the correct answer would almost surely be 65 miles per hour. Indeed, it is but the work of a moment for a mouse to sneak on board a truck or train and thus make Boston in a few days.

Hantavirus Outbreak Yields to PCR

Science, **262** (1993) 832–836.
Eliot Marshall

Researchers have yet to isolate the virus that has killed 26 people in the United States this year, but they have its genes, they know where it hides, and they are working out its modus operandi.

Fall weather is bringing chilly nights and sparkling, crisp days to the southwestern United States. It may be a relief from the blistering summer, but for Gary Simpson, New Mexico's chief infectious disease officer, the change of seasons is accompanied by a nagging concern. The cold is driving small rodents indoors, and Simpson fears that, as the animals hunker down next to people, they could trigger a second outbreak of a rodent-borne hantavirus disease that spread panic throughout the state earlier this year. New cases, if they appear this fall, would come smack in the middle of flu season, when it would be hard to distinguish patients who need critical care from those who should be sent home with aspirin. "We really have no idea" what to expect this autumn, says Simpson, since the disease was unknown until a few months ago.

As Simpson makes contingency plans, researchers around the country are busy taking stock of the virus itself. The fact that they have a culprit to study at all is the result of an extraordinary bit of detective work—as James Hughes, Stuart Nichol, and their colleagues at the Centers for Disease Control and Prevention (CDC) in Atlanta outline in a perspective and report in this issue (see pages 850 and 914). Just 30 days after the first reported death from a mysterious pulmonary disease in New Mexico, CDC had identified the perpetrator. Using polymerase chain reaction (PCR) to amplify viral genes from victims' tissue, Nichol's team pinned the blame on a previously unknown strain of hantavirus—a member of a family of viruses long known in Asia and Europe where they have been associated with hemorrhagic fevers and renal disease (see page 835).

Researchers at CDC and the U.S. Army Medical Research Institute of Infectious Diseases (US-AMRIID) in Frederick, Maryland, are now furiously trying to culture the virus in cells but, like other members of its family, it is proving hard to isolate. Once the organism is grown in vitro, researchers will be better able to study its modus operandi, develop cheap diagnostic tests, and work on therapies and vaccines. But even without a "tame" virus, CDC researchers, in an impressive demonstration of the power of modern genetic techniques, have extracted much of the organism's genome from infected tissue (see box, p. 361). Using immunological tests and these genetic data, they have developed probes and shown that the organism is carried primarily by deer mice, whose habitat includes most of North America. These probes have also enabled CDC to link 42 cases and 26 deaths to the virus. Most were in the Southwest, where an explosion in the deer mouse population early this year appears to have been a key factor, but some have been found as far afield as North Dakota and California.

Like other hantaviruses, which are named after a prototype identified near Korea's Hantaan River in the 1950s, this virus does not appear to be passed directly from one person to another. But the new strain—initially dubbed the Four Corners strain* after the region of New Mexico, Utah, Arizona, and Colorado, where the first cases occurred— differs from other known hantaviruses in three key respects: Its effects are more rapid and more lethal—apparently killing about two-thirds of those it infects compared to 5% to 20% for the Hantaan strain—and it destroys the lungs rather than the kidneys.

The big question researchers and public officials like Simpson are now grappling with is: Just how serious a threat does this virus pose? Researchers are now using PCR to scour old tissue samples for clues to how long the virus has been in the deer mouse population and whether it is genetically stable. Few virologists are yet willing to stick their necks out, but many agree with CDC epidemiologist James Childs, who says, "I think we're looking at a very well-established virus-rodent relationship in which the virus has been a parasite of these mice for a very long time." If he's right, the organism has had plenty of time to cause human disease, yet major outbreaks have not been observed.

The virus may, however, cause occasional deaths that are lost in the background noise. They could easily get lost: Depending on whether you prefer CDC or the National Institutes of Health (NIH) estimates, the number of people in the United States who die each year of unexplained respiratory disease is somewhere between 50,000 and 150,000. "I think it's a fairly rare illness, but when it occurs the results can be devastating," says Jay Butler, a CDC medical epidemiologist.

Yet the behavior of any organism such as this, notes former CDC virologist Karl Johnson, is hard to predict. Johnson is a senior figure in hantavirus research, part of the team that isolated Hantaan. He considers the Four Corners strain a classic "emerging virus" (*Science*, 6 August, p. 680). If deer mice are chronically infected with it, he says, "I would be very surprised if we don't find it everywhere" we find deer mice. Does that mean the disease will expand? Not necessarily, Johnson says. Perhaps the virus has made only a fleeting appearance as it "boiled up from the bottom of the cauldron" of pathogens existing in the wild. It is too soon to tell whether it will sink back down or resurface in regular cycles.

Early warning

One researcher who has long been convinced that hantaviruses are endemic in the United States and that they might be causing disease is Richard Yanagihara, a virologist and longtime hantavirus hunter in the lab of D. Carleton Gajdusek at NIH. His conviction stems from Gajdusek's four decades of studying Asian strains of hantavirus—work that

---

*Tradition calls for naming a new virus after the place of discovery, but to avoid offending local people, CDC is now calling this one Pulmonary Syndrome Hantavirus.

provided some of the basis for the rapid identification of the Four Corners strain.

The Asian form of the virus, as Gajdusek pointed out in the early 1960s, is related to other viruses in Russia and Scandinavia (the Puumala group), which are carried by voles and cause a mild kidney disease. Although there had been no reports of such disease in the United States, Gajdusek and Yanagihara thought it likely that North American rodents carried a Hantaan-like virus, and they set out to find it.

Yanagihara recalls how he, Gajdusek, and 10 of Gajdusek's adopted children from New Guinea set out small rodent traps on Gajdusek's property, called Prospect Hill, in Frederick, Maryland, in the early 1980s. They tested blood from voles, noting a positive reaction with Hantaan and Puumala virus. In 1982, they announced the first domestic hantavirus—the Prospect Hill strain. But they couldn't link it to any human disease.

Yanagihara continued looking. In 1983, he bled 203 mammalogists at their convention in Florida, and found that four had an antibody response to Prospect Hill strain. That led Yanagihara and his colleagues to warn in 1984 that some types of hemorrhagic virus "may go unnoticed in the United States" and that "persons having frequent contact with wild rodents ... might constitute a high-risk group." They called for further studies on the subject.

Scientists did do more studies, financed by an outfit with pockets deep enough and a mission broad enough to cover such an arcane topic—USAMRIID. Childs, then at Johns Hopkins University, was among the most active grantees. He surveyed the rats of Baltimore and found that many were infected with a variant of the Seoul virus. Rats in other cities tested positive, too, and after scouring public health clinics, Childs has found three mild cases of hemorrhagic disease.

At the cellular level, USAMRIID virologist Connie Schmaljohn made an important contribution at this time: In 1986 she was the first to sequence Hantaan's genetic code. She also sequenced part of Prospect Hill, and Mark Parrington and Yang Kang at the University of Ottawa obtained the full sequence in 1989. CDC researchers credit this work as being a key to the quick identification of the Four Corners strain as a new hantavirus.

Ironically, USAMRIID has recently been the focus of public suspicion rather than approbation for its hantavirus research, thanks to its former involvement in classified biological defense work. When the new hantavirus appeared this year in the Southwest, some people suspected it might have escaped from the Army's labs. The speculation intensified when it became clear that the virus whose genetic structure most closely resembled that of the lethal southwestern version was the Prospect Hill strain, discovered practically in USAMRIID's back yard in Frederick, Maryland. *Scientific American* reported on the speculation this month in an article headlined: "Were Four Corners Victims Biowar Casualties?" Those who know the virus well say such speculation is wrong-headed. "Totally ridiculous," scoffs Yanagihara. "It disturbs me a great deal that people have this Andromeda strain mentality ... Even some of my scientific colleagues seem to think a monstrous mutational event" was needed to make the virus pathogenic.

Yanagihara doesn't think any monstrous mutation was needed to make the Four Corners strain of hantavirus deadly. In fact, he thinks the virus has been around for a long time in pathogenic form. His theory will be put to the test over the next few months, as researchers begin to recheck blood and tissue samples they collected and filed away long ago. For example, they will tap into decade-old deer mouse collections in New Mexico and Pennsylvania. Yanagihara himself has already reanalyzed samples of lung tissue from deer mice trapped in the early 1980s from Mammoth Lakes, California, near the spot where a young field biologist contracted fatal respiratory disease this year. In a recent letter to Lancer, Yanagihara reports finding hantaviral genes in these samples, and he believes that this 10-year-old virus will turn out to be essentially the same as the one that caused deaths in the Four Corners area.

Attacking the virus

Now that the characteristics of U.S. hantaviral disease are coming into focus, a key task is to find the means of combating it. The virus's attack is insidious. At first, it causes ordinary flu-like aches and pains. But within a few days it begins to wreak havoc in the lungs, causing capillaries to leak. After about a week, says Frederick Kosrer, a physician who saw 12 victims last summer at the University of New Mexico Hospital in Albuquerque, people enter a "crisis phase." Over a span of hours, they find it harder and harder to breathe. He recalls one

Virology Without a Virus

"We didn't know what we had on our hands when we started getting samples in," says Stuart Nichol, a virologist at the Centers for Disease Control and Prevention (CDC) in Atlanta. "All we knew was that people were dying of something." That was on 21 May. Just a week earlier, an observant physician in New Mexico, Bruce Tempest, had alerted state officials about some unusual deaths. A young Navajo long-distance runner had died in the clinic of sudden pulmonary failure after attending the funeral of his fiancée. The fiancée had died similarly a few days earlier. No one could determine why they had died.

Tempest began calling around and quickly identified three more cases that looked similarly suspicious. He thought it might be plague, but tests showed that the plague bacterium wasn't the cause. Nor could New Mexico officials identify a cause. They sent blood and tissue specimens to CDC, where researchers discussed strategies for a week, then began processing samples on 31 May in the high-security containment labs. CDC virologists Thomas Ksiazek and Pierre Rollin ran a battery of tests on blood samples, looking for antibodies to known pathogens. The results were negative, except for signals for the Puumala hantavirus—which causes a relatively mild disease in Europe that affects the kidneys. CDC scientists then were at a critical juncture, says CDC medical epidemiologist Jay Butler: "We asked ourselves, Do we try to focus in on hantavirus, or is that a red herring?"

That weak signal, however, turned out to be anything but a red herring. Relying on research done by the Army, the National Institutes of Health, and by CDC itself, CDC pathologist Sherif Zaki probed the tissue samples with monoclonal antibodies that react with other hantaviruses. At the same time, Nichol synthesized pieces of shared gene sequences of two known hantaviruses—Prospect Hill and Puumala—to use as "primers" for polymerase chain reaction (PCR). His aim was to amplify traces of hantaviral genes that might be present in tissues of disease victims. Using molecular and immunological techniques, by 9 June, CDC proved that a new hantavirus was killing people in the Southwest.

Nichol used the PCR process to pull out information from tissue samples about the unique genetic structure of the virus. As more of the sequence came to light, Nichol incorporated it into targeted PCR primers, enabling him eventually to get pieces of the new virus's genome.

Nichol's group inserted the new genetic material into bacteria, causing them to produce viral proteins that could be used as probes for antibodies specific to this new virus. After exposing disease tissues to staining antibodies, Zaki began looking for signs of infection. He found that lung tissues were full of antigens, especially along the capillary walls. He and other CDC researchers conclude that the virus targets endothelial cells. While Asian strains of hantavirus seem to focus on endothelial cells in the kidney, the U.S. strain goes to the lungs.

Scientifically, the identity of a virus usually remains in limbo until it has been grown in the lab and its effects replicated. But this hantavirus, like other members of the hantavirus family, has proven uncooperative, and it has not yet been isolated in cell cultures. Even in the absence of a cultured virus, however, the PCR method has firmly established the identity of this virus. One remarkable example of its specificity came to light as CDC was reviewing the case of a man who died in Snowflake, Arizona. When Nichol studied the genetic sequence of the virus that killed this man, he noticed that it was slightly different from others in Arizona and more like sequences from Colorado. Did CDC have the wrong address for the patient? A second check revealed that the victim had been in Arizona only 2 weeks, and that he lived in Colorado. The lab then tested a deer mouse caught near his home in Colorado. It was infected with hantavirus. Even more striking, the viral sequence from the mouse was identical to the sequence obtained from the man. In the age of PCR, a lot of virology can clearly be conducted without a virus. —E. M.

12-year-old boy who came to the clinic with what looked like ordinary flu, then went home. His father brought him back again, and within hours, the boy stopped breathing. Even if assisted by a ventilator, many die from blood loss.

It's a frightening syndrome, made worse because researchers don't know how to stop the infection once it begins. The CDC has obtained permission from the Food and Drug Administration for "open label" use of the broad anti-viral agent ribavirin (a nucleoside). It amounts to compassionate use, with no statistical results anticipated. One study of ribavirin published in 1991 found it effective in reducing deaths from Hantaan virus. But the drug must be given to patients early to prevent hantaviruses from doing catastrophic tissue damage, and the Four Corners strain may move too rapidly for ribavirin to be effective. Few experts are optimistic about ribavirin. Johnson says that prescribing it is just an "emotional" gesture. CDC's Butler notes,

A Rogues' Gallery of Hantaviruses

The hantavirus nailed as the cause of 26 deaths in the United States this year belongs to a broad family of viruses responsible for widespread illness in Asia and Europe. National Institutes of Health virologist Richard Yanagihara estimates that each year diseases associated with hantaviruses strike some 200,000 people (half of them in China) and kill between 4000 and 20,000.

Western interest in hantaviruses dates to 1951, when United Nations troops in the Korean war began to come down with an illness characterized by high fever, severe headache, muscle pain, vomiting, and hemorrhaging that ends in renal failure and death in about 7% of cases; in 3 years some 3000 troops were diagnosed with Korean hemorrhagic fever with renal syndrome (HFRS). As soldiers were falling ill in Korea, Western virologists began to realize that similar hemorrhagic fevers were prevalent outside Southeast Asia. They saw parallels between HFRS and illnesses such as Tula fever in Russia and a mild form of HFRS in China. In addition, in 1953 NIH virologist D. Carleton Gajdusek, then with the U.S. Army, proposed a kinship between HFRS and nephropathia epidemica, a mild kidney illness first seen in Scandinavia.

Recognized Hantaviruses			
Virus	**Disease**	**Principal Reservoir**	**Distribution**
Hantaan	HFRS	striped field mice	Asia
Seoul	HFR5	rats	Cities worldwide
Belgrade	HFRS	yellow-necked mice	Yugoslavia
Puumala	nephropathia epidemica	bank voles	Europe
Prospect Hill	none described	meadow voles	United States
Thottapalsyam	none described	shrews	India
Thailand	none described	bandicoots	Thailand
Hemorrhagic fever with renal symptoms. SOURCE Ted Tsai			

Efforts to identify the agents responsible for these diseases were stymied until 1976, when virologist Ho Wang Lee of Korea University College of Medicine in Seoul isolated a virus from striped field mice (*Apodemus agrarius*) captured near the Hantaan River in South Korea. All viruses of this genus have resisted culturing in the lab; it wasn't until 1981 that Army virologist George French, Lee, and colleagues got Hantaan into a continuous cell line. Researchers quickly flushed out other hantaviruses. In 1982, Gajdusek's team isolated a hantavirus, which they called Prospect Hill, from meadow voles (*Microtus pennsylvanicus*) in Frederick, Maryland; they couldn't link it to any human disease, however. Later that year, Lee isolated a hantavirus from rats (*Rattus norvegicus*) in Seoul that causes a mild form of HFRS. This finding led researchers to believe that rats on trading ships had dispersed hantaviruses worldwide. Indeed, the Seoul serotype appears to cause much of the HFRS in China, and in 1985 a group led by microbiologist Ted Tsai of the Centers for Disease Control and Prevention found a virus similar to Seoul—which they labeled Tchoupitoulas—in a New Orleans rat. After that, Tsai says, "virtually everywhere we looked—West Coast, East Coast, Cincinnati, New York City—we found infected rats."

Later in 1985, a research team led by virologist Joel Dalrymple at the U.S. Army Medical Research Institute of Infectious Disease (USAMRIID) isolated from bank voles (*Clethrionomys glareolus*) in Sweden the Puumala serotype (tentatively identified 5 years earlier by Finnish researchers) that causes nephropathia epidemica. Other hantaviruses, still poorly characterized, appear to have been isolated from a patient in Greece, a house mouse in Texas, house shrews (*Suncus murinus*) in India, bandicoots (*Bandicota indica*) in Thailand, and yellow-necked mice (*Apodemus flavicollis*) in Yugoslavia. "This is a moving train in terms of our knowledge base—I think we're going to find new hantaviruses as time goes on," says Ernest Takafuji, USAMRIID's chief.

"So far, we don't have any evidence it changes the course of disease."

It might be possible to develop a vaccine against the Four Corners strain, once the virus has been cultured in the lab or its entire genome extracted. For the past decade, Army microbiologist Schmaljohn has been developing a vaccine to combat the Hantaan virus in Asia, and she plans to begin a large field trial in China next year. But developing a similar weapon against the Four Corners strain could take years.

Meanwhile, two researchers at the University of

New Mexico—Brian Hjelle and Steven Jenison—have been working on a quick diagnostic test based on "Western blot" technology to distinguish flu sufferers from hantavirus victims in the clinic. They began work on it in a rush in June, "as soon as we heard about hantavirus." After downloading viral sequence data by computer from federal databanks, they got help from CDC and learned to pull pieces of viral genome from patient tissue. They spliced these genes into bacteria, which expressed viral proteins that could be used to test human serum for antibodies to the virus. They claim their test can provide a reliable identification of hantavirus infection within 26 hours. But CDC staffers feel the test is still too complex to be used on a mass scale.

Attacking the vector

If researchers can't yet cure those who are infected, they aren't entirely helpless against this new viral threat. The best route of attack against the virus is preventing it from being transmitted from rodents to people. CDC researchers have concluded that it is most likely transmitted by aerosol from the rodents' urine and feces. The evidence, says CDC immunologist Pierre Rollin is anecdotal and is derived from experience with other hantaviruses. He recalls the case of the doctor who made a short visit to a barn in France where infected rodents lived, saw no animals, but came down with a fatal infection of Puumala. C. J. Peters, chief of CDC's special pathogens branch, recalls the case of a Korean restaurateur who used a broom to beat a rat to death in his bedroom, then got sick. No one knows exactly how the U.S. victims acquired their infections, but many lived around rodents, and transmission by aerosol

seems likely, says Peters. For that reason CDC is telling people to "avoid contact" with deer mice by keeping them out of homes.

Public health officials in New Mexico have already run a massive mouse roundup, and some officials have advised people to trap rodents and disinfect them with Lysol. CDC staffers aren't keen on the idea of having people handle rodents, however, since they often excrete fluids around the trap. But for now, "We don't know if it's safer" to trap or not to trap, Childs says.

Most people are hoping that 1993 will prove to have been an extraordinary year, and that the virus will not return. "But we would be remiss not to be prepared" for a resurgence, Simpson says, as he braces for the unknown. He's not alone in his concern. "We're worried that there's a good possibility we'll see more hantavirus infections in the area this fall," says Childs. One reason to worry is that the Hantaan and Puumala viruses exhibit two epidemic peaks: one in summer and a second in late fall between November and January.

One hopeful sign is that the deer mouse population in New Mexico declined in late summer, although it is still above 1992 levels. But just in case the Four Corners strain mimics Hantaan, New Mexico has set up an emergency team involving the University of New Mexico and the CDC to sort emergency cases from ordinary flu sufferers. The aim is to perform rapid triage if necessary. It's possible that none of this planning will be needed. In spite of the rapid progress made so far in identifying the culprit, developing initial diagnostic tests, and providing advice to the local populace, even CDC's experts cannot predict what course the virus will take next. —Eliot Marshall

---

US Braces for Hantavirus Outbreak

Science, **280** (1998) 993.
C. Holden

Health authorities in the southwestern United States are raising the alarm that—courtesy of El Niño—conditions there are ripe for an outbreak of a deadly Hantavirus strain called Sin Nombre. The virus was first identified in 1993 when it killed 20 people in the Four Corners region. So far this

spring, four people in three states have been infected by the virus, and two have died.

Doctors believe that most people catch Sin Nombre, which has proven fatal in about half the cases, through contact with droppings and urine from infected deer mice. Abundant rain and a mild winter, brought to the region by El Niño, have made for booming rodent populations, says mammalogist Terry Yates of the University of New Mexico, Albuquerque. Although deer mice are usually able to reproduce twice a year, he says, last year they turned out three litters, and the population is about 10 times normal. Furthermore, it appears that higher proportions of deer mice might be infected: The usual rate is 6% to 10% of the population, but in a sample this month near Gallup, seven of 30 mice tested positive, says virologist Brian Hjelle of the University of New Mexico School of Medicine.

Officials hope to head off an epidemic by warn-

ing residents to be careful when entering poorly ventilated spaces where mice might have been. A blood test developed by Hjelle since the last outbreak can identify the presence of the virus within hours, and authorities are encouraging area doctors to be vigilant. Although there is still no drug against the virus, Hjelle says, doctors can treat symptoms like low blood pressure and can reduce fluid buildup in beleaguered lungs. "In the last outbreak, patients were referred on death's door," he says. "But if you get to a good intensive care unit, then you have a pretty good chance."

Exercises for Chapter 21

1. For each of the following differential equations, make the traveling wave substitution $u(t, x) = f(x - ct)$ with c being a constant and so derive a differential equation in one variable for the function f.

(a) $\frac{\partial}{\partial t} u = 3 \frac{\partial^2}{\partial x^2} u + 2(u - \sin(u))$

(b) $\frac{\partial}{\partial t} u = \mu \frac{\partial^2}{\partial x^2} u - 6 \frac{\partial}{\partial x} u + u^3$

(c) $\frac{\partial}{\partial t} u = -5 \frac{\partial}{\partial x} u + e^u$

(d) $\frac{\partial^2}{\partial t^2} u = \mu \frac{\partial^2}{\partial x^2} u - 25 u^2$

2. In Exercises 1(a), 1(b), 1(c), and 1(d), you derived a differential equation in one variable for a function f. Four each of these four differential equations, do the following:

(a) Call the variable s so that f is a function of s; then write p for $\frac{d}{ds} f$ and rewrite the differential equations for f as a pair of equations for f and p that involve at most one derivative of each.

(b) Sketch the phase plane diagram for the resulting differential equation for the unknowns p and f. Be sure to label the null clines and the equilibrium points, and mark the null clines with arrows to indicate the direction of the trajectories that cross them.

3. Aside from not taking topography into account, list some ways in which the model represented by Equation (1) in Chapter 21 may be oversimplifying the problem of predicting the behavior of the proportion of infected mice as a function of time and space.

22

Traveling Wave Velocities

Recall from the previous chapter that we were looking for a solution $u(t, x)$ to the differential equation

$$\frac{\partial}{\partial t} u = \frac{\partial^2}{\partial x^2} u + ru(1 - u),$$ (1)

where $r > 0$ is a constant. In addition, we were looking for a solution to (1) that is valid for all values of t and x and that obeyed

(a) $0 \le u \le 1$.

(b) As x gets very negative, then $u(t, x) \to 1$.

(c) As x gets very positive, then $u(t, x) \to 0$. (2)

    We then considered the possibility that there might be a solution to (1) and (2) that maintains its shape over time. This would be a traveling wave where the function $u(t, x)$ has the form $u(t, x) = f(x - ct)$, where f is a function of one variable and where $c > 0$ is a constant.

    The constant c is the speed at which the wave described by f travels from left to right along the x-axis. From the point of view of our applications, it would be of great import were we able to predict the value of c. [Remember that we were considering (1) and (2) in the context of predicting the fraction of mice at time t and position x across the United States that are infected with the Four Corners hantavirus.]

You should recall that if $u(t, x)$ is to satisfy (1) and (2), then the function $f(s)$ must obey

$$-c\frac{d}{ds}f = \frac{d^2}{ds^2}f + rf(1-f)$$

(3)

subject to the auxilliary conditions:

(a) $0 \le f \le 1$.

(b) $f \to 1$ as s gets very negative.

(c) $f \to 0$ as s gets very positive.

(4)

We analyzed (3) by writing it as a pair of equations for $(f(s), p(s))$ obeying

$$\frac{d}{ds}f = p$$

(5a)

$$\frac{d}{ds}p = -cp - rf(1-f).$$

(5b)

We showed by phase plane analysis that (5) is solvable subject to the conditions in (4) whenever $c^2 > 4r$.

Problem: Resolve the following paradox: According to the preceding analysis, there are traveling wave solutions to (1) and (2) that cross the country at arbitrarily high rates of speed. Yet our intuition tells us that this is not a reasonable conclusion.

22.1   The Resolution

Here is a resolution of the preceding paradox: Consider the function $\frac{d}{ds}f = p(s)$. Suppose that $p(s)$ has a maximum at some s that I will call s_0. At this point, $\frac{d}{ds}p = 0$. Therefore, according to (5b),

$$cp(s_0) = -rf(s_0)(1 - f(s_0)).$$

(6)

Since f is constrained to lie between 0 and 1, Equation (6) tells us that p can never be positive. (It's maximum value is nonpositive!) This shouldn't be so surprising, since f is supposed to be nearly 1 at large negative x and to be nearly zero at large positive x, so we expect, on the whole, that $\frac{d}{ds}f$ should be nonpositive.

Let us now see just how negative $p(s) = \frac{d}{ds}f$ can be. Use s_1 to denote the value of s where $p(s)$ is most negative. Here, again, $\frac{d}{ds}p = 0$ so that (5b) at $s = s_1$ rearranges to give

$$-cp(s_1) = rf(s_1)(1 - f(s_1)).$$

(7)

Now, $f(s_1)$ is constrained to lie between 0 and 1, and this means that $f(s_1)(1 - f(s_1))$ is no bigger than $\frac{1}{4}$. [The parabola $y = x(1 - x)$ has its maximum at $x = \frac{1}{2}$ where $y = \frac{1}{4}$.] Therefore,

$$\max \left| \frac{d}{ds} f \right| = -p(s_1) \leq \frac{r}{4c}. \tag{8}$$

Equation (8) needs some interpretation: It says that a solution to Equations (3) and (4) with c large must have everywhere gentle slope. See Figure 22.1.

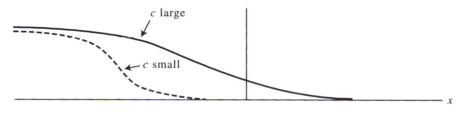

Figure 22.1

Indeed, Equation (8) asserts that the velocity of the traveling wave can be a priori bounded if we know the maximum of the absolute value of the slope of the wave. Put succinctly, a high-speed wave will have very slow falloff in x, but a wave traveling slowly will have a large falloff with x.

22.2   The Mouse/Virus Infestation

In the context of the mouse/virus problem, the preceding remarks can be turned around to tell us how we might obtain an upper bound for the speed at which the infection travels across the country: Indeed, the maximum slope, as a function of x, of the fraction of mice infected at time zero is the number

$$\max \left| \frac{\partial}{\partial x} u(0, x) \right|. \tag{9}$$

If we can determine this number (from field data) then, with knowledge of the coefficient r in (1), we can predict an upper bound to the speed at which the infection travels across the country.

Do you think it is reasonable that slow waves should be steeper than fast ones? Can you justify this conclusion directly from the biological assumptions given at the beginning of Chapter 21 about the behavior of the mice?

22.3 Lessons

Here are some key points from this chapter:

- In some cases, the speed of a traveling wave solution of a reaction-diffusion equation is determined by the wave's shape.

- *Always* critically question the conclusions from a mathematical model. Ask whether the results are consistent with your scientific intuition. If they are not, then one or more of the following is most likely the case:

    (a) You made a mathematical mistake.

    (b) Your model is inappropriate.

    (c) Your intuition for the underlying biology is mistaken.

Cases (b) and (c) are most useful, for in those cases, the mathematics has pinpointed weak parts of your understanding.

READINGS FOR CHAPTER 22

READING 22.1

Travelling Waves in Vole Population Dynamics

Commentary: This article reports on evidence of traveling waves in the population dynamics of voles in Finland. The article reports on the observation of the local vole population number and how it changes with time and space as if it were a moving wave. Whether the population density $u(t, x)$ is characteristic of a classical wave, $u(t, x) = f(x - ct)$, or whether the behavior of u is more complicated cannot be discerned from the article.

Travelling Waves in Vole Population Dynamics

Nature, **390** (1997) 456.
Esa Ranta, Veijo Kaitala

Spatial self-organization patterns in population dynamics have been anticipated [1-3], but demonstrating their existence requires sampling over long periods of time at a range of sites. Voles cause severe economic damage and are therefore extensively monitored, providing a source of the required data. Using two long-term data sets [4-6] we now report the existence of travelling waves in vole population numbers.

Information about the intensity of annual vole damage to young forests was available for the years 1972-83 for 19 forestry board districts of Finland [5, 6]. The geographical distribution of the most damaged areas fluctuates (Fig. 1a, b) and the timing of damage is asynchronous among the different areas. Locally, the damage peaks at intervals

---

Author affiliations: Esa Ranta: Department of Ecology and Systematics, University of Helsinki. Veijo Kaitala: Department of Biological and Environmental Science, University of Jyvaskyla.

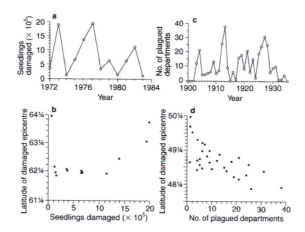

Fig. 1 Extent and severity of vole damage. **a,** The number of seedlings destroyed by voles on study plots in young forest plantations in Finland. **b,** The dynamics are characterized by movement of the annual damage epicentre from south to north as total damage increases. **c,** Geographical extent of French vole plagues; and **d,** annual epicentre of plagues plotted against the number of plagued departments.

of three to four years, matching the cyclic dynamics of Scandinavian voles [7]. This is in agreement with simulation studies [2] showing that spatial order is associated with the presence of spatial waves showing a certain degree of periodicity.

Correlation between the latitude of the "centre of mass" and the annual severity of the damage (Fig. 1a, b) is $r = 0.84$ ($P < 0.001$; for the longitude $r = 0.13$). Thus, with increasing volume of vole damage the damage epicentre moves north. There was also a negative correlation between intensity of damage and distance between provinces ($r = -0.47$; $P < 0.001$) suggesting that nearby localities have similar damage levels, and providing another indication of spatial structure [8] in vole damage.

We also analysed records of vole plagues in 53 departments of France for the years 1900–35 [4] (Fig. 1c, d). There were rapid changes in the extent of the plagued area, but there were no clear periodicities like those seen in the Scandinavian voles [7]. Again there was a correlation between the latitude of the annual location of the centre of mass and the geographical extent of the damage ($r = -0.67$, $P < 0.001$; for the longitude $r = 0.06$). Accordingly, the French vole plagues spread from north to south, whereas in Finland they spread from south to north.

A variable is said to be spatially autocorrelated when it is possible to predict the values of this variable in one site from the known values at nearby sampling sites [9]. We performed autocorrelation analyses for irregularly spaced interval (Finland) or nominal scales (France) [10].

For the Finnish data, we computed annual autocorrelation coefficients [9–11] for the distance class of 200 km assuming binary links. We detected significant positive spatial autocorrelation in six of the twelve years. For the French data we computed annual spatial autocorrelations using binary weights and assuming a link between sites that were less than 120 km apart. We found a significant spatial autocorrelation present in the data in 14 of the 33 years. Both results indicate that spatial structure exists in these populations.

By identifying spatial autocorrelation in vole populations and an annually moving epicentre of vole plagues as a function of damage extent, we conclude that travelling waves, or pulses, in the dynamics of vole populations exist, a phenomenon predicted [1–3] by theoretical population ecology.

REFERENCES

1. M. P. Hassell, H. N. Comins, and R. M. May, *Nature* **353**, pp. 255–258 (1991).
2. P. Rohani and O. Miramontes, *Proc. R. Soc. Lond. B* **260**, pp. 335–342 (1995).

3. J. Bascompte and R. Sole, *Trends Ecol. Evol.*
   10, pp. 361–366 (1995).

4. C. Elton, *Voles, Mice and Lemmings. Problems in Population Dynamics* (Clarendon, Oxford, 1942).

5. T. Teivainen, *Folia Forestalia* **387**, pp. 1–23, (1979).

6. T. Teivainen, *Metsäntutkimuslaitoksen Tiedonantoja* **145**, pp. 1–12 (1984).

7. L. Hansson and H. Henttonen, *Oecologia* **67**,

pp. 394–402 (1985).

8. E. Ranta, V. Kaitala, J. Lindström and H. Linden, *Proc. R. Soc. Lond. B* **262**, pp. 113–119 (1995).

9. R. R. Sokal and Oden, F. M. L. S., *Biol. J. Linn. Soc.* **10**, pp. 199–228 (1978).

10. P. Legendre and M.-J. Fortin, *Vegetatie* **80**, pp. 107–138 (1989).

11. P. A. P. Moran, *Biometrica* **37**, pp. 17–23 (1950).

READING 22.2

Control of Spiral-Wave Dynamics in Active Media by Periodic Modulation of Excitability

Commentary: Wavelike patterns are commonly observed now in excitable systems. Of particular interest are spiral waves that appear in many examples. This article reports on attempts to affect the behavior of the spiral waves in a light-sensitive version of the Belousov-Zhabotinsky reaction. The authors modeled the observed behavior using a diffusion equation for two density functions, u and v. The equation that they use appears in the fifth paragraph after the abstract and is called the Oregonator model. They run the model on a computer and compare the computer-generated behavior of spiral waves with the observed chemical waves.

Control of Spiral-Wave Dynamics in Active Media by Periodic Modulation of Excitability

Nature, **366** (1993) 322–324.
Oliver Steinbeck, Vladimir Zykov, Stefan C. Müller

Excitable media exhibit a wide variety of geometrically complex spatio-temporal patterns, perhaps the most striking of which are rotating spiral waves. Spiral waves have now been observed in many excitable systems, including heart muscle [1], aggregating slime-mould cells [2], retinae [3], CO oxidation on platinum [4] and oscillatory chemical systems such as the Belousov-Zhabotinsky (BZ) reaction [5, 6]. In the last case, the spiral cores trace out circular or hypocycloidal trajectories, depending on the specific reaction conditions [7–9]. In addition, if the excitability of the BZ reaction is light-sensitive [10–13], constant illumination has been shown to influence the dynamics of spiral waves [14, 15]. Here we investigate the effect of illumination that is periodically modulated in time. We find that, for a single set of reaction conditions, the motion of the spiral cores can be forced to describe a wide range of open and closed hypocycloidal trajectories, in phase with the applied modulation frequency. Numerical simulations using a modified version of the Oregonator model [16, 17] of the BZ reaction reproduce this behaviour. We suggest that the modulation of excitability with weak external forces might be used as a means for controlling the dynamics of other excitable media.

Author affiliations: Oliver Steinbeck, Vladimir Zykov, and Stefan C. Müller: Max-Planck-Institut für molekulare Physiologie.

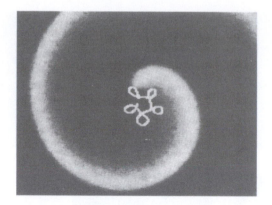

Fig. 1 Spiral wave in the ruthenium-catalysed BZ reaction at a steady level of light intensity (0.93 mW cm^{-2}). In the experiment bright bands are green, and dark bands are orange. The tip trajectory (overlaid white curve) is similar to a hypocycloid. Image size 3.8 × 3.0 mm.

In the light sensitive BZ reaction [10–11], a ruthenium-bipyridyl complex is used to promote the autocatalytic production of the reaction activator HBrO$_2$. This only occurs when the complex is in its reduced and electronically unexcited state. If the ruthenium complex is photochemically excited, it slowly catalyses the production of the inhibitor species, bromide. Thus externally applied illumination suppresses the excitability of the medium (for instance, it decreases the propagation velocity of excitation waves) and allows control of spiral wave parameters [12–15].

In our experiments Ru(bpy)$_3^{2+}$ (4mM) was immobilized in a silica-gel matrix [18] (thickness 0.7 mm, diameter 7 cm). The reactants and their concentrations (disregarding bromination of malonic acid) were: NaBr (0.09 M), NaBrO$_3$ (0.19 M), malonic acid (0.17 M) and H$_2$SO$_4$ (0.35 M). The temperature was kept constant at 23 ± 1 C. White light (halogen lamp, 150 W) illuminating the entire observation area was polarized by a rotating polarization filter and applied to the active medium through a tilted glass plate. Because of the rotation of the polarization vector the intensity of the reflected light was modulated sinusoidally with time (0.49–1.36 mW cm^{-2}). The transmitted light was detected by CCD (charge-coupled device) camera (Hamamatsu C3077) at 490 nm, stored on a video recorder and finally digitized by an image-acquisition card.

We created a spiral wave by using a thin laser beam to break a propagating circular wave (this creates two spirals) and moving one of the open ends

to the boundary of the dish [19]. The temporal trace of the wave tip was detected visually with a reticle in digitized images. Figure 1 shows a snapshot of the created spiral wave (constant light intensity, 0.93 mW cm^{-2}) with the overlaid trajectory of its tip. The trajectory is almost a five-lobed hypocycloid with the wave period $T_0 = 24.5$ s at the centre of the meandering pattern. The change in the stationary intensity level of illumination in the range from 0.49 to 1.36 mW cm^{-2} leads only to small variations of the spatio-temporal parameters of the spiral wave. For the minimum and maximum intensities applied here, we observed four- and six-lobed trajectories, respectively.

Periodic modulation of the applied light intensity within the same range causes dramatic changes in the dynamics of spiral rotation. The small variations of the trajectory shape that occur during each period of the modulation accumulate with time. As a result, the modulation forces the tip to follow trajectories that differ significantly from those observed at constant intensities. The shape of trajectories depends strongly on the modulation period T_m (see Fig. 2). The trajectories of Fig. 2b–d are members of an entrainment band ($T_m \approx 20 - 35$ s) of closed hypocycloids with one lobe corresponding to one external period. The number of lobes continuously increases from 3 to more than 12 with increasing values of T_m. Figure 2a shows a deformed five-lobed pattern with one lobe described during two modulation periods. Figure 2e depicts a surprising trajectory with alternating distances be-

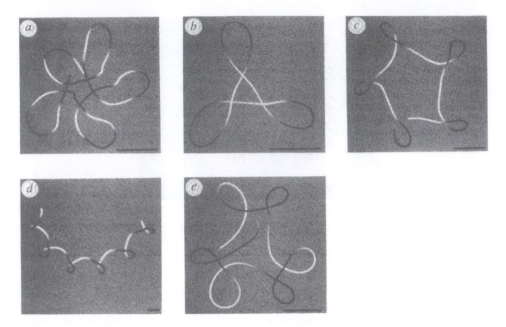

Fig. 2 Sequence of tip trajectories measured under sinusoidal modulation of light intensity with a period T_m of: 17.0 s (a), 26.2 s (b), 30.4 s (c), 34.5 s (d) and 52.2 s (e). The shading of the traces is due to the time-varying intensity of illumination and shows the number of lobes per modulation period. Scale bar, 0.2 mm.

tween neighbouring lobes. In this small frequency range the spiral tip describes a pair of lobes during one external modulation. For modulation periods between those of Fig. 2d and e we observed irregular motion with epicyclic segments of the trajectories. In all examples of Fig. 2 the tip motion is synchronized by the external rhythm.

We complemented the experiments by numerical simulations of the observed process of synchronization. For this purpose we used an Oregonator model [16, 17], which was extended by a term ϕ describing the light-induced bromide production [11].

$$\frac{\partial u}{\partial t} = \Delta u + \frac{1}{\epsilon}\left[u - u^2 - (fv + \phi)\frac{u - q}{u + q}\right]$$

$$\frac{\partial v}{\partial t} = u - v$$

The variables u and v describe the time-space evolution of the $HBrO_2$ and catalyst concentration, respectively. While the parameters $\epsilon = 0.05$, $q = 0.002$ and $f = 2.0$ are constant, the term $\phi(t) = 0.01 + A\sin(2\pi t/T_m)$ describes the periodic modulation of bromide production, and hence excitability.

In our calculations we studied the trajectories of the spiral tip for different values of the modulation period T_m and amplitude A. The unperturbed trajectory ($A = 0$) is similar to a five-lobed hypocloid (see Fig. 1) and the excitation period measured at the symmetry centre is $T_0 = 2.8$. Our calculations show that the motion of the spiral tip can be synchronized by external modulation. In Fig. 3 the most pronounced entrainment bands are shown (labelled 1 : 3, 1 : 2, 1 : 1 and 2 : 1), with the broadest band observed around the modulation period $T_m = T_0$. For this band one loop is completed during one modulation period. Each modulation period for the furthest right (2 : 1) band in Fig. 3 contains two loops. On the left side of the diagram two bands (1 : 3 and 1 : 2) are shown which correspond to five-lobed trajectories but deformed with respect to the unperturbed one. Each loop of these meandering pictures is described during two (1 : 2 band) or three modulation periods (1 : 3 band). Within the entrainment bands the tip trajectory is sometimes rather complex but regular, in that the motion is still phase-locked. Between the bands, irregular behaviour apparently dominates. The sim-

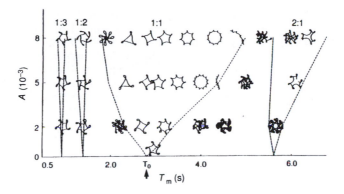

Fig. 3 Tip trajectories calculated as a response to sinusoidal modulation of bromide production. Trajectories (not to scale) are shown in the plane spanned by amplitude A and period T_m of modulation. Dashed lines indicate boundaries of entrainment bands related to different ratios $n : m$, where n is the number of lobes per m periods of the modulation. The vertical arrow on the x-axis shows T_0, the intrinsic period of the unperturbed meandering spiral.

ulation results for $A = 0.005$ are in good qualitative agreement with the experimental data given in Fig. 2. One can see the wide entrainment band ($T_m \approx T_0$) in which the lobe number increases from three up to infinity (as in Fig. 2b-d), the deformed five-lobed trajectory for $T_m \approx T_0/2$ (as in Fig. 2a) and characteristic pictures with pairs of petals for $T_m \approx 2 \times T_0$ (see Fig. 2e).

The nature of the described synchronization phenomenon is similar to entrainment in different kinds of oscillating systems in nonlinear physics [20] or chronobiology [21], but cannot be reduced to the case of zero spatial dimension. The resulting family of very unusual spatio-temporal patterns, including the transition from localized to infinite motion, has no close analogies with other examples of entrainment. The observed effects also differ essentially from the resonance drift induced by periodic forcing of an active medium [12] which occurs without synchronization and leads to displacement of the spiral-wave core as a whole, without the pronounced deformations of the tip trajectory [22] characteristic of the described phenomenon. This new way to deform and translate a spiral wave by controlling excitability with even weak external forces could ultimately prove relevant to our understanding of pattern formation in the early stages of morphogenesis [2], or to the origin and possible control of abnormalities of heart rhythm [1].

REFERENCES

1. J. M. Davidenko, A. V. Pertsov, R. Salomonsz, W. Baxter and J. Jalife, *Nature* **355**, pp. 349–351 (1992).
2. F. Siegert and C. J. Weijer, *Cell Sci* **93**, pp. 325–335 (1989).
3. N. A. Gorelova and J. J. Bures, *Neurobiol* **14**, pp. 353–363 (1983).
4. S. Jakubith, H. H. Rotermund, W. Engel, A. von Oertzen and G. Erth, *Phys. Rev. Lett* **65**, pp. 3013-3016 (1990).
5. A. T. Winfree, *Science* **175**, pp. 634–636 (1972).
6. R. J. Field and M. Burger (eds) *Oscillations and Traveling Waves in Chemical Systems* (Wiley, New York, 1985).
7. W. Jahnke, W. E. Skaggs and A. T. Winfree, *J. phys. Chem* **93**, pp. 740-749 (1989).
8. Th. Plesser, S. C. Müller and B. Hess, *J. phys. Chem.* **94**, pp. 7501-7507 (1990).
9. G. S. Skinner and H. L. Swinney, *Physica* **D40**, pp. 1–16 (1991).
10. L. Kuhnert, *Naturwissenschaften* **73**, pp. 96–97 (1986).
11. H. -J. Krug, L. Pohlmann and L. Kuhnert, *J. phys. Chem.* **94**, pp. 4862–4866 (1990).
12. K. I. Agladze, V. A. Davydov and A. S. Mikhailov, *JETP Lett.* **45**, pp. 767–770 (1987).
13. L. Kuhnert, K. I. Agladze, and V. I. Krinsky, *Nature* **337**, pp. 244–247 (1989).

14. M. Markus, Nagy-Ungverai and B. Hess, *Science* **257**, pp. 225–226 (1992).
15. M. Braune and H. Engel, *Chem. Phys. Lett.* (in press).
16. R. J. Field and R. M. Noyes, *J. phys. Chem.* **60**, pp. 1877–1884 (1974).
17. W. Jahnke and A. T. Winfree, *Int. J. Bifurcation and Chaos* **1**, pp. 445–466 (1991).
18. T. Yamaguchi, L. Kuhnert, Zs. Nagy-Ungvarai, S. C. Müller and B. Hess, *J. phys. Chem.* **95**, pp. 5831–5837 (1991).
19. O. Steinbock and S. C. Müller, *Physica* **A188**, pp. 61–67 (1992).
20. H. G. Schuster, *Deterministic Chaos* (VCH, Weinheim, 1989).
21. G. Hildebraudt, Ch. Gutenbrunne and R. Moog (eds) *Chronobiology and Chronomedicine* (Peter Lang, Frankfurt, 1922).
22. V. A. Davydov, V. S. Zykov and A. S. Mikhailov, *Soviet. Phys. Usp.* **34**, pp. 665–684 (1991).

Exercises for Chapter 22

1. Imagine a hypothetical mouse/virus situation in which, in 1993, mice west of the Great Divide (the central Rocky Mountains) are almost 100% infected by a harmless virus, but mice on the Great Plains and eastward are virus free. In a page or so, describe those measurements that I could make (over some years time) to determine if a traveling wave model is valid for predicting the rate of spread to the east of the hypothetical viral infection.

2. Roughly sketch out how I could determine the appropriate values for the coefficients r and μ in my model diffusion equation

$$\frac{\partial}{\partial t}u = \mu \frac{\partial^2}{\partial x^2}u + ru(1 - u).$$

3. Given the analysis in Chapter 22 for the speed of wave propagation, design a strategy that will slow the spread of the epidemic. Consider, for example, which of the following makes more sense:

    (a) Inoculate mice with a hantavirus vaccine in states already infected.

    (b) Inoculate mice with a hantavirus vaccine in states east of the line of infection.

    (c) Inoculate mice with a hantavirus vaccine in border states where some percent of the mice are infected.

   (No inoculation program is 100% successful, especially no animal inoculation program where the inoculation must be made through the use of "baited" food airdropped or otherwise spread over a large region. Given a large area of concern, inoculation might well be hopeless.)

4. Find a nonmathematical explanation for the fact that the steeper waves travel slower than the faster ones. (*Hint:* Turn this around and ask yourself why slow waves are going to be steep.)

23

Periodic Solutions

In this and the subsequent chapters, I return to the milieu of the first 12 chapters, in which only dependence on time was at issue. However, whereas the first 12 chapters considered equilibrium issues almost exclusively, this chapter and the remaining chapters consider nonequilibrium phenomena. Here, the story is amazingly complicated and there is no sense in which it can be said that the interesting questions are all solved. Indeed, the complicated nonequilibrium dynamics that arise even from very simple models are still a wealthy source of very interesting mathematics. Meanwhile, similar appearing dynamics appears in real biological systems and the underlying causes are a subject of intense investigation.

I shall start by describing a predator-prey model that has a stable, time-dependent, periodic solution. The model presented provides one mathematical explanation for cyclic behavior. However, the model itself is not the main point of this chapter. Rather, you should focus on those aspects of the model that guarantee the existence of cyclic solutions, for those aspects are found in many other models. That is, there are certain generic properties of a differential equation that a priori imply that there are periodic solutions. In particular, the properties to notice are the existence of a basin of attraction in the phase plane that has inside a certain type of unstable equilibrium point.

The discussion in this chapter follows that of the same model in Beltrami's book, *Mathematics for Dynamic Modeling*.

23.1   The Model

Let $p(t)$ denote the number of prey at time t, and let $q(t)$ denote the number of predators at time t measured in some convenient units (say thousands for the prey and hundreds for the predator). Consider the following differential equations for $p(t)$ and $q(t)$:

$$\frac{d}{dt}p = \frac{2}{3}p\left(1 - \frac{p}{4}\right) - \frac{pq}{1+p} \tag{1a}$$

$$\frac{d}{dt}q = sq\left(1 - \frac{q}{p}\right). \tag{1b}$$

Here, $s > 0$ is a real number that could, in principle, be determined by some measurement. (As I remarked previously, this model and the following analysis come from Beltrami's book *Mathematics for Dynamic Modeling*.)

Before starting the analysis, I ask you to consider the following justification for this model. First, I admit that the various numerical constants were chosen solely for the purpose of simplifying subsequent computations. For example, there is no biological justification for the factor of 2/3 that appears in equation (1a). Next, consider the form of (1a). In particular, if there are no predators (so $q = 0$), (1a) becomes a standard logistics equation for p with its stable equilibrium point at $p = 4$. The existence of predators decreases the rate of change of p with the term $\frac{pq}{1+p}$. The thing to notice about this term is that when p is much less than 1, it is approximately equal to pq, which has been our standard predator-prey interaction term. The justification for the latter, as you may recall, is that the death rate of prey due to predation is, to a first approximation, proportional to the number of prey times the number of predator. However, this is just a first approximation, and it doesn't take into account the fact that certain kinds of predators can get satiated. To put it bluntly, a single lion can only eat so many antelope in any given day. The factor of $(1 + p)$ in the denominator is a crude way of modeling predator satiation, for when p is large, it makes the death rate due to predation approximately proportional to the number of predators, which is to say that each predator can eat only so many prey per day.

Now consider equation (1b). If p were a consant, this would be a standard logistics equation for q with the stable equilibrium point equal to p. Thus, this equation models the fact that the carrying capacity of the ecosystem for the predator is proportional to the number of prey. (Of course, you can argue whether the proportionality factor should be equal to 1, but as I said previously, the constants taken here are not meant to have biological relevance.)

Do you think this is a reasonable model for prey and predators that can be satiated?

23.2   The Phase Plane

Figure 23.1 is a diagram of the phase plane for the equation in (1).

The p null clines are given by line $p = 0$ and the parabola $q = \frac{2}{3}(1 - \frac{p}{4})(1 + p)$. Meanwhile, the q null clines are given by the line $q = 0$ and the line $q = p$. The equilibrium points where $p > 0$ are

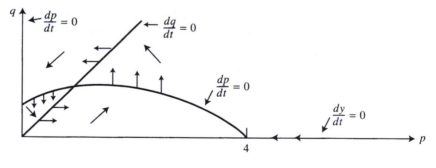

Figure 23.1

1. $p = 1, q = 1$

2. $p = 4, q = 0$

23.3   Stability

The stability of these equilibrium points can be analyzed by computing the matrix of derivatives, **A**.

1. At $p = 1, q = 1$: $\mathbf{A} = \begin{pmatrix} 1/12 & -1/2 \\ s & -s \end{pmatrix}$

2. At $p = 4, q = 0$: $\mathbf{A} = \begin{pmatrix} -2/3 & -4/5 \\ 0 & s \end{pmatrix}$

We are interested in the critical point at $p = 1, q = 1$. Observe that

1. $\text{trace}(\mathbf{A}) = \frac{1}{12} - s$

2. $\det(\mathbf{A}) = \frac{5}{12} s$ $\qquad\qquad\qquad\qquad\qquad\qquad$ (2)

Thus, this is a stable equilibrium point if $s > 1/12$. If $s < 1/12$, the equilibrium point is unstable.

23.4   A Repelling Equilibrium Point

Say that an equilibrium point is **repelling** if the following is true: Whenever a non-equilibrium solution to (1) is very close to the equilibrium point at time t, it then moves further away. [Stated differently, an equilibrium point is repelling if there are no nonconstant solutions to (1) starting very close to the point and not moving away.]

The equilibrium point $p = 1$, $q = 1$ is repelling when $s < \frac{1}{12}$. In fact, if $p(t)$ and $q(t)$ are both near 1 (their equilibrium values), then the vector is almost a solution to the equation

$$\frac{d}{dt}\begin{pmatrix} p - 1 \\ q - 1 \end{pmatrix} = \begin{pmatrix} (p - 1)/12 - (q - 1)/2 \\ s(p - 1) - s(q - 1) \end{pmatrix}. \tag{3}$$

The solutions to the preceding equation will grow exponentially fast with time. In particular, any nonequilibrium solution to (1) that is very close to $\begin{pmatrix} 1 \\ 1 \end{pmatrix}$ will move away from this point.

In general, an equilibrium point is repelling when the matrix $\mathbf{A}$ obeys

$$\text{trace}(\mathbf{A}) > 0 \quad \text{and} \quad \det(\mathbf{A}) > 0. \tag{4}$$

[Remember that stability of an equilibrium point required $\text{trace}(\mathbf{A}) < 0$ and $\det(\mathbf{A}) > 0$. Thus, the difference between stability and repelling is only in the sign of the trace of $\mathbf{A}$.]

23.5   Basin of Attraction

A *basin of attraction*, or *trapping region*, is a region V in the (p, q) plane that has the following property: No solution $\begin{pmatrix} p(t) \\ q(t) \end{pmatrix}$ of Equation (1) that enters V ever leaves V. For example, the square region

$$V = \{(p, q) : 0 < p < 4, 0 < q < 4\} \tag{5}$$

is a basin of attraction. Indeed, $\frac{d}{dt}p \leq 0$ on the edge where $p = 4$, so no solution can exit V through that edge. Also, $\frac{d}{dt}q \leq 0$ on the edge $q = 4$ (remember that $p < 4$), so no solution can exit V through the $q = 4$ edge. On the edge where $q = 0$, we have $\frac{d}{dt}q = 0$, so no solution can exit that edge. Finally, the edge $p = 0$ has $\frac{d}{dt}p = 0$, so no solution can hit this last edge unless it starts on it.

If you are worrying about the corners, please note the following: If a solution starts in V near the point $(0, 0)$, then $p > 0$ and $\frac{d}{dt}p > 0$, so it moves away from this corner. Likewise, if it starts in V near the corner point $(0, 4)$, then $q < 4$ and $\frac{d}{dt}q < 0$, so it moves away from $(0, 4)$. If it starts in V near the point $(4, 4)$, then $p < 4$ and $\frac{d}{dt}p < 0$, so it moves away. Finally, if it starts in V near $(4, 0)$, the $q > 0$ and $\frac{d}{dt}q > 0$, so it moves away from this corner, too.

Note that $p = 1$, $q = 1$ is the only equilibrium point in this basin of attraction. As noted previously, this equilibrium point is repelling in that it has the property that all nonequilibrium solutions to (1) that are very near it at some time $t = t_0$ will move away from it as t increases.

23.6   Poincaré-Bendixson Theorem

A basic fact (known as the Poincaré-Bendixson theorem) is the following:

Theorem.   Consider the equation

$$\frac{d}{dt}\begin{pmatrix} p \\ q \end{pmatrix} = \begin{pmatrix} f(p,q) \\ g(p,q) \end{pmatrix} \tag{6}$$

for functions $p(t)$ and $q(t)$. Here, f and g are some given pair of functions on the (p,q) plane. Suppose that a region V is a basin of attraction in the (p,q) plane for this equation. [That is, no solution to (6) that enters V ever leaves V.] Also, suppose that V has, inside it, a single equilibrium point that is repelling. [That is, any non-equilibrium solution to (6) that is very close to the equilibrium point at some time t will move away from the equilibrium point.] Then there is a periodic solution to (6) that is inside V for all times.

23.7   Periodic Solutions to Equation (1)

With the Poincaré-Bendixson theorem understood, we can conclude that (1) has a periodic solution in the case where $s < \frac{1}{12}$ since we have a basin of attraction with one equilibrium point, which is a repelling one.

The justification for the Poincaré-Bendixson theorem in the case of (1) for $s < \frac{1}{12}$ is not complicated. The basic point is that every nonequilibrium solution to (1) that starts in V will wind around the equilibrium point. You can confirm this by considering the regions where $\frac{d}{dt}p$ and $\frac{d}{dt}q$ are positive and negative.

Because the equilibrium point is repelling, the solutions to (1) that start near the equilibrium point will spiral out. Meanwhile, because V is a basin of attraction, the solutions to (1) that start near the boundary of the square will spiral *in*. Somewhere in between, there must be a solution that spirals neither in nor out (i.e., a periodic solution). [Remember that trajectories for (1) can never cross each other because the solutions to (1) are uniquely determined by their starting points.] See Figure 23.2.

23.8   Stability

The periodic solution here is stable in the following sense: If you start near it and evolve according to Equation (1), then the orbit in the (p,q) plane will spiral around, getting ever closer to the periodic solution. If you start inside the periodic orbit, then you spiral out to it, and if you start outside the periodic orbit, then you will spiral into it.

One last remark: This periodic orbit is stable against changing the parameters slightly in Equation (1), since a small change in parameters will not destroy the basin of attraction, nor will it change the repelling nature of the critical point inside.

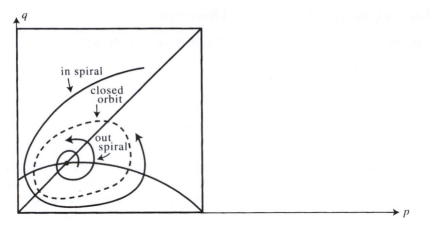

Figure 23.2

23.9   Lessons

Here are some key points from this chapter:

- A differential equation for two unknown functions of time can have time-periodic solutions.

- The time-periodic solutions must occur if there is a basin of attraction that has inside it a single equilibrium point that is unstable and repelling.

- In particular, this time-periodic solution will move in the basin of attraction and it is stable in the sense that nearby trajectories spiral to it.

READINGS FOR CHAPTER 23

READING 23.1

Wolves, Moose and Tree Rings on Isle Royale

Commentary: This article is also discussed in Commentary 6.2, where it is reproduced. This commentary summarizes the issues that are of concern here.

Isle Royale is a large island in Lake Superior and a national park. The island has long had both a moose population and a wolf population. This article presents data on the time variation of the populations of the wolves and moose on Isle Royale. Also presented are data on tree ring width of fir trees on Isle Royale. The latter is taken as a proxy for the effect of moose forage on the trees. The authors conclude from the data that the populations of moose and wolves and the health of the fir population vary in a cyclic fashion. The authors used their data to conclude further that the three-level interaction (wolves-moose-fir trees) is controlled by the two-level interaction of wolves and moose. (Do the authors' arguments convince you?)

READING 23.2

Snowshoe Hare Populations Squeezed from Below and Above;
Impact of Food and Predation on the Snowshoe Hare Cycle

Commentary: The first article is a discussion and synopsis of the second. The latter reports on experiments to determine the cause of the observed cyclic variation of the snowshoe hare population in northern Canada. The authors of the second article consider the possibility that the cyclic behavior is due to the interaction among the hares, their predators (lynx, coyote, and owls), and their food supply (willow and birch). In particular, the second article reports on experiments that are designed (in principle) to detect whether the hare population dynamics is due to the interaction of two, three, or more variables. The basic idea is to consider the hare population $h(t)$ as a function of time t in years. The philosophy here is that if the dynamics are controlled by three unknowns, then knowledge of $h(t)$ at three different times, say $(h(t), h(t + 1), h(t + 2))$, will allow us to compute $h(t + 3)$ with little error. Likewise, if there are only two variables that control the dynamics, then we might expect the pair $(h(t), h(t + 1))$ to determine $h(t+2)$ for all t. If there is one variable, then $h(t)$ should determine $h(t+1)$. The point is that $h(t)$ serves as a "proxy" for the one variable.

Say you wanted to determine if the dynamics were controlled by one variable. In this case, you would expect $h(t + 1)$ to be determined by $h(t)$, so if you plot $y = h(t + 1)$ versus $x = h(t)$, you should see a smooth curve graphed. For example, if the controlling equation is $\frac{d}{dt}h = rh$, then $h(t) = h(0)e^t$ and $h(t+1) = h(0)e^{t+1} = eh(t)$. So your plot would find the straight line $y = ex$ with slope e. If the dynamics were controlled by the equation $\frac{d}{dt}h = -rh^2$, the solution has $h(t) = h(0)(1 + rh(0)t)^{-1}$, and setting $y = h(t + 1)$ and $x = h(t)$ you would find that $y = x(1 + rx)^{-1}$. Anyway, if you find a nice curve for the y versus x plot, then you can say that the dynamics are controlled by one variable. If you find a scatter of points with no obvious graph shape, then you probably have more than one variable controlling the dynamics, as the diagram in Figure 23.3 illustrates.

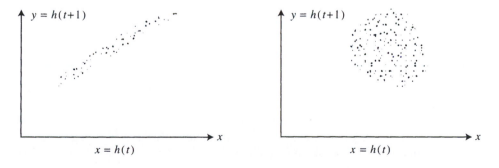

1-variable dynamics   More complicated dynamics

Figure 23.3

In the case of three variables, you plot $w = h(t+3)$ versus $x = h(t)$, $y = h(t+1)$,

$z = h(t+3)$. If the result is a graph, then you know that the dynamics are controlled by three variables. If the result is a scatter plot, then the dynamics are more complicated.

Of course, it is one thing to conclude that the dynamics are controlled by three variables and quite another to sort out which, of all the variables in a hare's environment, are the right three. This article attempts to do just that via judiciously designed experiments. After reading it, see if you are convinced.

Of independent interest is the observation of Krebs: "monitoring of populations is politically attractive but ecologically banal unless it is coupled with experimental work to understand the mechanisms behind the system changes." (Compare with the approach in the article "Wolves, Moose and Tree Rings on Isle Royale," by McLaren and Peterson, *Science* **266** (1994) 1555–1558.)

Snowshoe Hare Populations: Squeezed from Below and Above

Science, **269** (1995) 1061–1062.
Nils Christian Stenseth

All ecologists favor long-term studies and in this respect differ from chemists, physicists, most other biologists, and all politicians. But, like other scientists, ecologists prefer to do experimental work ... We must combine these two approaches to solve the major ecological questions [1, p. 3].

The cyclical population density of the Canadian snowshoe hare (*Lepus americanus*), with high densities occurring every 9 to 11 years, is a classic example of the multiannual cycles in many vertebrate populations of the boreal zone [2–4]. Although the cycles in the hare population are cited in almost all introductory biology texts, their cause has been obscure. Food, predators, disease, and sunspots have all been put forward as essential, but there has been no agreement as to which factors best explain the cycles. Now, Krebs and his co-workers, in the next article, report a technically unique and important experiment indicating that the hare population cycle results from a food-hare-predator interaction.

The new results show that the effects of food and predation on population density are nonadditive. Food augmentation and exclusion of mammalian predators separately caused about a twofold increase in the abundance of individual hares, whereas combined addition of food and reduction of predation increased the population density by a factor of 10. Krebs and co-workers therefore argue that neither plant-herbivore nor predator-prey interactions are by themselves sufficient for cycling.

Although this notion is consistent with the earlier suggestion of Keith [4] that the periodic fluctuations in the hare population are due to a dynamic interaction between predators and food shortage during winter, the new results do not necessarily support Keith's proposed sequential two-level interaction that assumes food shortage to be temporarily followed by predation.

The validity of the proposed three-trophic-level hypothesis may be tested independently by examining long-term monitoring data of snowshoe hare populations [6]. If a three-level interaction is truly responsible for generating and maintaining the observed dynamics, the relevant time series should exhibit dimension [7] three or higher. Indeed, the structure of the hare time series is consistent with dimension three [8] and is therefore consistent with the proposed three-level hypothesis. In theory, a three-dimensional structure in the time series could also arise because of several other three-factorial explanations; however, the extraordinary consistency between the experimental and the time-series data greatly strengthens the plant-hare-predator hypothesis. This consistency is emphasized by the facts that (i) the experimental and the time series data are independent sources of information; (ii) by combining these two data sources, insights derived from experimental manipulation can reinforce insights from statistical analyses of observed patterns; and (iii) the new experimental

---

Author affiliation: Nils Christian Stenseth: Department of Biology, University of Oslo.

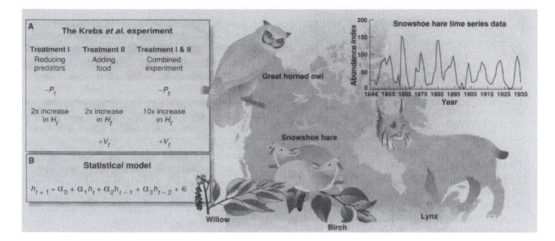

Fig. 1 Understanding population dynamics by combining experimental and modeling approaches. The structure and dynamics of ecological systems (such as the snowshoe hare of the Canadian boreal forest) may be deduced through a pluralistic approach. (**A**) Experimental manipulations like those reported by Krebs *et al.* in the following article [5]. Assuming H_t to be the abundance of the hare at time t, the statistical modeling has been done using $h_t = \ln[H_t]$. (**B**) Statistical modeling of long-term data [6, 8] (as well as mathematical modeling [7]). The snowshoe hare cycle may result from a dynamic interaction between the food supply of the hares [willow (*Salix glauca*) and birch (*Betula glandulosa*)], the hares, and the mammalian predators on the hares [lynx (*Lynx canadensis*), coyote (*Canis latrans*), and great horned owls (*Buteo virginianus*)] [9]. If the food supply, the hare population (h), and the predators each can be modeled as one entity, a system of three difference equations is a plausible mathematical description of the system, implying an expected three-dimensional structure of long-term time series data on the hare, which in fact is observed [8]. [In the statistical modeling $h_t = \ln(H_t)$ has been used as the transformed variable in the analysis.]

results allow a biological interpretation of the estimated dimensional structure of the time series.

Indeed, this synthesis is an example of the integrated, dual approach to ecology advocated by Krebs in 1991 [1]. I agree with Krebs (1, p. 3) that "monitoring of populations is politically attractive but ecologically banal unless it is coupled with experimental work to understand the mechanisms behind system changes." Further, the utility of experimentally deduced mechanisms, like the new work of Krebs *et al.* [5], is greatly enhanced if these mechanisms can be shown, as I have attempted above, to generate the patterns they are supposed to explain.

Many Northern microtines—lemmings and voles—also exhibit periodic fluctuations in their population densities [10]. The estimated dimension of the time series for small rodents is typically two [11], suggesting that the microtine cycle may be caused by fewer processes than are involved in the snowshoe hare cycle. Krebs favors a structurally

simpler hypothesis—the so-called Chitty hypothesis [12]—for the microtine cycle, consistent with the estimated lower dimensionality of the microtine time series. This hypothesis assumes that some population-intrinsic factor by itself causes the density cycle. But much experimental and theoretical evidence suggests that the Chitty hypothesis cannot explain the cycle [10, pp. 70-73]; extrinsic factors also seem essential [10, 13]. Perhaps either food or predators may be responsible for the microtine cycle, accounting for the estimated dimensionality of two of the small rodent time series. This two-trophic-level hypothesis for microtines awaits experimental testing, such as has now been provided for the snowshoe hare.

The hare cycle and the microtine cycle will continue to fascinate ecologists as they have since Elton's classic 1924 paper [2]. Thanks to the new study by the Krebs team, ecologists are now able to pose their questions more sharply, and some pieces in the jigsaw puzzle are falling into place.

REFERENCES

1. C. J. Krebs, *Ibis* **133** (suppl. 1), 3 (1991).
2. C. S. Elton, *Br. J. Exp. Biol.* **2**, 119 (1924).
3. ———, *Voles, Mice and Lemmings* (Clarendon, Oxford, 1942).
4. L. B. Keith, *Wildlife's Ten-Year Cycle* (Univ. of Wisconsin Press, Madison, WI, 1963); *Proc. Int. Congr. Game Biol.* **11**, 17 (1974); *Oikos* **40**, 385 (1983); *Curr. Mammal.* **2**, 119 (1990).
5. C. J. Krebs *et al.*, *Science* **269**, 1112 (1995).
6. Data from 1844 to 1904 are from the Hudson Bay Company fur records; those from 1905 to 1935 are from trapper questionnaires; both sets are from D. A. MacLulich [*Fluctuations in the Numbers of the Varying Hare (Lepus americanus)* (Univ. of Toronto Press, Toronto, 1937)] and A. R. E. Sinclair *et al.* [*Am. Nat.* **141**, 173 (1993)]. This combined time series is from a different location than the data used for the experimental study in [5], but because the snowshoe hare cycles are synchronized throughout most of the Canadian boreal forest, the two sets of data are assumed to represent the same general phenomena [[4]; A. R. E Sinclair *et al.*, *Am. Nat.* **141**, 173 (1993), but see C. H. Smith, *Can. Field Nat.* **97**, 151 (1983)].
7. This refers to the embedding dimension [B. Cheng and H. Tong, *Philos. Trans. R. Soc. London Ser. A* **348**, 325 (1994); P. Turchin and J. A. Millstein, *EcoDyn/RSM: Response Surface Modelling of Nonlinear Ecological Dynamics* (Applied Biomathematics, New York, 1994)] and represents the number of lags d needed to describe the dynamics with a model $N_{t+1} = N_t \exp[f(N_t, N_{t-1}, \ldots, N_{t-d}) + \epsilon]$ [T. Royama, *Analytical Population Dynamics* (Chapman & Hall, London, 1992); *Ecol. Monogr.* **51**, 473 (1981)], where f is some appropriate (linear or nonlinear) function, N_t is the abundance at time t, and ϵ is a sequence of martingale differences with constant variance and is assumed to represent the environmental stochasticity [D. Goodman, *Viable Populations for Conservation*, M. E. Soul, Ed. (Sinauer, Sunderland, MA, 1987), p. 11]. If the underlying dynamics are linear, a model with k factors (for example, k trophic levels) will result in an embedding dimension k. If the dynamics are highly nonlinear, the embedding dimension may be higher, but not lower, than k [D. S. Broomhead and R. Jones, *Proc. R. Soc. London Ser. A.* **423**, 103 (1989)]. Many mechanisms other than k trophic interactions can cause lagged dynamics and an embedding dimension of order k; the process implies the dimension, but the dimension does not imply the process.
8. N. C. Stenseth, O. Bjørnstad, W. Falck, unpublished results. We have analyzed the hare time series [6] both as one 92-year-long time series and as two separate time series; both analyses provide analogous results. Details for the combined time series are as follows: The dimension of the square root-transformed time series was estimated by cross-validation with (i) a linear autoregressive model, and (ii) a nonparametric kernel autoregressive model. In (i), maximum likelihood estimates were calculated with a Kalman filter applied to the state space representation of the likelihood [R. Kohn and C.F. Ansley, *J. Am. Stat. Assoc.* **81**, 751 (1986)]. In (ii), nonparametric conditional means were calculated with a Gaussian product kernel [7]. The optimal dimension was five for both methods. However, that dimension gave only marginally better predictions (< 0.4 percent units) than that of dimension three. Models of dimension two were inferior in either case (by > 3.5 percent units). The result is analogous to that obtained with standard information theoretic approximations based on Akaikes information criterion [C. M. Hurvich and C.-L. Tsai, *Biometrika* **76**, 297 (1989)]. Tests for nonlinearity were also carried out. The null hypothesis of linearity could not be rejected. A hypothesis of three trophic interactions responsible for the most of the population dynamics is consistent with the time series data: Components of the various trophic levels presumably contribute marginally to the overall dynamics of the snowshoe hare population.
9. S. Boutin *et al.*, *Oikos*, in press.
10. N. C. Stenseth and R. A. Ims, *The Biology of Lemmings* (Academic Press, London, 1993), pp. 75–79.
11. O. N. Bjørnstad, W. Falck, N. C. Stenseth, *Proc. R. Soc. London Ser. B*, in press.
12. C. J. Krebs, *Arc. Inst. N. Am. Tech. Pap.* **15**, 1 (1964); C. J. Krebs *et al.*, *Science* **179**, 35 (1973); C. J. Krebs and J. H. Myers, *Adv. Ecol. Res.* **8**, 267 (1974); C. J. Krebs, *Can. J. Zool.* **56**, 2463 (1978); in [10], p. 247; ———, R. Boonstra, A. J. Kenney, *Oecologia*, in press; D. G. Reid,

C. J. Krebs, A. J. Kenney, *Oikos*, in press; D. Chitty, *Cold Spring Harbor Symp. Quant. Biol.* **22**, 277 (1958); *Can. J. Zool.* **38**, 99 (1960); *Proc. Ecol. Soc. Aust.* **2**, 51 (1967). For a review, see C. J. Krebs, *Can. J. Zool.* **56**, 2463 (1978).

13. M. J. Taitt and C. J. Krebs, *Am. Soc. Mammal. Spec. Publ.* **8**, 567 (1985); G. O. Batzli, *Wildlife*

2001, D. R. McCullough and R. H. Barrett, Eds. (Elsevier, New York, 1992), pp. 831–850.

14. I thank O. N. Bjørnstad and W. Falck for discussions and contributions to the information in [7] and [8] and E. Framstad, R. Anker Ims, C. J. Krebs, S. L. Pimm, H. Steen, H. Tong, and J. Wolff for comments on an earlier draft and for discussions.

Impact of Food and Predation on the Snowshoe Hare Cycle

Science, **269** (1995) 1112–1115.

C. J. Krebs, S. Boutin, R. Boonstra, A. R. E. Sinclair, J. N. M. Smith, M. R. T. Dale, K. Martin, and R. Turkington

Snowshoe hare populations in the boreal forests of North America go through 10-year cycles. Supplemental food and mammalian predator abundance were manipulated in a factorial design on 1-square-kilometer areas for 8 years in the Yukon. Two blocks of forest were fertilized to test for nutrient effects. Predator exclosure doubled and food addition tripled hare density during the cyclic peak and decline. Predator exclosure combined with food addition increased density 11-fold. Added nutrients increased plant growth but not hare density. Food and predation together had a more than additive effect, which suggests that a three-trophic-level interaction generates hare cycles.

The 10-year cycle of snowshoe hare populations and those of their predators is one of the dominant perturbations of the boreal forests of North America. Predation and food shortage have been postulated as the major factors causing these fluctuations [1]. Because in all cyclic populations many factors will change in a manner correlated with population density, necessary conditions can be recognized only by experimental manipulations [2]. From 1976 to 1984, we manipulated food supplies of snowshoe hares (*Lepus americanus*) in the southern Yukon and showed that the cyclic decline could not be prevented by either artificial or natural food

addition [3]. Single-factor manipulations have been criticized in field ecology because they may miss important interactions between factors [4]. For the past 8 years, we have carried out large-scale experiments on nutrients, supplemental food, and predation in the Yukon to untangle the causes of the hare cycle and the consequences the hare cycle has for the vertebrate community. By crossing a predator reduction manipulation with food addition we estimated interaction effects caused by the failure of factors to combine additively.

We chose 1-km^2 blocks of undisturbed boreal forest near Kluane Lake, Yukon, as our experimental units [5]. The boreal forest in this region is dominated by white spruce (*Picea glauca*) and was not disturbed by logging, fire, or extensive fur trapping during our studies. We used a factorial design to untangle the effects of food and predation on hares. Three areas were used as controls [6]. Two experimental areas were provided with ad lib supplemental food year-round. We excluded mammalian predators by building one electric fence in the summer of 1987. In the summer of 1988, we built a second electric fence to use for the combined predator reduction-food addition treatment [7]. Since January 1989, the electric fences have worked effectively to prevent mammalian predators from entering the two areas. The fences are permeable

Author affiliations: C. J. Krebs, A. R. E. Sinclair, J. N. M. Smith: Department of Zoology, University of British Columbia. S. Boutin and M. R. T. Dale: Department of Biological Sciences, University of Alberta. R. Boonstra: Department of Life Sciences, Scarborough Campus, University of Toronto. K. Martin: Department of Forest Sciences, University of British Columbia. R. Turkington: Department of Botany, University of British Columbia.

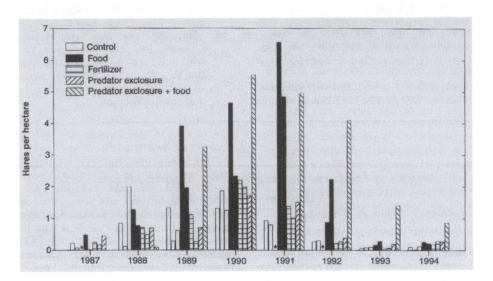

Fig. 1 Spring densities of snowshoe hares in three control and six treatment areas, Kluane Lake, Yukon, 1987 to 1994. Densities were estimated from mark-recapture live trapping for 4 to 5 days in late March and early April each year with the use of the jackknife estimator in the program Capture [10]. Asterisk indicates one of the three control areas that was not trapped in 1987, 1991, and 1992.

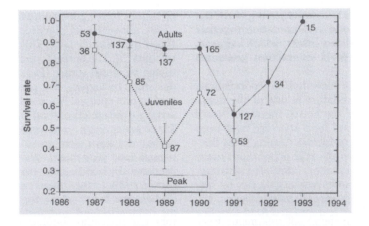

Fig. 2 Survival rates of adult (circles) and juvenile (squares) snowshoe hares in the most intensively studied control area (control area 1). Juvenile survival is already deteriorating in the late increase phase (1988) and is low in the peak and decline phases of the cycle. We could not estimate juvenile survival in 1992 because few of the 13 juveniles caught were ever recaptured. We presume juvenile survival was poor in 1992. Survival rates (per 28 days) were estimated from mark-recapture data with the use of the Jolly-Seber Model B [10]. Juvenile survival refers to trappable juveniles only, which are those more than 6 to 8 weeks old. Sample sizes of hares are given next to data points.

to snowshoe hares. Beginning in 1987, we added nitrogen-potassium-phosphorus (NPK) fertilizer to two blocks of forest to increase plant growth [8]. We chose to manipulate a few large areas rather than many small areas because of the failure of most field experiments to address large-scale issues [9]. We captured, marked, and released snowshoe hares every March and October and estimated densities with the robust design [10].

Snowshoe hares in the control areas increased from a low in the mid-1980s to a peak in 1989 and 1990 (Figure 1). The increase phase from 1986 to 1988 showed considerable variation among the three control populations, but from the peak phase onward all the controls were similar in their year-to-year dynamics. The cyclic decline began in autumn and winter 1990 and continued until the spring of 1993 when hares had reached low numbers of approximately one hare per 15 ha. Population increase in snowshoe hares is stopped both by increased mortality and by reduced reproductive output [11]. This previously described syndrome of demographic changes was consistent over the cycle we observed. Juvenile mortality increased while the population was still in the increase phase of the cycle (Figure 2), whereas adult losses did not become severe until the decline phase. The decline phase in 1991 and 1992 was characterized by poor survival of both juveniles and adults and by reduced reproductive output by females through restriction or elimination of their second and third summer litters [11].

The impact of our experimental treatments can be measured in several ways. We concentrate here on changes in the population density of hares in the treated areas and on the survival rates of radio-collared hares. Density effects can be most simply expressed as ratios of the density in the treated area to the density in the control areas. We estimated these each spring and autumn for all treatments (Figure 3). Effects were small during the increase phase in 1987 and 1988 because the treatments were just being established. All treatments were effective by spring 1989. The food addition effect was always positive and produced densities ranging from 1.5- to 6-fold over control levels during the peak and decline phases (Fig. 3B). The predator exclosure effect was negligible in the peak phase from 1989 to 1990 but became pronounced in the late decline and low phases, producing densities ranging from 1.4- to 6-fold over control levels (Fig. 3A).

The largest effect was shown by predator exclosure and food treatment combined, particularly in the late decline phase when densities exceeded control levels by 36-fold (Fig. 3C). Averaged over both the peak and the decline phase, predator exclosure approximately doubled the density of hares, food addition approximately tripled density, and the combined treatment increased density 11-fold.

In contrast to the strong effects shown by manipulation of predation and food supply, the addition of nutrients had virtually no effect on snowshoe hare numbers. In spite of increased growth of herbs, grasses, shrubs, and trees [8], the fertilized plots contained virtually the same number of hares as did the control plots (Fig. 1). Fertilized vegetation in the boreal forest cannot duplicate either the quantity or quality of the artificial food that we added in our experiments, and for this reason fertilization is a relatively ineffective method of food addition for hares.

Survival rates can be estimated from mark-recapture methods or from radio-telemetry [12]. Treatments had little impact on survival rates during the peak phase of the hare cycle. Monthly adult survival rates were greater than 90% in the peak phase, leaving little room for improvement. The major effects of the treatments on survival were visible in the decline phase (Figure 4). The probability of a hare living for 1 year in the control areas during the decline was 0.7%. In fertilized areas, this probability was 1.9%, which is slightly but not significantly higher than in the control areas. This probability improved to 3.7% in the food addition grids and to 9.5% in the predator exclosure areas. The best chances of survival occurred in areas treated with a combination of mammalian predator reduction and food, where the probability of survival was 20.8% for 1 year during the decline. The effects of food and predation on survival during the decline phase were nearly additive and showed no sign of an interaction. The addition of food by itself was not sufficient to prevent large losses to predators, and the rapid population collapse in the food areas from 1991 to 1992 (Fig. 1) was due to heavy predation.

The numbers of both avian and mammalian predators follow the hare cycle, but with a 1- to 2-year time lag [1]. Virtually all snowshoe hares in our study area die from predator attack in the immediate sense. From 1989 to 1993, we found that 83% of the deaths of all radio-collared hares were due

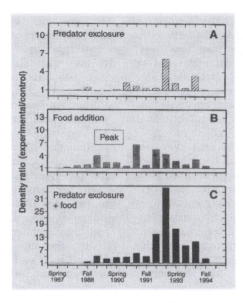

Fig. 3 Ratio of population densities for the three treatments to average control population densities at the same time. If there is no treatment effect, we expect a ratio of 1.0. During the peak and decline phases, the predator exclosure (**A**) doubled density on average, food addition (**B**) tripled density, and the combined treatment (**C**) increased density 11-fold.

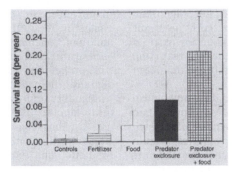

Fig. 4 Annual survival rates for snowshoe hares with radio collars during the decline phase from autumn 1990 to autumn 1992. Ninety-five percent confidence limits are shown. Sample sizes for the estimates are (in order): 278, 206, 197, 246, and 262 hares. Radio collars were placed only on fully grown animals and thus are used to measure adult mortality rates.

to predation and only 9% were attributed to starvation [13]. We presume that hares suffering from food stress will be more susceptible to death from predation.

Because we used live trapping as our primary technique of study, we have less data on the re-

productive output of snowshoe hares in relation to these three treatments. During the peak phase, the food treatment areas have the same reproductive output as do the control areas [14]. Table 1 gives the total production of live young hares at birth for a female over the summer breeding pe-

Table 1

Year	Control areas [no. of young ± SE (n)]	Food areas [no. of young ± SE (n)]	Predator exclosure + food area [no. of young ± SE (n)]
1988	16.4 ± 0.44 (10)	—	—
1989	13.7 ± 0.39 (21)	14.1 ± 0.43 (36)	—
1990	13.7 ± 0.40 (33)	15.1 ± 0.43 (36)	—
1991	7.8 ± 0.37 (18)	—	16.3 ± 0.73 (15)
1992	3.3 ± 0.25 (4)	—	17.1 ± 0.31 (50)

Total production of live young by female snowshoe hares over the summer breeding season during the late increase (1988), peak (1989 and 1990), and decline (1991 and 1992) phases of the hare cycle. Litter sizes at birth were not available for all treatments in all years. We assume that total production in the predator exclosure plus food treatment area would have been the same as in the food areas in the peak years of 1989 and 1990. Numbers in parentheses indicate number of females sampled.

riod for the control areas, the food areas, and the predator exclosure plus food grid. The collapse of reproduction in the control areas was prevented in the combined treatment area. Because we do not know the reproductive schedule for the food or the predator exclosure treatment areas during the decline phase, we cannot assess the separate contributions of food and predation to the reproductive output of females.

There are three possible explanations of these differences in reproductive output. Reproductive changes may be driven by food limitations in the decline period. Alternatively, hares may respond to predation risk in the decline phase by altering their habitat use so that they cannot achieve adequate nutrition [15]. In both of these cases, this reproductive curtailment is due to food shortage, but in the first case it is absolute food shortage and in the second case it is relative food shortage caused as an indirect effect of predation. The third possibility is that these reproductive effects are a direct result of stress and the physiological derangement associated with stress [16]. We cannot yet determine which of these explanations is correct. Absolute winter food shortage does not necessarily occur during the peak or decline phase [3], and the weight of evidence is against the first explanation. Behavioral evidence suggests that there may be a relative shortage of food [15].

If both food and predation are together sufficient to explain population cycles in snowshoe hares, why were we not able to prevent the decline entirely in the combination treatment area?

Hare densities fell from about seven per hectare in 1989 to about one per hectare in 1994 in the combination treatment area, even though hare density in this area remained at or above normal peak densities for 7 years from 1988 to 1995. The combination treatment delayed the decline but did not prevent it. There are three possible reasons for this, and they illustrate one difficulty of largescale experiments. First, hares could move freely into and out of the predator exclosures. Individuals were often killed by predators when they moved out, but others could emigrate into nearly unoccupied landscape outside the fence in the low years of 1992 to 1994 [17]. Second, we could not prevent raptor predation inside the predator exclosure. Our monofilament treatment covered only a small fraction of the predator exclosure and was ineffective in preventing raptors such as goshawks from invading the area. Because raptors and owls cause about 40% of the predation mortality in our hare populations, the effect of the exclosures was to reduce total predation losses, not eliminate them. Great horned owls and goshawks continued to kill hares inside the predator exclosures during the decline and low phases. Third, another factor in addition to food and predation may be sufficient to cause the decline.

These results support the view that population cycles in snowshoe hares in the boreal forest are a result of the interaction between food supplies and predation. They do not support either the plant-herbivore model or the predator-prey model for cycles but suggest that hare cycles result from a three-trophic-level interaction [18]. Our experimental results are consistent with the general ideas of Keith [19] and Wolff [20] that both food and predation play a role in generating hare cycles, but they do not support Keith's sequential two-factor model that states that food shortage effects are followed by predation effects in causing cyclic declines. Further work will be required to determine if the nutritional effects on hares are an indirect effect of predation that is explicable in terms of the hares' behavioral responses to predation risk. Our studies have provided little data on the causes of the low phase of the hare cycle, which can persist for 3 to 4 years. Food supplies recover quickly after the peak has passed, and predator numbers collapse during the hare decline. Whether the direct or indirect effects of predation can also explain the low phase remains an open question.

REFERENCES

1. L. B. Keith, *Wildlife's Ten Year Cycle* (Univ. of Wisconsin Press, Madison, WI, 1963); ——, J. R. Cary, O. J. Rongstad, M. C. Brittingham, *Wildl. Monogr.* **90**, 1 (1984); L. B. Keith, *Proceedings of the World Lagomorph Conference*, Guelph, Ontario, 12 to 16 August 1979, K. Myers and C. D. MacInnes, Eds. (Univ. of Guelph, Ontario, Canada, 1981), pp. 395-440; M. R. Vaughan and L. B. Keith, *J. Wildl. Manage.* **45**, 354 (1981); J. P. Finerty, *The Population Ecology of Cycles in Small Mammals* (Yale Univ. Press, New Haven, CT, 1980).

2. C. J. Krebs, *Oikos* **52**, 143 (1988); D. Chitty, *Can. J. Zool.* **38**, 99 (1960).

3. C. J. Krebs, B. S. Gilbert, S. Boutin, A. R. E. Sinclair, J. N. M. Smith, *J. Anim. Ecol.* **55**, 963 (1986); A. R. E. Sinclair, C. J. Krebs, J. N. M. Smith, S. Boutin, *ibid.* **57**, 787 (1988); J. N. M. Smith, C. J. Krebs, A. R. E. Sinclair, R. Boonstra, *ibid.*, p. 269; C. J. Krebs, S. Boutin, B. S. Gilbert, *Oecologia* **70**, 194 (1985).

4. E. A. Desy and G. O. Batzli, Ecology 70, 411 (1989); G. O. Batzli, in *Wildlife 2001: Populations*, D. R. McCullough and R. H. Barrett, Eds. (Elsevier, London, 1992), pp. 831-850.

5. Blocks were spaced at least 1 km apart. Within each block, we surveyed checkerboard grids of 20-by-20 points with 30.5-m spacing and used these grids for snowshoe hare live trapping. Two experimental areas were provided with supplemental food (commercial rabbit chow, 16% protein) year round. In

the summer of 1987, we built one electric fence around 1 km^2 to exclude mammalian predators, and over the following year we covered 10 ha with monofilament to reduce avian predation. The monofilament was never effective in preventing avian predation inside the electric fences, and consequently we did not rely on it as a part of the treatment. In the summer of 1988, we built a second electric fence around 1 km^2 to use for the combined predator reduction-food addition treatment. We modified the design of the electric fences in 1988 to make them more effective, and since then they have worked effectively to prevent mammalian predators from entering the area. The fences are permeable to snowshoe hares. We could not replicate either the predator reduction or the predator reduction-food addition treatment because of maintenance costs and the difficulty of maintaining electric fences in the Yukon winter with −45°C temperatures. The fences had to be checked every day during winter. From 1976 to 1985, we trapped hares in six areas within the main study region and found that their population trajectories were very similar [3]. We thus have no reason to suspect strong area effects on the unreplicated predator reduction plots.

6. We used three control areas but were not able to trap hares in all of them every year. We have more detailed data on hares from control area 1. The three control areas had quite different histories during the increase phase from 1986 to 1988. Control area 3 reached its greatest hare density in 1988 and remained at a plateau until 1990. Control area 2 reached its peak density in 1990, and control area 1 reached its peak in 1989. By the late peak in 1990 and during the decline phase, the control areas were much more similar to each other in hare densities.

7. The electric fence was 10-stranded, 2.2 m in height, and carried 8600 V. Snow tracking of mammalian predators meeting the fence illustrated its effectiveness. We excluded mammalian predators virtually continuously from January 1989 onward. Our attempts to use monofilament fishing line as a deterrent to birds of prey was largely ineffective because ice formation and snow accumulation on the lines in winter caused them to break or collapse to the ground. We used

monofilament on 10 ha of the predator ex-closure but did not attempt to use it on the combination treatment area. The predator exclosures thus were mammalian predator exclosures and were still subject to avian pre-dation.

8. We fertilized two 1-km² blocks of forest with commercial fertilizer. In May 1987, we used ammonium nitrate at 25 g/m². In May 1988, we switched to NPK fertilizer and used 17.5 g of N/m², 5 g of P/m², and 2.5 g of K/m². In 1989, we used half this amount, and in the years 1990 to 1994 we used the full amount as in 1988. The fertilizer was spread aerially and we did ground checks to make sure it was uniformly spread. We do not present the data here to show the plant growth responses, but all elements of the flora responded dramati-cally to the added nutrients (C. J. Krebs *et al.*, unpublished data).

9. D. Tilman, in *Long-Term Studies in Ecology*, G. E. Likens, Ed. (Springer-Verlag, New York, 1989), pp. 136–157; J. F. Franklin, *ibid.*, pp. 3–19.

10. K. H. Pollock, J. D. Nichols, C. Brownie, J. E. Hines, *Wildl. Monogr.* **107**, 1 (1990); K. H. Pol-lock, *J. Wildl. Manage.* **46**, 752 (1982). Con-fidence limits were typically plus minus 15% of the population estimates. Snow tracking in winter confirmed density estimates.

11. L. B. Keith, *Curr. Mammal.* **2**, 119 (1990).

12. S. Boutin and C. J. Krebs, *J. Wildl. Manage.* **50**, 592 (1986); K. H. Pollock, S. R. Winterstein, C. M. Bunck, P. D. Curtis, *ibid.* **53**, 7 (1989).

13. D. Hik, *Wildl. Res.* **22**, 115 (1995); S. Boutin *et al.*, unpublished data.

14. M. O'Donoghue and C. J. Krebs, *J. Anim. Ecol.*

61, 631 (1992); J. R. Cary and L. B. Keith, *Can. J. Zool.* **57**, 375 (1979).

15. D. Hik, unpublished data; R. Boonstra and G. R. Singleton, *Gen. Comp. Endocrinol.* **91**, 126 (1993).

16. R. Boonstra, *Evol. Ecol.* **8**, 196 (1994); ____, D. Hik, G. R. Singleton, in preparation.

17. In snowshoe hares, dispersal occurs mainly in the juvenile stage; adult hares are mostly sedentary [S. Boutin, B. S. Gilbert, C. J. Krebs, A. R. E. Sinclair, J. N. M. Smith, *Can. J. Zool.* **63**, 106 (1985)]. Artificial food addition causes immigration of adult hares [S. Boutin, *Oecolo-gia* **62**, 393 (1984)], and population changes in food addition areas are more affected by movements than are changes at other treat-ment or control sites.

18. L. Oksanen, in *Perspectives on Plant Consump-tion*, D. Tilman and J. Grace, Eds. (Academic Press, New York, 1990), pp. 445–474; S. D. Fretwell, *Oikos* **50**, 291 (1987); S. R. Carpen-ter, J. F. Kitchell, J. R. Hodgson, *Bioscience* **35**, 634 (1985).

19. L. B. Keith, *Oikos* **40**, 385 (1983).

20. J. O. Wolff, *Ecol. Monogr.* **50**, 111 (1980).

21. We thank the Natural Sciences and Engineer-ing Research Council of Canada for support-ing this research program through a Collab-orative Special Project grant; the Arctic In-stitute of North America, University of Cal-gary, for the use of the Kluane Lake research station; and all the graduate students, tech-nicians, and undergraduates who have as-sisted in this project over the past 9 years, in particular V. Nams, S. Schweiger, and M. O'Donoghue.

Exercises for Chapter 23

1. This exercise considers equations for functions $p(s)$ and $q(s)$, of a parameter s that have the following form:

$$\frac{d}{ds}\begin{pmatrix} p \\ q \end{pmatrix} = \begin{pmatrix} f(p, q) \\ g(p, q) \end{pmatrix}.$$

Here f and g are suitable functions on the (p, q) plane. In answering the following questions, remember that a ***basin of attraction*** or ***trapping region*** for (1) is a region, V, in the (p, q) plane from which trajectories never exit. (Think of V as a black hole; trajectories can enter V, but they can never leave V.) You can check whether or not a given set V is a basin of attraction by considering the direction of the flow along the boundary of V.

For which f and g given is the square $\{(p, q) : 0 \leq p \leq 1, 0 \leq q \leq 1\}$ a basin of attraction?

(a) $f(p, q) = q - p$ and $g(p, q) = p - q$

(b) $f(p, q) = q(p - \frac{1}{2})$ and $g(p, q) = -pq$

(c) $f(p, q) = p(q - \frac{1}{2})$ and $g(p, q) = -q$

(d) $f(p, q) = \cos(\pi(p + q/8))$ and $g(p, q) = p^2 - q^2$

(e) $f(p, q) = qp(p - \frac{1}{2})$ and $g(p, q) = pq - 1$

[For example, in Exercise 1a, a hypothetical trajectory can point out of the square along the side, where $p = 0$ and $0 \leq q \leq 1$ only if $p(s)$ is decreasing when the trajectory hits this side. However, $p(s)$ cannot be decreasing when $p = 0$ and $0 \leq q \leq 1$ because $\frac{d}{ds}p = f(p, q) = q \geq 0$, where $p = 0$ and where $0 \leq q \leq 1$. Likewise, trajectories cannot point out of the square along the side where $p = 1$ and $0 \leq q \leq 1$ because $\frac{d}{ds}p = f(p, q) = q - 1 \leq 0$ along this side.]

2. The following are some matrices. If each were the stability matrix at an equilibrium point, classify the equilibrium point as being stable (so $\det > 0$ and $\mathrm{tr} < 0$) or repelling (so $\det > 0$ and $\mathrm{tr} > 0$) or neither.

(a) $\begin{pmatrix} 1 & 2 \\ 1 & 3 \end{pmatrix}$

(b) $\begin{pmatrix} -1 & 2 \\ 2 & 3 \end{pmatrix}$

(c) $\begin{pmatrix} 1 & 2 \\ -2 & -3 \end{pmatrix}$

(d) $\begin{pmatrix} -2 & 4 \\ -3 & 3 \end{pmatrix}$

24

Fast and Slow

We saw in the previous chapter some criteria for a differential equation for two unknown functions of time to have a periodic solution. But we only proved that there were stable periodic solutions, and we did not discuss the speed of movement. This chapter discusses speed issues.

In particular, this chapter discusses a model system that swings rapidly between two or more approximate equilibrium positions. In this chapter, the point is not so much the model, but the way in which the rapid swings between different approximate equilibria occur. Indeed, the phenomenon of rapid switching between approximately constant states occurs in many different contexts, so it is important to realize that there are reasonably general properties of certain kinds of differential equations that guarantee this switching behavior. In particular, with the model discussed here, focus on the fact that two systems are interacting; one is fundamentally slow moving while the other can move much faster. The resulting dynamics of the interacting system sometimes mirror the motion of the slow system and sometimes the fast one.

The simple model discussed here describes the switching behavior for a muscle that controls a valve in the heart. I learned of the model and learned the analysis from Beltrami's book *Mathematics for Dynamical Modeling*.

The model is much too simple to be biologically reasonable. However, we shall see that it exhibits some remarkably lifelike behavior. In particular, the model experiences periodic behavior that is characterized by the fact that the muscle spends most of the time moving slightly and very slowly near one of two positions. I shall call these the "rest positions." They are not true equilibrium positions of the system, but they are approximate equilibria in that the muscle spends most of its time near one or the other of them. However, there are short time intervals during which the muscle moves rapidly from one rest position to the other. (You may think of this rapid motion as an analog to the time during a heartbeat cycle when a valve slams shut or swings open. The rest positions correspond to the open valve position and to the closed valve position.)

24.1 The System

Suppose that the position of the muscle in question at time t is described by a function $x(t)$. The concentration at time t of a chemical stimulus above or below some fixed concentration is described by a function $\alpha(t)$. Suppose that the dynamics of x and α are controlled by the following system of differential equations:

$$\frac{d}{dt}x = -\frac{x^3}{3} + x + \alpha \tag{1a}$$

$$\frac{d}{dt}\alpha = -\varepsilon x. \tag{1b}$$

Here, $\varepsilon > 0$ is a parameter; its inverse will be seen to estimate roughly the time that x spends near one or the other of the rest positions. I shall suppose that the number ε is very small.

24.2 Slow and Fast Subsystems

As a prelude to the subsequent analysis of (1), let me point out two crucial features of (1): First, as long as $|x|$ never gets too big (say $|x| < 10$), then the speed with which α changes is never more than 10ε. Thus, if ε is small, then α always moves slowly. In particular, you should view the function $\alpha(t)$ as representing an inherently slow-moving subsystem. Second, Equation (1a) has no ε in it, and thus its solutions can move relatively quickly. Indeed, if you forget (1b) for α and take α in (1a) to be constant, then as long as $x(t)$ starts away from one of the equilibrium points of (1a), it will move relatively quickly. With this fact understood, you should view the function $x(t)$ as representing a subsystem that has the potential to move relatively quickly.

It is in the preceding sense that the system described by (1) should be viewed as one in which an inherently slow subsystem is interacting with a subsystem that is not inherently slow.

24.3 A Heuristic Picture

Here is a picture of the behavior of the system in (1): Suppose that we start with $\alpha = 2/3$ and with $x = 2$. Note that $x = 2$ is a stable equilibrium point for the equation

$$\frac{d}{dt}x = -\frac{x^3}{3} + x + \frac{2}{3}. \tag{2}$$

But α will not stay at 2/3 since Equation (1b) asserts that α will decrease slowly. Now, as α decreases from 2/3 (but $\alpha > -2/3$), the equation

$$\frac{d}{dt}x = -\frac{x^3}{3} + x + \alpha \tag{3}$$

has a stable equilibrium point at positive x, but no longer at $x = 2$. Rather, this stable equilibrium point is at some value of x between 2 and 1. Call this equilibrium point x_α since its position depends on the value of α. See Figure 24.1.

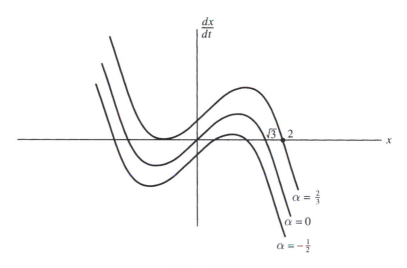

Figure 24.1

Since α moves very slowly, we might expect that x will also move very slowly to track the moving equilibrium point. That is, we might guess that $x(t)$ should stay close to the equilbrium point, $x_\alpha(t)$, of (3) for the case where α has value $\alpha(t)$. Indeed, as long as $\alpha(t)$ changes very slowly [which is the case when ε in (1) is tiny], then a good first approximation takes α in the first line of (1) and in (3) as a constant. In this case, the dynamics of (3) push $x(t)$ toward $x_{\alpha(t)}$ as long as $x(t)$ is near $x_{\alpha(t)}$ and the latter is a stable equilibrium point to the $\alpha = \alpha(t)$ version of (3). Of course, in reality, $\alpha(t)$ is slowly changing, but even so, (3) will still keep $x(t)$ near $x_{\alpha(t)}$ when the latter is a stable equilibrium point. In particular, we should expect that $x(t)$ tracks the slowly moving equilibrium point $x_{\alpha(t)}$ until α decreases to $-2/3$ when this equilibrium point becomes unstable. The $\alpha = -2/3$ version of Figure 24.1 is pictured in Figure 24.2.

You can see that $x_{-2/3} = 1$ is an equilibrium point, but it is not stable. Furthermore, when $\alpha(t) < -2/3$, then (3) does not have a positive equilibrium point at all. See Figure 24.3.

So when $\alpha(t)$ drops below $-2/3$, Equation (3), which is governing the motion of x, has no positive equilibrium points. This means that $x(t)$ must travel from where it is nearly 1 to the remaining stable equilibrium point, where $x < -2$. Notice that

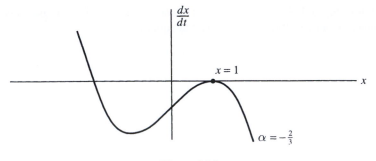

Figure 24.2

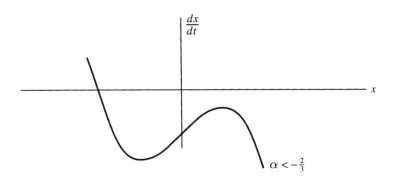

Figure 24.3

this motion is governed by (3) and so occurs at a speed that is governed by (3). Equation (3) sees no sign of the parameter ε, so this transition for $x(t)$ from where x is approximately 1 to where $x < -2$ should occur rapidly, at least with respect to the time it took for x to decrease from 2 to 1. (The time for x to decrease from 2 to 1 is controlled by the speed at which α goes from 2/3 to $-2/3$, and this speed is determined by ε.)

When x has switched to nearly -2, then Equation (1b) asserts that α will start increasing. However, this increase will occur slowly (as determined by ε). As long as $\alpha < 2/3$, Equation (3), which governs the behavior of x, has a stable equilibrium point, call it x_α, which lies between -2 and -1. Here, the subscript α is to remind you that this point depends on the value of α in (3). It moves as α changes. We might expect that $x(t)$ should now hug $x_{\alpha(t)}$ as long as $\alpha(t) < 2/3$. (This is the mirror image of the behavior that occurred before when x hugged $x_{\alpha(t)}$.) See Figure 24.4.

However, when $\alpha(t)$ reaches the value of 2/3, then $x_{2/3} = -1$ is no longer a stable equilibrium point. And when $\alpha(t) > 2/3$, then there is no negative stable equilibrium point for (3). This means that $x(t)$ will be forced to move from near -1 to the positive equilibrium point of (3), which is near 2. This motion will be governed by (3) and so will not see the value of ε. Thus, the motion here can be relatively fast

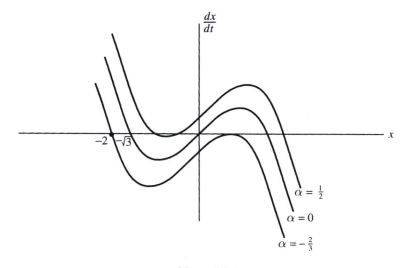

Figure 24.4

with respect to the motion of $x(t)$ as it followed the negative equilibrium point $x_{\alpha(t)}$.

When $x(t)$ returns to the vicinity of 2, the cycle is complete and can start again.

24.4   The Existence of a Cycle

I will now argue that (1) has a stable, periodic solution. I will use the Poincaré-Bendixson method that I used previously. That is, I will show that the origin in the (x, α) plane is a *repelling* equilibrium point, and I will show that the origin is also contained in a basin of attraction.

To begin, Figure 24.5 shows the phase diagram for (1). Note that the origin is the only equilibrium point.

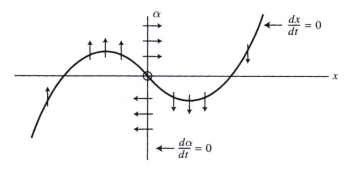

Figure 24.5

The stability matrix for (1) at the origin is

$$\mathbf{A} = \begin{pmatrix} 1 & 1 \\ -\varepsilon & 0 \end{pmatrix} \tag{4}$$

whose determinant equals ε and whose trace equals 1. Since both trace and determinant are positive, the equilibrium point at the origin is repelling. (This means that trajectories that start near this equilibrium point will move away from it.)

I shall now argue that the domain pictured in Figure 24.6 is a basin of attraction for (1) in the sense that no trajectories can exit this domain.

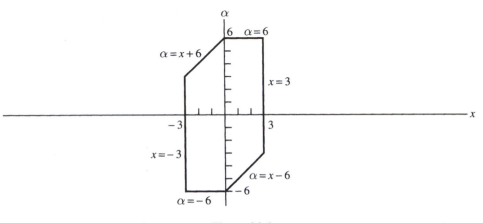

Figure 24.6

Indeed, as long as $\alpha \le 6$, then no trajectory can pass through the wall at $x = 3$ because there, $\frac{d}{dt}x = -6 + \alpha \le 0$. Likewise, as long as $\alpha \ge -6$, no trajectory can pass through the wall $x = -3$ because there, $\frac{d}{dt}x = 6 + \alpha \ge 0$. As for the wall where $\alpha = 6$ and $x \ge 0$, no trajectory passes out because $\frac{d}{dt}\alpha = -\varepsilon x < 0$ there. Likewise, trajectories cannot pass out of the wall where $\alpha = -6$ and $x \le 0$. (I leave the proof to you.)

Now consider the wall where $\alpha = x+6$ and $x \le 0$. Inside this wall, $\alpha - x - 6 < 0$, and outside this wall we have $\alpha - x - 6 > 0$. Therefore, no trajectory will pass out of this wall if we can show that

$$\frac{d}{dt}(\alpha - x - 6) \le 0 \quad \text{where } -3 \le x \le 0 \text{ and where } \alpha - x - 6 = 0. \tag{5}$$

Remember that we are trying to show that (5) is true. We demonstrate the veracity of (5) by using (1) to substitute for the derivatives. Thus, (5) is true if

$$-\varepsilon x + \frac{x^3}{3} - x - \alpha \le 0 \quad \text{where } -3 \le x \le 0 \text{ and where } \alpha - x - 6 = 0. \tag{6}$$

Thus, we are trying to prove that (6) is a true statement. We do this by eliminating α by using the equation $\alpha = x + 6$. Thus, (6) is true if

$$(-\varepsilon - 2)x + \frac{x^3}{3} - 6 \le 0 \quad \text{where} - 3 \le x \le 0. \tag{7}$$

When $\varepsilon \le 1$, we can check that the maximum of $-(\varepsilon + 2)x + \frac{x^3}{3} - 6$ where $x \le 0$ occurs where $-(\varepsilon + 2) + x^2 = 0$, that is, where $x = -(\varepsilon + 2)^{1/2}$. At this value of x, (7) has the value $\frac{2}{3}(\varepsilon + 2)^{3/2} - 6$, which is negative, as required for $\varepsilon \le 1$. (Check with a calculator if you can't prove this yourself.)

The case for the wall where $\alpha = x - 6$ and $x \ge 0$ is left for you to do.

Thus, we have established that Figure 24.6 is a basin of attraction for (1). Together with the fact that the origin is the only equilibrium point in this basin, and the origin is repelling, we can conclude that (1) has a stable, periodic solution that stays inside the basin of attraction for Figure 24.6.

24.5   Fast and Slow

Now that we know that (1) has a solution, we can consider estimating the amount of time that this solution stays near various parts of its orbit in the (x, α) plane. In particular, I will verify the fast/slow dichotomy that is exhibited by the solution. Suppose that at some time t_0, the value of x is positive and the value of α is 0. Then, at some later time t_1, the value of α is $-1/3$. We want to estimate $\Delta t = t_1 - t_0$. To make this estimate, we integrate both sides of (1b) from t_0 to t_1:

$$-\int_{t_0}^{t_1} \frac{d\alpha}{dt} dt = \varepsilon \int_{t_0}^{t_1} x(t) \, dt. \tag{8}$$

Since x is no larger than 3 (because Figure 24.6 is a basin of attraction), we have

$$-\int_{t_0}^{t_1} \frac{d\alpha}{dt} dt \le 3\varepsilon \int_{t_0}^{t_1} dt. \tag{9}$$

Use the fundamental theorem of calculus to evaluate the integrals in (9): The result is

$$\alpha(t_0) - \alpha(t_1) \le 3\varepsilon(t_1 - t_0). \tag{10}$$

Now, plug in the value 0 for $\alpha(t_0)$ and plug in $-1/3$ for $\alpha(t_1)$ to find that

$$\frac{1}{3} < 3\varepsilon \Delta t, \tag{11}$$

which implies that $\Delta_t > 1/(9\varepsilon)$. Thus, if ε is very near zero, it takes a very long time for α to move from 0 to $-1/3$.

By comparison, we should ask for the time taken for x to move from $1/2$ to $-3/2$ when α has value less than $-1/2$. Here, we can use (1a) to estimate this elapsed time. Indeed, suppose that $x(t_2) = 1/2$, that for some $t_3 > t_2$, $x(t_3) = -3/2$, and that

$\alpha < -1/2$. We want to estimate $\delta t = t_3 - t_2$. To make this estimate, use the fact that $-\alpha \geq 1/2$ in (1a) to find that

$$-\frac{d}{dt}x \geq \frac{1}{3}\left(x^3 - 3x + \frac{3}{2}\right). \tag{12}$$

Now, when $-3/2 < x < 1/2$, then the right side of (12) has a local maximum at $x = -1$ but takes on its smallest size at $x = 1/2$, where it equals $1/24$. Therefore, where $\alpha \leq -1/2$ and also where $-3/2 \leq x \leq 1/2$, we have

$$-\frac{d}{dt}x \geq \frac{1}{24}. \tag{13}$$

Integrate both sides of this last equation from t_2 to t_3 to find that

$$x(t_2) - x(t_3) \geq \delta t/24. \tag{14}$$

Since $x(t_2) - x(t_3) = 2$, this means that $\delta t \leq 48$.

Likewise, we can show that when $\alpha \geq 1/2$, the time spent for x to move from $-1/2$ to $3/2$ is less than 48.

24.6 Summary

When ε is small, then the elapsed time for x to move between $1/2$ and $-3/2$ when $\alpha < -1/2$ and the elapsed time for x to move between $-1/2$ and $3/2$ when $\alpha > 1/2$ are both much less than the time spent for α to move between 0 and $-1/3$ (or from 0 to $1/3$). This is because the ratio of $\delta t/\Delta t$ is less than $(48)(9)\varepsilon$, which is very small when ε is very small. That is, only a small fraction of the total period time is taken up by the motion of x between $1/2$ and $-3/2$ or between $-1/2$ and $3/2$. Most of the time, the value of x must be either greater than $1/2$ or else less than $-1/2$. This verifies that there is, in fact, a fast/slow dichotomy in our solution to (1).

24.7 Lessons

Here are some key points from this chapter:

- There are some fairly simple differential equations for two unknown functions of time that exhibit cyclic motion that is relatively fast in one part of the cycle and relatively slow in another. Stated differently, some cyclic phenomena, even those whose cycle has very fast and very slow parts, are governed by very simple differential equations.

- The fast/slow dichotomy can arise in a system that has two interacting parts, one always slow moving and the other with the potential for much faster movement. (Chapter 26 discusses such fast/slow interactions.)

- The motion of the system as a whole moves slowly as long as the fast subsystem tracks a stable equilbrium point whose position is determined by the slow subsystem. The motion speeds up when the motion of the slow subsystem makes the equilibrium point in the fast subsystem either disappear or become unstable.

- Learn how to use the governing equations to estimate the elapsed time in various parts of a cycle. (Chapter 25 discusses this subject.)

READINGS FOR CHAPTER 24

READING 24.1

Disparate Rates of Molecular Evolution in Cospeciating Hosts and Parasites

Commentary: This article describes a conjectured example of a "fast" system interacting with a "slow" one. Here, the fast variable should be associated with the parasite (a chewing lice), while the slow variable should be associated with the pocket gopher. The point is that the parasite has a number of generations per year, while the gopher has only one. Thus, the parasite should be able to adjust quickly to changes in the gopher physiology. From the point of view of the parasite, the gopher should serve as an essentially fixed background to its evolution and so the parasite should, at almost all points in time, evolve to virtual perfection with respect to the particulars of the gopher.

The fact that the parasite should quickly adapt to any change in the gopher population suggests the hypothesis that the speciation tree for the parasite should closely resemble that of its host, the gopher. Among other things, this article presents favorable evidence for the latter hypothesis.

Disparate Rates of Molecular Evolution in Cospeciating Hosts and Parasites

Science, **265** (1994) 1087–1090.

Mark S. Hafner, Philip D. Sudman, Francis X. Villablanca, Theresa A. Spradling, James W. Demastes, Steven A. Nadler

DNA sequences for the gene encoding mitochondrial cytochrome oxidase I in a group of rodents (pocket gophers) and their ectoparasites (chewing lice) provide evidence for cospeciation and reveal different rates of molecular evolution in the hosts and their parasites. The overall rate of nucleotide substitution (both silent and replacement changes) is approximately three times higher in lice, and the rate of synonymous substitution (based on analysis of fourfold degenerate sites) is approximately an order of magnitude greater in lice. The difference in synonymous substitution rate between lice and gophers correlates with a difference of similar magnitude in generation times.

Chewing lice of the genera *Geomydoecus* and *Thomomydoecus* are obligate ectoparasites of pocket gophers (Fig. 1). Because the entire life

Author affiliations: M. S. Hafner, P. D. Sudman, F. X. Villablanca, T. A. Spradling, J. W. Demastes: Museum of Natural Science and Department of Zoology and Physiology, Louisiana State University. S. A. Nadler: Department of Biological Sciences, Northern Illinois University.

Fig. 1 Chewing lice (*Geomydoecus texanus*, inset) are wingless insects that are obligate ectoparasites of pocket gophers (*Thomomys bottae* is shown here). The entire life cycle of the chewing louse occurs in the fur of these fossorial rodents.

cycle of these lice occurs exclusively in the fur of the host, and because different host species rarely interact, each species of louse is normally restricted to a single host species [1]. As a result, there is close correspondence between gopher taxonomic boundaries and louse taxonomic boundaries [2]. When viewed over large geographic and temporal scales, this restricted distributional pattern of chewing lice on pocket gophers has resulted in phylogenetic histories of lice and gophers that are remarkably similar [3–5].

Although well-documented cases of host-parasite cospeciation are rare [3, 6], they are of interest because they permit comparative study of organisms with a long history of parallel evolution. The temporal component of parallel phylogenesis (in which lineages of hosts and their parasites speciate repeatedly at approximately the same time) permits examination of relative rates of evolution in the two groups by comparison of the amount of change each has undergone during their parallel histories. Because the life histories of hosts and their parasites are often profoundly different, studies of molecular evolution in host-parasite assemblages can help answer a broad spectrum of questions relating to the possible effects of generation time, metabolic rate, and other life history parameters on rates of mutation and evolutionary change.

We examined DNA sequence variation in 14 species of pocket gophers and their chewing lice [7] to test for cospeciation and to investigate rates of molecular evolution in this host-parasite assemblage. We sequenced and compared homologous regions of the gene encoding the mitochondrial cytochrome c oxidase subunit I (COI) in both groups [8]. Of the 379 nucleotides sequenced for each taxon, 134 positions were variable in pocket gophers and 178 positions were variable in chewing lice (Table 1).

The cospeciation hypothesis predicts that the branching structure of the host and parasite phylogenies will be similar to a degree beyond that expected to occur by chance. This prediction can be evaluated statistically [3, 4]. For any particular host-parasite assemblage, confidence in the test of cospeciation can be no stronger than confidence in the phylogenies under comparison. Thus, it is essential that the host and parasite phylogenies accurately estimate the evolutionary history of each group. There are many methods for estimating phylogenies from sequence data [9], each of which uses a different model of nucleotide evolution and potentially yields a phylogenetic hypothesis (a tree) that differs from that estimated with other methods [10]. To consider the effects of different evolu-

Table 1   Observed percent of difference (mean ±1 SD) in various elements of the COI nucleic acid sequence from pocket gophers and their ectoparasitic chewing lice.

	Percent of sequence difference (±1 SD)*	
	Gophers	Lice
First position transitions	1.43 (0.49)	2.09 (0.62)
First position transversions	0.00 (0.25)	0.46 (0.34)
Second position transitions	0.24 (0.18)	0.38 (0.37)
Second position transversions	0.00 (0.00)	0.16 (0.17)
Third position transitions	8.74 (1.65)	9.59 (1.51)
Third position transversions	5.01 (1.67)	7.68 (1.99)
Total difference	15.64 (3.20)	20.75 (2.31)
Silent nucleotide differences	14.75 (3.09)	17.66 (2.32)
Replacement nucleotide differences	0.89 (0.71)	2.67 (1.05)
Amino acid differences	2.40 (1.83)	6.85 (2.62)

*Mean and standard deviation based on all pairwise comparisons.

tionary models on our estimates of phylogeny, we applied multiple methods of analysis [11, 12] to our sequence data. In cases where different methods yielded different results, we retained all host and parasite trees for topological comparison in order to determine whether the inference of cospeciation is warranted and, if so, whether the inference is sensitive to the method of analysis.

All analyses of the pocket gopher sequence data (using different models of DNA sequence evolution) yielded trees that were very similar in overall branching structure. For example, phylogenetic analysis [11] of the COI sequence data for pocket gophers yielded two most-parsimonious trees of equal length (1423 steps). One of these trees (Fig. 2A) was topologically identical to the tree generated by a maximum-likelihood analysis of the same data [12]. The other most-parsimonious tree showed only minor differences [13] from the tree shown in Fig. 2A. The general structure of the gopher parsimony tree (Fig. 2A) also was supported by Fitch-Margoliash (Fig. 2B) and neighbor-joining [14] analyses of genetic distances [12]. Differences among the trees gener-

ated by the parsimony, maximum-likelihood, Fitch-Margoliash, and neighbor-joining analyses of the pocket gopher data were judged nonsignificant by a likelihood ratio test [15]. Accordingly, all four trees were retained for topological comparison with the parasite trees. The basic structure of these trees and, in particular, relations within the genera *Orthogeomys* and *Geomys*, also are supported by independent phylogenetic studies based on morphology, allozymes, comparative immunology, karyology, and nucleotide sequence data [3, 5, 16–18].

All analyses of the chewing louse sequence data likewise yielded trees with similar branching structures. Phylogenetic analysis of the louse COI data yielded three most-parsimonious trees of equal length (4208 steps). One of these trees (Fig. 2A) was topologically identical to the tree generated by a maximum-likelihood analysis of the same data. The two remaining parsimony trees showed only minor differences (involving one species in each case) from the tree in Fig. 2A [19]. The Fitch-Margoliash analysis (Fig. 2B) and the neighbor-joining analysis [20] yielded louse trees very sim-

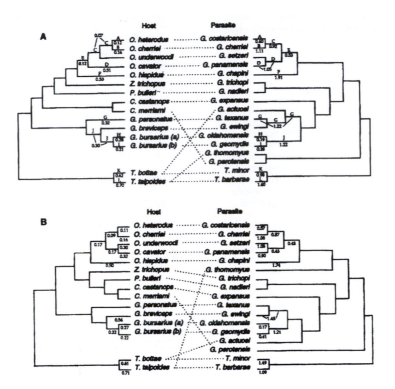

Fig. 2 Phylogenies of pocket gophers (left) and chewing lice (right) generated by analysis of DNA sequences of the gene encoding COI. In each set of trees, coexisting hosts and parasites are linked by dotted lines (each parasite examined was taken directly from the fur of the host individual examined). (A) Host and parasite trees generated by parsimony and maximum-likelihood analyses of the sequence data. (B) Trees generated by Fitch-Margoliash analyses of the host and parasite data. Pocket gopher taxa examined include the genera *Orthogeomys*, *Zygogeomys*, *Pappogeomys*, *Cratogeomys*, *Geomys*, and *Thomomys*. Chewing louse taxa include the genera *Geomydoecus* and *Thomomydoecus*. The louse *G. setzeri* also has been reported from *O. cherriei* in areas where the range of *O. cherriei* abuts that of *O. underwoodi*. Relative rates of molecular evolution were investigated by comparison of expected numbers of substitutions per site for hosts and parasites. Maximum-likelihood branch lengths (×100) based on all nucleotide substitutions are indicated for one of four possible branch combinations in Fig. 2A and for one of the two possible combinations in Fig. 2B [24]. Letters above branches indicate branches compared in Fig. 3. Comparisons were restricted to parasites whose phylogenetic history is topologically identical to that of their hosts. Because most of the uncertainty in the phylogenetic analyses involved branches near the base of the trees, only terminal and subterminal branches were compared between gophers and lice.

ilar to those generated by the parsimony analysis. Differences among the trees generated by the parsimony, maximum-likelihood, Fitch-Margoliash, and neighbor-joining analyses of the louse data were judged nonsignificant by a likelihood ratio test [15]. Accordingly, all five louse trees were retained for topological comparison with the host trees. The basic structure of the louse trees and, in particular, relations among lice hosted by species of

Orthogeomys and *Geomys*, also are supported by independent phylogenetic studies of allozymes [3, 5, 21].

The COMPONENT program [22] determined if the fit between observed parasite and host trees was significantly better than the fit between the parasite tree and trees drawn at random from the set of all possible host trees [4]. For each of 20 pairwise comparisons (four host trees and five par-

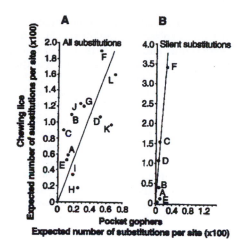

Fig. 3 Comparison of rates of molecular change in a 379-bp region of the gene encoding COI in cospeciated pocket gophers and chewing lice. Letters refer to branches labeled in Fig. 2A. (**A**) Comparison of maximum-likelihood branch lengths (based on all nucleotide substitutions) for the phylogeny shown in Fig. 2A (one of six possible combinations of cospeciating taxa) [24]. Slopes of Model II regressions (through the origin) ranged from 2.60 to 2.83 (with a mean of 2.74), which indicates that the overall rate of nucleotide substitution in lice is approximately three times higher than in gophers [27]. (**B**) Comparison of maximum-likelihood branch lengths based solely on nucleotide substitutions at fourfold degenerate sites for *Orthogeomys* gophers and their lice. Slopes of Model II regressions (through the origin) ranged from 9.88 to 11.83 (with a mean of 11.04), which indicates that the rate of synonymous substitution in this gene region is approximately an order of magnitude greater in chewing lice than in pocket gophers.

asite trees), the observed degree of fit between the gopher and louse trees was significantly better ($P < -0.01$) than the fit between the louse tree and 10,000 randomized gopher trees [23]. These results, which are robust to the method of phylogenetic inference and to the evolutionary models used, falsify the null hypothesis of chance similarity between the host and parasite trees. Although this evidence is consistent with the hypothesis of cospeciation, the concordant phylogenies might instead result from dispersal, extinction, or incomplete sampling of closely related taxa [4]. However, only the cospeciation hypothesis predicts temporal congruence of host and parasite speciation events, which (given roughly time-dependent molecular change in each group) would result in a significant relation between measures of molecular differentiation in the host and parasite trees. We demonstrate below that our molecular data are consistent with this prediction. This finding, which requires no assumptions about rate similarity between hosts and parasites [3], corroborates inde-

pendent evidence for cospeciation in several genera of pocket gophers (*Orthogeomys*, *Geomys*, and *Thomomys*) and their lice [3–5].

Given evidence for cospeciation, it is possible to test the null hypothesis that pocket gophers and chewing lice have undergone equivalent amounts of genetic differentiation during their parallel histories. It is appropriate that this test be restricted to hosts and parasites that have cospeciated, because time since divergence can be assumed to be equal only for host-parasite pairs that show cospeciation (9 host-parasite pairs in Fig. 2A and 10 host-parasite pairs in Fig. 2B) [24]. We first compared maximum-likelihood distance matrices for cospeciating gophers and lice using Mantel's test [25], which showed a highly significant ($P < 0.01$) association between genetic distances in corresponding hosts and parasites. This test demonstrates that evolutionary rates in gophers and lice are significantly correlated, regardless of tree structure. To test for equality of rates between gophers and lice, we compared maximum-likelihood branch lengths for all possible combinations of cospeciating taxa [24, 26]. In all cases, Wilcoxon sign-rank tests showed that louse branches were significantly longer than gopher branches ($P < 0.003$ in each case). Given this significant difference, we used Model II regression analysis (through the origin) and determined that the slopes of the regressions (Fig. 3A) ranged from 2.60 to 2.83 (with a mean of 2.74). This indicates that the overall rate of nucleotide substitution in lice is approximately three times higher than in gophers [27].

To estimate rates of synonymous substitution in gophers and lice, we restricted our analysis of the COI sequences to fourfold degenerate sites (sites at which all base substitutions are silent). We focused on the largest group of closely related species (*Orthogeomys* species and their lice) because these species are sufficiently closely related to ensure that corrections for multiple mutations are effective [28]. The numbers of variable fourfold degenerate sites in gophers and lice were approximately equal (67 and 69, respectively). We used a maximum-likelihood model of evolution to infer branch lengths based solely on substitutions at these sites [24]. The model included corrections for observed transitional bias (a maximum 4:1 bias in gophers; 10:1 bias in lice) and for significantly different nucleotide compositional biases in the two data sets ($P < 0.05$) [29]. These corrections are necessary to estimate evolutionary rates [28], and spurious results should not occur simply because of differences in the number of characters compared or differences in mutational dynamics (and resultant saturation levels) in the two groups being compared. In theory, rates of synonymous substitution are proportional to mutation rates [30], but our estimates cannot be considered direct measures of mutation rates because we have not controlled for possible constraints on the translational apparatus, such as codon bias and secondary structure of mRNAs.

If nucleotide substitutions at fourfold degenerate sites are selectively neutral, then change at these sites should fit a molecular clock model [30]. Accordingly, we tested all possible combinations of cospeciating gophers (*Orthogeomys* only) and their lice [24] for significant departure from clocklike behavior, using the log-likelihood ratio test [12]. In all cases, the data were consistent with molecular-clock assumptions, which indicates that substitutions within gophers and within lice accumulate in

a roughly time-dependent fashion. Wilcoxon sign-rank tests showed that louse branches were significantly longer than gopher branches in four of the six possible comparisons ($P < 0.05$ in each case). We used Model II regression analysis (through the origin) to quantify the relation between gopher and louse branch lengths. Slopes of the regressions (Fig. 3B) ranged from 9.88 to 11.83 (with a mean of 11.04), which indicates that the estimated rate of silent substitution for this gene region is approximately an order of magnitude greater in chewing lice than in pocket gophers. Evidence for a higher rate of substitution in lice appears to be independent of the evolutionary model employed, although the magnitude of the rate difference is sensitive to certain parameters of the model [31].

Viewed together, the analysis of all nucleotide substitutions (Fig. 3A) and the analysis of substitutions at fourfold degenerate sites (Fig. 3B) provide insight into the dynamics of molecular evolution for this gene region in the species studied. The analysis of all substitutions indicates that the overall rate of evolutionary change is approximately three times greater in chewing lice than in their hosts (Fig. 3A). Likewise, the means of all pairwise replacement differences for nucleotides and amino acids are approximately three times greater in lice than in gophers (Table 1). In contrast, the analysis of nucleotide substitutions at fourfold degenerate sites indicates that rates of silent substitution in this gene region are roughly 11 times greater in lice than in gophers (Fig. 3B). The fact that this 11-fold rate difference is not evident when all substitutions are considered is probably the result of selective constraints on replacement substitutions. High levels of functional constraint on the COI enzyme have been reported in other organisms [28].

The 11-fold difference in rates of synonymous substitution in *Orthogeomys* gophers and their lice (Fig. 3B) cannot be explained by transition bias or nucleotide frequency bias. Because silent substitutions at the fourfold degenerate sites show clocklike behavior, it is likely that they are neutral or nearly neutral [30]. Several possible mechanisms could account for this rate difference, including mutation rate differences caused by possible differences in gene order that affect vulnerability to mutation [32], differences in metabolic rate or general metabolic physiology, generation-time differences, or other factors correlated with body size [33]. Alternatively, this rate difference could be caused by mechanisms that are independent of mutation rate, such as codon bias and other constraints on the translational apparatus, or differences in DNA repair efficiency. It is perhaps important that this 11-fold rate difference is accompanied by a similar difference in generation time between gophers and lice (approximately 1 year in gophers and 40 days in lice) [34]. If the observed rate difference results from an underlying difference in mutation rate, then generation time may explain this difference. However, mutation rates are more likely to be influenced directly by nucleotide generation time than by organismal generation time [33]. As such, our study suggests that each organismal generation is equivalent to equal numbers of nucleotide generations in pocket gophers and chewing lice. If the 11-fold rate difference reflects a similar difference in mutation rate, then these findings are consistent with the neutral theory of molecular evolution [30], because once the data are corrected for the difference in generation time, they suggest equal rates of mutation per generation in distantly related groups of animals.

REFERENCES

1. S. A. Nadler, M. S. Hafner, J. C. Hafner, D. J. Hafner, *Evolution* **44**, 942 (1990).
2. R. D. Price and K. C. Emerson, *J. Med. Entomol.* **8**, 228 (1971).
3. M. S. Hafner and S. A. Nadler, *Syst. Zool.* **39**, 192 (1990); *Nature* 332, 258 (1988).
4. R. D. M. Page, *Syst. Zool.* **39**, 205 (1990).
5. J. W. Demastes and M. S. Hafner, *J. Mammal.* **74**, 521 (1993).
6. R. D. M. Page, *Int. J. Parasitol.* **23**, 499 (1993).
7. Voucher specimens are deposited in the Museum of Natural Science, Louisiana State University (LSUMZ) or the New Mexico Museum of Natural History (NMMNH) and are as follows: *Orthogeomys underwoodi* (LSUMZ 29493), *O. hispidus* (LSUMZ 29231), *O. cavator* (LSUMZ 29253), *O. cherriei* (LSUMZ 29539), *O. heterodus* (LSUMZ 29501), *Geomys breviceps* (LSUMZ 33940), *G. personatus* (LSUMZ 31460), *G. bursarius halli* (LSUMZ 31463; designated "a" in Fig. 2), *G. b. majusculus* (LSUMZ 31448; designated "b" in Fig. 2), *Cratogeomys castanops* (LSUMZ 31455), *C. merriami* (LSUMZ 34343), *Pappogeomys bulleri* (LSUMZ 34338), *Zygogeomys*

trichopus (LSUMZ 34340), *Thomomys bottae* (LSUMZ 29320 and 29569), and *T. talpoides* (NMMNH 1634 and 1637).
8. Pocket gophers were trapped, killed, and immediately brushed to recover lice. Gopher tissues and lice were stored at negative 70°Celsius. DNA extractions from gophers followed the phenol-chloroform technique [D. M. Hillis, A. Larson, S. K. Davis, E. A. Zimmer, in *Molecular Systematics*, D. M. Hillis and C. Moritz, Eds. (Sinauer, Sunderland, MA, 1990), pp. 318–370]. DNA extractions from lice followed a modification of the protocol described by H. Liu and A. T. Beckenbach [*Mol. Phylogenet. Evol.* **1**, 41 (1992)] and used two lice from each gopher. One μl of the extraction solution was used for DNA amplification in a 50-μl reaction. A 379-base pair (bp) region of the mitochondrial COI gene was amplified by polymerase chain reaction (PCR) with two degenerate primers: L6625 (5'-CCGGATCCTTYTGRTTYTTYGGNCAYCC-3') and H7005 (5'-CCGGATCCACNACRTARTANGTRTCRTG-3'). Primer names refer to the 3' position of each primer relative to the human mitochondrial genome [S. Anderson *et al.*, *Nature* 290, 457 (1981)]. Double-stranded amplifications were done by four cycles of 1 min of denaturation (95°Celsius), 1 min of annealing (45°Celsius), and 1 min of extension (72°Celsius), followed by 30 cycles at reduced denaturation temperatures (93°Celsius) and increased annealing temperatures (60°Celsius). Methods for production of single-stranded templates and for sequencing of single-stranded products are given elsewhere [18]. Because no two sequences were identical either within or between lice and gophers, cross-contamination is not suspected. Gopher and louse sequences are available through GenBank (accession numbers L32682 to L32696 for gophers and L32665 to L32681 for lice).
9. D. L. Swofford and G. J. Olsen, in *Molecular Systematics*, D. M. Hillis and C. Moritz, Eds. (Sinauer, Sunderland, MA, 1990), pp. 411–501; D. M. Hillis, M. W. Allard, M. M. Miyamoto, *Methods Enzymol.* **224**, 456 (1993).
10. J. P. Huelsenbeck and D. M. Hillis, *Syst. Biol.* **42**, 247 (1993).
11. We used the PAUP program (Phylogenetic Analysis Using Parsimony, 3.1.1, D. L. Swof-

ford, Illinois Natural History Survey) for phylogenetic analyses of the sequence data. Of the 134 variable sites in pocket gophers (178 in lice), 113 were phylogenetically informative (142 in lice). To compensate for the effects of multiple mutations at nucleotide positions, we used a parsimony analysis in which transitions at first and third codon positions were designated as more likely than transversions (with a 10:1 step matrix for gophers and a 17:1 step matrix for lice). These empirical estimates of transition bias represent maximal values, which were observed between taxa with the most similar sequences (8.7% sequence divergence between the most similar gophers; 4.7% between the most similar lice). Because the effects of multiple mutations should be least for closely related taxa, we consider these to be reasonable approximations of the actual transition bias. For each database (gophers and lice) we conducted 100 heuristic searches, varying the addition order of taxa (using the random addition option of PAUP) and using alternate branch-swapping algorithms (using the TBR and SPR options). All gopher and louse parsimony trees contained significant phylogenetic signal (g_1 < −0.54, P < 0.01) [D. M. Hillis and J. P. Huelsenbeck, *J. Hered.* **83**, 189 (1992)]. Gophers in the genus *Thomomys* and lice of the genus *Thomomydoecus* were designated as outgroups. The outgroup status of these taxa is supported by previous morphological and molecular analyses [16, 17] [R. A. Hellenthal and R. D. Price, J. Kans. *Entomol. Soc.* **57**, 231 (1984); S. A. Nadler and M. S. Hafner, *Ann. Entomol. Soc. Am.* **82**, 109 (1989)].

12. We used PHYLIP [Phylogeny Inference Package, 3. 5c, J. Felsenstein, Department of Genetics, University of Washington, Seattle] for maximum-likelihood, Fitch-Margoliash [W. M. Fitch and E. Margoliash, *Science* **155**, 279 (1967)], and neighbor-joining [N. Saitou and M. Nei, *Mol. Biol. Evol.* **4**, 406 (1987)] analyses of the sequence data. Empirical base frequencies and maximal observed transition biases (10:1 for gophers and 17:1 for lice) were used in the maximum-likelihood analysis. The Fitch-Margoliash and neighbor-joining analyses were based on distance matrices corrected with the Kimura two-parameter model [substituting maximal observed transition bias values for the default (2:1) value M. Kimura, *J. Mol. Evol.* **16**, 111 (1980)]. We repeated each analysis at least 10 times, varying the input order of taxa using the jumble option of PHYLIP.

13. In the alternative parsimony tree for pocket gophers, the phylogenetic positions of *P. bulleri* and *Z. trichopus* were reversed and the two species of *Cratogeomys* were not depicted as sister taxa.

14. The neighbor-joining tree differed from the Fitch-Margoliash tree (Fig. 1B) only in the placement of *P. bulleri* (immediately basal to the five-taxon clade containing *O. hispidus*) and *Z. trichopus* (basal to the eight-taxon clade containing *C. castanops*, *C. merriami*, and *P. bulleri*).

15. H. Kishino and M. Hasegawa, *J. Mol. Evol.* **29**, 170 (1989), as used in PHYLIP [12].

16. R. S. Russell, *Univ. Kans. Publ. Mus. Nat. Hist.* **16**, 473 (1968); R. L. Honeycutt and S. L. Williams, *J. Mammal.* **63**, 208 (1982); M. S. Hafner, *ibid.* **72**, 1 (1991).

17. M. S. Hafner, Z. *Zool. Syst. Evolutionsforsch.* **20**, 118 (1982).

18. P. D. Sudman and M. S. Hafner, *Mol. Phylogenet. Evol.* **1**, 17 (1992).

19. In the second parsimony tree, the louse *G. actuosi* was linked with the *oklahomensis-geomydis* clade, rather than the *texanus-ewingi* clade (as in Fig. 2A). The third tree showed *G. thomomyus* near the root of the tree between the outgroups *T. minor* and *T. barbarae*.

20. The neighbor-joining tree was identical to the Fitch-Margoliash tree (Fig. 2B), except that *G. thomomyus* and *G. perotensis* were linked as sister taxa (as in the parsimony and maximum-likelihood trees, Fig. 2A) and positioned basal to the *G. chapini* branch.

21. S. A. Nadler and M. S. Hafner, *Int. J. Parasitol.* **23**, 191 (1993).

22. COMPONENT 2.0, R. D. M. Page, Natural History Museum, London; R. D. M. Page, *Syst. Biol.* **43**, 58 (1994).

23. The test was based on the criterion of minimum number of independent losses necessary to reconcile the host and parasite trees [22]. Sixteen of the 20 pairwise comparisons were also found to be statistically significant (P < 0.05) according to the criterion of "number of leaves added" [22]. Number of leaves added is equivalent to one-half the number of "items of error" [G. Nel-

son and N. Platnick, *Systematics and Biogeography: Cladistics and Vicariance* (Columbia Univ. Press, New York, 1981)]. The four comparisons that were not statistically significant (according to the criterion of number of leaves added) all involved the louse parsimony tree that depicted the outgroup *Thomomydoecus* as polyphyletic. The branching structure of this tree is challenged by evidence from previous morphological and allozyme studies [11] supporting monophyly of the louse genus *Thomomydoecus*.

24. The trees (Fig. 2) were reduced to the largest number of taxa that showed identical branching patterns in the hosts and parasites. All possible combinations of cospeciating taxa were compared. For example, there are four possible combinations of nine cospeciating taxa in Fig. 2A [that is, the nine-taxon gopher tree can include either *O. underwoodi* or *O. cavator* (but not both) and *G. breviceps* or *G. personatus* (but not both)]. There are 2 possible combinations of 10 cospeciating taxa in Fig. 2B (that is, the tree can include either *G. breviceps* or *G. personatus*, but not both). Because most of the uncertainty in the phylogenetic analyses involved branches near the base of the trees, only terminal and subterminal branches were compared between gophers and lice.

25. N. Mantel, *Cancer Res.* **27**, 209 (1967). We calculated distance matrices using the maximum-likelihood model [12]. For the 9 cospeciating taxa in Fig. 2A, $t = 4.073$, $P < 0.01$. For the 10 cospeciating taxa in Fig. 2B, $t = 4.289$, $P < 0.01$. Degrees of freedom were adjusted to $n - 1$, where $n =$ equals number of taxa compared [3].

26. Trees for cospeciating taxa (24) were input as user trees into PHYLIP [12] in order to estimate maximum-likelihood branch lengths. Empirical estimates of transition bias, nucleotide frequency bias, and positional bias (first:second:third codon position: 5:1:40 for gophers and 5:1:21 for lice) were incorporated into the maximum-likelihood model [12]. We recognize that the parameters used in maximum-likelihood models will influence rate estimates [K. Fukami-Kobayashi and Y. Tateno, *J. Mol. Evol.* **32**, 79 (1991)]. Our models use empirical estimates of these parameters (rather than arbitrary or constant values) to obtain a more realistic approximation of the actual mutational dynamics of the gene region being surveyed [12, 28].

27. We used Model II regression analysis (reduced major axis model), which does not permit conventional tests of significance [R. R. Sokal and F. J. Rohlf, Biometry (W. H. Freeman, San Francisco, 1969)]. However, the non-parametric Mantel's test documents a significant relation between genetic distances in gophers and lice [25], and the non-parametric Wilcoxon sign-rank test shows that louse branches are significantly longer than gopher branches. When the regression analyses were not constrained through the origin, none of the y-intercepts was significantly different from zero.

28. R. Kondo, S. Horai, Y. Satta, N. Takahata, *J. Mol. Evol.* **36**, 517 (1993); R. A. Capaldi, *Annu. Rev. Biochem.* **59**, 569 (1990). Total sequence differences among the five species of *Orthogeomys* range from 4 to 11.9%. Sequence differences among the parasites of *Orthogeomys* range from 11.9 to 20.3%.

29. Differences in nucleotide composition were tested using the B-statistic [A. P. Martin, G. J. P. Naylor, S. P. Palumbi, *Nature* **357**, 153 (1992)].

30. M. Kimura, *The Neutral Theory of Molecular Evolution* (Cambridge Univ. Press, London, 1983); K. H. Wolfe, P. M. Sharp, W.-H. Li, *Nature* **337**, 283 (1989).

31. The slope of this relationship is affected very little by use of different nucleotide frequency estimates in the maximum-likelihood model. The slopes are more sensitive, however, to different estimates of transition:transversion bias. Our model used empirical estimates of transition:transversion bias (4:1 in gophers; 10:1 in lice) because such estimates are more likely to reflect the actual mutational dynamics of the gene regions being surveyed [12, 28].

32. G. G. Brown and M. V. Simpson, *Proc. Natl. Acad. Sci. U.S.A.* **79**, 3246 (1982); [28].

33. A.P. Martin and S. R. Palumbi, *ibid.* **90**, 4087 (1993); C.-I. Wu and W.-H. Li, *ibid.* **82**, 1741 (1985).

34. R. W. Rust, *Oecologia* **15**, 287 (1974); A. G. Marshall, *The Ecology of Ectoparasitic Insects* (Academic Press, London, 1981).

35. We thank D. R. Wolstenholme, R. Okimoto, and C. T. Beakley for providing PCR primer sequences and protocols. Photography was

provided by R. J. Bouchard. D. H. Clayton, S. V. Edwards, D. A. Good, D. J. Hafner, B. D. Marx, R. D. M. Page, S. Scheiner, and two anonymous reviewers provided helpful advice and comments on this manuscript. Supported by NSF grant BSR-8817329 and NSF-Louisiana Education Quality Support Fund grant ADP-02.

Exercises for Chapter 24

1. Consider an equation for a function of time, $x(t)$, of the form $\frac{dx}{dt} = f(x) + c$, where c is a constant and where $f(\cdot)$ is a function. For some values of c, there will be three equilibrium points and for others there will be one. In each of the following cases, the function $f(x)$ is given and your task is to first find the value of c where the number of equilibrium points changes from three to one.

   [Here is how to find this value of c: The number of equilibrium points changes when the graph of the function $x \to f(x) + c$ has a local maximum or minimum on the x-axis, as illustrated in Figure 24.7.

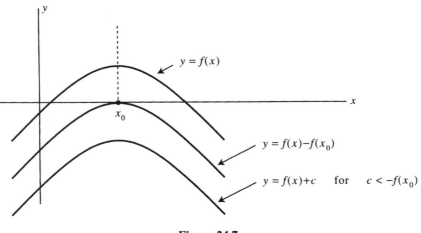

Figure 24.7

   Because the maxima and minima of the function $x \to f(x) + c$ occur at the zeros of the function $x \to \frac{df}{dx}(x)$, the strategy is to solve for the points x_0 that make $\frac{df}{dx}(x_0) = 0$ and then set c equal to $-f(x_0)$.]

   (a) $f(x) = x^3 - 3x$

   (b) $f(x) = 2x^3 - 3x^2 - 12x$

   (c) $f(x) = 2x^3 + 9x^2 + 12x + 7$

   (d) $f(x) = x^3 - 12x + 1$

2. Suppose that f is a function of variables x and y, but we are interested in solutions to the equation $\frac{dx}{dt} = f(x, y)$, where y is fixed and treated as a parameter. Here, we don't allow all functions $f(x, y)$, but just those with the following property: For all values of y, the function $f(x, y)$ with y fixed goes to $+\infty$ as $x \to \infty$ and it goes to $-\infty$ as $x \to -\infty$. [Here is an example of a function with this desired property: $f(x, y) = x^5 + yx^2$. A function that is not allowed is $f(x, y) = x^2 + y$.) Having chosen f (subject to our constraints), you will find that as y changes, the number of equlibrium points can change. However, consider the following assertion:

The number of equilibrium points where $\frac{df}{dx}$ is positive minus the analogous number of equilibrium points where $\frac{df}{dx}$ is negative is always equal to 1.

That is, for all values of y, and no matter the function f (subject to the constraints about its behavior as x tends to $\pm\infty$), the difference between the numbers of these two kinds of equilibrium points is 1. Do you believe this outlandish assertion?

Part 1 of this exercise is to verify that the preceding assertion holds in the following four cases (which are taken from Exercise 1).

(a) $f(x, y) = x^3 - 3x + y$

(b) $f(x, y) = 2x^3 - 3x^2 - 12x + y$

(c) $f(x, y) = 2x^3 + 9x^2 + 12x + 7 + y$

(d) $f(x, y) = x^3 - 12x + 1 + y$

Part 2 of this exercise asks you to outline a justification for this outlandish assertion (if you believe it) or else find an example of $f(x, y)$ that proves this assertion false.

25

Estimating Elapsed Time

To use differential equations, it is important to learn how to estimate. Creative estimation is the key to getting useful information from a differential equation. (Even if you are a programming whiz, it still pays to estimate, for estimates allow you to check your program and to develop error bounds for your program's answers.)

25.1 Estimating Time

Suppose that s is a function of time t that evolves according to a differential equation of the form

$$\frac{ds}{dt} = f(s), \tag{1}$$

where f is a function of a single variable. Suppose also that at time $t = 0$, the value of $s(0)$ is some known quantity s_0. Finally, imagine that we wish to estimate the amount of time it takes $s(t)$ to move from its value at 0, s_0, to some new value s_1. Here is how to get upper and lower bounds for this time in the case where $f(s)$ is strictly positive for $s_0 \leq s \leq s_1$.

Step 1: Suppose that $t = t_1$ is when s first hits the value s_1. (For the sake of argument, we take $s_1 > s_0$.) Let f_{max} denote the maximum value of $f(s)$ when s is between s_0 and s_1. Likewise, let f_{min} denote the minimum value of $f(s)$ when $s_0 \leq s \leq s_1$.

Step 2: It follows from (1) that

$$f_{\min} \leq \frac{ds}{dt} \leq f_{\max}. \tag{2}$$

Step 3: Integrate $\frac{ds}{dt}$ against dt from $t = 0$ to $t = t_1$. According to the fundamental theorem of calculus,

$$\int_0^{t_1} \frac{ds}{dt}\, dt = s_1 - s_0. \tag{3}$$

Step 4: Replace $\frac{ds}{dt}$ in (3) by $f_{\max}$. According to (2), this just makes the integral bigger. The integral of this replacement is equal to $f_{\max} \cdot t$. Therefore,

$$f_{\max} \cdot t_1 \geq s_1 - s_0, \tag{4}$$

which implies that

$$t_1 \geq \frac{s_1 - s_0}{f_{\max}}. \tag{5}$$

Step 5: Replace $\frac{ds}{dt}$ in (3) by $f_{\min}$. According to (2), this just makes the integral smaller. The integral of this replacement is equal to $f_{\min} \cdot t$. Therefore,

$$f_{\min} \cdot t_1 \leq s_1 - s_0. \tag{6}$$

This last equation implies the upper bound

$$t_1 \leq \frac{s_1 - s_0}{f_{\min}}. \tag{7}$$

25.2   An Estimation Technique for Functions

When estimating the maximum or minimum of a function on an interval $[a, b]$, the following strategies can be helpful:

Strategy 1: If $f(s)$ is the function, find the points where $\frac{d}{ds} f = 0$. These, plus possibly the endpoints $\{a, b\}$, will be the locations of maxima or minima.

For example, suppose I wish to find the maximum and minimum values of the function $f(s) = 3s^{10} + 4s^2 + 2$ on the interval $[0, 1]$. I compute $\frac{d}{ds} f$ to equal $30s^9 + 8s$, which I see to be nonnegative on $[0, 1]$. Thus, f is increasing on $[0, 1]$, its minimum must be at $s = 0$, and its maximum must be at $s = 1$. That is, $\min(f) = 2$ and $\max(f) = 9$.

For a second example, suppose I want to find the maximum and minimum of the function $f(s) = 2s^4 - s^2 + 2$ on the interval $[0, 1]$. I compute its derivative to be $8s^3 - 2s$. This is equal to $2s(4s^2 - 1)$ and so vanishes in my interval at $s = 0$ and $s = 1/2$. Thus, the possible points where the maximum and minimum are taken are

$s = 1/2$ and, as always, the endpoints 0 and 1. I can plug these values of s into f to find $f(0) = 2$, $f(1/2) = 15/8$ and $f(1) = 3$. Thus, the maximum of f on the interval $[0, 1]$ is 3 and the minimum is $15/8$.

Strategy 2: If the function $f(s)$ is a sum of two functions, $f(s) = g(s) + h(s)$, then $\max(f)$ can be no larger than $\max(g) + \max(h)$ and $\min(f)$ can be no smaller than $\min(g) + \min(h)$.

For example, to estimate the maximum and minimum of the function

$$f(s) = 1 + s^3 + \cos(\pi s) \tag{8}$$

on $[0, 1]$, set $g(s) = 1 + s^3$ and set $h(s) = \cos(\pi s)$. Then $\max(g) = 2$ and $\max(h) = 1$ so we find that $\max(f) \leq 3$. Also, $\min(g) = 1$ and $\min(h) = -1$, so $\min(f) \geq 0$.

Strategy 3: The preceding strategy often results in rather crude estimates. To refine the strategy, we can split the interval $[a, b]$ by selecting a number c between $[a, b]$. Then use the fact that the maximum of f on $[a, b]$ is equal to the maximum of two numbers, the first being the maximum of f on $[a, c]$ and the second being the maximum of f on $[c, b]$. Likewise, $\min(f)$ is equal to the minimum of two numbers, the first being the minimum of f on $[a, c]$ and the second being the minimum of f on $[c, b]$.

For example, to estimate the maximum and minimum of the function $f(s)$ in (8) on $[0, 1]$, consider the strategy with $c = 1/2$. Then we can find (using Strategy 2) that $\max(f|_{[0,1/2]}) \leq 17/8$ and $\max(f|_{[1/2,1]}) \leq 2$, so that $\max(f|_{[0,1]}) \leq 17/8$. Also, $\min(f|_{[0,1/2]}) \leq 1$ and $\min(f|_{[1/2,1]}) \geq 1/8$, so that $\min(f|_{[0,1]}) \geq 1/8$.

Note that we could break up the interval into more than two regions and get, presumably, even better estimates.

With regard to (1), if the given function f is strictly negative, then $s_0 > s_1$ and (5) and (7) apply with f_{min} and f_{max} (now, both negative) interchanged.

25.3   Lessons

Here are some key points from this chapter:

- Learn the rule for bounding elapsed time from above and from below.

- Learn the methods given here for obtaining an upper bound and a lower bound for the values of some function.

Exercises for Chapter 25

In Exercises 1–6, the behavior of $x(t)$ as a function of time is governed by the given differential equation. If $x(0) = 0$, find upper and lower bounds for the time t at which $x(t) = 1$.

    1. $\frac{dx}{dt} = 3x + x^2 + 3$

2. $\frac{dx}{dt} = x^3 + 5$

3. $\frac{dx}{dt} = x - x^2 + 1$

4. $\frac{dx}{dt} = x + x^2 + 1$

5. $\frac{dx}{dt} = 2x - x^2 + 1$

6. $\frac{dx}{dt} = 2x + x^2 + 3$

26

Switches

The fast and slow analysis of the heart model from Chapter 24 can be applied in a more general context. In particular, the fast-slow analysis applies when two systems (or populations) interact and where the reaction time of one is much faster than that of the other.

To illustrate how this works, consider first an example that is suggested by the article from Reading 24.1: The example considers the coevolution of pocket gophers and their mite parasites. The mites can be considered to be the fast system and the gophers the slow system, inasmuch as the generation time for mites can be as short as a few days or weeks, while gophers typically have one or at most two litters per year. The point is that the fast generation time of the mite allows the latter to adapt quickly (relatively speaking) to environmental changes, while the adaptation time of gophers is much slower. To be slightly more specific, take $g(t)$ to denote some property of gophers under consideration at time t, and take $m(t)$ to denote some property of mites. [For example, $g(t)$ might represent the fraction of gophers with some particular blood type, and $m(t)$ might denote the fraction of mites that preferentially feed on that blood type.] We can then model the time evolution of $g(t)$ and $m(t)$ by some system of differential equations that can be written schematically as

$$\frac{d}{dt}g = F(g, m)$$
$$\frac{d}{dt}m = H(g, m). \tag{1}$$

416

Here, F and H are some specific functions of two variables. The fast-slow dichotomy between mites and gophers can be expressed by the requirement that $|F(g, m)|$ should be much smaller than $|H(g, m)|$ except near those values of (g, m) where $H = 0$.

The requirement that $|F|$ should generally be much less than $|H|$ implies that for most values of (g, m), the function $g(t)$ is changing at a much slower rate than is the function $m(t)$. In particular, for any given initial value for m, a good first approximation to (1) takes $g =$ constant in the second line of (1). That is, to a first approximation, the function g in the second line of (1) should be viewed as a constant—simply a parameter in the equation for m. With this approximation understood, then the second line of (1) predicts that $m(t)$ will be found to equal one of the solutions to the equation

$$H(g, m) = 0 \qquad (2)$$

where

$$\frac{\partial}{\partial m} H(g, \cdot)|_m < 0. \qquad (3)$$

[The condition in (2) says that m is an equilibrium point to the equation in the second line of (1), and the condition in (3) ensures that this equilibrium point is a stable one. Note that in all of this, g is taken to equal a constant so our one-variable analysis from the first few chapters can be applied.]

In reality, g will not, in general, be truly constant as its motion is controlled by the first line in (1). That is, g will change slowly and thus the conditions in (2) and (3) that depend on g will change slowly, and thus m will change slowly even as it stays close to obeying (2) for each value of t. See the graphs in Figures 26.1 and 26.2 for a hypothetical function $H(g, m)$ with selected values for g chosen.

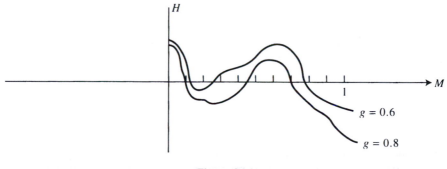

Figure 26.1

With regard to the slow change in m, I made one crucial oversight in the assertion of the preceding paragraph. This one exception occurs when g [as a parameter in (2)] evolves so as to make one of the stable equilibria of the m equation disappear. Figure 26.2 provides an example.

For $g > 0.5$, there are solutions to (2) and (3) with m approximately 0.1 and also for m approximately 0.7, but when $g < 0.5$, there is only the solution with m

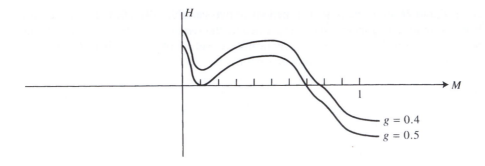

Figure 26.2

approximately 0.7. Thus, if $m(t)$ is found to be tracking the m approximately 0.1 solution to (2) and (3) for $g > 0.5$, then as g approaches 0.5 from above and decreases below 0.5, there is no equilibrium point near 0.1 for m to track. The m equation in (1) then forces m to jump quickly to the next accessible stable equilibrium point, which in this case is the 0.7 equilibrium point. This quick jump (with time interval governed by the m equation) will be viewed by us as a sudden and drastic switching of the nature of the interacting system. (In some circles, these sudden switches are called "catastrophes").

Sudden and drastic changes of behavior happen in many natural systems, and this fast-slow analysis provides a simple mathematical model of its appearance.

A truly fundamental application of these fast-slow ideas can be seen in the article *Thresholds in Development* by Lewis, Slack, and Wolpert (in *Journal of Theoretical Biology* **65** (1977) 579–590; see Reading 26.1). The point of the article is to present (and give evidence) for a model that explains how nearest-neighbor cells in an embryo might naturally develop in drastically different ways. For example, after development is complete, a cell on the outside of the heart is heart muscle, but the next cell out is drastically different. Even so, these two cells are descendants from cells that were almost identical and perhaps even nearest neighbors in the undifferentiated embryo. In particular, Lewis, Slack, and Wolpert were looking for a mechanism that was compatible with the notion that development is determined by the relative concentrations of ambient chemicals (i.e., morphogens). The historical context here is that at the time of the writing of this article, the proposed explanations for such catastrophic differences in offspring fate for embryo cells had two fundamental flaws:

- They required drastic and very unrealistic changes in the size of the morphogen concentration over very small distances.

- They couldn't explain how cells "remember" morphogen signal after the morphogen dissipates.

(The article refers to the size of the morphogen "gradient." The **gradient** of a function of space variables is the vector whose components are the partial derivatives of that function. The gradient of a function f is often denoted ∇f. For example, the gradient of a function f on the xy-plane is the vector

$$\nabla f = \begin{pmatrix} \frac{\partial f}{\partial x} \\ \frac{\partial f}{\partial y} \end{pmatrix} \tag{4}$$

In any event, large gradient means rapid changes in at least some directions for the function in question.)

Lewis, Slack, and Wolpert proposed the following model as a prototype for solving this morphogen gradient conundrum: They consider the activation of a gene G by a signaling substance S. The amount of G's product at time t is denoted by $g(t)$, and the amount of S at time t by $S(t)$. They suppose that the rate of change of g depends linearly on the amount of S, but that there are "feedbacks" so that relatively small concentrations of g promote g's growth, while large concentrations inhibit it. In particular, they consider the following equation for g:

$$\frac{dg}{dt} = k_1 S + \frac{k_2 g^2}{k_3 + g^2} - k_4 g, \tag{5}$$

where k_1, k_2, k_3, and k_4 are constants.

Lewis, Slack, and Wolpert now consider the behavior of g for different values of S. As they report, the plot of the right-hand side of (5) as a function of g for different values of S has different numbers of stable equilibrium points. Such a plot is pictured in Figure 26.3.

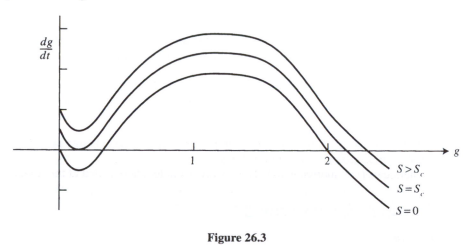

Figure 26.3

As can be seen, when $S < S_c$, there are two stable equilibria, one near $g = 0$ and one with g considerably larger than zero. There is also one unstable equilibrium point between the two stable ones. Thus, as S increases toward S_c, the small g stable equilibrium point cancels against the unstable one so that when $S > S_c$, there is only one stable equilibrium point, and this one is where g is relatively large.

The graph in Figure 26.3 explains how two adjoining cells can have widely different values of g even when they are very close together: All we need is that S is near S_c at these two cells but with S slightly greater than S_c at one cell and slightly less at the other. The result will be that the former cell has g near zero, while the other will have g relatively large. Moreover, if S subsequently decreases to zero (because the signaling cells are no longer active), then the drastic difference in g output by these two nearby cells still remains.

The explanation by Lewis, Slack, and Wolpert for reaction of embryonic cells to the gradients of signaling molecules now plays a major role in current thinking in developmental biology. See, for example, the discussion in Chapter 6 of *Cells, Embryos, and Evolution*, by J. Gerhart and M. Kirschner, which was published in 1997 by Blackwell Science Inc. Also, take a look at Reading 18.4

26.1   Lessons

Here are some key points from this chapter:

- When an inherently slow-moving subsystem interacts with a relatively fast-moving subsystem, then, to a first approximation, the dynamics can be analyzed by viewing the unknowns for the slow subsystem as fixed parameters in the equations for the unknowns of the fast subsystem.

- The fast subsystem may track a stable equilibrium point as determined by the "parameters" given by the slow subsystem's unknowns. However, as these unknowns slowly change their values under the direction of the differential equation, the fast subsystem's equilibrium points will move slowly.

- Rapid and drastic changes in the properties of the fast subsystem can occur when the slowly moving parameters cause a stable critical point of the fast system to disappear.

- The disappearance of stable equilibrium points as parameters change provides a simple explanation for switching behavior.

READINGS FOR CHAPTER 26

READING 26.1

Thresholds in Development

Commentary: As remarked in this chapter, the switching model proposed in this article to explain developmental thresholds has proved to be extremely influential to developmental biologists and it is still a commonly cited reference. Indeed, when written, the work was quite speculative and it is only recently that molecular biology techniques have advanced to the stage where such a model can be fully tested. In this regard, I refer you back to the articles in Reading 18.4, which describe such tests. Here, I refer to the article 'Direct and continuous assessment by cells of their position in a morphogen gradient' by J. B. Gurdon, A. Mitchell, and D. Mahony in *Nature* **376** (1995) 520–521, and to the article 'Activin signalling and responce to a morphogen gradient' by J. B. Gurdon, P. Harger, A. Mitchell and P. Lemaire in *Nature* **371** (1994) 487–492.

By the way, the third section of the paper contains a nice application of the diffusion equation. The latter is cleverly used by the authors to obtain numerical output to make some simple tests of their model.

Thresholds in Development

Journal of Theoretical Biology, **65** (1977) 579–590.
J. Lewis, J. M. W. Slack, L. Wolpert

The interpretation of gradients in positional information is considered in terms of thresholds in cell responses, giving rise to cell states which are discrete and persistent. Equilibrium models based on co-operative binding of control molecules do not show true thresholds of discontinuity, though with a very high degree of co-operativity they could mimic them; in any case they do not provide the cells with any memory of a transient signal. A simple kinetic model based upon positive feedback can account both for memory and for discontinuities in the pattern of cell states. The model is an example of a bistable control circuit, and transitions from one state to another may be brought about not only by morphogenetic signals, but also by disturbances in the parameters determining the kinetics of the system. This might explain some aspects of transdetermination in insects.

An attempt is made to analyse the precision with which a spatial gradient of a diffusible morphogen could be interpreted by a kinetic threshold mechanism, in terms of the length of the field, the steepness of the concentration gradient, and the intrinsic random variability of cells. It is concluded that it would be possible to specify as many as 30 distinct cell states in a positional field 1 mm long with a concentration span of 10^3. Mechanisms for reducing the positional error are considered.

1. Introduction

Many pattern-forming processes in development may be viewed in terms of a mechanism involving positional information. Two main steps are envisaged: the cells have their position specified with respect to certain boundary regions, and then they interpret this positional information by an appropriate choice of cell state. The choice of cell state may correspond to a change in determination or initiate a course of cytodifferentiation. Much more attention has been given to how positional information may be supplied than to the process of interpretation. A classic mechanism for specifying position depends on a gradient in some morphogen acting as a positional signal, as in the insect epidermis (Lawrence, Crick & Munro, 1972), the insect egg (Sander, 1975), hydra (Wolpert, Clarke & Hornbruch, 1972) and the chick limb (Tickle, Summerbell & Wolpert, 1975). Possible molecular mechanisms for setting up such gradients have been considered by Gierer & Meinhardt (1972). A gradient in a positional signal is not the only way of providing positional information, as is shown by the phase-shift model of Goodwin & Cohen (1969) and the progress zone model of Summerbell, Lewis & Wolpert (1973). Nevertheless, it seems reasonable to assume that ultimately the positional value of the cell will be governed by the concentration of one or more chemical compounds.

The first step in the interpretation of positional information is to choose amongst a set of different states, discrete choices being associated with thresholds in the response of the cells: for concentrations just above a threshold, the cells will adopt one state, and for concentrations just below it, another state. If the positional signal is graded, the spatial boundaries between cells in different states will correspond to critical concentrations of the signal substance, as first pointed out by Dalcq & Pasteels (1937). Spemann, indeed, evidently saw here a serious objection to the gradient concept of Child (1941) when he wrote: "Another difficulty seems to lie in the fact that the gradient ... must be conceived of as continuous, whereas the series of formations whose differentiation would be determined by that gradient ... is absolutely discontinuous" (Spemann, 1938).

We have to meet this objection if we are to believe that a smooth concentration gradient of a morphogen can provide positional information which the cells can then interpret to give the spatial pattern of cellular differentiation (Wolpert, 1971). The rather abstruse analysis provided by catastrophe theory (Thom, 1975) leaves some important biological questions unanswered. We need to know how many discrete distinctions could plausibly be specified by such a gradient, and how reliably, given the natural variability of cells. Crucial as the problem is, it has received virtually no attention. We wish here to suggest some simple mechanism, and to discuss the fidelity with which they could generate patterns.

Thresholds are intimately related to another feature of cell determination: its persistence. In many cases the differences set up in a positional field remain long after the field-like properties have disappeared together with the positional signal. The cells must remember the effect of the signal (Wolpert & Lewis, 1975). The threshold and the memory are, we believe, two aspects of a single process: both depend on a positive feedback loop in the intracellular control system.

The detailed chemistry of thresholds in embryonic development is unknown. Elsewhere in biology, however, certain threshold phenomenon have been analysed thoroughly, and the role of positive feedback has been made plain; as, for example, in the triggering of the nerve action potential (Hodgkin, 1967) and in the induction of the lac permease in *E. coli* (Kennedy, 1970). In other cases, such as the very abrupt and striking response of certain enzyme activities in plants to the amount

of phytochrome activated by light (Mohr & Oelze-Karow, 1976), something of the chemistry is known, but the mechanism is as yet only half-elucidated.

2. Equilibrium Control and Kinetic Control

Thresholds are often dismissed as being easily accounted for by co-operative binding of control molecules to an allosteric enzyme. The threshold, on this conception, consists simply in a steep shift in the level of an equilibrium between different forms of the enzyme as the concentration of the control molecule changes (Fig. 1). The threshold is thus viewed essentially as an equilibrium phenomenon. We wish to contrast this with a view of thresholds which is essentially kinetic.

Although the cell as a whole is in no sense at thermodynamic equilibrium, it is possible to isolate sets of metabolites whose relationships are very near those of equilibrium (Newsholme & Start, 1973). One may thus consider the effect of variations in a signal substance on a subsystem where all the other substances assume their equilibrium concentrations passively. Consider, for example, the case where the observed response of the cell is directly proportional to the degree of saturation of a receptor enzyme, E, by the morphogen S. Suppose that E binds n molecules of S with very high co-operativity. Under these conditions the equilibrium degree of saturation, Y, will show a sigmoid dependence on S with a Hill coefficient approaching n (Fig. 1):

$$E + nS \rightleftharpoons ES_n \text{ with dissociation constant } K \quad (1)$$

$$Y \approx \frac{[ES_n]}{[E_{\text{total}}]} \approx \frac{[S]^n}{K + [S]^n}$$

The equilibrium dependence of Y upon S is always continuous. One might admittedly mimic the effect of a true threshold of discontinuity, if the dependence of Y on the signal concentration S were so steep that there was a large difference between Y in adjacent cells in the graded signal field. But some rather extreme conditions would have to be met. Suppose, for example, that Y changed from 10% to 90% of maximal. With a Hill coefficient of 4 this would require a difference of S of three-fold between adjacent cells. At this rate, the concentration of S would change by 3^{10} fold across ten cell diameters and by 3^{100} fold across 100 cell diameters. Such very steep concentration gradients do not seem plausible (see Fig. 1). Thus any equilibrium chemical control of this sort which could

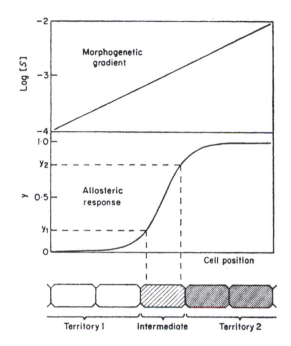

Fig. 1 A simple equilibrium model for a "threshold". The concentration gradient of S spans two decades across five cells. The lower curve depicts the degree of saturation of an allosteric enzyme obeying the Hill equation with $K = 10^{-12}$, $n = 4$. The intermediate response of the cells is arbitrarily drawn to occur between $Y_1 = 0.2$ and $Y_2 = 0.8$.

produce a steep enough response to mimic a threshold would need a receptor with a number and a co-operativity of binding sites much greater than those of most known allosteric enzymes. But even then, a crucial requirement for cell determination would be missing: the cells would have no memory of the effect of the signal. They would all revert to the same state after the signal gradient was gone. Admittedly, true discontinuities may occur in equilibrium systems undergoing phase transitions, and may be associated with hysteresis, as in the supercooling of water or the orientation of domains in a magnet. But phase transitions depend on the establishment of long-range order in very large (strictly speaking, in infinite) assemblies of molecules, and are an unlikely means of producing the subtle and multifarious thresholds occurring in animal cells. The memory of a change of determination that is heritable from one cell generation to the next must be a non-equilibrium phenomenon.

Let us therefore consider the cell as a kinetic sys-tem, in which chemicals react far from equilibrium at rates depending on their concentrations. Such a system can jump discontinuously from one sort of steady state to another as external control parameters are changed (Kacser, 1960; Prigogine Nicolis, 1971; Edelstein, 1972; Thornley, 1972). We illustrate this sort of threshold phenomenon by a simple and biochemically plausible model, whose basic formulation was suggested to us by Graeme Mitchison. The model has interesting properties which are not intuitively obvious, and which it shares with a much more general class of kinetic systems.

We consider the activation of a gene G by a signal substance S. Transcription of the gene G is assumed to be promoted in a linear fashion by S and in a sigmoidal fashion by its own product g, giving a positive feedback g is in addition degraded at a rate proportional to its concentration. If (we) eliminate various intermediate chemical variables from the system of equations describing transcription, translation and metabolism, the behavior of this

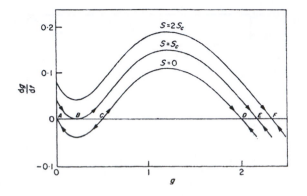

Fig. 2 The rate of change, dg/dt, of the concentration of the gene product as a function of the instantaneous concentration g, for three different fixed concentrations of the signal substance S. The curves are calculated assuming

$$\frac{dg}{dt} = k_1 S + \frac{g^2}{1 + g^2} - 0 \cdot 4g.$$

The arrows on the curves point in the direction of increasing g where $dg/dt > 0$, and of decreasing g where $dg/dt < 0$. A steady state occurs where a curve cuts the g-axis; it is stable if the arrows converge on it, and unstable if the arrows diverge from it. When $S = 0$, there are stable steady states at the points A [where $g = g_0(0) = 0$] and D [where $g = g_1(0)$] and there is an unstable state at C. When S equals the critical value S_c, there is one stable steady state at E [where $g = g_1(S_c)$] and a singular unstable state at B [where $g = g_0(S_c)$]. When $S = 2S_c$, there is only one steady state, at F [where $g = g_1(2S_c)$] and it is stable.

subsystem when subject to slowly changing conditions may be summed up by a single equation for g. To be precise, let the net rate of change of g be

$$\frac{dg}{dt} = k_1 S + \frac{k_2 g^2}{k_3 + g^2} - k_4 g$$

where the k's are constants. For a given value of S, there are steady states at the values of g such that $dg/dt = 0$ (Fig. 2). A steady state is stable if any small departure from it is self-correcting: that is, if dg/dt is negative for slightly higher values of g, and positive for slightly lower values of g. The stable steady states for a given S can be read directly from the graph of dg/dt as a function of g.

Figure 2 shows the graphs for three different concentrations of the signal substance S. When S is low, there are two stable steady states, at $g_0(S)$ and $g_1(S)$. When S is high, there is only one, at $g_1(S)$. The critical intermediate signal concentration S_c, divides the regime where there are two possible stable steady states from the regime where there is only one. Suppose that a cell starts with the gene G inactive, that is with $g = 0$, and is exposed to a signal concentration that increases gradually from zero. The concentration of the gene product g will stay close to the lower stable value $g_0(S)$ until S reaches the threshold S. When S becomes greater than S_c, the first downward dip of the dg/dt curve fails to meet to the zero line, and there ceases to be a steady state in the neighbourhood of the point $g_0(S_c)$ to which the cell has been brought. The concentration of g will thereupon rise autonomously until it reaches the upper stable value $g_1(S)$. Thus there is a discontinuous transition from one steady state to another as S is increased past its threshold value S_c. In principle, an infinitesimal increment from $S_c - dS$ to $S_c + dS$ will bring this transition about. Our kinetic system stands here in contrast with the equilibrium systems, where no such discontinuity occurs.

The kinetic system, furthermore, has a memory. If the signal concentration S, having been raised above threshold, is then brought down again to zero, the system will not return to the lower stable state $g_0(0)$; it will instead be left in the upper stable state $g_1(0)$. The gene will, in effect, have been turned on permanently. Much of the argument can be carried over into a far more general case. Discrete alternative stable modes are a natural feature of chemical kinetic systems, and the discontinuous transitions from one mode to another can be triggered by transient control signals.

Our kinetic system is an example of the type of bistable control circuit postulated by Kauffman (1975) to explain determination and transdetermination in the imaginal discs of *Drosophila*. It shows clearly why transitions between alternative steady states may occur more readily in one direction than in the other: in our particular model, the upper stable steady state g is stable for example even against large fluctuations in S, whereas the lower stable steady state g_0 is not. Random disturbances of other concentrations affecting the values of k_1, k_2, k_3 and k_4, could likewise cause transdetermination preferentially in one direction. A fluctuation in the concentration of a small molecule, which might pass from cell to cell through gap junctions, could cause transdetermination of a group of cells simultaneously, as is observed (Gehring, 1972).

A related feature of the system is that the cells in an embryonic field may continue to be sensitive, say, to an increase in the concentration of a signal substance, even after they have ceased to be affected by a reduction. One might thus understand why, for example a graft of ectopic polarizing tissue in a chick limb bud can cause an almost complete reduplication of digits, whereas excision of polarizing tissue at the same stage causes only slight deficiencies (Tickle, Summerbell & Wolpert, 1975).

3. Positional Signalling and Positional Precision

To give the body its spatial pattern, developmental signals must specify the positions where cells cross thresholds. Most developmental fields are about 1 mm or 100 cell-diameters in maximum linear dimension, and specification takes place over a period of hours. As Crick (1970) has pointed out, this makes diffusion a plausible means to set up a positional signal, and there is some experimental evidence from Hydra (Wolpert, Clarke & Hornbruch, 1972), insects (Sander, 1975; Lawrence, Crick & Munro, 1972) and the vertebrate limb (Tickle, Summerbell & Wolpert, 1975) which is consistent with this. The local concentration of a substance diffusing across the field may serve as the signal S which controls the choice of cell state. Thresholds in the

response to S would appear as a spatial pattern of discontinuities demarcating different cell states. There is no reason in principle why there should not be many thresholds in the response to a single signal, depending on different receptor circuits, so that many distinctions are simultaneously determined in the one field.

There are, however, some limitations imposed by the random variability of cells. Given our threshold mechanism, we can relate the random errors in the placing of thresholds in the body to the random errors in the control of chemical concentrations inside cells. The errors in the placing of thresholds will in turn set a limit to the number of thresholds that can be drawn in a reliable spatial sequence across one developmental field.

There are some rather scanty data bearing on these points. The lengths of most parts of the body in higher animals seem to be determined with a precision of about 3% or better (Wolpert, 1972; Maynard-Smith, 1960). The random variability of intracellular concentrations will be discussed more fully below: it may be much greater, perhaps of the order of 20% for many substances. The number of distinct states typically specified per embryonic field is hard to establish. There is some suggestion that all the 14 segments of the insect body may be rendered distinct by a single axial field acting round the time of blastoderm formation (Sander, 1975); and it is possible that the number of territories specified during primary induction along the main body axis in vertebrates may be of the same order of magnitude. On the other hand, the subdivision of *Drosophila* imaginal discs into compartments, for example, seems to proceed simply by binary distinctions (Garcia-Bellido 1975; Crick & Lawrence, 1975).

Concentration gradients could be established in various ways. Diffusion is perhaps the simplest conceptually and we shall assume it in this discussion. But most of our general conclusions do not depend on this specific assumption. Consider, then, a field a hundred cell-diameters long, in which a signal gradient is set up by diffusion from a source at one end. Could the gradient serve to specify, say, ten distinct regions? Could it determine lengths to a precision of 3%? Let us assume that the signal substance is kept at a constant concentration S_0 at the source, and that elsewhere it is degraded uniformly at a rate proportional to its concentration. The steady-state distribution will then be of exponential form. The concentration S at a distance x from the source will be given by

$$S = S_0 e^{-\alpha x}$$

or equivalently

$$c = \tfrac{1}{\alpha} \ln S_0/S \qquad (2)$$

where α is a constant depending on the diffusion constant and the rate of degradation.

If a threshold occurred invariably at $S = S_c$, there would be a transition in the spatial pattern precisely at

$$c = \frac{1}{\alpha} \ln S_0/S_c.$$

If the cells are variable, however, with their individual thresholds ranging from $S_c - \Delta S_c$ to $S_c + \Delta S_c$, then the position of the transition will be variable by $\pm \Delta x_c$, where

$$\Delta x_c = \frac{1}{\alpha} \Delta S_c/S_c.$$

We can express the positional error as a fraction of the length L of the field using equation (1):

$$\frac{\Delta x_c}{L} = \frac{1}{\alpha L} \frac{\Delta S_c}{S_c} = \frac{\Delta S_c}{S_c} \bigg/ \ln \frac{S_0}{S_L}.$$

Thus the positional error can be reduced in two ways: by reducing the error in the setting of the threshold, $\Delta S_c/S_c$, and by increasing the concentration span of the gradient, S_0/S_L.

How big a concentration span is plausible? The upper extreme of contentration which one might plausibly expect for a small molecule in the soluble pool is about 10^{-2} M. If for the lower extreme we take ten molecules per cell, i.e., about 10^{-10} M, we have a maximum concentration span of 10^8. Consideration of the flux to be drawn from each source cell, however, and of the, homeostasis of sources, sets a rather more stringent limit of 10^5 for the concentration span.

The positional error $\Delta x_c/L$ would then be less than the error in the threshold $\Delta S_c/S_c$ by a factor of $\ln 10^5$, that is, about 12. For a concentration span of only 10, the positional error would be only five times larger than for a concentration span of 10^5.

The value of $\Delta S_c/S_c$ is the crucial unknown. It depends on the random variability of chemical concentrations in cells, and we have found it remarkably difficult to obtain information on this basic

point; it is also hard to distinguish experimentally between intrinsic variability in single cells, and random errors in the techniques of measurement. A cytochemical study of mouse fibroblasts (Killander & Zetterberg, 1965) showed that the RNA content had a coefficient of variation of 22% and the dry mass of 11% even for sister cells at the same stage of the division cycle, with a DNA content variable by no more than 2%. Measurements of chloride concentration in squid giant axons (Keynes, 1963) of resting potential in frog muscle fibres (Adian, 1956) and of pH in crab muscle fibres (Aickin & Thomas, 1975) have standard deviations corresponding to coefficients of variation in ionic concentrations of about 8%, 12% and 30% respectively; these can be regarded as upper limits on the variation intrinsic to the cells.

From rather inadequate evidence such as this we would guess that a variability of 20% would be typical of many intracellular concentrations. Several different substances must take part in the threshold reactions, and the error in each may be expected to contribute to the error in the definition of the threshold concentration of S. In the same way, errors in many different metabolite concentrations will contribute to the error in the concentration of any one given metabolite. The cell's system of homeostatic controls, however, presumably keeps the error at almost every point of the reaction network within standard limits, avoiding disastrous accumulations of error. In the first instance let us therefore suppose that the threshold setting error $\Delta S_c/S_c$ is of the same order as most concentration errors in the cell, say 20%. Then we can find the positional error from equation (2): for concentration spans of 10^6, 10^3 and 10, it will be roughly 1.5%, 3% and 9%, respectively. Thus our arguments lead to an estimate which is perfectly compatible with the observations of positional errors of about 3%. As many as 30 distinct states could be specified in reliable serial order by one positional signal, although its gradient would need to be rather steep.

4. Modification of the Error

The size of the threshold setting error $\Delta S_c/S_c$ might in any case be reduced in various ways. The concentrations of substances involved in the threshold reactions might be controlled with special accuracy. If DNA is involved, as seems likely, its concentration may be standardized by confining the action of the signal to a particular phase of the cell cycle, such as G_1. Fluctuations in the concentrations of small molecules may be reduced by their exchange through low-resistance junctions between the cells (Loewenstein, 1973; Furshpan & Potter, 1968; Slack & Warner, 1975). The signal substance might act through a reaction of high order in S, so that a 20% error in the level of the immediate reaction product required for threshold would correspond to a much smaller percentage error in the required threshhold level of S itself. But we recognize that we are here glossing over complex problems raised by variability and fluctuations in control systems.

Because of the errors in the thresholds of the individual cells, the boundary between tissues composed of cells in different states may not be shared. There may be a transition region, of width Δx_c, where cells above and below their individual thresholds are mixed together. The position of the center of the transition region will be defined by an average over many cells, and will have an accuracy better than Δx_c. The effect of cell variability can be much reduced if the cells are free to move a little. The cells in the transition region may then sort out if, for example, those in the same state of determination stick together more strongly than those in opposite states (Steinberg 1976). Lawrence (1975) has suggested such a mechanism for maintaining the sharp and precisely placed boundaries at compartment borders in *Drosophila*.

5. General Discussion

The kinetic model for thresholds given above shows how cells may change their state discontinuously at particular concentrations of a signal substance and how this change of cell state may persist when the signal is withdrawn. Although our argument has been presented specifically in terms of chemical concentration gradients, thresholds must be an essential feature of any theory of development, if it is to explain the origin of qualitative differences between cells. This holds true irrespective of whether the differences are from cytoplasmic differences in the egg, or from signals from other cells.

If a signal can supply positional information only over a small field, and yet is to determine a gross feature of a large animal, it must do its work when the embryo is still small; it must assign different states of determination to the cells, and they must remember the assignments when the embryo has grown big, and the signal is no longer

there. Mesoderm must remain mesodermal, thoracic somites must help to make a thorax, and the leg bud must develop into a leg. Thus in the pattern of differentiation of the body, all but the finest details must depend on temporary signals that set up lasting distinctions; and discontinuities of cell character are characteristic of that process. Such perpetuated differences represent discrete alternative steady mode of the dynamical system of reacting chemicals in the cell, and cannot in general be continuously graded. On the other hand, where a morphogenetic signal is permanently present there is no need for such discontinuities: a maintained graded signal could maintain a smooth gradation of cell character and an equilibrium model for interpretation might be appropriate, as perhaps in hydra. A distinction should be drawn between two different ways in which a smooth gradation between two types of tissue might occur. In the one there could be continuous variation in the character of the constituent cells; in the other, the tissues might be composed of discrete types of cell, mixed in smoothly variable proportions.

The analysis above has shown that a concentration gradient across a small field could easily define the position of a discontinuity of cell state with a precision of 3% of the length of the field. In a field of 100 cell diameters it should be possible to specify as many as 30 distinct cell states in a reliable sequence. A basic factor determining the precision of such a pattern is the random variability of the individual cells. This is probably an important source of variability between organisms which is neither genetic nor environmental.

This work is supported by the Medical Research Council.

REFERENCES

1. Adrian, R. H. (1956). *J. Physiol.* **133**, 631.
2. Aikin, O. C. & Thomas, R. C. (1975). *J. Physiol.* **252**, 803.
3. Crick, C. M. (1941). *Patterns and Problems in Development.* Chicago: University of Chicago Press.
4. Crick, F. H. C. (1970). *Nature, Lond.* **225**, 420.
5. Crick, F. H. C. & Lawrence, P. A. (1975). *Science, N. Y.* **189**, 340.
6. Dalcq, A. & Pasteels, J. (1937). *Arch. Biol.* **48**, 669.
7. Edelstein, B. B. (1972). *J. theor. Biol.* **37**, 221.

8. Furshipan, E. J. & Potter, D. D. (1968). *Curr. Top. Devel. Biol.* **3**, 95.

9. Garcia-Bellido, A. (1975). In *Cell Patterning*, Ciba Foundation Symp. **29**, 161. Amsterdam: Associated Scientific Publishers.

10. Gehring, W. (1972). in *Biology of Imaginal Disks* (H. Ursprung & R. Nothiger, eds) p. 35. Berlin: Springer.

11. Gierer, A. & Meinhardt, H. (1972). *Kybernetik* **12**, 30.

12. Goodwin, B. & Cohen, M. H. (1969). *J. theor. Biol.* **25**, 49.

13. Hodgkin, A. L. (1964). *The Conduction of the Nervous Impulse.* Liverpool: University Press.

14. Kacser, H. (1960). *Symp. Soc. exp. Biol.* **14**, 13.

15. Kauffman, S. (1975). In *Cell Patterning*, Ciba Foundation Symp. **29**, 201. Amsterdam: Associated Scientific Publishers.

16. Kennedy, E. P. (1970). In *The Lactose Operon* (J. Beckwith & D. Zipser, eds), Cold Spring Harbor Lab.

17. Keynes, R. D. (1963). *J. Physiol.* **169**, 690.

18. Killander, D. & Zetterbero, A. (1965). *Expl. Cell Res.* **38**, 272.

19. Lawrence, P. A. (1975). In *Cell Patterning*, Ciba Foundation Symp. **29**, 2. Amsterdam: Associated Scientific Publishers.

20. Lawrence, P. A., Crick, F. H. C. & Munro, M. (1972). *J. cell Sci.* **11**, 815.

21. Loewenstein, W. R. (1973). *Fedn. Proc. Fedn. Am. Socs. exp. Biol.* **32**, 60.

22. Maynard-Smith, J. (1960). *Proc. R. Soc. B.* **152**, 397.

23. Mohr, H. & Oelze-Karow, H. (1976). In *Light and Plant Development* (H. Smith, ed) Kent: Butterworth.

24. Newsholme, E. A. & Start, C. (1973). *Regulation in Metabolism.* London: John Wiley & Sons Ltd.

25. Prigogine, I. & Nicolis, G. (1971). *Q. Rev. Biophys.* **4**, 107.

26. Sander, K. (1975). In *Cell Patterning*, Ciba Foundation Symp. **29**, 241. Amsterdam: Associated Scientific Publishers.

27. Slack, C. & Warner, A. E. (1975). *J. Physiol.* **248**, 97.

28. Spemann, H. (1938). In *Embryonic Development and Induction*, p. 331. Yale: Yale University Press.

29. Sternberg, M. S. (1970). *J. expl. Zool.* **173**, 395.

30. Summerbell, D., Lewis, J. H. & Wolpert, L. (1973). *Nature, Lond.* **244**, 492.

31. Thom, R. (1975). *Structural Stability and Morphogenesis.* Reading, Mass.: Benjamin.

32. Thornley, J. H. M. (1972). *An. Bot.* **36**, 861.

33. Tickle, C., Summerbell, D. & Wolpert, L. (1975). *Nature, Lond.* **254**, 199.

34. Wolpert, L. (1971). *Curr. Top. devel. Biol.* **6**, 183.

35. Wolpert, L., Clarke, M. R. B. & Hornbruch, A. (1972). *Nature, Lond.* **239**, 101.

36. Wolpert, L. & Lewis, J. H. (1975). *Fedn. Proc. Fedn. Am. Socs. exp. Biol.* **34**, 14.

Exercises for Chapter 26

1. In (a)–(c), decide which (if any) of the functions of time, $x(t)$ or $y(t)$, represent "slow subsystem" unknowns and which represent "fast subsystem" unknowns. Assume that your interest in these equations concerns, only those values for x and y that lie between -10 and 10.

(a) $$\frac{d}{dt}x = 0.1x^3 - 3xy,$$

$$\frac{d}{dt}y = 0.1y - 0.001xy.$$

(b) $$\frac{d}{dt}x = 0.1x - .05xy,$$

$$\frac{d}{dt}y = 0.1y + 0.03xy.$$

(c) $\dfrac{d}{dt}x = 0.1x - 0.03xy,$

$\dfrac{d}{dt}y = y + 0.007xy.$

2. In each of the cases (a)–(d), c is a parameter. For each case, there are three parts to the problem.

Part 1: Find those values of c at which an equilibrium point in the given equation for the function $x(t)$ disappears.

Part 2: On the x versus t plane, sketch the graphs of typical solution curves $t \rightarrow x(t)$ for the differential equation using a value of c that is slightly less than that found in Part 1.

Part 3: On the x versus t plane, sketch the graphs of typical solution curves $t \rightarrow x(t)$ for the differential equation using a value of c that is slightly greater than that found in Part 1.

(a) $\frac{d}{dt}x = 5x^2 + c$

(b) $\frac{d}{dt}x = 2\sin(x) + c$

(c) $\frac{d}{dt}x = e^{-x} + cx$

(d) $\frac{d}{dt}x = e^{-x} + 1 - c$

3. This is an open-ended research problem: Find examples of biological systems that exhibit switchlike behavior in their development. By "switchlike behavior," I mean that various subsystems that start in seemingly similar circumstances end up, at the end of the development, to have very different appearance or function. In your examples, say a few words about whether you think that a (complicated) variant of the Lewis-Slack-Wolpert model in Commentary 26.1 might be appropriate.

27

Testing for Periodicity

It is important to have a model that predicts periodic behavior, but this sort of prediction is moot if the data don't support such a hypothesis. More often than not, any given data, say a measurement $m(t)$ as a function of time t, will not exhibit true periodicity. There are many possible reasons for this. For example, even if the phenomenon in question is truly periodic, any measurement has some uncertainty in it, and this uncertainty will tend to obscure the periodicity in the data function $m(t)$.

Often, the phenomenon in question is not truly periodic but is the result of many (sometimes a lot and sometimes a few) confluent factors, where each factor is itself periodic, but where the different factors have different periods. For example, if you were studying the behavior of horseshoe crabs on the New England coast, you would have to consider effects that vary on a daily cycle (for example, light and dark), a monthly cycle (for example, tides), and a yearly cycle (for example, the seasons). Thus, you would expect any measured function of time $m(t)$ for horseshoe crabs to be a sum of (at least) three functions, $m(t) = m_d(t) + m_m(t) + m_y(t)$, which are each periodic, but with periods equal to 24 hours, 29 days, and 365 days, respectively. Moreover, there may be longer period effects which add a fourth term to $m(t)$. For example the solar flux varies over an 11 year cycle, and so one might see $m(t) = m_d(t) + m_m(t) + m_y(t) + m_s(t)$, where $m_s(t)$ is a function with an 11 year period.

Another example concerns the daily fluctuations $T(t)$ in the average temperature of the surface waters of the North Atlantic (for example, at the latitude of Southern Greenland). Here, we expect $T(t) = T_1(t) + T_{11}(t) + \cdots$ to be a sum of functions where the first, T_1, is periodic with period one year, and the second is periodic with

430

period 11 years. But there are evidently other terms with much longer periods, on the order of thousands and hundreds of thousands of years.

27.1   Testing for a Term with Period p

Suppose that $m(t)$ is a measured function of time that we have some how obtained. We shall suppose that we know $m(t)$ for t between $-\tau_0$ and $-\tau_0 + \tau$, where τ_0 and τ are known. (The number τ_0 represents the amount of time before the present at which our data begins, and τ is the time span for which we have accumulated data.) Suppose we ask, Does our data $m(t)$ have a component that is periodic with period p? Of course, if the hypothetical period p is larger than τ, we shall never know, for one period will not have passed in the time represented by our data. However, if p is much less than τ, then there is a way to detect the presence in $m(t)$ of a term that is periodic with period p.

Indeed, to detect a period p component to $m(t)$, we need only compute three integrals, namely

$$a \equiv \frac{1}{\tau} \int_{-\tau_0}^{-\tau_0+\tau} m(t)\cos(2\pi t/p)\,dt,$$

$$b \equiv \frac{1}{\tau} \int_{-\tau_0}^{-\tau_0+\tau} m(t)\sin(2\pi t/p)\,dt,$$

$$\sigma \equiv \frac{1}{\tau} \int_{-\tau_0}^{-\tau_0+\tau} m(t)^2\,dt. \tag{1}$$

With the numbers a, b and σ computed, we then compute

$$f = (a^2 + b^2)/\sigma. \tag{2}$$

A significant component of the original function $m(t)$ with period p is signified by a large value for f. In this regard, it is important to note that f is always between zero and one, so *large* means that f is a significant fraction of 1.

27.2   The Power Spectrum Function

In general, we may not have an a priori guess as to the periods of the components of m (if any are periodic). In this case, a standard approach is to compute the f using (1) and (2) for many values of p and see which values give large f. That is, we think of the numbers a and b in (1) and f in (2) as functions of the variable p. Actually, it is customary to change variables from the period p to the frequency,

$$\nu \equiv 1/p \tag{3}$$

because $1/p$ appears in (1). Then both a and b are considered to be functions of ν:

$$a \equiv \frac{1}{\tau} \int_{-\tau_0}^{-\tau_0+\tau} m(t)\cos(2\pi \nu t)\,dt,$$

$$b \equiv \frac{1}{\tau} \int_{-\tau_0}^{-\tau_0 + \tau} m(t) \sin(2\pi \nu t)\, dt. \tag{4}$$

Likewise, f is considered to be a function of $\nu = 1/p$ given by

$$f(\nu) = (a(\nu)^2 + b(\nu)^2)/\sigma. \tag{5}$$

The function $f(\nu)$ is called the **power spectral density** function. A significant peak in the function $f(\nu)$ at some value ν can signify that the original data function $m(t)$ has a significant component that has period p roughly equal to $1/\nu$. To be precise, if the peak occurs at ν where the resulting period $1/\nu$ is much less than τ (thus, $\nu \gg 1/\tau$), then we can conclude that $m(t)$ has a significant component that has period p that is nearly $1/\nu$.

The computation of the functions $a(\nu)$ and $b(\nu)$ is a common operation on data—the name for this operation is **Fourier transforming**. The functions $a(\nu)$ and $b(\nu)$ are called the **Fourier coefficients** of the original function $m(t)$.

27.3   An Example

Suppose that the function $m(t)$ is exactly periodic with period q. More precisely, suppose that we can write $m(t)$ as a sum

$$m(t) = \alpha \cos(2\pi t/q) + \beta \sin(2\pi t/q), \tag{6}$$

where α and β are numbers. We can then compute the Fourier coefficients $a(\nu)$ and $b(\nu)$ via the integrals in (4) and then the function $f(\nu)$ as indicated in (5). The relevant integrals involve trigonometric functions which are given in Equations (8) and (9). (You can also look them up in any standard calculus book.) The upshot is that when q and $1/\nu$ are much less than τ, then $f(\nu)$ looks as pictured in Figure 27.1.

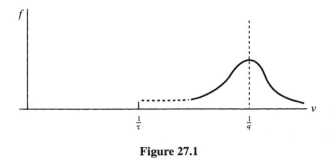

Figure 27.1

Note that the width of the peak at $\nu = 1/q$ is of the order of magnitude of $1/\tau$ when q is much less than τ. Meanwhile, the height of the peak is proportional to $\alpha^2 + \beta^2$.

More generally, suppose that $m(t)$ is a sum of some number, say N, of sines and cosines with periods $\{q_j\}_{1 \le j \le N}$ as follows:

$$m(t) = \alpha_1 \cos(2\pi t/q_1) + \beta_1 \sin(2\pi t/q_1) + \alpha_2 \cos(2\pi t/q_2) + \beta_2 \sin(2\pi t/q_2) + \cdots .$$
(7)

In this case, the graph of the function $f(v)$ has the form pictured in Figure 27.2.

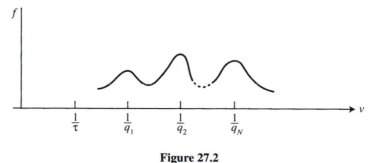

Figure 27.2

Note that there are peaks near the inverse periods, $v = 1/q_1, \dots, 1/q_N$. The heights of the peak near any given $1/q_j$ are determined both by $\alpha_j^2 + \beta_j^2$ and by the distance of q_j from the other q_k's as measured on a scale where τ is equal to 1.

27.4   Trigonometric Integrals

In the preceding examples, the Fourier coefficients $a(v)$ and $b(v)$ can be computed exactly in closed form. Should you wish to do so, you can use the following indefinite integrals: First, when $A \ne B$, then

$$\int \cos(At)\cos(Bt)\,dt$$

$$= 2^{-1}(A-B)^{-1}\sin((A-B)t) + 2^{-1}(A+B)^{-1}\sin((A+B)t)$$

$$\int \sin(At)\cos(Bt)\,dt$$

$$= -2^{-1}(A-B)^{-1}\cos((A-B)t) - 2^{-1}(A+B)^{-1}\cos((A+B)t)$$

$$\int \sin(At)\sin(Bt)\,dt$$

$$= 2^{-1}(A-B)^{-1}\sin((A-B)t) - 2^{-1}(A+B)^{-1}\sin((A+B)t). \quad (8)$$

In addition, the $A = B$ cases of the integrands in (8) are

$$\int \cos^2(At)\,dt = t/2 + 4^{-1}A^{-1}\sin(2At)$$

$$\int \sin(At)\cos(At)\,dt = -4^{-1}A^{-1}\cos(2At)$$

$$\int \sin^2(At)\,dt = t/2 - 4^{-1}A^{-1}\sin(2At). \qquad (9)$$

27.5   Lessons

Here are some key points from this chapter:

- A significant, periodic component of a function of time can be identified using integrals of the products of the function with $\cos(2\pi t/p)$ and with $\sin(2\pi t/p)$.

- Remember how to compute the power spectrum function for any given function of time.

- Peaks in the graph of the power spectrum function occur near frequencies that are inverses of the periods of significant periodic components of the original function. The heights of these peaks correlate with the sizes of these periodic components.

READINGS FOR CHAPTER 27

READING 27.1

A Pervasive Millennial-Scale Cycle in North Atlantic Holocene and Glacial Climates

Commentary: With evidence from North Atlantic deep ocean cores, the authors identify evidence for a cyclic component with period $\sim 1400 \pm 500$ years to climate variation in the North Atlantic. Their technique is to produce a 'power spectrum', $f(\nu)$, as a function of frequency, and then identify peaks of this function. Their power spectrum curve appears as Figure 7C in the article.

A Pervasive Millennial-Scale Cycle in North Atlantic Holocene and Glacial Climates

Science, **278** (1997) 1257–1266.
Gerard Bond, William Showers, Maziet Cheseby, Rusty Lotti, Peter Almasi, Peter deMenocal,
Paul Priore, Heidi Cullen, Irka Hajdas, Georges Bonani

Evidence from North Atlantic deep sea cores reveals that abrupt shifts punctuated what is conventionally thought to have been a relatively stable Holocene climate. During each of these episodes, cool, ice-bearing waters from north of Iceland were advected as far south as the lati-

---

Author affiliations: G. Bond, M. Cheseby, R. Lotti, P. Almasi, P. deMenocal, P. Priore, and H. Cullen: Lamont-Doherty Earth Observatory of Columbia University. W. Showers: Department of Marine, Earth and Atmospheric Sciences, North Carolina State University. I. Hajdas and G. Bonani: AMS14 C Lab, ITP Eidgenossische Technische Hochschule (ETH) Honeggerberg.

tude of Britain. At about the same times, the atmospheric circulation above Greenland changed abruptly. Pacings of the Holocene events and of abrupt climate shifts during the last glaciation are statistically the same; together, they make up a series of climate shifts with a cyclicity close to 1470 ± 500 years. The Holocene events, therefore, appear to be the most recent manifestation of a pervasive millennial-scale climate cycle operating independently of the glacial-interglacial climate state. Amplification of the cycle during the last glaciation may have been linked to the North Atlantic's thermohaline circulation.

More than 20 years ago, Denton and Karlén [1] made two provocative suggestions about the climate of our present interglacial or Holocene period. Having found what appeared to be synchronous advances of mountain glaciers in North America and Europe, they concluded that Holocene climate was much more variable than implied by broad trends in pollen and marine records. On the basis of radiocarbon chronologies of the glacial advances, they further suggested that the climate variations were part of a regular millennial-scale pattern, which when projected backward coincided with climate shifts of the preceding glaciation and when projected forward predicted a progressive warming over the next few centuries.

Despite its important implications, Denton and Karlén's concept of a predictable, millennial-scale climate rhythm has not been widely accepted, partly because it has been difficult to find corroborating evidence in other climate records. For example, measurements of oxygen isotopes, methane concentrations, and snow accumulation in Greenland ice cores reveal no evidence of millennial-scale fluctuations during the Holocene, except perhaps for a brief cooling about 8200 years ago [2]. Many researchers now view the Holocene climate as anomalously stable [3]. Recent evidence from deep sea sediments in the Nordic Seas [4] suggests that Holocene climate there was even more stable than climate during the last interglaciation (Eemian).

In 1995, however, O'Brien et al. [5] demonstrated from measurements of soluble impurities in Greenland ice that Holocene atmospheric circulation above the ice cap was punctuated by a series of millennial-scale shifts. The most prominent of those shifts appeared to correlate with Denton and Karlén's glacial advances. Encouraged by the findings of O'Brien and her colleagues, we launched an investigation of deep sea Holocene sediments in the North Atlantic, anticipating that the shifts in atmospheric circulation above the ice cap were part of a much larger climate pattern that left its imprint in the deep sea record. Here, we report the results of that investigation, showing that the North Atlantic's Holocene climate indeed exhibits variability on millennial scales, and we then compare the Holocene variations with climate shifts of the last glaciation.

We analyzed Holocene sediment in two cores from opposite sides of the North Atlantic (Figure 1), VM 28-14 (64°47'N, 29°34'W; 1855 m of depth) and VM 29-191 (54°16'N, 16°47'W; 2370 m of depth). High-resolution accelerator mass spectrometer radiocarbon datings of planktonic foraminifera demonstrate that both cores have thick and nearly complete Holocene sections (Table 1 and Figure 2). Core top ages are less than 1000 years (all ages are in calendar years B.P. unless otherwise indicated). Sedimentation rates in both cores exceeded 10 cm per 1000 years, more than sufficient to resolve millennial-scale variability, and the rates were nearly constant (Figure 2). We sampled both cores at intervals of 0.5 to 1 cm (equivalent to a resolution of 50 to 100 years), and in each sample we measured nine proxies [6].

The Holocene signal: Episodes of ice-rafting. The most conspicuous evidence of variations in the North Atlantic's Holocene climate comes from the same three proxies that we used to document ice-rafting in the North Atlantic during the last glaciation [7]. One proxy is the concentration of lithic grains, defined as the number of grains with diameters greater than 150 μm in 1 g of core. At both sites we found a series of increases in grain concentrations, which, although of much smaller magnitude than those of the last glaciation, are distinct and reach peak values several times that of the ambient grain concentrations (Figure 2).

The other two proxies are petrologic tracers, defined as the percentages of certain types of lithic grains. One of the tracers is fresh volcanic glass, which comes from Iceland or Jan Mayen, and the other is hematite-stained grains, mostly quartz and feldspar, that come from sedimentary deposits containing red beds (Figure 3) [7]. During each of the lithic events, both tracers display prominent increases well above the 2σ counting error (Figure 2) [8].

These lithic/petrologic events demonstrate that

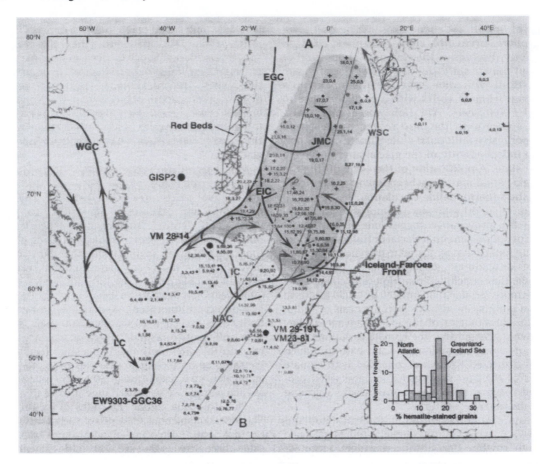

Fig. 1 Location of cores we analyzed and principal surface currents in the North Atlantic and Nordic Seas. The green dots and green line are locations of COADS temperature estimates and the profile in Figure 4A. The line from A to B is the line of the cross section of petrologic data shown in Figure 4B. Small dots and plus signs are locations of core tops analyzed for the two petrologic tracers; numbers (from left to right) are percentage of hematite-stained grains, percentage of Icelandic glass, and core locator number (core locations and site numbers can be obtained from the first author; see also [17]). Shaded red area encloses core tops with ≥10% hematite-stained grains. Dashed black line encloses core tops with ≥15% Icelandic glass. Histogram insert summarizes core top data for hematite-stained grains showing contrast in percentages north and south of the Denmark Strait and Iceland-Faeroes frontal systems, indicated by blue shading. Locations of red beds in East Greenland and Svalbard are from [7]. Surface currents: EGC, East Greenland Current; JMC, Jan Mayen Current; WSC, West Spitsbergen Current; EIC, East Iceland Current; WGC, West Greenland Current; LC, Labrador Current; NAC, North Atlantic Current; IC, Irminger Current.

ice-rafting episodes also occurred during the Holocene. Age differences of lithic/petrologic maxima between the two sites are within the 2σ calendar age error (Figure 2 and Table 1); hence, the events cannot be local in origin. Regional changes in carbonate dissolution or in winnowing of fine sediment could not have produced the petrologic changes we measured, and in neither core could we find evidence of current laminations or cross-bedding, making it unlikely that the events were produced by strong bottom currents (see also caveats in [9]).

Table 1   Table 1. Accelerator mass spectrometer radiocarbon measurements and calibrated (calendar) ages.

Depth (cm)	Species	Corrected radiocarbon age*	Calendar age†	2σ age range
		VM 29-191		
f1	G. bulloides	990 ± 55	868	752– 983
f21	G. bulloides	2,825 ± 60	2,928	2,779– 3,077
f41	G. bulloides	5,355 ± 75	6,136	5,984– 6,288
f61	G. bulloides	7,600 ± 80	8,334	8,168– 8,499
f81.2	G. bulloides	8,895 ± 80	9,714	9,651– 9,776
f91	G. bulloides	9,175 ± 95	10,166	9,979– 10,353
≠111	Mixed planktonic	10,150 ± 55	10,931	10,854– 11,007
≠121	N. pachyderma (s.)	10,950 ± 50	12,459	12,281– 12,637
≠125	Mixed planktonic	11,450 ± 45	12,962	12,829– 13,094
		VM 28-14		
¡8	G. bulloides	1,643 ± 60	1,516	1,396– 1,635
f19.5	G. bulloides	2,895 ± 60	3,019	2,862– 3,175
¡32	G. bulloides	3,540 ± 53	3,809	3,687– 3,931
¡44	G. bulloides	4,318 ± 59	4,931	4,816– 5,046
f52	G. bulloides	4,895 ± 70	5,659	5,561– 5,757
¡60	G. bulloides	5,515 ± 51	6,337	6,264– 6,410
f74	G. bulloides	6,680 ± 80	7,507	7,390– 7,623
f76.7	G. bulloides	7,100 ± 75	7,854	7,707– 8,001
f82	G. bulloides	8,010 ± 80	8,788	8,558– 9,017
‡90	N. pachyderma (s.)	8,810 ± 140	9,749	9,494– 10,004
‡95	N. pachyderma (s.)	10,290 ± 130	11,906	11,329– 12,483
‡103	N. pachyderma (s.)	10,880 ± 130	12,797	12,521– 13,072
‡108	N. pachyderma (s.)	11,040 ± 130	12,960	12,684– 13,235
		VM 23-81		
‡157	N. pachyderma (s.)	10,880 ± 90	12,651	12,597– 12,704
‡160	N. pachyderma (s.)	10,980 ± 150	12,900	12,586– 13,213
‡172	N. pachyderma (s.)	11,580 ± 110	13,531	13,240– 13,822
f205	N. pachyderma (s.)	12,180 ± 100	14,243	13,882– 14,603
f210	N. pachyderma (s.)	13,440 ± 120	16,068	15,689– 16,447
f219	N. pachyderma (s.)	13,630 ± 100	16,325	16,009– 16,641
f221	N. pachyderma (s.)	14,150 ± 110	16,965	16,658– 17,272
f223	N. pachyderma (s.)	14,330 ± 100	17,176	16,894– 17,457
f227	N. pachyderma (s.)	14,770 ± 110	17,669	17,378– 17,960
f229	N. pachyderma (s.)	15,040 ± 110	17,744	17,674– 17,814
f251	N. pachyderma (s.)	15,650 ± 150	18,552	18,222– 18,881
f263	N. pachyderma (s.)	17,280 ± 130	20,511	20,003– 21,018
f274	N. pachyderma (s.)	17,875 ± 150	21,305	20,774– 21,835
¡287.5	N. pachyderma (s.)	18,269 ± 163	21,820	21,299– 22,340
¡H2-328		20,780 –	24,000	–
¡H3-381		26,270 –	30,000	–
		VM 19-30		
f100	Mixed planktonic	12,430 ± 90	14,571	14,201– 14,940
f150	Mixed planktonic	17,500 ± 130	20,808	20,307– 21,308
f198.5	Mixed planktonic	22,730 ± 190	26,562	–
f249.5	N. pachyderma (d.)	28,090 ± 250	30,249	–
f297.4	Mixed planktonic	31,310 ± 390	34,847	–
f350	Mixed planktonic	37,640 ± 610	42,845	–
		GGC36		
f1	G. bulloides	2,540 ± 70	2,087	2,220– 1,954
¡15	G. bulloides	5,422 ± 75	5,850	5,997– 5,703
f30	G. inflata	7,720 ± 100	8,070	8,218– 7,922
¡50	G. bulloides	8,146 ± 90	8,526	8,725– 8,327
f60	G. inflata	8,720 ± 100	9,268	9,453– 9,082
¡87.5	G. inflata	9,052 ± 88	9,662	9,871– 9,452
f90	G. inflata	8,580 ± 95	9,166	9,387– 8,945
f120	G. inflata	9,165 ± 85	9,725	9,924– 9,525
¡140	G. inflata	10,555 ± 138	11,665	12,264– 11,065
f150	G. inflata	10,750 ± 95	12,110	12,526– 11,694

*Corrected by −400 years for the age of the surface ocean reservoir, except for VM 19-30 for which the correction is −500 years.   †Radiocarbon ages to 18,800 years calibrated according to (46). f from ETH, Zurich; ≠ from (61); ‡ from (62); ¡, from Arizona; !, interpolated radiocarbon ages from (7) and calibrated according to (24). Of the three Barbados corals dated beyond 18,000 radiocarbon years, one has a radiocarbon age of 26,160 ± 520 years and a U-Th age of 30,230 ± 160 years (46), consistent with the estimate of H3 calibrated age from ice core–marine correlation (48). This age is also supported by ^{14}C–U–Th datings of aragonite and organic material from Lake Lisan (49).

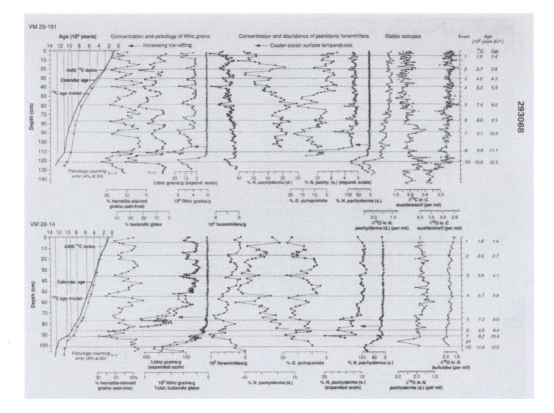

Fig. 2 Summary of proxy measurements in cores VM 29-191 and VM 28-14. Dashed lines are approximate peaks of the Holocene events, taking into account all of the records. In the lithic and *N. pachyderma* (s.) records, the Holocene events become evident only after expanding the horizontal scales. The record from VM 28-14 lacks the oldest event, most likely because of anomalously low sedimentation rates. Hematite-stained grain percentages are always given on an ash-free basis and are thus independent of variations in the material coming solely from Iceland. Petrologic counting error is from [8]. Radiocarbon age- and calendar age-depth models are based on data in Table 1 and were produced by assuming linear sedimentation rates between dated intervals. After ~8200 years, $\delta^{13}C$ values in *C. wuellerstorfi* fall close to the modern value of ~1 per mil at the nearby GEOSECS site 23 [45].

Hence, contrary to the conventional view, the North Atlantic's Holocene climate must have undergone a series of abrupt reorganizations, each with sufficient impact to force concurrent increases in debris-bearing drift ice at sites more than 1000 km apart and overlain today by warm, largely ice-free surface waters of the North Atlantic and Irminger currents. The ice-rafted debris (IRD) events exhibit a distinct pacing on millennial scales, with peaks at about 1400, 2800, 4200, 5900, 8100, 9400, 10,300 and 11,100 years ago.

Ocean forcing of the Holocene signal. We

argue that the immediate cause of the Holocene ice-rafting events was a series of ocean surface coolings, each of which appears to have been brought about by a rather substantial change in the North Atlantic's surface circulation. The most consistent evidence of ocean surface coolings is the succession of prominent increases in *Globigerina quinqueloba* (Figure 2), a species that today dominates the planktonic foraminiferal populations in cool Arctic waters north of Iceland [10]. Corroborating that evidence are increases in abundances of *Neogloboquadrina pachyderma* (s.), the polar planktonic

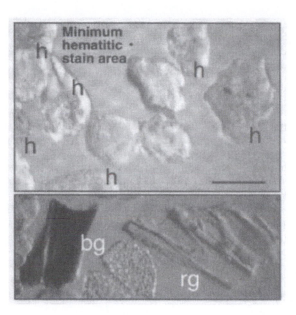

Fig. 3 Hematite-stained grains (h); lower panel, Icelandic glass [clear rhyolitic (rg) and dark basaltic (bg) varieties]. Hematite-stained grains are defined as grains with at least one stained area the size of the small black dot, the minimum size at which hematite can be identified (the dot is set in the ocular of the microscope). Although the definition may seem at first to overdefine the grain type, 60 to 80% of non-Icelandic grains have no visible hematite staining. By counting all grains with identifiable hematite stain, therefore, we avoid rectifying the signal. Scale bar, 120 μm.

foraminifera, during the first four events (events 5 to 8), and marked decreases in abundances of the warm-water planktonic species *N. pachyderma* (d.) during middle and late Holocene events (Figure 2). Although some of the faunal shifts are not large, all are defined by more than one species, and they are correlative at two widely separated sites. Moreover, because the foraminiferal concentrations increased markedly during most events (Figure 2), it is unlikely that the assemblages were modified by carbonate dissolution.

Analysis of stable isotopes in *G. bulloides* and *N. pachyderma* (d.) in the two cores produced isotopic evidence of cooling at the level of the Younger Dryas event—as much as 5°C if the isotopic enrichment was entirely the result of a temperature change [11]—but revealed little variability during any of the Holocene events at either coring site (Figure 2). The cooler surface waters may have been fresher, thereby offsetting temperature-driven δ^{18}O enrichment, or cooler surface waters may have forced planktonic species to

shift growth seasons or depth habitats, as recently documented for *N. pachyderma* (s.) in the North Atlantic [12]. Whatever the reason, at least in the subpolar North Atlantic, planktonic isotopic compositions apparently are not robust indicators of variability during present-day interglacial climates.

The magnitudes of the ocean surface coolings probably did not exceed 2°C, at least in the eastern North Atlantic; this conclusion is based on a comparison of planktonic abundances in VM 29-191 and recent measurements of modern (core top) assemblages in the Nordic Seas (Figure 4A). We chose not to further quantify ocean surface temperatures with transfer function methods because their errors are relatively large [13], and in an earlier effort at VM 28-14, results were uncertain because of no-analog problems [14]. If this upper limit is correct, the Holocene coolings did not exceed 15 to 20% of the full Holocene-to-glacial temperature change [15].

The strongest evidence for changes in ocean surface circulation during the ice-rafting events comes

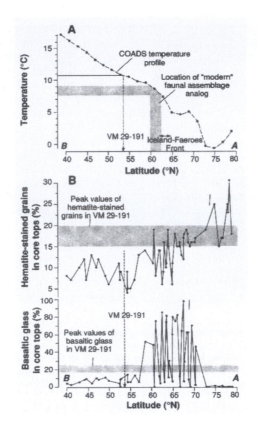

Fig. 4 (A) COADS temperature profile [63] as located in Figure 1 and the latitudinal position of VM 29-191. This diagram provides an estimate of how much cooling might have occurred during each of the Holocene events at VM 29-191. The rationale is as follows: Within the Nordic Seas, there is only one place where the core top or "modern" abundances of *N. pachyderma* (s.), *G. quinqueloba*, and *N. pachyderma* (d.) are within the ranges of their values at the IRD peaks in VM 29-191. That is a rather narrow area just south of the Iceland-Faeroes Front, between ~60° and 62°N [10]. On the basis of COADS temperature data averaged over the last 150 years, comparable to the duration of a single point in our records, the mean summer and winter temperature difference between the "modern analog" location and the location of VM 29-191 is between 1.5° and 2°C. If we assume, as the simplest alternative, that the temperature-faunal relation within the analog location can be transferred directly to the coring site, and that ambient temperatures at VM 29-191 were comparable to those of today, then probably 2°C of cooling at most occurred there during each Holocene event. (B) Cross section of the two petrologic tracers from section A-B (Figure 1). All core top data from within the area of A-B were projected to the center along the green line. The cross sections demonstrate the difference in percentages of the two tracers north and south of the Iceland-Faeroes frontal system and the relation of those changes to the peak values of the tracers in our record from VM 29-191, as in Figure 2.

from VM 29-191. Today, most of the ice entering the subpolar North Atlantic is calved in southern and western Greenland, circulates through the southwestern part of the subpolar gyre in the Labrador Sea, and enters the North Atlantic near Newfoundland (Figure 1) [16]. If present-day circulation remained unchanged during the Holocene, IRD at VM 29-191 and in Labrador Sea ice should have similar compositions. However, in the broad area south of Iceland where Labrador Sea ice is dispersed [16], percentages of the two tracers in modern (core top) IRD are consistently lower than even their mean values in VM 29-191 (Figure 1; see also [17]). In core GGC36, which contains debris from icebergs exiting the Labrador Sea, percentages of the two tracers exhibit similar low values throughout almost the entire Holocene (Figure 1 and Figure 5).

We find instead that the sources of IRD at VM 29-191 must have been much farther to the northeast in the Greenland-Iceland seas. There, percentages of the two tracers in core top sediments consistently reach or even exceed their peak values in VM 29-191 (Figure 1 and Figure 4B). The Icelandic material probably was erupted onto drifting ice (see also [18]), and hematite-stained grains may have come from along coastal East Greenland and Svalbard (Figure 1) and perhaps from farther north around the Arctic Ocean, where red beds also are present. The abrupt eastward and southward drops in tracer percentages almost certainly result from melting of ice along frontal systems where warm and cool surface waters mix (Figure 1 and Figure 4B). The low tracer percentages in Labrador Sea ice, therefore, likely reflect melting of tracer-rich ice in the Denmark Strait frontal system and the dearth of hematite-bearing sources of IRD around the Labrador Sea [7].

Hence, the North Atlantic's surface circulation must have alternated between two modes. At times of minimum concentrations of IRD and warmer sea surface temperatures at the coring sites, circulation probably was similar to today's, such that small amounts of tracer-poor ice from the Labrador Sea reached both coring sites. During the ice-rafting events, export of ice from the Labrador Sea may have increased, but surface waters of the Greenland-Iceland seas must have been advected much farther southward or southeastward than they are today, carrying tracer-rich ice to both coring sites (Figure 1).

The changes in faunal assemblages during each ice-rafting event at VM 29-191 further support our circulation reconstruction and appear to restrict the IRD sources to surface waters in the southerly parts of the Greenland-Iceland seas near or within the frontal systems (Figure 4A). Surface waters in the frontal systems tend to be quite productive [19], and their southward shifts, therefore, are consistent with prominent increases in foraminiferal concentrations associated with most of the IRD events at both sites (Figure 2). The large magnitudes of the circulation changes are especially evident in the

eastern North Atlantic where cool, ice-bearing surface waters shifted across more than 5° of latitude, each time penetrating well into what is at present the core of the warm North Atlantic Current.

Thus, with a single mechanism—an oscillating ocean surface circulation—we can explain at once the synchronous ocean surface coolings, changes in IRD and foraminiferal concentrations, and changes in petrologic tracers at both sites. Although we cannot rule out an increase in iceberg discharge from tidewater glaciers during the ice-rafting events, it is not required by the data in hand.

The deep southward advections of cooler and fresher surface waters into the core of the North Atlantic Current imply links with the North Atlantic's thermohaline circulation. For example, at VM 29-191, $\delta^{13}C$ values in the benthic foraminifera *Cibicidoides wuellerstorfi* decrease by ~0.5 to 0.6 per mil (Figure 2), corroborating previous evidence that North Atlantic deep water (NADW) production was reduced during that climatic event [20, 21].

In the subsequent Holocene record, $\delta^{13}C$ values decrease by 0.4 to 0.5 per mil during event 7 at ~10,300 years (Figure 2), suggesting reduced NADW production at that time as well. The absence of depletions in $\delta^{13}C$ values exceeding background noise (that is, depletions greater than ~0.3 per mil) during the other events does not necessarily rule out corresponding changes in NADW production; shifts in thermohaline circulation at those times may be recorded only in deeper, more sensitive sites near the present boundary between NADW and Southern Ocean water [22].

Correlations with the Greenland ice core. We also found evidence of a close link between the shifts in Holocene climate and atmospheric circulation above Greenland. In particular, the marine records exhibit a good correlation with the flux of non-sea salt K, one of the several constituents of atmospheric chemistry dissolved in Greenland ice and regarded as evidence of changes in loadings of terrigenous dust [5]. With one exception, at 1400 years, the North Atlantic's Holocene events fall either on or close to peaks in that record (Figure 6). The 1400-year event correlates with a prominent increase in the flux of another member of the geochemical series, sea salt Na, an indicator of storminess and entrainment of sea spray into the atmosphere [5]. The Little Ice Age episode, which is clearly reflected in the geochemical series, is absent from our marine records because the core tops

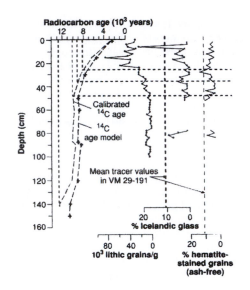

Fig. 5 Radiocarbon age- and calendar age-depth models for EW9303-GGC36 and our measurements of lithic grain concentrations and the two petrologic tracers (water depth, 3980 m). Mean values for the two tracers in VM 29-191 are from data in Figure 2.

slightly predate that event.

O'Brien *et al.* [5] argued that the increases in soluble impurities in Holocene ice from Greenland probably occurred at times of lowered atmospheric temperatures. Their conclusion was based on evidence that certain of those shifts were associated with well-documented climatic deteriorations, such as the Little Ice Age, the 8200-year cooling event, and the Younger Dryas [2, 23]. Our correlations with the ice core records, therefore, imply that during the Holocene, just as during the last glaciation, the North Atlantic's ocean surface and the atmosphere above Greenland were a coupled system undergoing recurring shifts on short, millennial scales.

A millennial-scale climate cycle. Despite a profound reorganization of climate at the onset of the Holocene, there is no statistical difference between pacings of the Holocene climate events and the rapid climate shifts that dominated the last glaciation. The evidence comes from a composite climate record we constructed by extending our measurements of IRD concentrations and petrologies back to ~32,000 years, the present limit of radiocarbon age calibrations. The glacial portion of the composite record is from VM 23-81 (Figure 1) in which

earlier work documented a series of fast-paced ice-rafting events [7]. We increased the resolution of those events with additional measurements of lithic grain concentrations and their petrologies. The glacial records were placed on a calendar time scale, as described in Table 1 and [24], and joined to the Holocene record of VM 29-191, which is only a few kilometers away (Figure 1) at the midpoint of the Younger Dryas detrital carbonate-lithic peak, dated at 10,880 radiocarbon years in both records (Figure 6).

Thus, the composite record is based on the same proxies throughout. Moreover, all of the events reflect the same sense of climate change-that is, coolings [25]. The three events between interstadials 1 and 2, equivalents of which have recently been identified in the Labrador Sea [25], demonstrate that the climate signal persists through the last glacial maximum where ice core $\delta^{18}O$, dissolved Ca (Figure 6), and methane records [26] reveal no evidence of comparable climate variability. Why this is the case is unclear, but perhaps air masses carrying the signals were blocked from the interior of the ice cap by expansion of polar atmospheric circulation [23].

Using the procedure described in [27], we measured the pacing of the Holocene-glacial climate

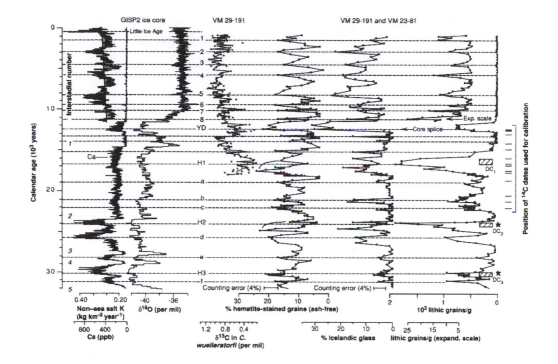

Fig. 6 Glacial-Holocene record of IRD, petrology, and $\delta^{13}C$ values in *C. wuellerstorfi*, placed on a calendar time scale and compared with GISP2 Ca, $\delta^{18}O$, and flux of non-sea salt K [5, 47]. $\delta^{13}C$ values in *C. wuellerstorfi* are from VM 29-191; IRD and petrology are from the composite VM 29-191 and VM 23-81 record, spliced as indicated. The calendar time scale for the composite marine record was constructed as in [24]; stars (at right) are ages for H2 and H3 used to extend the age model beyond the radiocarbon calibrations (see also Table 1). Calibration of the age of H3 is based on multiple criteria [24] and is ~4000 years younger than estimated by McIntyre and Molfino [64]. Hatched boxes give positions of layers rich in detrital carbonate (DC). Interstadial numbers in ice core records are from [65]. Holocene data are from VM 29-191; glacial data are from VM 23-81; the cores are joined at the core splice at the midpoint of the Younger Dryas event. Depletions in benthic $\delta^{13}C$ values associated with IRD events increase into the last glaciation, suggesting that the glacial amplification of the millennial-scale oscillator was linked to increasingly stronger decreases in NADW production.

shifts in the most consistent and robust component of the composite record, the series of peaks in hematite-stained grains (Figure 7A). For the glacial interval, the mean pacing is 1536 ± 563 years, essentially the same as a 1450-year cycle identified in the glacial portion of the Greenland Ice Sheet Project 2 (GISP2) geochemical series [28]. For the Holocene interval, the mean pacing is 1374 ± 502 years. The standard deviations are from the means of the event pacings. Calendar age errors have been estimated back to 22,000 years, and for that interval, which makes up two-thirds of our total record, the errors are smaller than the standard deviations

(compare Table 1 and Figure 7A).

Thus, the pacings of the Holocene and glacial events are the same statistically, and together the two series constitute a cyclic signal centered on ~1470 ± 532 years. The signal persists across at least three major climate transitions: the Younger Dryas-Holocene transition, the deglaciation, and the boundary between marine isotope stages 2 and 3, which we have dated at ~30,000 years (Figure 7A and Table 1). The implication of this finding is clear: The millennial-scale variability in our records reflects the presence of a pervasive, at least quasiperiodic, climate cycle occurring inde-

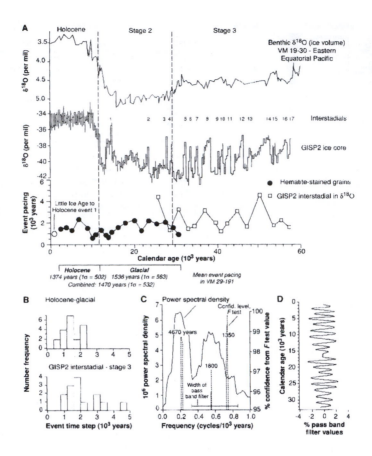

Fig. 7 (A) Time series of pacings of Holocene-glacial events from VM 29-191 and VM 23-81 (•), compared with pacings of numbered interstadials in GISP2 $\delta^{18}O$ (□) and benthic $\delta^{18}O$ in VM 19-30 [66]. Pacings were calculated as in [27]. The pacings of the Dansgaard/Oeschger cycles were obtained by measuring the time steps between numbered interstadials and placing the value at the midpoints between them. The time step to the Little Ice Age (○) is also shown for comparison with older event pacings. Mean values of event pacings are from the composite record of hematite-stained grains in Figure 6. The calibrated time scale for VM 19-30 is from [25]. On the basis of radiocarbon ages and their calibrations, the stage 2-stage 3 boundary is almost 6000 years older than given by SPECMAP tuning [67]. (B) Histogram of Holocene-glacial IRD events and GISP2 interstadial events. (C) Multi-taper spectral analysis (7 tapers) of time series of hematite-stained grains. (D) Filtered record of time series of hematite-stained grains.

pendently of the glacial-interglacial climate state.

Our composite record further suggests that the Holocene and glacial event pacings are nearly the same as those of the prominent Dansgaard/Oeschger $\delta^{18}O$ shifts of the last glaciation, especially in marine isotope stage 3 where they are best developed (Figure 7, A and B). That finding implies that Dansgaard/Oeschger cycles were not

forced by internal instabilities of large ice sheets during glaciations, as some have suggested [3], but instead originated through processes linked to the climate cycle. Similarly, the consistent association of Heinrich events with cold phases of the cycle (Figure 6) supports arguments by Bond and Lotti [7] that climate plays a fundamental role in their origin.

Finally, spectral analysis of the time series of hematite-stained grains by the multitaper method of Thompson [29] reveals that power is concentrated in two broad bands. One is centered at ~1800 years, near the mean of Holocene-glacial events, and the other is centered at ~4700 years (Figure 7C). Cycles close to both have been noted previously in spectra from other paleoclimate records from the last glaciation [30]. In addition, F variance ratio tests reveal lines with >95% probability at 4670, 1800, and 1350 years (Figure 7C). Further corroboration of cyclicity close to the mean of the IRD events is given by applying a broad Gaussian bandpass filter to the record of hematite-stained grains centered on 1800 years (Figure 7D).

Discussion. Our conclusion that millennial-scale shifts of Holocene and glacial climates originated fundamentally from the same forcing mechanism, a quasiperiodic climate cycle, has implications for the debate over the origin of abrupt climate changes and how those changes may influence future climates. The stability of the present climate in the North Atlantic must be considered in the context of millennial-scale variability in addition to interannual and interdecadal fluctuations, such as the North Atlantic Oscillation [31]. For example, as originally suggested by Denton and Karlén [1], the Little Ice Age, a widespread climatic deterioration culminating about 300 years ago [32], was not an isolated event. Our records demonstrate that the time step separating the main phase of the Little Ice Age from the previous Holocene event (our event 1) is ~1100 years, falling well within the pacings of events in the composite record (Figure 7A). The Little Ice Age, therefore, appears to have been the most recent cold phase in the series of millennial-scale cycles.

Whether the climatic amelioration since the Little Ice Age marks the onset of a warm phase of the cycle (for example, [33]) is, however, unclear. The series of climate events we have identified were not strictly periodic, and brief warmings often punctuated the cold phases of the millennial-scale fluctuations.

The 8200-year cooling, which is well documented by the only distinct Holocene $\delta^{18}O$ shift in Greenland ice, has received much attention because of its prominence in a growing number of other records [2]. In our records, what is almost certainly the correlative of that cooling, event 5 (Figure 6), is clearly part of the recurring series of climate shifts. We argue, therefore, that the origin of the 8200-year cooling is linked to the climate cycle and that its large amplitude in climate records reflects a mechanism that in some way amplified the climate signal at that time.

Finally, we are able to directly compare in the North Atlantic two prominent series of abrupt climate shifts from completely different climate states: the Holocene climate episodes and the marine imprint of the Dansgaard/Oeschger cycles. In addition to their similar pacings, events in both series document southward shifts of cooler, ice-bearing surface waters deep into the subpolar North Atlantic and a coupling of ocean surface circulation with atmospheric circulation above Greenland. As was the case for the glacial events, the Holocene shifts were abrupt, switching on and off within a century or two at least and probably faster, given the likely blurring of event boundaries by bioturbation (Figure 2). Sudden, recurring reductions in NADW production accompanied the glacial cycles [34], and the same appears to have occurred during at least one of the Holocene events.

Hence, at least with respect to the surface circulation in the subpolar North Atlantic, it seems entirely consistent with the evidence to view the Holocene events as mini-Dansgaard/Oeschger cycles, and to regard the surface North Atlantic as a hydrographic system that shifts persistently in a Dansgaard/Oeschger-like mode, even when ice volumes are small. If that is correct, then the much larger amplitudes of the Dansgaard/Oeschger cycles relative to those of the Holocene reflect an amplification of the cyclic signal by a mechanism unique to the glaciation.

The amplifying mechanism is uncertain, but it may have been linked to thermohaline circulation in the North Atlantic. At VM 29-191, for example, large depletions of $\delta^{13}C$ values in *C. wuellerstorfi* during glaciation coincide with relatively large increases in IRD (Figure 6). In cores from south of Iceland, similar shifts in $\delta^{13}C$ values in *C. wuellerstorfi* coincide with cold phases of Dansgaard/Oeschger cycles [34] and, by correlation, coincide with increases in IRD in VM 23-81 (Figure 6) [7]. During glaciation, the 1470-year cycle may have regulated iceberg discharges into the North Atlantic, thereby amplifying the signal through the impact of recurring increases in fresh water (icebergs) on rate of NADW production [35]. An alternative mechanism is suggested by recent models demonstrating that a large-amplitude oscillatory mode can be induced in thermohaline circulation simply by increasing fresh water fluxes to the ocean, as would occur with growth of Northern Hemisphere ice sheets [36]. The frequency of the oscillator could perhaps lock onto the frequency of a weak but persistent external forcing, such as the 1470-year cycle, in effect an amplification through entrainment or frequency locking. Regardless of the exact amplifying mechanism, the results of our study implicate ocean circulation as a major factor in forcing the climate signal and in amplifying it during the last glaciation.

We know too little thus far to identify the origin of the 1470-year cycle. Its constant pacing across major stage boundaries, especially the last glacial termination, almost certainly rules out any origin linked to ice sheet oscillations. Rather, the close correlation of shifts in ocean surface circulation with changes in atmospheric circulation above Greenland is consistent with a coupled ocean-atmosphere process. Coupled ocean-atmosphere modes of variability on decadal scales have been inferred from observational records in the North Atlantic [37], but those records are too short to assess longer, millennial-scale phenomena. Millennial-scale climate cycles may arise from harmonics and combination tones of the orbital periodicities, but cycles currently thought to fall within those bands are longer than the cycle we have identified [38]. Forcing of millennial-scale climate variability by changes in solar output has also been suggested, but that mechanism is highly controversial, and no evidence has been found of a solar cycle in the range of 1400 to 1500 years [39].

In any case, if we are correct that the 1470-year climate cycle is a pervasive component of Earth's climate system, it must be present in previous glacial-interglacial intervals. If that turns out to be true, the cycle may well be the pacemaker of rapid climate change.

REFERENCES

1. G. H. Denton and W. Karlén, *Quat. Res.* **3**, 155 (1973).
2. R. B. Alley *et al.*, *Geology* **25**, 483 (1997).
3. W. S. Broecker, *Nature* **372**, 421 (1994).
4. T. Fronval and E. Jansen, *Paleoceanography* **12**, 443 (1997).
5. S. R. O'Brien *et al.*, *Science* **270**, 1962 (1995).
6. Core samples were collected with precision cutting tools to remove material in narrow

slots at centimeter intervals. We avoided sampling the core's outer rind and obvious disturbances such as burrowing. All samples (dry) were weighed, then washed to separate the fine (<62 μm) from the coarse sediment fraction. All petrologic analyses were done in the 63- to 150-μm size range, which consistently contains at least a few hundred lithic grains. A portion of that grain fraction was placed on a glass slide (with glycerin as a mounting medium) and 300 to 500 grains were counted per sample, using a line rather than point counting method. A key factor in our ability to measure certain tracers accurately is a specially prepared microscope. We placed a white reflector on the condenser lens and illuminated the sample slide from above with a halogen light source. By moving the substage reflector up and down, a position can be found that creates a strong impression of relief and brings into striking view details of surface textures and coatings on grains. The technique is especially useful for identifying hematite-stained grains, even when the stain is quite small (Figure 3). Counting of lithic grain concentrations was done in the >150 μm fraction. To avoid errors introduced by splitting small grain populations, we did not split samples for this procedure. Counting of planktonic foraminiferal concentrations and abundances was done on splits of the >150 μm fraction. For the abundance measurements, 300 to 500 individuals were counted. Species in the >150 μm fraction were selected for planktonic and benthic isotopic measurements. All isotopic measurements were made at North Carolina State University by W.S. The samples were crushed and washed with reversed-osmosis water and airdried at 65°C, and then 30- to 100-μg sample aliquots were isotopically analyzed in a Kiel Autocarbonate device attached to a FMAT 251 RMS. NBS-19, NBS-18, and the NCSU CM-1 marble standard were analyzed with every sample run for standard calibration and sample correction. During the period these samples were analyzed, for samples in the 10- to 100-μg size range the standard reproducibility was 0.07 per mil for ^{13}C and 0.08 per mil for ^{18}O. Planktonic isotopic analyses for VM 28-14 were made on 18 to 20 individuals per sample and were replicated at a number of depths. For VM 29-191, 10 replicate analyses were done on two or three individuals from each centimeter depth with the objective of investigating small-sample variability over time.   Because those results showed no coherent relation to our other proxies, we combined all measurements from each centimeter with a mass balance Equation using the sample size (micromolar) and the sample isotopic composition per mil, the results of which are shown in Figure 2. The final result is equivalent to isotopically analyzing 20 planktonic individuals at once.

7. G. C. Bond and R. Lotti, *Science* **267**, 1005 (1995).

8. The counting error is the 2σ standard deviation for the mean of 20 replicate counts in each of two representative samples.

9. That we find any measurable record of drift ice as far south as 54°N is not as extraordinary as it might seem. Between about 1880 and the mid-1960s there were more than 20 iceberg sightings in the eastern North Atlantic between 45°and 60°N, and icebergs have been reported from as far south as Bermuda and the Azores [40]. Even so, keeping in mind that the lithic grain concentrations are very low and that each sample integrates 50 to 100 years of time, the amounts of drifting ice reaching either site, even at peak concentrations, must have been small, and certainly much less than during the glaciation.

10. T. Johannessen, E. Jansen, A. Flatoy, A. C. Raveo, in *Carbon Cycling in the Glacial Ocean: Constraints on the Ocean's Role in Global Change*, R. Zahn et al., Eds. (NATO ASI Series, Vol. I 17, Springer-Verlag, Berlin, 1994), pp. 61-85.

11. S. R. Epstein, R. Buchsbaum, H. A. Lowenstam, H. C. Urey, *Geol. Soc. Am. Bull.* **64**, 1315 (1953).

12. M. Sarnthein et al., *Paleoceanography* **10**, 1063 (1995); G. Wu and C. Hillaire-Marcel, *Geochim. Cosmochim. Acta* **58**, 1303 (1994).

13. D. W. Oppo, M. Horowitz, S. J. Lehman, *Paleoceanography* **12**, 51 (1997).

14. T. B. Kellogg, in *Climatic Changes on a Yearly to Millennial Basis*, N.-A. Morner and W. Karlén; Eds.   (Reidel, Dordrecht, Netherlands, 1984), pp. 123-133.

15. L. Labeyrie et al., *Philos. Trans. R. Soc. London* **348**, 255 (1995).

16. B. E. Viekman and K. D. Baumer, *International Ice Patrol Technical Report 95-03* (Department of Transportation, U.S. Coast Guard, International Ice Patrol, 1995).

17. Most of the cores (indicated by solid dots in Figure 1) have Holocene sections, documented on the basis of radiocarbon dating, isotopic measurements, or the presence of ash layer 1 (~10,000 years) at depths of tens of centimeters. The existence of Holocene sections in the remainder, indicated by plus signs, is inferred from distinct color patterns and bulk carbonate measurements [41]. Core top ages likely vary from close to the present to perhaps a few thousand years; on average, then, the core top petrologic data reflect distributions of the tracers over time as well as over a large geographic area. The petrographic analyses of each core top lithic sample were done in the same way as for VM 29-191 and VM 28-14.

18. There probably were no tidewater glaciers in Iceland, at least during the middle and late parts of the Holocene, ruling out transport of debris to the sites by Icelandic icebergs; direct transport in eruption clouds is also ruled out by the large size of clasts (coarsest grains typically are 0.5 to 1 mm) and by evidence that only two of the six largest Holocene eruptions in Iceland are correlative with the petrologic events [42]. The only plausible mechanism, therefore, is eruption of Icelandic volcanic material onto nearby sea ice (and probably glacier ice as well) and subsequent transport of that ice in surface currents to the coring sites. Explosive eruptions from volcanoes such as Hekla, Katla, and Grimsvotn occur once every 20 to 30 years or so in Iceland [43], and the Icelandic lithic grains at both sites are dominantly (70 to 95%) clear rhyolitic glass shards, the characteristic product of explosive eruptions in silicic volcanoes. The importance of our interpretation of the Icelandic IRD is that its increases at VM 28-14 require increases in drift ice within reach of eruption clouds to the east. Hence, changes in surface hydrography favoring preservation and transportation of drift ice in the western North Atlantic must have taken place, at the very least, from Denmark Strait to the vicinity of Iceland or Jan Mayen.

19. S. Sathyendranath, A. Longhurst, C. M. Caverhill, T. Platt, *Deep Sea Res. Pt. 1*, **42**, 1773 (1995).

20. The carbon isotopic composition of
C. wuellerstorfi appears to record the [13]C
of total CO_2 of sea water, and therefore is
taken as a proxy of the nutrient content of
that water [44]. Today, VM 29-191 lies within
NADW, which is nutrient-depleted and hence
δ^{13}C-enriched; below the depth of the core
are modified waters of Southern Ocean origin, which are nutrient-enriched and δ^{13}C-
depleted [45]. Variations in the δ^{13}C of that
foraminifera are related to shifts in the rate
of production of NADW and hence in the rate
of convective overturning and thermohaline
circulation in the North Atlantic [34, 44].

21. E. Boyle, *Philos. Trans. R. Soc. London* **348**, 243
(1995).

22. M. S. McCartney, *Prog. Oceanogr.* **29**, 283
(1992).

23. P. A. Mayewski *et al.*, *Science* **263**, 1747
(1994).

24. The calendar age model in the composite
record for the interval between the Younger
Dryas and 18,000 radiocarbon years was constructed by linear interpolation between radiocarbon dates from VM 23-81 calibrated
following Bard *et al.* [46] (Table 1). For the
remainder of the record, we transferred the
GISP2 ages of Heinrich events 2 and 3 (H2
and H3) [47] into their equivalent levels in VM
23-81 (Table 1) [48] and interpolated linearly
between the two ages. Our estimate of the
age of H3 is corroborated by calibrated radiocarbon measurements in Barbados corals
(Table 1) and by paired U-Th-^{14}C measurements in aragonite from Lake Lisan in the
Dead Sea Rift [49]. Detrital carbonate at the
Younger Dryas level was deposited rapidly
and is widespread, extending from the eastern North Atlantic to sources in Hudson Strait
and Cumberland Strait [50]; it serves as an
excellent chronostratigraphic marker. Calibrations of radiocarbon ages in VM 19-30 (Table 1) at 100 and 150 cm are from [46]. Remaining calibrations are from the age model
for the Holocene-glacial composite record as
described above.

25. The Heinrich events are tied to well-
documented lowerings of atmospheric and
ocean surface temperatures and to large
reductions in the rate of NADW formation
(Figure 6) [34]. The two distinct ice-rafting
peaks between interstadial 1 and the Younger
Dryas, dated at ~13,200 and ~14,000 calen-

dar years (~11,300 and 12,000[14]C years), are
accompanied by increases in *N. pachyderma*
(s.) measured in the same core [51], and they
are coincident with prominent decreases in
δ^{18}O in Greenland ice (Figure 6). Evidence
of coolings at about the same times have
been found in other deep sea cores from the
North Atlantic [52] and the Nordic Seas [53]
and in pollen and glacial records in Europe
and North America [53]. From 23,000 calendar years to the end of the record, the IRD-
petrologic episodes between Heinrich events
also occur at times of maxima in abundances
of *N. pachyderma* (s.) [7] and at times of
prominent cold phases in the ice core δ^{18}O
records (Figure 6). We also find distinct IRD
events punctuating prolonged stadials, such
as between interstadials 2 and 3 and interstadials 4 and 5. There, the IRD-petrologic
peaks correspond to prominent increases
in dissolved Ca (Figure 6) and in the polar
circulation index [23], suggesting a close association of the marine events with expansion of polar circulation above the ice cap.
Within the long interval between interstadials 1 and 2, we have identified four other
IRD-petrologic events in addition to H1. The
N. pachyderma (s.) abundances are saturated
through the entire interval, preventing reliable estimates of changes in sea surface temperatures. The youngest of the four events,
dated at ~15,200 calendar years (~12,900[14]C
years), however, is correlative with glacial advances in the U.S. midwest [for example, Port
Huron Stade [54]], with a glacial readvance on
the Scotian shelf [55], and with a Heinrich-like
cooling and ice-rafting event identified on the
Scotian Slope [56]. The other three events, a,
b, and c, previously identified by Bond and
Lotti [7] and dated at about 19,000, 21,000,
and 22,000 calendar years (~16,500, 17,500,
and ~18,600[14]C years), are present in IRD
records from a number of other sites in the
subpolar North Atlantic and GIN Seas [57],
and they also appear in magnetic susceptibil-
ity records from the Labrador Sea [58]. Two
of these widespread events, a and c, are correlative with prominent coolings identified
in pollen records from a long lacustrine sequence in northern Norway [59]. There is evidence that the two older events, b and c, may
even be correlative with glacial advances in
the Southern Hemisphere, implying a bihemi-

spheric symmetry of the abrupt changes [60].

26. E. J. Brook, T. Sowers, J. Orchardo, *Science* **273**, 1087 (1996).

27. We define event pacing as the time separating adjacent peaks placed at the midpoints between the peaks (Figure 6). We define peaks as maxima with at least two points and amplitudes equal to or exceeding the 2σ counting error (4% [8]); the measure of separation is made between maxima in peak percentage. At the levels of H1 and H2, the sharp decreases in percentages of the two petrologic tracers reflect dilution by massive amounts of IRD discharged from the Labrador Sea [7], and we place the peaks for those two events at maxima in the lithic grain concentrations (Figure 6).

28. P. A. Mayewski *et al.*, *J. Geophys. Res.*, in press.

29. D. J. Thompson, *Proc. IEEE* **70**, 1055 (1982).

30. P. Yiou and J. Jouzel, *Geophys. Res. Lett.* **22**, 2179 (1995); L. D. Keigwin and G. A. Jones, *J. Geophys. Res.* **99**, 12397 (1994); E. Cortijo, P. Yiou, L. Labeyrie, M. Cremer, *Paleoceanography* **10**, 911 (1995).

31. J. C. Rogers, *J. Clim.* **3**, 1364 (1990).

32. J. M. Grove, *The Little Ice Age* (Methuen, London, 1988).

33. L. D. Keigwin, *Science* **274**, 1504 (1996).

34. D. W. Oppo and S. J. Lehman, *Paleoceanography* **10**, 901 (1995).

35. S. Rahmstorf, *Nature* **378**, 145 (1995); S. Manabe and R. J. Stouffer, *Paleoceanography* **12**, 321 (1997); T. F. Stocker, D. G. Wright, W. S. Broecker, *ibid.* **7**, 529 (1992).

36. K. Sakai and W. R. Peltier, *J. Clim.* **10**, 949 (1997); E. Tziperman, *Nature* **386**, 592 (1997); A. J. Weaver and T. M. C. Hughes, *ibid.* **367**, 447 (1994).

37. R. T. Sutton and M. R. Allen, *Nature* **388**, 563 (1997); Y. Kushnir, *J. Clim.* **7**, 141 (1994).

38. P. Pestiaux, I. Van der Mersch, A. Berger, J. C. Duplessy, *Clim. Change* **12**, 9 (1988); P. Yiou *et al.*, *J. Geophys. Res.* **96**, 20365 (1991).

39. M. Stuiver and T. F. Braziunas, *The Holocene* **3**, 289 (1993); E. A. Frieman, *Solar Influences on Global Change* (National Academy Press, Washington, DC, 1994).

40. *Oceanographic Atlas of the North Atlantic* (U.S. Naval Oceanographic Office, Washington, DC, 1968); J. E. Murry, in *Ice Seminar*, V. E. Bohme, Ed. (Canadian Institute of Mining and Metallurgy, Washington, DC, 1969), pp. 3–18.

41. T. B. Kellogg, thesis, Columbia University (1973).

42. A. J. Dugmore, G. Larsen, A. J. Newton, *The Holocene* **5**, 257 (1995); S. Thorarinsson, in *Tephra Studies*, S. Self and R. S. J. Sparks, Eds. (Reidel, Dordrecht, Netherlands, 1981), pp. 109–134; R. A. J. Cas and J. V. Wright, *Volcanic Successions* (Allen and Unwin, Winchester, MA, 1987); H. J. Gudmundsson, *Quat. Sci. Rev.* **16**, 81 (1997).

43. T. Simkin *et al.*, *Smithsonian Institution Volcanoes of the World* (Hutchinson Ross, Stroudsburg, PA, 1981).

44. E. A. Boyle and L. D. Keigwin, *Nature* **330**, 35 (1987).

45. P. Kroopnick, *Earth Planet. Sci. Lett.* **49**, 469 (1980).

46. E. Bard, M. Arnold, R. G. Fairbanks, B. Hamelin, *Radiocarbon* **35**, 191 (1993); M. Stuiver and J. Reimer, *ibid.*, p. 215.

47. M. J. Bender *et al.*, *Nature* **372**, 663 (1994).

48. G. C. Bond *et al.*, *ibid.* **365**, 143 (1993).

49. S. Goldstein, personal communication.

50. J. T. Andrews *et al.*, *Paleoceanography* **10**, 943 (1995).

51. S. J. Lehman and L. D. Keigwin, *Nature* **356**, 757 (1992).

52. H. P. Sejrup, H. Haflidason, D. K. Kristensen, *J. Quat. Sci.* **10**, 385 (1995).

53. N. Koc, E. Jansen, M. Hald, L. Labeyrie, in *Late Quaternary Palaeoceanography of the North Atlantic Margins*, J. T. Andrews, W. E. N. Austin, H. Bergsten, A. E. Jennings, Eds. (Geological Society Special Publication 111, 1996), pp. 177–185.

54. G. H. Denton and T. J. Hughes, in *The Last Great Ice Sheets* (Wiley, New York, 1981).

55. R. R. Stea, R. Boyd, O. Costello, G. B. J. Fader, D. B. Scott, in [53], pp. 77–101.

56. L. D. Keigwin and G. A. Jones, *Paleoceanography* **10**, 973 (1995).

57. R. Stein, S.-I. Nam, H. Grobe, H. Hubberten, in [53], pp. 135–151; T. Fronval and E. Jansen, *Nature* **383**, 806 (1996); H. Haflidason, H. P. Sejrup, D. K. Kristensen, S. Johnsen, *Geology* **23**, 1059 (1995).

58. J. S. Stoner, J. E. T. Channell, C. Hillaire-Marcel, *Paleoceanography* **11**, 309 (1996).

59. T. Alm, *Boreas* **22**, 171 (1993).

60. T. V. Lowell *et al.*, *Science* 269, 1541 (1995).

61. I. M. Lagerklint, thesis, University of Maine (1995).

62. E. Bard *et al.*, *Earth Planet. Sci. Lett.* **126**, 275 (1994).

63. S. D. Woodruff, R. J. Slutz, R. L. Jenne, P. M. Steurer, *Bull. Am. Meteorol. Soc.* **68**, 521 (1987).

64. A. McIntyre and B. Molfino, *Science* **274**, 1867 (1996).

65. W. Dansgaard *et al.*, *Nature* **364**, 218 (1993).

66. N. J. Shackleton, J. Imbrie, M. A. Hall, *Earth Planet. Sci. Lett.* **65**, 233 (1983).

67. D. G. Martinson *et al.*, *Quat. Res.* **27**, 1 (1987).

68. We thank G. Kukla, J. Lynch-Stieglitz, and A.

Van Geen for comments on the manuscript. Supported in part by grants from NSF and the National Oceanic and Atmospheric Administration. Support for the core curation facilities of the Lamont-Doherty Earth Observatory Deep-Sea Sample Repository is provided by NSF through grant OCE94-02150 and the Office of Naval Research through grant N00014-96-I-0186. This is L-DEO contribution 5714.

Exercises for Chapter 27

In Exercises 1–5, compute the functions $a(v)$, $b(v)$ and the power spectrum $f(v)$ for the following cases of $m(t)$. Then try to figure where $f(v)$ has its maximum by sketching a graph of this function. For exercises 1–5, assume that $m(t)$ is defined for all t, so your integrals range over $-\infty < t < \infty$.

1. $m(t) = e^{-|t|}$

2. $m(t) = e^{-|t+a|}$, where a is some fixed integer

3. $m(t) = e^{-|t+1|} + e^{-|t|} + e^{-|t-1|}$

4. $m(t) = \cdots + e^{-|t-2|} + e^{-|t+1|} + e^{-|t|} + e^{-|t-1|} + e^{-|t-2|} + \cdots$

5. $m(t) = 0$ if $|t| > 1$ and $m(t) = 1$ if $|t| < 1$

   Hint: You can simplify your work considerably if you use the fact that if m is written as a sum of functions of t (as in Exercises 3 and 4), then both $a(v)$ and $b(v)$ will be the sum of the respective a's and b's.

6. Explain why the preceding hint is true.

28

Causes of Chaos

In the previous chapters, I talked about the differential equation

$$\frac{dv}{dt} = f(v) \tag{1}$$

only in the case where v is either a function of time or a two-component vector function of time. In this chapter, I discuss the case where v has three or more components, and I focus on some new and bizarre behavior that can appear. Here, I will tell you that solutions to the three-component version of (1) can behave in a very chaotic fashion. That is, they can stay in a bounded region of the three dimensional version of the phase plane—the phase space—yet wind through that region along a path that is incredibly convoluted and, for all practical purposes, completely unpredictable.

In this regard, I should remind you that even when v has three or more components, Equation (1) is predictive in the same sense as for the two- or one-component cases. That is, if you specify a starting value of v, then there is precisely one solution to (1) that, at the chosen starting time, has your chosen starting value.

Even so, I can still say that the behavior is, *for all practical purposes*, unpredictable, by which I mean the following: The trajectory taken by a solution may be so sensitive to the precise starting value that no individual solution can be used to model real data reliably. The point here is that real data always has some inherent uncertainty; so starting values taken from real data are never known precisely. Moreover, there are versions of (1) where there are three or more components for which a certain *hyper* unpredictability is present. In these versions of (1), no matter how *long* you watch a trajectory, you will not be able to predict its future behavior. That is, no amount of past information will allow you to make a reasonable bet on the future behavior. (This hyper unpredictability is one of the signatures of what is often called "chaotic" behavior. However, you should know that there does not seem to be a universaly agreed on technical definition of either chaos or chaotic.)

So much for the bad news. The good news is that not even the hyper unpredictability destroys the scientific value of any given version of (1). The point is that there is often a sense in which the statistical properties of a suite of solutions to (1) give valuable information even though no individual solution is helpful. The point here is that you must be careful about the sorts of questions that you ask of (1). Anyway, nature doesn't seem to care about our difficulties with the multicomponent versions of (1) and is not shy about using them.

There is a tremendous amount of fascinating research on the chaotic versions of (1), both in mathematics and in applications to real biological systems. See, for example, the readings for this chapter. However, as the discussion of this research goes beyond the scope of this book, I won't dwell on it further.

My goal for this chapter is to present a plausibility argument for the appearance of hyper unpredictability in some of the three- (or more) component versions of (1).

28.1   A Concrete Example

What follows is a famous example of a three-component version of (1). It is called the Lorenz equations after its discoverer. These equations were devised as an approximation to model certain weather-related phenomena, and it was subsequently realized that for certain values of the parameters, the trajectories of the solutions were incredibly convoluted and effectively unpredictable. In any event, the unknowns here consist of three functions of time, $x(t)$, $y(t)$, and $z(t)$; the equations are

$$\frac{d}{dt}x = -\sigma x + \sigma y,$$
$$\frac{d}{dt}y = rx - y - xz,$$
$$\frac{d}{dt}z = -bz + xy. \tag{2}$$

Here, σ, r, and b are constants. For certain values of these constants, the trajectories are both crazy and extremely sensitive to their starting positions.

28.2   Equilibrium Points When v Has Two Components

One heuristic explanation for the hyper unpredictability of some three-component versions of (1) can be found by considering the nature of equilibrium points for equations such as (1). In this regard, it is convenient to start with a review of the situation when

v has two components. In this case, you should recall that we first met the stable equilibrium points of (1). Such an equilibrium point is characterized by the condition that nearby trajectories move into the equilibrium point. See Figure 28.1.

Figure 28.1 Trajectories near a stable equilibrium point.

We also met repelling equilibrium points. (Remember the discussion of limit cycles and the Poincaré-Bendixson theorem in Chapter 23.) An equilibrium point is repelling when nearby trajectories move away. See Figure 28.2.

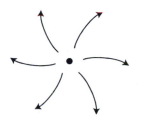

Figure 28.2 Trajectories near a repelling equilibrium point.

There is a third kind of equilibrium point that we can meet in the generic version of (1). This third kind of equilibrium point is called ***hyperbolic*** for a reason that will be explained shortly. This equilibrium point has some nearby trajectories (two, precisely) moving into it, and two others moving out. All other trajectories pass nearby, but eventually leave. See Figure 28.3.

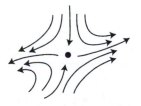

Figure 28.3 Trajectories near a hyperbolic equilibrium point.

Notice that a trajectory that does not crash into the equilibrium point traces out, at least in some rough sense, a branch of the hyperbola $xy = $ constant. It is for this reason that such equilibrium points are called hyperbolic.

The trajectories that (in the limit of large t) end on the equilibrium point form what is called the "stable manifold" for the equilibrium point. The trajectories that seem to emanate from the equilibrium point as t runs toward ∞ form what is called the "unstable manifold" for the equilbrium point. Notice how the trajectories that don't hit the equilibrium point seem to converge onto the unstable manifold (but they never actually touch it).

28.3   Equilibrium Points When v Has Three Components

In an equation such as (1) when v has three components, there can still be stable equilibrium points. Here, as in the two-component case, these are characterized by the condition that nearby trajectories move in. The local picture is, more or less, a three-dimensional version of Figure 28.1. Likewise, there can also be repelling equilibrium points. Here, all nearby trajectories are required to move out from the equilibrium point. The local picture here is more or less a three-dimensional version of Figure 28.2.

In a three-component equation, there can also be the analog of hyperbolic equilibrium points. Here, again, some trajectories move in and some move out, but most nearby trajectories trace out a path that resembles a part of a hyperbola. In the case where there are three components to v in (1), there are two types of such hyperbolic equilibrium points. The first type has its in-moving trajectories lying in a piece of surface that passes through the equilibrium point. (This surface is almost planar and, to a first approximation, can be thought of as being a flat plane.) This surface of in-moving trajectories is called the stable manifold. In this case, there are two out-moving trajectories that move in opposite directions on a piece of a curve through the equilibrium point. These two out-moving trajectories make up the unstable manifold. Figure 28.4 shows the picture.

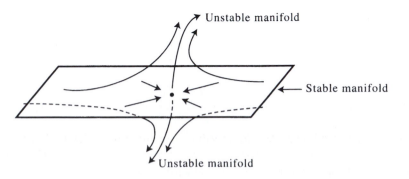

Figure 28.4 Trajectories near a hyperbolic equilibrium point with a two-dimensional stable manifold.

Note that the nearby trajectories that miss the equilibrium point come very close to the unstable manifold.

The other type of hyperbolic equilibrium point is the more interesting one for someone who is interested in chaos. This equilibrium point has only two in-spiraling trajectories; they lie on a piece of a curve through the equilibrium point (the stable manifold). In this case, the out-spiraling trajectories form a piece of an almost planar surface through the equlibrium point.

Figure 28.5 shows a picture of the nearby trajectories for this sort of hyperbolic equilibrium point.

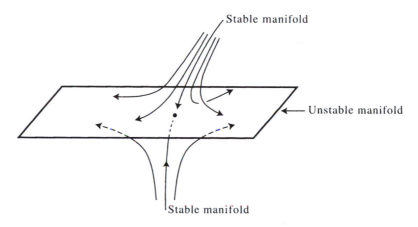

Figure 28.5 Trajectories near a hyperbolic equilibrium point with a curve for a stable manifold.

Notice how the trajectories that miss the equilibrium point in Figure 28.5 congregate along the unstable manifold (the piece of surface in the picture).

28.4 Throwing the Dice

Consider now the behavior of trajectories that pass near an equilibrium point as depicted in Figure 28.5: A trajectory coming in near the stable manifold (say from the top) gets shot out near the unstable manifold, the almost planar surface. But observe that two trajectories can be very close to each other at the start (when they are at the top of the picture) yet end up, a short time later, going off in opposite directions. (See Figure 28.6.)

All it takes for Figure 28.6 to be valid is for the trajectories to lie on opposite sides of the stable curve. Note that the initial points can be arbitrarily close together and yet still lie on opposite sides of the stable curve and so end up going in different directions.

Moreover, any given trajectory may, after passing close to this equilbrium point once, wind around and pass it again and again. Each time past the equilibrium point,

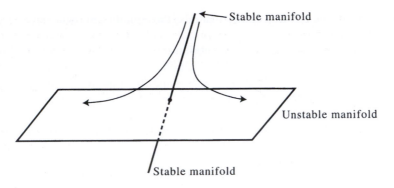

Figure 28.6 Two trajectories that start close ending up far away.

the same dichotomy of nearby trajectories can arise. For example, two trajectories can lie on the same side of the stable curve the first time past the equilibrium point. However, they may wind close to this equilbrium point a second time, in which case they may get sent off in opposite directions even though they were sent off in similar directions the first time past. Or, they may be sent off in essentially the same direction the first 10 or 12 or 12 million times they pass close to this equilibrium point, only to find that on the eleventh, or thirteenth, or twelfth million and first time through, they lie just enough on opposite sides of the stable curve that they get sent out in opposite directions.

Thus, the passage of any trajectory through a neighborhood of an equilibrium point as in Figure 28.5 introduces an unavoidable unpredictability to the dynamics governed by Equation (1). Furthermore, this unpredictability can be repeated any number of times (even infinitely many times), since there is nothing, in principle, that keeps a trajectory (or family of trajectories) from passing close to such an equilibrium point any number of times.

Thus, if you run the system governed by (1) and it has equilibrium points as in Figure 28.5, you will not, in principle, be uable to predict the future behavior of any trajectory if you don't know its starting position *exactly*. Furthermore, *no matter how long you watch the trajectory*, you will not be able to predict its long-term behavior. This is because the trajectory can pass close to the equilibrium point any number (even infinitely many) times.

28.5   Unpredictability for Two-Component Systems

A system of equations as in (1) where v has just two components can also have a certain level of unpredictability. For example, one with a hyperbolic equilibrium point as in Figure 28.3 is unpredictable in the sense that two trajectories can start very close to the stable manifold in Figure 28.3 (this being the in-coming trajectories), but lie on different sides of the stable manifold, and thus get shunted off in opposite directions.

However, this sort of unpredictability can only occur once along a trajectory (once for each hyperbolic equilibrium point). The point is that trajectories cannot cross each other [because there is a unique evolution governed by (1) for every starting value]. This means that if two trajectories initially lie on the same side of the stable manifold in Figure 28.3, then they must lie on the same side *every* time they pass the equilibrium point. For if, on different approaches, they were to lie on different sides, then one of the trajectories would have to cross the stable manifold trajectory, and as I said, crossing trajectiories is forbidden. (A non-self-intersecting loop on the plane separates the plane into two pieces and you can, if you watch long enough, decide which piece your trajectory is on.) See the illustration in Figure 28.7.

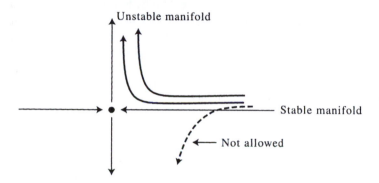

Figure 28.7 Crossing trajectories is forbidden.

The dynamics of a two-component version of (1) are predictable in the sense that if you watch the behavior for some length of time (enough to know which side of the stable manifold a trajectory is on), then you will be able to predict roughly the behavior for all subsequent times.

As remarked previously, this last level of predictability can fail for three-component versions of (1) because a line in three-dimensional space does not separate the space into two pieces. (You need a plane for that.)

28.6 Lessons

Here are some key points from this chapter:

- Although completely predictive in principle, an equation such as $\frac{dv}{dt} = f(v)$ for a vector function of time might not allow predictions in practice if the behavior of solutions is extremely sensitive to their starting points. This is because real data always has some experimental uncertainty so the starting point cannot be known to an arbitrarily high level of precision.

- If v has three or more unknowns, a hyper level of unpredictability can occur with the property that *no matter how long you observe the solution, you still can't make bets on its long-time behavior.*

READINGS FOR CHAPTER 28

READING 28.1

Chaos, Persistence, and Evolution of Strain Structure in Antigenically Diverse Infectious Agents

Commentary: Certain infectious agents mutate quickly (such as the HIV virus) and thus an infected individual soon carries a suite of related, but genetically diverse strains of infectious agents. Meanwhile, the immune response attacks these agents, but is hampered by the fact that it must mount a separate response to different strains if their 'genetic' distance is large. Of interest here are the circumstances under which the number of strains attains an equilibrium, and more generally, to predict the shape of the 'attractors' when there are no equilibria. The authors make some progress towards this end with a fairly simplistic model of the interactions involved. In particular, they find circumstances where their model has stable equilibria, limit cycles, and also 'chaotic' behavior. (For the uninitiated: An attractor is a part of the phase space where sets of trajectories congregate. Attractors can be extremely convoluted and bizarre for differential equations with three or more components.) By the way, the model's equations are given in the fourth reference citation.

Chaos, Persistence, and Evolution of Strain Structure in Antigenically Diverse Infectious Agents

Science, **280** (1998) 912–915.
Sunetra Gupta, Neil Ferguson, Roy Anderson

The effects of selection by host immune responses on transmission dynamics was analyzed in a broad class of antigenically diverse pathogens. Strong selection can cause pathogen populations to stably segregate into discrete strains with nonoverlapping antigenic repertoires. However, over a wide range of intermediate levels of selection, strain structure is unstable, varying in a manner that is either cyclical or chaotic. These results have implications for the interpretation of longitudinal epidemiological data on strain or serotype abundance, design of surveillance strategies, and the assessment of multivalent vaccine trials.

New epidemics of an infectious disease can be triggered by the evolution of a novel antigenic type or strain that evades the acquired immunity within the host population created by its predecessors. The most studied case is the influenza virus, where major shifts in the structure of surface antigens can often trigger worldwide pandemics of the novel variant [1, 2]. The antigens that are most likely to exhibit diversity are those under strong selection by host immune responses. These polymorphic antigens may be ranked by the degree to which the associated immune response reduces the reproductive or transmission success of the pathogen. We demonstrated previously that those antigens that elicit the strongest immune response, which in turn have the strongest impact on transmission success, may be organized by immune selection acting within the host population into sets of nonoverlapping variants [3]. For example, in the case of two antigens each encoded by a distinct locus with two alleles, namely *a* and *b* at one locus and *c* and *d* at the other, four genotypes exist (*ac*, *ad*, *bc*, and *bd*). One set of nonoverlapping variants is *ac* and *bd* (a discordant set) and the other is *bc* and

---

Author affiliations: Sunetra Gupta, Neil Ferguson, and Roy Anderson: Wellcome Trust Centre for the Epidemiology of Infectious Disease, Zoology Department, University of Oxford.

ad. The pathogen population may exhibit a discrete strain structure where one set of nonoverlapping variants exists at much greater frequency than the other. For this pattern to emerge and be stable over time, the intensity of acquired immunity to a specific variant antigen (encoded by a given allele) within the host population must reduce considerably the transmission success or fitness of all subsequent infections by genotypes possessing that allele. Pathogen populations may therefore be categorized into discrete "strains" or serotypes according to the genetic loci that encode antigens eliciting immune responses with the greatest effect on the transmission success of the infectious agent [3].

Here, we show that pathogens that possess antigens that do not elicit an immune response that is strong enough to induce a discrete stable strain structure may exist as a set of strains that exhibit cyclical or chaotic fluctuations in frequency over time. They may still be organized by immune selection into discrete groups of variants, where all the members in a given group have different alleles at every locus, but the dominance of the group, relative to that of other groups, may fluctuate widely over time, either cyclically or chaotically. Pathogen antigens that elicit immune responses that have little effect on transmission success will not be organized to express a discrete nonoverlapping strain structure. In this case, all possible allele combinations will be maintained at abundances commensurate with their individual transmission success or fitness.

We define the conditions under which each of these three outcomes arises, in terms of the biological characteristics of both the pathogen and the immune response of the host to the various antigens of the infectious agent, using a model in which a pathogen strain is defined by the m alleles that exist at each of n loci. We take no account of other important biological complications such as mutation, seasonality in transmission, time delays between the acquisition of infection and infectiousness to susceptible hosts, or genetic diversity in the host population, influencing the immune response to particular antigenic variants. These exclusions are deliberate, because we wish to assess the impact of selection imposed by acquired immunity in the host population on temporal trends in the frequencies of different variants and the evolution of the associated strain structure in the pathogen population.

Strains that do not share any alleles (hereafter referred to as a discordant set) are assumed not to interfere with each others' transmission success or fitness as mediated by host immune responses. The degree to which infection with a given strain limits the ability of another strain that shares any alleles to infect the same host is defined by a cross-protection or cross-immunity parameter γ. If $\gamma = 0$, then the strains do not induce cross-protective responses, whereas if $\gamma = 1$, then there is complete cross-protection. It is also assumed that immunity to a given strain i does not prevent infection by any other strain, but only reduces the probability of transmission of a nondiscordant strain j by a factor $(1 - \gamma)$. This implies that the process of the acquisition of immunity to any one strain is independent to that of its acquisition to any other strain. These simple but realistic biological assumptions can be expressed by a system of differential equations representing the changes in the abundances of each of the m^n strains over time in a genetically homogeneous host population [4].

The model system generated three distinct dynamical behaviors (Figure 1). (i) When the degree of cross-immunity is below a threshold value, γ_L, all the strains coexist in the host population with stable abundance, and no strain structure (NSS) is apparent. Increasing the degree of cross-protection in this dynamical domain acts to decrease strain abundance because of increased immunological interference between strains. (ii) When the degree of cross-immunity exceeds an upper threshold, γ_U, one discordant set of strains dominates in terms of their prevalence. This situation represents the presence of stable discrete strain structure (DSS). (iii) For $\gamma_L < \gamma < \gamma_U$, no stable strain structure occurs and the relative proportions of the different strains exhibit very complex and often chaotic dynamics with marked fluctuations over time. This is referred to as cyclical or chaotic strain structure (CSS).

Figure 1 plots the parameter regions for which these different dynamical behaviors and associated strain structures pertain. The upper threshold, γ_U, is given by $\gamma_U = 1 - 1/(2R_0)$ if all strains have identical transmission success as defined by the case reproductive rate R_0 (the average number of secondary infections generated by one primary infection in a susceptible host population [5]). The lower threshold value γ_L is determined by the ratio σ/μ [where $1/\mu$ is host life expectancy and $1/\sigma$

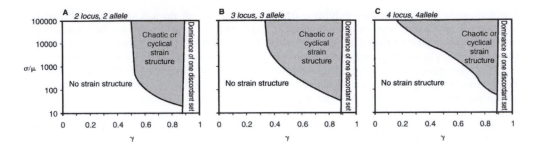

Fig. 1 Strain structure type is shown as a function of the degree of cross-protection (y) and the ratio of host life-span to pathogen infectious period (σ/μ) for pathogens with (A) two antigenically active loci, each with two alleles; (B) three loci, each with three alleles; and (C) four loci, each with four alleles ($R_0 = 4$ for all strains).

is the average duration of infection (equal to infectiousness)], R_0, and the number of antigen loci and alleles. For pathogens in developed countries, with human life expectancies of around 70 years and infectious periods of between 4 days to 1 year ($1/\sigma = 0.01$ to 1), σ/μ is between 70 to 7000. The large size of the parameter space generating cyclical or chaotic behavior (Figure 1), for relevant parameter assignments for the duration of infection and host life expectancy, implies that such complex dynamics may be the norm for many antigenically variable infectious agents that induce moderate cross-protective immune responses in the human host.

Figure 2 portrays the wide range of complex dynamical behaviors generated by the model in the CSS parameter domain. Initially the strain prevalences follow simple limit cycles (Figure 2B), but as the value of y is increased, chaotic intermittency is seen (Figure 2C), followed by increasingly large-amplitude chaos (Figure 2, D and E). Several important trends are apparent from extensive numerical studies. First, as y increases in value, the amplitude and period of the epidemic cycles rise. Second, when y is close to the two boundaries of the CSS region, complete coherence is seen between the epidemics of strains within discordant sets (that is, the abundances of all strains within a set are identical). This is a direct result of competitive exclusion between genotypes that share alleles where the competition is created by herd immunity. Third, for a large band of cross-protection values in the center of the CSS region, this coherence begins to break down (Figure 2D), and the trajectories of the prevalence of individual members of discordant sets can

be distinguished. The exact dynamical mechanism driving this symmetry breakdown remains unclear, but the behavior appears to be associated with a crossover effect as the system moves from the low-y region of short-period cycles (a few years) to the large-y region where long-period (many years) chaos dominates. In the crossover region, both time scales are apparent, with "generation" cycles (with period $\approx 1/\mu$ = host life expectancy) being modulated by shorter period oscillations, the periods of which are determined by the average infectious period ($1/\sigma$) and the transmission success of the pathogen (R_0). In other words, the long periods are determined by host demography and the short periods by the variables that determine the typical course of infection in the host and the transmission dynamics of the infectious agent. As the duration of the infectious period is decreased, the period of the low-y regime limit cycles decreases, and chaotic behavior increasingly dominates the entire CSS region. Similarly, discordant set decoherence increases as the number of antigenically active loci and alleles increase. Decoherence also increases markedly as the infectious period decreases.

Other theoretical studies indicate that complex dynamics are likely to be common for any pathogen populations that are organized in such a manner that there is cross-immunity within certain strain subsets and none between these subsets. For instance, Andreasen *et al.* [2] modeled influenza by assuming that interstrain cross-immunity exists only between nearest neighbors in a one-dimensional phenotypic space (that is, immunity to strain *i* affects the transmission success of

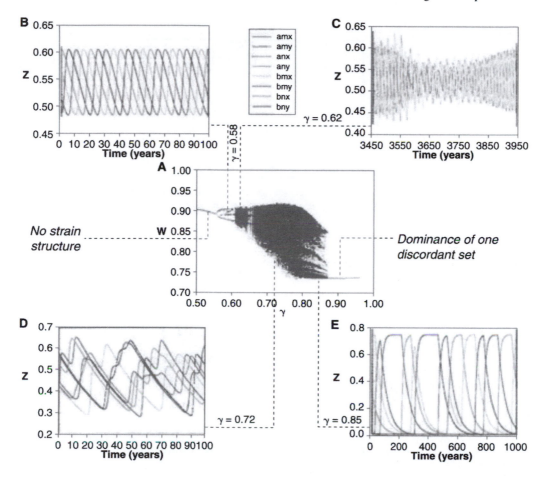

Fig. 2 Long-term population dynamics of pathogens with three loci, each with two alleles. $R_0 = 4$ for all strains, $\sigma = 10$ years^{-1}, and $\mu = 0.02$ years^{-1}. (A) Bifurcation diagram shows the location of local maxima of W_{amx} (the fraction of the population exposed to strain amx or any strain sharing alleles with amx) as a function of the degree of cross-protection. (B through D) Time-series of the prevalences, z_i, of each strain in the population: (B) for $\gamma = 0.58$ (nonchaotic limit cycle), (C) for $\gamma = 0.62$ (intermittent chaotic episodes), (D) for $\gamma = 0.72$ (chaotic short-period cycles with decoherence between discordant set members), and (E) $\gamma = 0.85$ (large-amplitude chaos).

strains $i - 1$ and $i + 1$) and demonstrated that this form of population structuring can also generate stable limit cycles in systems with four or more strains. Other population-structuring mechanisms may also give interaction matrices that generate complex nonlinear dynamics, but it is particularly relevant that such a fundamental biological property of pathogens as the sharing of antigen variants yields this structure.

A key question arising from these analyses is whether we can expect to see deterministic chaos in epidemiological data for common viral, bacterial, and protozoan antigenically variable pathogens in human communities. Although stochastic effects in small populations may interfere with the persistence of long-period cycles [6], we may expect to see the following broad patterns. For polymorphic antigens that elicit weak immune responses, the abundances of the different strains (as defined by different combinations of alleles) will be deter-

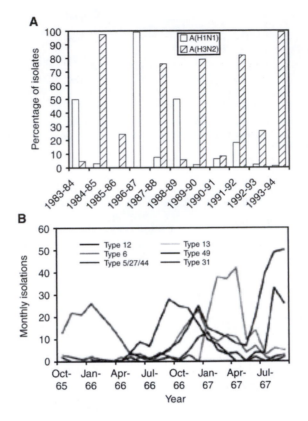

Fig. 3 Examples of dynamical changes in serotype frequencies of two major human pathogens: (A) influenza cases in the United States from 1983–94 [20] and (B) group A streptococcal infections in Minnesota, United States, from 1965–1967 [15].

mined by their respective transmission successes or fitnesses. In the cyclical or chaotic region, where the degree of cross-protection associated with the antigen is moderate, irregular epidemic cycles will be observed with average periods set by the mean duration of infectiousness in the host (short-period cycles of a few years for infectious periods of a few days to very long periods of many years for infectious periods approaching a month or longer). Concomitantly, there will be a high degree of correlation in the prevalence or incidence time-series of strains within the same discordant set and a low degree of correlation between strains or serotypes in different discordant sets. For antigens that elicit very strongly protective immune responses, the correlation between strains within the same discordant set will be extremely high, manifesting as DSS:

These theoretical predictions argue for considerable caution in the interpretation of observed patterns in longitudinal epidemiological data sets that are stratified by strain type or serotype based on molecular or immunological taxonomic characteristics. Although the limited time-series data currently available show significant fluctuations in strain incidences for a variety of pathogens (Figure 3), better quality, longer term data will be required to accurately assess the precise strain structure of a given antigenically variable pathogen population. Essential to this process is the cloning of the relevant antigen genes; for example, the recent cloning of the *Plasmodium falciparum* gene encoding PfEMP-1 [7–9] is crucial to the validation of the hypothesis that PfEMP-1 is the antigen that structures *P. falciparum* into discrete strains [10]. The analysis we present suggests that it is also worth monitoring other *P. falciparum* antigens such as the merozoite surface proteins (MSP), which may not have as significant a role in protective immunity as PfEMP-1, because these may be more likely to exhibit cyclical or chaotic fluctuations. A recent study in Senegal [11] involving the typing of successive clinical malaria isolates during a 4-month period of intense transmission shows a large degree of genetic diversity in MSP1 and MSP2 combinations. However, long-term large-scale studies are required to elucidate the precise dynamics of these allele combinations. For bacteria, such as *Neisseria meningitidis*, large-scale nucleotide sequencing studies of genes encoding surface antigens are essential to understand the population structure of these organisms [12, 13] and, in particular, to elucidate whether the strong association observed between VR1 and VR2 epitopes of the PorA protein [2] forms the basis for the fluctuations in lineages of *N. meningitidis* [14]. Serotypes of group A streptococci also exhibit rapid fluctuations (Figure 3) [15]. These serotypes are defined by an antiphagocytic cell-surface molecule known as M protein, which appears to occur in nonoverlapping combinations with another variable element known as the serum opacity factor [16], and thus, may exemplify a two-locus multiallele system in a state of CSS.

Such information on pathogen population structure is crucial in many different contexts, including the assessment of vaccine trials where the candidate vaccine contains only a subset of the antigens expressed by the pathogen population [17] such as those currently planned for *N. meningitidis* [18] and *Streptococcus pneumoniae* [19]. Attributing observed changes in strain structure following mass immunization to the intervention may be problematic, given the complex nonlinear dynamics suggested by our analyses. The intricate behavior of these multistrain systems is a consequence of the selective pressures imposed on the pathogen population by the profile of herd immunity in the host population. This profile is, in turn, conditioned by the prevailing antigenic structure of the pathogen population. It is the subtle interplay between these two factors that leads to the unstable evolutionary dynamics we describe.

REFERENCES

1. A. D. Cliff, P. Haggett, J. K. Ord, *Spatial Aspects of Influenza Epidemics* (Pion, London, 1986).
2. V. Andreasen, J. Lin, S. A. Levin, *J. Math. Biol.* **35**, 825 (1997).
3. S. Gupta *et al.*, *Nature Med.* **2**, 437 (1996).

4. The proportion immune to a given strain i, z_i is simply given by

$$\frac{dz_i}{dt} = (1 - z_i)\lambda_i - \mu z_i$$

Here, λ_i represents the force of infection of strain i. We assume, for simplicity, that immunity is lifelong. We then define an additional compartment, w_i, which represents those immune to any strain j that shares alleles (at the relevant polymorphic loci) with strain i (including i itself). The dynamics of w_i are then given by (where $j \sim i$ means j shares alleles with i)

$$\frac{dw_i}{dt} = (1 - w_i)\sum_{j \sim i}\lambda_i - \mu w_i$$

Individuals who have never been exposed to any strain sharing alleles with strain i (that is, $1 - w_i$) are completely susceptible to strain i. However, those that have been exposed to a strain sharing alleles with i, but not exposed to strain i itself (that is, $w_i - z_i$) will become infectious with a probability $1 - y$ when they are infected by strain i. With σ being the rate of loss of infectiousness of the host, the dynamics of the proportion of the population infectious for strain i may therefore be represented as

$$\frac{dy_i}{dt} = [(1 - w_i) + (1 - y)(w_i - z_i)]\lambda_i - \sigma y_i$$

The impact of genetic exchange on the population structure of infectious disease agents may be examined within this framework by modifying the force of infection term, λ, to include the assumption that the progeny of parasites within hosts infectious for two or more strains will consist of defined fractions, Ω_{ijk}, of the various combinations of the different strains, j and k, that may generate strain i through recombination. Because the proportions of infectious hosts are very small, the force of infection of strain i may be approximately represented as $\lambda_i = \beta_i(y_i + \sum \Omega_{ijk}y_j y_k)$, where β_i is a combination of parameters affecting the transmission of strain i. The behavior of the model is largely unaffected by the inclusion or precise functional form of the recombination term.

5. R. M. Anderson and R. M. May, *Infectious Diseases of Humans: Dynamics and Control* (Oxford Univ. Press, Oxford, 1991).

6. B. M. Bolker and B. T. Grenfell, *Philos. Trans. R. Soc. London Ser. B* **348**, 308 (1995); N. M. Ferguson, R. M. May, R. M. Anderson, in *Spatial Ecology: The Role of Space in Population Dynamics and Interspecific Interactions*, D. Tilman and P. Kareiva, Eds. (Princeton Univ. Press, Princeton, NJ, 1997).

7. D. I. Baruch *et al.*, *Cell* **82**, 77 (1995).

8. X.-z. Su *et al.*, *ibid.*, p. 89.

9. J. D. Smith *et al.*, *ibid.*, p. 107.

10. S. Gupta, K. Trenholme, R. M. Anderson, K. P. Day, *Science* **263**, 961 (1994).

11. H. Contamin *et al.*, *Am. J. Trop. Med. Hyg.* **54**, 632 (1996).

12. M. C. J. Maiden and I. M. Feavers, in *Population Genetics of Bacteria*, S. Baumberg, P. Young, J. Saunders, E. Wellington, Eds. (Cambridge Univ. Press, Cambridge, 1995), pp. 269-293.

13. I. M. Feavers, A. J. Fox, S. Gray, D. M. Jones, M. C. J. Maiden, *Clin. Diagn. Lab. Immunol.* **3**, 444 (1996).

14. D. A. Caugant *et al.*, *J. Infect. Dis.* **162**, 867 (1990).

15. B. F. Anthony, E. L. Kaplan, L. W. Wannamaker, S. S. Chapman, *Am. J. Epidemiol.* **104**, 652 (1976).

16. J. V. Rakonjac, J. C. Robbins, V. A. Fischetti, *Infect. Immun.* **63**, 622 (1995).

17. S. Gupta, N. M. Ferguson, R. M. Anderson, *Proc. R. Soc. London Ser. B*, in press.

18. P. van der Ley and J. T. Poolman, *Infect. Immun.* **60**, 3156 (1992).

19. D. A. Watson, D. M. Musher, J. Verhoef, *Eur. J. Clin. Microbiol. Infect. Dis.* **14**, 479 (1995).

20. *Morb. Mortal. Wkly. Rep.* **41**, 3 (1992); *ibid.*, p. 5; *ibid.* **42**, 1 (1993); *ibid.* **46**, 1 (1997).

21. S.G. and R.A. thank the Wellcome Trust, and N.F. thanks the Royal Society for grant support. We thank M. Maiden and J. Mathews for their helpful comments.

READING 28.2

Controlling Chaos in the Brain

Commentary: This article reports on evidence for the existence of an underlying chaotic dynamical system that governs the firing of neurons in an *in vitro* brain slice preparation. In this case, the slices come from rat brains; they are stimulated by an abnormally high potassium concentration. The nature of the dynamics is investigated here by plotting pairs ($x = I_{n-1}$, $y = I_n$) on the xy-plane, where I_n is the nth interburst interval. If the interburst intervals were constant, then all of the points plotted for $n = 1, 2, \ldots$, would give the same point on the xy-plane. The fact is that we obtain a scatter plot on the xy-plane, which demonstrates that I_n cannot be predicted from knowledge of I_{n-1}. If this were the case, we would expect that the points would trace out a graph in the xy-plane. The evidence for chaotic behavior comes from the way successive points plot out on the plane. For example, if there were a cycle, say of period j, so that $I_n = I_{n-j}$, then you would see the successively plotted points hit only j points on the plane. If this cycle were unstable, you could see approximately periodic behavior, with the approximation getting worse each time around as the dynamics moves away from the cycle in an unstable direction. In fact, this sort of behavior is seen in Figure 2 of the article, where successive plotted points move away from the $x = y$ line along an orthogonal line.

Controlling Chaos in the Brain

Nature, **370** (1994) 615–620.
Steven J. Schiff, Kristin Jerger, Duc H. Duong, Taeun Chang, Mark L. Spano, William L. Ditto

In a spontaneously bursting neuronal network in vitro, chaos can be demonstrated by the presence of unstable fixed-point behaviour. Chaos control techniques can increase the periodicity of such neuronal population bursting behaviour. Periodic pacing is also effective in entraining such systems, although in a qualitatively different fashion. Using a strategy of anticontrol such systems can be made less periodic. These techniques may be applicable to in vivo epileptic foci.

Following the recent theoretical prediction that chaotic physical systems might be controllable with small [1, 2] perturbations, there has been rapid and successful application of this technique to mechanical systems [3], electrical circuits [4], lasers and chemical reactions [6, 7]. Following the demonstration of the control of chaos in arrhythmic cardiac tissue [8], there are no longer any technical barriers to applying these techniques to neural tissue.

One of the hallmarks of the human epileptic brain during periods of time in between seizures is the presence of brief bursts of focal neuronal activity known as interictal spikes. Often such spikes emanate from the same region of brain from which the seizures are generated but the relationship between the spike patterns and seizure onsets remains unclear [9, 10]. Several types of *in vitro* brain slice preparations, usually after exposure to convulsant drugs that reduce neuronal inhibition, exhibit population burst-firing activity that in many ways seems analogous to the interictal spike [11]. One of these preparations is the high potassium concentration ($[K^+]$) model, where slices from the hippocampus of the temporal lobe of the rat brain (a frequent site of epileptogenesis in the human) are exposed to artificial cerebrospinal fluid containing 6.5–10 mM $[K^+]$ [12]. After exposure to high $[K^+]$, spontaneous bursts of synchronized neuronal activity originate in a region known as the third part of the cornu ammonis or CA3 [13]. Impulses from

the CA3 bursts are propagated through a recurrent collateral fibre tract (the Schaffer collateral fibres) from CA3 to CAl, where electrographic seizure-like discharges can frequently be observed. A detailed computer model of the high $[K^+]$ burst discharges in CA3 successfully replicates many of the experimental findings [15] (reviewed in [16]). Although it is difficult to identify determinism in long time series of such bursting activity using nonlinear prediction techniques, some evidence of determinism was recently identified [17, 18]. We sought to determine whether such neuronal bursting activity was amenable to control.

There have been substantial efforts to influence neuronal activity with electric fields and currents. Regarding the in vitro hippocampal slice, brief direct current (d.c.) currents from nonpolarizable electrodes in the perfusion bath can influence the evoked excitability of pyramidal cells in normal $[K^+]$ when the electric fields are suitably oriented [19]. Apparently similar effects can be achieved with brief currents from monopolar polarizable microelectrodes placed directly into the tissue [20, 21]. To our knowledge the use of electrical stimulation to entrain spontaneous burst discharges from CA3 in high $[K^+]$ has not been attempted, whether indirectly with electric fields or with direct stimulation of the nerve cells themselves. Direct stimulation can be accomplished for CA3 neurons by stimulating the Mossy fibre neurons that synapse on the CA3 pyramidal cells (orthodromic stimulation) or by stimulating branches of the CA3 pyramidal cell axons (Schaffer collateral fibres) and allowing the action potentials to propagate into the pyramidal cells in a retrograde fashion (antidromic stimulation).

With the observation that it was feasible to entrain burst discharges from CA3 with both orthodromic and antidromic stimulation (K.J. and S.J.S., manuscript in preparation), we were in a position

Author affiliations: Steven J. Schiff, Kristin Jerger, Duc H. Duong, Taeun Chang: Department of Neurosurgery, Children's National Medical Center and The George Washington University School of Medicine. Mark L. Spano: Naval Surface Warfare Center, White Oak Laboratory. William L. Ditto: School of Physics, Georgia Institute of Technology.

to ask the following questions: (1) is there evidence for deterministic chaotic behaviour in this preparation; and (2) could such activity be controlled?

If one observes the timing of events from a chaotic physical system, those events are aperiodic. The timing of events evolves from one unstable periodicity to another. Furthermore the approach to these unstable periodicities shows recurring patterns which can be quantitatively understood by examining the relationship between the timing of sequential events. This can be visualized using a plot which is a type of a return map. Such a map plots the present interval between events, I_n, versus the jth previous interval, I_{n-j}. Periodicities (of period j) are revealed on such a plot as intersections with the line of identity, $I_n = I_{n-j}$. In a chaotic system these intersections are known as unstable fixed points. These unstable fixed points are deterministically approached from a direction called the stable direction or manifold and exponentially diverge from these points along the unstable direction or manifold. This type of local geometry has the shape of a saddle. Chaos control [1, 3, 8] consists of the identification and characterization of these saddles, followed by a control intervention which exploits the local geometry of the saddle to increase the periodic behaviour of the system.

We demonstrate here three separate approaches to control: periodic pacing, an implementation of chaos control theory, and the inverse of chaos control which we term anticontrol.

Brain slice preparation. Hippocampal slices were prepared using standard techniques [18]. Glass micropipette electrodes in CAI and CA3 were used to record neuronal activity, and a computer calculated and delivered appropriately timed control pulses. The anatomy of the transverse hippocampal slice and arrangement of electrodes are illustrated in Fig. 1a, and details of the experimental preparation are summarized in the figure legend.

Chaos identification and control. A key feature of chaotic systems is that they contain an infinite number of unstable periodic fixed points [22]. That these spontaneously active neuronal networks are chaotic was supported by evidence of unstable fixed points in the return maps with the characteristics shown in Fig. 2a, b. These candidate unstable fixed points met four criteria. First, the sequence of points approaches the unstable fixed-point candidate along a stable direction and diverges from it along an unstable direction.

Second, the departing trajectory must be locally linear. Third, multiple approaches to the same candidate along the same stable direction with the corresponding departures along the same unstable direction must be detected. Last, the distances of the departing points from the candidate fixed point must increase exponentially, thereby demonstrating the sensitivity to initial conditions that is the defining feature of chaos. Although short single approaches to candidate unstable fixed points can be occasionally observed in random data, multiple approaches fitting these stringent criteria will not be seen.

In the implementation of these criteria, we excluded higher period behaviour (period 2, period 3, and so on), insisted that the geometry about the fixed point candidates be a flip saddle, and set criteria for the minimum acceptable linearity of the unstable manifold. We also set criteria for limits on the rates of approach and divergence of the trajectories. Because on these return plots it is easily shown that the eigenvalues (local rates of expansion and contraction along the respective manifolds) are simply the slopes of the respective manifolds, it is important to note that we found that the stable manifolds had negative slopes with magnitudes less than 1 and that the unstable manifolds had negative slopes with magnitudes greater than 1, further supporting the presence of chaos in these data (see Fig. 2).

Our chaos control technique begins with a learning phase consisting of the identification of unstable fixed points and the performance of local linear least-square fits to obtain the stable and unstable directions along with the rates of approach and divergence along them. The control phase consisted of waiting until the system executed a close approach to the unstable fixed point (within a small radius ε) along the stable direction, followed by an intervention that modified the timing of the predicted next interval in order to place it back onto the stable manifold. In this way we use the saddle structure inherent in the chaotic dynamics to bring the system back to the desired unstable fixed point with minimal intervention.

In the following experiments, for periodic pacing, we used the same pulse interval as our calculated unstable fixed point. For anticontrol, we chose an interval that placed the next point on a line that was completely off the manifolds. We somewhat arbitrarily chose a direction that was the mir-

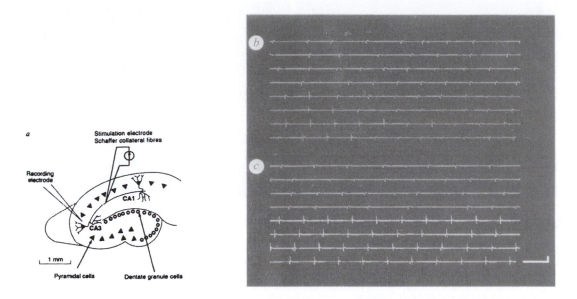

Fig. 1 a, Schematic diagram of the transverse hippocampal slice, and arrangement of recording electrodes. Female Sprague-Dawley rats weighing 125–150 g were anaesthetized with diethyl-ether and decapitated. Transverse slices 400 μm thick were prepared from the hippocampus with a tissue chopper and placed in an interface-type perfusion chamber at 32–35 °C. Slices were perfused with artificial cerebrospinal fluid (ACSF) flowing at 2 ml min^{-1} and composed of 155 mM Na$^+$, 136 mM Cl$^-$, 3.5 mM K$^+$, 1.2 MM Ca^{2+}, 1.2 mM Mg^{2+}, 1.25 MM PO$_4^{2-}$, 24 mM HCO$_3^-$, 1.2 MM SO$_4^{2-}$, and 10 mM dextrose. After 90 min of incubation, slices were tested for viability by recording a greater than 2 mV unitary population spike in the stratum pyramidale of CAI, in response to stimulation of Schaffer collateral fibres in the stratum radia- turn with 100 μs constant current 50–150 μA square-wave pulses delivered at 0.1 Hz through tungsten microelectrodes. Recordings were made with 2–4-MΩ glass needle electrodes filled with 150 mM NaCl. With confirmation of viability, the perfusate was switched to ACSF containing 8.5 mM [K$^+$] and 141 mM [CI$^-$]. After 15–20 min of high [K$^+$] perfusion, spontaneous burst firing could be recorded from CA3a or CA3b. Recordings were digitized across 12 bits at 5 kHz with a Digidata 1200 analogue to digital converter (Axon Instruments), and stored on a personal computer using Axotape 2.0 (Axon Instruments). CA3 interburst intervals were measured with Datapac II (Run Technologies). For control purposes, a separate computer system, digitizing across 16 bits at 1 kHz, identified spontaneous bursts from CA3 using a threshold and peak amplitude detection strategy. This computer triggered a stimulator (Model S8800, Grass Corp.) linked to a photoelectric stimulus isolation unit (Model SIU7, Grass Corp.), to deliver 100-μs constant current square-wave pulses to the Schaffer collateral fibres through a tungsten microelectrode. At times, double pulses consisting of pairs of 100-μs pulses with 150-μs interpulse intervals were used. b. Shown are 100 s of recording from an extracellular electrode within CA3 following expo- sure to 8.5 mM [K$^+$]. The upper four traces in red show the burst discharges occurring at irregular intervals. The lower four traces in blue show the effect of turning on chaos control. When control of chaos pulses are delivered to the Schaffer collateral fibres, large stimulation artefacts are seen at the onset of a burst. Note that each control pulse evokes a synchronized discharge in CA3. As control becomes more stable, the bursts become increasingly periodic, as seen most clearly on the bottom trace. Note that the control pulses are delivered only intermittently. The plot of the full series of these intervals before and during control is seen in the first half of the plot in Fig. 3. c, Shown are 100 s of recording from the same experiment before (red) and after (yellow) the initiation of periodic pacing. Note that periodic pulses immediately entrain the population bursts in the fifth tracing (10/10 pulses), but by the sixth tracing. there is occasional escape behaviour evident when the population burstsjust before the delivered pulse (seen in 3/10 bursts in tracing 6, 2/11 in tracing 7, and 2/10 in tracing 8). The plot of the full series of these intervals before and during periodic pacing is seen in the second half of the plot in Fig. 3. The calibration mark indicates 5 mV ordinate and 1 s abscissa.

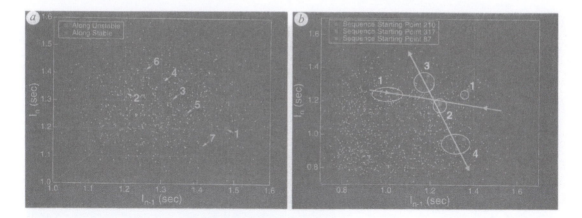

Fig. 2 a, Return plots of interburst intervals I_n versus the previous interval. I_{n-1} without control. Seven sequential points are colour coded and numbered 1-7. Note that as the trajectory crosses the line of identity along a particular direction from 1 to 2, the next points take a peculiar sequence that starts close to the line of identity at point 3, and then alternates on either side of the line of identity and progressively diverges from it along a nearly straight line for points 3-7. The points coloured in green (1, 2), define a stable direction or manifold, whereas the points in red (3-7) define an unstable manifold. The intersection of these manifolds with the line of identity defines the unstable fixed point. Note that, as for other physical and biological systems [3, 8], the saddle observed in these preparations is a flip saddle; that is, while the distances of successive state points from the fixed point increase in an exponential fashion along the unstable manifold, the state points alternate on opposite sides of the line of identity. b, Return plot showing multiple colour-coded trajectories that did not follow each other in time. The starting point for each sequence, numbered 1 in each, began at burst numbers 87 (blue), 210 (red) and 317 (green), out of a total series of 320 bursts. The stable manifold is shown with arrows pointing towards the unstable fixed point, and the unstable manifold has arrows drawn in the direction away from the unstable fixed point. Note that each sequence starts at a point roughly the same distance from the unstable fixed point in a region close to the stable manifold. Each trajectory follows roughly the same path, and after closely approaching the unstable fixed point for the points labelled 2 (red, green and blue), there evolves for each trajectory a sequence of exponentially diverging jumps (on alternating sides of the line of identity along the unstable manifold). Although the trajectories are clustered along the unstable manifold for four bursts, the fifth bursts are widely scattered (not shown). This is the manifestation of the sensitivity to initial conditions typical of chaotic systems. These plots show clear evidence for nonrandom structure in these trajectories, and their patterns bear the hallmarks of deterministic chaos.

ror image of the unstable manifold about the line of identity. This anticontrol technique proved effective in diverting the state of the system away from the stable direction and therefore from the unstable fixed point.

Experimental results. We did 91 experimental trials on 22 hippocampal slices from 9 rats. These trials consisted of combinations of chaos control, periodic pacing and anticontrol using both single and double stimulation pulses (Table 1). Double pulses were used at times to increase the effectiveness of stimulation.

Figure lb shows an example of the synchronized burst discharges recorded extracellularly in CA3. The upper four traces show the burst discharges occurring at irregular intervals before control. On the fifth tracing chaos control pulses are delivered to the Schaffer collateral fibres, seen as large artefacts at the onset of a burst. Note that each control pulse can evoke a synchronized discharge in CA3, but control pulses are only given intermittently. As control becomes more stable, as seen most clearly on the bottom trace, the spontaneous bursts become increasingly periodic. We examined the relationship between burst intensity (duration, integrated amplitude and root mean square deviation) and interburst interval but found no correlations. The plot of the full series of these intervals before and during control is seen in the first half of the plot in Fig. 3.

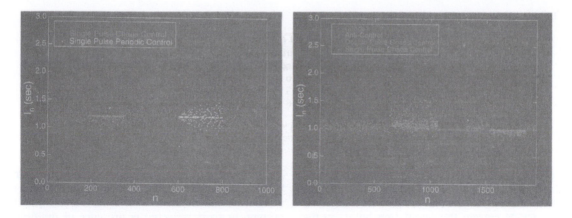

Fig. 3 (left) Plot showing chaotic (red) interburst intervals 1. before and after single-pulse chaos control (blue) and single-pulse periodic pacing (yellow). Burst number is indicated by n. Note the qualitative differences in the control achieved with each method. Raw data from the onsets of chaos control and periodic pacing are shown in Fig. 1b, c.

Fig. 4 (right) Comparison of anticontrol (pink), double-pulse chaos control (green) and single-pulse control (blue). Note how anticontrol reduces the periodicity of the preparation. Also note that double-pulse control is more effective than single-pulse control, as was generally the case (see Table 1).

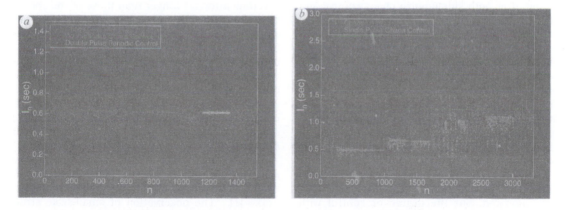

Fig. 5 Comparison of double-pulse chaos control (green) with double-pulse periodic pacing (orange) in an exceptional experiment where double-pulse chaos control was not very tight, and where double pulse periodic pacing was a better control method. Note the frequent escapes and recaptures of control with the chaos control method. This example helps illustrate how the chaos control method differed from overdrive periodic pacing. b, Experimental run where a wide range of control behaviours were observed. There are 5 sequential attempts to identify unstable fixed points and learn the manifold structure. Each control run is coded in blue. The first control run demonstrates what we classify as good control in Table 1. After relearning, the second attempt at control was also good, but at a higher fixed point (where the period of the underlying periodic orbit of the fixed point is larger). The third trial, despite learning nearly the identical fixed point as the previous trial, chose different manifolds and the control was not as good. We call such control runs, where intervals I_n larger than the fixed point are selectively eliminated partial control; note that this is a manifestation of the manifold selection and not just a function of the mean frequency of stimulation. This point is further clarified in the fourth and fifth control runs. The fourth control run selected a larger fixed point than the previous runs but control was a failure. The fifth and final control run of this sequence, despite selection of the highest fixed point of this series, achieved partial control with more adequate manifold and fixed-point selection.

Table 1

	Summary of experiments					
	Chaos control		Periodic pacing		Anticontrol	
	Single	Double	Single	Double	Single	Double
Good	10	4	6	2	4	1
Partial	15	2	5	0	—	—
Bad	21	0	6	0	16	0
Total	46	6	17	2	20	1

Figure 1c shows 100 s of recording from the same experimental run before and after the initiation of periodic pacing. Note that now pulses immediately entrain the population bursts in the fifth tracing, but by the sixth tracing there is occasional escape behaviour evident when the population burstfires just before the delivered pulses. The plot of the full series of these intervals before and during periodic pacing is seen in the second half of the plot in Fig. 3.

Figure 3 illustrates burst intervals before, during and after both single-pulse chaos control and single-pulse periodic pacing. Note that there is a qualitative difference in the control achieved with each method, as was seen from the traces of raw data shown in Fig. 1b, c.

Figure 4 compares the effects of anticontrol, double-pulse chaos control and single-pulse chaos control for another experiment. Here, the effect of anticontrol in reducing the periodicity of the preparation is clearly seen. We also demonstrated an increase in the effectiveness of double over single pulsing for this experiment.

Figure 5a illustrates an experiment where double-pulse chaos control and periodic pacing were compared, but the degree of control with chaos control was not as good as with periodic pacing. This experiment was exceptional, because most of the double-pulse experiments achieved very tight control (see Fig. 4, and summary in Table 1). Note that even with the increased stimulation of double pulsing, chaos control was not simply overdrive pacing, in that there were frequent escapes and recaptures of control consistent with the interactive control hypothesis.

Figure 5b illustrates the full gamut of results obtained with single-pulse chaos control. The first control trial demonstrated good control. Note that during this control sequence, the quality of the control improved. We then allowed the algorithm to re-learn the fixed point and it chose one with a longer interval, again with good control. Whether this is a manifestation of the non-stationary nature of the system, or truly reflects the presence of multiple unstable period 1 fixed points is unclear. Relearning a third time, the algorithm selected different manifolds without substantially changing the value of the fixed point, but the control is not as good. This third control run exhibited what we call partial control in the results shown in Table 1; in partial control, manifold selection was not optimal, and time intervals longer than those of the chosen fixed-point interval appear eliminated. Relearning a fourth time, a fixed point is chosen at a longer interval than the previous three attempts. This was a control failure. Nevertheless, relearning a fifth time with an even longer fixed-point interval gave partial control. This sequence demonstrates that the quality of control is not simply a function of the frequency chosen, the control is critically dependent on the quality of the fixed points and manifolds selected.

Discussion. This is the second attempt at achieving control of a chaotic biological system [8] with a derivative method of Ott, Grebogi and Yorke [1]. The observation of small-scale structure and the identification of stable and unstable manifolds near unstable fixed points for many of these burst-firing slices demonstrated the presence of deterministic chaos in this simple neuronal system. For this preparation, complicated control theory is not required just to achieve relatively fast periodic behaviour. Above certain frequencies, periodic pacing was effective in entraining the spontaneous burst discharges in CA3. However, the quality of control with periodic pacing was not equivalent with the chaos control method; furthermore, the chaos control method has the advantage over overdrive periodic pacing in terms of its ability to identify and track fixed points over time. In addition, the control of chaos strategy offers the ability to break up fixed-point behaviour with anticontrol. The anticontrol method used here uses a minimal number of stimuli needed to prevent periodic behaviour.

Although it was easy to observe the unstable manifolds from this preparation, it was difficult to place accurately the stable manifold. This is because the system has many degrees of freedom (that is, is high dimensional), and the expectation of fully disentangling its dynamics with a two-dimensional embedding is simplistic. In addition, increasing amounts of noise in a system will increase the frequency of escapes from control and eventually render control impossible [1]. Despite these difficulties, good or partial control could frequently be achieved with our implementation of chaos control. The application of new theoretical techniques recently developed for the control of high dimensional systems [23] could improve the reliability of control for such neuronal preparations.

We experimented with several variants of stimulation delivery. In addition to single pulses, double pulses were used and were often more effective in achieving higher quality control. We also explored limiting the delivery of control pulses by prohibiting consecutive control pulses without an intervening spontaneous burst. This latter method was less effective than permitting consecutive pulsing.

Because this neuronal preparation shares similar characteristics with epileptic interictal spike foci, we believe these methods may be applied to such foci. Although it is impossible to predict what effect increasing the periodicity of epileptic foci will have, the opposite effect of breaking up fixed-point periodic behaviour with anticontrol could be a more useful intervention.

REFERENCES

1. Ott, E., Grebogi, C. & Yorke, J. A. *Phys. Rev. Lett.* **64** 1196-1199 (1990).
2. Shinbrot, T., Grebogi, C., Ott, E. & Yorke, J. A. *Nature* **363**, 411-417 (1993).
3. Ditto, W. L., Rauseo, S. N. & Spano, M. L. *Phys. Rev. Lett.* **65**, 3211-3214 (1990).
4. Hunt, E. R. *Phys. Rev. Lett.* **67**, 1953-1955 (1991).
5. Roy, R., Murphy, T. W., Maier, T. D. & Gills, Z. *Phys. Rev. Lett.* **68** 1259-1262 (1992).
6. Petrov, V., Gaspar, V., Masere, J. & Showalter, K. *Nature* **361**, 240-243 (1993).
7. Rollins, R. W., Parmananda, P. & Sherard, P. *Phys. Rev.* E**47**, R780 (1993).
8. Garfinkel, A., Spano, M., Ditto, W. L. & Weiss, J. *Science* **257**, 1230-1235 (1992).
9. Gotman, J. *Can. J. Neurol. Sci.* **18** 573-576 (1991).
10. Katz, A., Marks, D. A., McCarthy, G. & Spencer, S. S. *Electroenceph. Clin. Neurophysiol.* **79**, 153-156 (1991).
11. Pedley, T. A. & Traub, R. D. in *Current Practice of Clinical Electroencephalography* 2nd edn (eds Daly, D. D. & Pedley, T. A.) 107-137 (Raven, New York. 1990).
12. Rutecki, P. A., Lebeda, F. J. & Johnston, D. *J. Neurophysiol.* **54** 1363-1374 (1985).
13. Korn, S. J., Giacchino, J. L., Chamberlin, N. L. & Dinglecline, R. *J. Neurophysiol.* **57**, 325-341 (1987).
14. Traynelis, S. F. & Dingledine, R. *J. Neurophysiol.* **59**, 259-276 (1988).
15. Traub, R. D. & Dingledine, R. *J. Neurophysiol.* **64**, 1009-1018 (1990).
16. Traub, R. D. & Miles, R. *Neuronal Networks of the Hippocampus* (Cambridge Univ. Press, Cambridge, 1991).
17. Chang, T., Schiff, S. J., Sauer, T., Gossaird, J.-P. & Burke, R. E. *Biophys. J.* **67** 671-683 (1994).
18. Schiff, S. J., Jerger, K., Chan, T., Sauer, T. & Aitken, P. G. *Biophys. J.* **67**, 684-691 (1994).
19. Jefferys, J. G. R. *J. Physiol.* **319**, 143-152 (1981).
20. Durand, D. *Brain Res.* **392**, 139-144 (1986).
21. Kayyali, H. & Durand, D. *Eypt Neurol.* **113**, 249-254 (1991).
22. Ott, E. *Chaos in Dynamical Systems* (Cambridge Univ. Press, Cambridge, 1993).
23. Auerbach, D., Grebogi, C., Ott, E. & Yorke, J. A. *Phys. Rev. Lett.* **69**, 3479-3482 (1992).

READING 28.3

Predicting and Producing Chaos;
Experimentally Induced Transitions in the Dynamic Behaviour of
Insect Populations

Commentary: The first article is a commentary on the second. The latter describes
a relatively simple mathematical model for the population as a function of time of
the flour beetle Tribolium. According to the model, changes in the adult mortality
produce drastic changes in the population dynamics, shifts from stable equilibrium
to stable cycles to aperiodic, unpredictable behavior. The adult mortality was then
manipulated accordingly in experiments with beetles to compare the predicted behavior
with the actual population dynamics. Of particular importance here is that the work
reported goes beyond simply observing a time series. Rather, the beetle populations
are experimentally manipulated to challenge the predictions of the model.

Predicting and Producing Chaos

Nature, **375** (1995) 189–190.
Peter Kareiva

One of the best reasons for building explicit demographic models is that they can help to predict
when subtle changes in the environment might pro-
voke profound changes in population dynamics.
On page 227 of this issue [1], Costantino and collaborators provide one of the first examples of a
model that actually lives up to this promise.

The authors use a nonlinear demographic model
to predict shifts in flour beetle dynamics from
stable equilibria to cycles to aperiodic (and even
chaotic) fluctuations simply as a result of modest
changes in beetle survival rates. They then validate
their model by manipulating beetle survival in elegant laboratory experiments and observing exactly
the shifts in dynamics that the theory predicts. In
particular, beetle fluctuations became highly irregular when adult mortality was experimentally increased (from 73 per cent to 96 per cent). This work
is an unusual blend of nonlinear dynamics theory,
statistics and experimentation—and the results are
of uncommon clarity for ecology.

The theory of nonlinear dynamics, which has
chaos theory as its most fashionable expression,
is one of this century's more important scientific
advances. The surprising richness of behaviour
possible from simple nonlinear feedback systems
has provided grist for Hollywood and spawned major academic research centres. One would have
thought that ecology, a cradle of chaos theory [2, 3],
would be especially illuminated by studies of nonlinear dynamics; after all, virtually all sensible ecological models are highly nonlinear in a way that
admits the possibility of complex dynamics, and
observations of population trajectories for animals
routinely reveal cycles and wild fluctuations. Yet
until recently the study of nonlinear dynamics in
ecology has been largely neglected by experimen-
talists.

The achievements of Costantino *et al.* stem from
the unique collaboration involved. Two of the four
contributors provided a command of flour beetle
biology that led to the design of practical experiments and the formulation of a population model
grounded in just the right amount of biological detail (with the key element being density-dependent
cannibalism, which provides the nonlinear feedback that drives the dynamics). The third applied
his expertise to the analysis of nonlinear dynamical
systems to spell out the possibilities for population
behaviour. And the fourth member analysed time-
series data to estimate model parameters, thereby
identifying exactly what sort of experimental
alterations of model parameters were needed to
produce dramatic shifts in population behaviour.

Author affiliation: Peter Kareiva: Department of Zoology, University of Washington.

Although a mix of models and experiments in ecology is nothing new, such ventures are rarely accompanied with rigorous parameter estimation, or careful statistics [4]—it is almost as though theoretical ecologists have felt that statistics is a subject for empiricists, not modellers.

It is also worth noting that most previous connections between the theory of nonlinear dynamics and natural populations have been aimed at simply establishing that chaos is evident in ecological time series [5]. Only within the past decade have researchers used their understanding of nonlinear dynamics to interpret key features of observed population fluctuations [6-9]. These reports have all been retrospective, however. In no case were predictions about exact boundaries in parameter space that demarcate different system behaviours tested by manipulating a population's parameters, with the object of seeing whether the dynamics actually did shift between stable equilibria and aperiodic cycles.

To a scientist mired in the mud of the real world, or concerned with only the most pragmatic of issues, it might seem curious to pay so much attention to aperiodic cycles and nonlinear theory. There is a reason, however. Nonlinear dynamics admits the possibility of surprise events occurring in response to subtle modification of a system's parameters. For instance, if we consider the flour beetle populations studied by Costantino et al., we might think that an increase in mortality of adult beetles caused by a pesticide application would simply reduce the populations affected. But an apt nonlinear model indicates that the increased mor-

tality could actually cause beetle populations to shift from a stable equilibrium to aperiodic fluctuations. In different nonlinear models for other organisms it could turn out that subtle decreases in mortality or increases in reproduction could cause chaotic dynamics.

The point is that simply documenting direct demographic effects of environmental change or environmental stress has little predictive power. Only with a solid understanding of underlying dynamics can ecologists ever hope to anticipate the consequences of environmental perturbation. Costantino et al. show us how the marriage of nonlinear models and experiments can help to accomplish the task.

REFERENCES

1. Costantino, R. F., Cushing, J. M., Dennis, B. & Desharnais, R. A. *Nature* **375**, 227-230 (1995).
2. May, R. M. *Science* **186**, 643-647 (1974).
3. Gleick, J. *Chaos: Making a New Science* (Penguin, New York, 1987).
4. Ludwig, D. *Ecology* **76**, 357 (1995).
5. Hastings, A. *et al. A. Rev. Ecol. Syst.* **24**, 1-33 (1983).
6. Murdock, W. W. & McCauley, E. *Nature* **316**, 628-630 (1985).
7. Grenfell, B. J., Price, O. J., Albon, S. D. & Clutton-Brock, T. H. *Nature* **355**, 823-826 (1992).
8. Hanski, I., Turchin, P., Korpimaki, E. & Henttonen, H. *Nature* **364**, 232-235 (1993).
9. Tilman, D. & Wedin, D. *Nature* **353**, 653-655 (1991).

Experimentally Induced Transitions in the Dynamic Behaviour of Insect Populations

Nature, **375** (1995) 227–230.

R. F. Costantino, J. M. Cushing, B. Dennis, R. A. Desharnais

Simple nonlinear models can generate fixed points, periodic cycles and aperiodic oscillations in population abundance without any external en- **vironmental variation. Another familiar theoretical result is that shifts in demographic parameters (such as survival or fecundity) can move a**

Author affiliations: R. F. Costantino: Department of Zoology, University of Rhode Island. J. M. Cushing: Department of Mathematics, University of Arizona. Brian Dennis: Department of Fish and Wildlife Resources, University of Idaho. Robert A. Desharnais: Department of Biology, California State University.

population from one of these behaviours to another [1–4]. Unfortunately, empirical evidence to support these theoretical possibilities is scarce [5–15]. We report here a joint theoretical and experimental study to test the hypothesis that changes in demographic parameters cause predictable changes in the nature of population fluctuations. Specifically, we developed a simple model describing population growth in the flour beetle *Tribolium* [16]. We then predicted, using standard mathematical techniques to analyse the model, that changes in adult mortality would produce substantial shifts in population dynamic behaviour. Finally, by experimentally manipulating the adult mortality rate we observed changes in the dynamics from stable fixed points to periodic cycles to aperiodic oscillations that corresponded to the transitions forecast by the mathematical model.

We modelled the relationship linking larval, pupal and adult numbers at time $t + 1$ to the number of animals at time t in the flour beetle (see [9]). The model is a system of three difference equations:

$$L_{t+1} = b A_t \exp(-c_{ea} A_t - c_{el} L_t)$$
$$P_{t+1} = L_t (1 - \mu_1) \tag{1}$$
$$A_{t+1} = P_t \exp(-c_{pa} A_t) + A_t (1 - \mu_a)$$

Here, L_t is the number of feeding larvae, P_t is the number of non-feeding larvae, and pupae and callow adults, and A_t is the number of mature adults, at time t; the unit of time (2 weeks) is taken to be the feeding larval maturation interval, so that after one unit of time a larva either dies or survives and pupates. This unit of time is also the cumulative time spent as a non-feeding larva, pupa and callow adult. The quantity $b > 0$ is the number of larval recruits per adult per unit of time in the absence of cannibalism. The fractions μ_1 and μ_a are the larval and adult probabilities, respectively, of dying from causes other than cannibalism. The exponential nonlinearities account for the cannibalism of eggs by both larvae and adults and the cannibalism of pupae by adults. The fractions $\exp(-c_{ea} A_t)$ and $\exp(-c_{el} L_t)$ are the probabilities that an egg is not eaten in the presence of A_t adults and L_t larvae. The fraction $\exp(-c_{pa} A_t)$ is the survival probability of a pupa in the presence of A_t adults.

For fitting to time-series data, the model was converted to a stochastic model with noise added

on a logarithmic scale:

$$L_{t+1} = b A_t \exp(-c_{ea} A_t - c_{el} L_t + E_{1t})$$
$$P_{t+1} = L_t (1 - \mu_1) \exp(E_{2t})$$
$$A_{t+1} = [P_t \exp(-c_{pa} A_t) + A_t (1 - \mu_a)] \exp(E_{3t}) \tag{2}$$

Here E_{1t}, E_{2t}, and E_{3t} are random noise variables assumed to have a joint multivariate normal distribution with means of zero. The noise variables represent the unpredictable departures of the observations from the deterministic model (1) due to environmental and other causes. Maximum-likelihood estimates of model parameters were calculated under the assumptions that the noise variables are correlated with each other but uncorrelated through time [16]. These assumptions were evaluated for all the data sets by standard diagnostic analyses of time-series residuals [17].

In the study reported here we experimentally set adult mortality rates at values suggested by the analysis of historical time-series data [16] (Figure 1(a)) to place the experimental cultures in regions of different asymptotic dynamics: $\mu_a = 0.04$, 0.27, 0.50, 0.73 and 0.96. There were also control cultures which were not manipulated. Cultures of *T. castaneum* (24 for each of two genetic strains, RR and SS) were initiated with 100 young adults, 5 pupae and 250 small larvae. Each population was contained in a half-pint (237 ml) milk bottle with 20 g of standard media and kept in a dark incubator at 31 degrees C. Every two weeks the L_t, P_t, and A_t stages were censused and returned to fresh media. This procedure was continued for 36 weeks. At week 12, four populations of each genetic strain were randomly assigned to each of the six treatments, and the imposition of adult mortalities began. Adult mortality was manipulated by removing or adding adults at the time of a census to make the total number of adults that died during the interval consistent with the treatment value of μ_a. To counter the possibility of genetic changes in life-history characteristics, beginning at week 12 and continuing every month thereafter, the adults returned to the populations after the census were obtained from separate stock cultures maintained under standard laboratory conditions.

For each genetic strain the four replicates of each treatment were randomly assigned into two groups. With half of the data, we fitted model (2) to the time series of the RR (Figure 2) and SS (Fig-

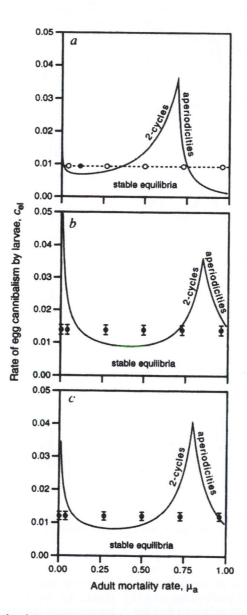

Fig. 1 Stability boundaries for the deterministic model (1) for three experiments. a, Sensitive strain [16] with $b = 11.68$, $c_{ea} = 0.011$, $c_{pa} = 0.017$, $\mu_l = 0.513$. The filled circle locates the population in parameter space. The broken line and open circles indicate the extrapolated predicted dynamic behaviour as a function of adult mortality. b, RR strain with $b = 7.88$, $c_{ea} = 0.011$, $c_{pa} = 0.004$, $\mu_l = 0.161$. c, SS strain with $b = 7.48$, $c_{ea} = 0.009$, $c_{pa} = 0.004$, $\mu_l = 0.267$. The filled circles locate the experimental populations in parameter space. In b and c the bar represents a 95% confidence interval for c_{el} based on the profile likelihood.

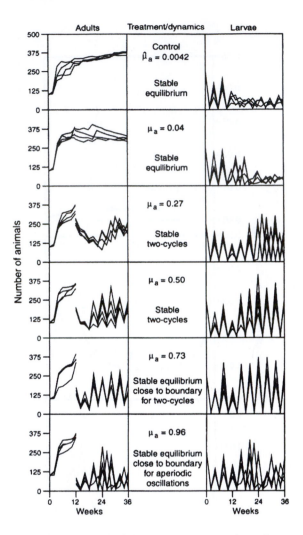

Fig. 2 Time-series data for individual replicates of the RR strain. At week 12 the experimental manipulation of adult mortality rate began. The center column lists the imposed adult mortality rates and the location of the cultures in parameter space.

ure 3) strains. The other half of the data were used for evaluating the model predictions. Analyses of time-series residuals [17] indicated that the stochastic model described the data quite well. Using the model and maximum-likelihood parameter estimates, based on the time series for weeks 12 to 36, we calculated the stability boundaries (Figure 1(b), (c)) and bifurcation diagrams (Figure 4) for each genetic strain.

The parameter estimates placed the control and

$\mu_a = 0.04$ treatments in the region of stable equilibria. In the $\mu_a = 0.27$ treatment there was a transition in the dynamics: regular, albeit small, fluctuations were found in the adult data, whereas large and sustained oscillations were noted in larval numbers. These populations were located in the two-cycle zone. At $\mu_a = 0.50$ both strains were in the two-cycle region and displayed sustained oscillations, most noticeably in larval numbers. With $\mu_a = 0.73$ the RR strain was close to the two-cycle

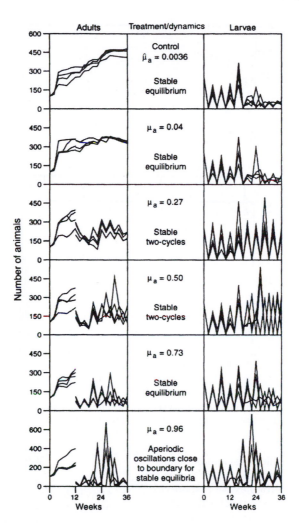

Fig. 3 Time-series data for individual replicates of the SS strain. At week 12 the experimental manipulation of adult mortality began. The centre column lists the imposed adult mortality rates and the location of the cultures in parameter space.

boundary but the SS strain underwent another transition and was clearly in the region of stable equilibria. Fluctuations in the RR strain were sustained (Figure 2) whereas the fluctuations in the SS strain appeared to dampen (Figure 3).

In the $\mu_a = 0.96$ treatment both strains underwent another transition in the dynamics. The RR strain was in the stable equilibria region and the SS strain was in the region of aperiodic oscillations; however, both were close to the boundary at which

a bifurcation to aperiodicities occurs (Figure 1(b), (c)). Being close to the aperiodic bifurcation boundary means that the transients of the RR strain are expected to appear aperiodic and only slowly decay into the equilibrium. Both genetic strains displayed the expected aperiodic fluctuations (Figure 2 and 3).

Oscillatory approaches to point equilibria, stable periodic oscillations and aperiodic oscillations are difficult to distinguish in short time series with model-free methods. Our model-based parametric

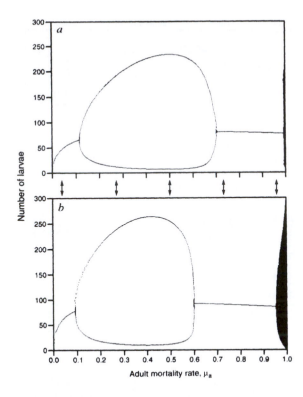

Fig. 4 Bifurcation diagrams for the model equations (1) for parameter values based on the experimental data. a, RR strain; b, SS strain. The double arrows indicate the adult mortality rate treatments.

approach to classifying dynamic behaviour is statistically more powerful than non-parametric regression methods [18] or parametric flexible-response-surface methods [19]. However, our classifications depend on the adequacy of the model (2). We minimized the risk of error by basing the model on detailed knowledge of a well-studied system [9] and by thorough diagnostic analyses of the resulting time-series residuals [16]. The data (Figures 2 and 3) also give a visual impression of dynamic behaviours consistent with our classifications. Other, less-studied systems require statistical approaches that are more robust to variations in model form [20].

Our joint theoretical and experimental study has documented the transitions in the asymptotic dynamics of laboratory cultures of *Tribolium*. This rigorous verification of the predicted shifts in dynamical behaviour provides convincing evidence for the relevance of nonlinear mathematics in population biology.

REFERENCES

1. May, R. M. *Science* **186**, 645–647 (1974).
2. May, R. M. *J. theor. Biol.* **51**, 511–524 (1975).
3. May, R. M. *Nature* **261**, 459–467 (1976).
4. May, R. M. & Oster, G. F. *Am. Nat.* **110**, 573–599 (1976).
5. Strong, D. R. in *Ecological Theory and Integrated Pest Management* (ed. Kogan, M.) 37–58 (Wiley, New York, 1986).
6. Kareiva, P. in *Perspectives in Ecological Theory* (ed. Roughgarden, J., May, R. M. & Levin, S. A.) 68–88 (Princeton University Press, Princeton, NJ, 1989).
7. Bartlett, M. S. *J. R. Statist. Soc. A* **52**, 321–347 (1990).
8. Logan, J. A. & Hain, F. (eds) *Chaos and Insect Ecology* (Virginia Polytechnic Institute and State University, Blacksburg, VA, 1991).
9. Costantino, R. F. & Desharnais, R. A. *Population Dynamics and the Tribolium Model: Ge-*

netics and Demography (Springer, New York, 1991).

10. Logan, J. A. & Allen, J. C. A. *Rev. Ent.* **37**, 455–477 (1992).

11. Hastings, A., Hom, C. L., Ellner, S., Turchin, P. & Godfray, H. C. J. A. *Rev. Ecol. Syst.* **24**, 1–33 (1993).

12. Murdock, W. W. & McCauley, E. *Nature* **316**, 628–630 (1985).

13. Grenfell, B. J., Price, O. J., Albon, S. D. & Clutton-Brock, T. H. *Nature* **355**, 823–826 (1992).

14. Hanski, I., Turchin, P., Korplmakl, E. & Henttonen, H. *Nature* **364**, 232–235 (1993).

15. Tilman, D. & Wedin, D. *Nature* **353**, 653–655 (1991).

16. Dennis, B., Desharnais, R. A., Cushing, J. M. & Costantino, R. F. *Ecol. Monogr.* (in press).

17. Tong, H. *Non-linear Time Series: a Dynamical System Approach* (Oxford Univ. Press, Oxford, 1990).

18. McCaffrey, D. J., Ellner, S., Gallant, A. R. & Nychka, D. W. *J. Am. statist. Ass.* **87**, 682–695 (1992).

19. Turchin, P. & Taylor, A. *Ecology* **73**, 289-305 (1992).

20. Turchin, P. *Oikos* **68**, 167-172 (1993).

Extra Exercises and Solutions

Extra Exercises for Chapters 1–12*

1. Consider the following differential equation for the function $x(t)$:

$$\frac{dx}{dt} = x(x+1)(x-1).$$

(a) What are the equilibrium points?

(b) Which equilibrium points are stable?

(c) If $x(0) = 2$, what happens to x as t gets very large?

(d) If $x(0) = -2$, what happens to x as t gets very large?

2. Consider the following differential equation for the function $x(t)$:

$$\frac{dx}{dt} = x(2-x) - c.$$

Here, c is a constant.

(a) For what values of c does this equation have equilibrium solutions?

(b) When $c > 0$, what happens to x as t gets very large if $x(0) = 0$?

---

*Answers are provided on page 484.

3. Is $x = 0$ a stable equilibrium solution for the equation $\frac{dx}{dt} = -x^3$? Justify your answer.

4. Consider the following differential equation for $\mathbf{v}(t) = \begin{pmatrix} x(t) \\ y(t) \end{pmatrix}$:

$$\frac{d}{dt}\begin{pmatrix} x \\ y \end{pmatrix} = \begin{pmatrix} x - y + 1 \\ y + 2x \end{pmatrix}.$$

(a) Draw and label the x null clines and the y null clines.

(b) On the drawing for part (a), label the equilibrium point(s).

(c) Decide whether any equilibrium points are stable. Justify your answers.

(d) On the drawing for part (a), label where $\frac{dx}{dt}$ is positive and where $\frac{dx}{dt}$ is negative. Do the same for $\frac{dy}{dt}$.

5. Consider the following differential equation for $\mathbf{v}(t) = \begin{pmatrix} x(t) \\ y(t) \end{pmatrix}$:

$$\frac{d}{dt}\begin{pmatrix} x \\ y \end{pmatrix} = \begin{pmatrix} x - 3y + 2 \\ -x + 2 \end{pmatrix}.$$

(a) Draw and label the x null clines and the y null clines.

(b) On the drawing for part (a), label the equilibrium point(s).

(c) Decide whether any equilibrium points are stable. Justify your answers.

(d) On the drawing for part (a), label where $\frac{dx}{dt}$ is positive and where $\frac{dx}{dt}$ is negative. Do the same for $\frac{dy}{dt}$.

(e) What happens to $\begin{pmatrix} x(t) \\ y(t) \end{pmatrix}$ as t gets very large if $x(0) = 4$ and $y(0) = -1$?

6. Consider the following differential equation for $\mathbf{v}(t) = \begin{pmatrix} x(t) \\ y(t) \end{pmatrix}$:

$$\frac{d}{dt}\begin{pmatrix} x \\ y \end{pmatrix} = \begin{pmatrix} x^2 - y^2 + 1 \\ -x \end{pmatrix}.$$

(a) Draw and label the x null clines and also the y null clines.

(b) On the drawing for part (a), label the equilibrium point(s).

(c) Decide whether any equilibrium points are stable. Justify your answers.

(d) On the drawing for part (a), label where $\frac{dx}{dt}$ is positive and where $\frac{dx}{dt}$ is negative. Do the same for $\frac{dy}{dt}$.

(e) What happens to $\begin{pmatrix} x(t) \\ y(t) \end{pmatrix}$ for small but positive t if $x(0) = 1$ and $y(0) = \sqrt{2}$?

7. Consider the following differential equation for $\mathbf{v}(t) = \begin{pmatrix} x(t) \\ y(t) \end{pmatrix}$:

$$\frac{d}{dt}\begin{pmatrix} x \\ y \end{pmatrix} = \begin{pmatrix} x - 3x^2 + xy \\ y - xy \end{pmatrix}.$$

(a) Draw and label the x null clines and the y null clines.

(b) On the drawing for part (a), label the equilibrium point(s).

(c) Decide whether any equilibrium points are stable. Justify your answers.

(d) On the drawing for part (a), label where $\frac{dx}{dt}$ is positive and where $\frac{dx}{dt}$ is negative. Do the same for $\frac{dy}{dt}$.

(e) Suppose that $x(t)$ and $y(t)$, when nonnegative, are meant to represent the populations of two kinds of animals at time t. Which is the predator and which is the prey? Does the predator have an alternate food source besides the prey?

(f) If $x(0) = 2$ and $y(0) = 1$, is x greater or less than 2 if t is positive but very small? Is y greater or less than 1 if t is positive but very small?

(g) If $x(0) = 1$ and $y(0) = 1$, is x greater or less than 1 if t is positive but very small? Is y greater or less than 1 if t is positive but very small? (*Hint:* In this case, first see what x does and then use that information to determine the behavior of y.)

8. Consider the following differential equation for $\mathbf{v}(t) = \begin{pmatrix} x(t) \\ y(t) \end{pmatrix}$:

$$\frac{d}{dt}\begin{pmatrix} x \\ y \end{pmatrix} = \begin{pmatrix} 1 - xy \\ y - 3xy \end{pmatrix}.$$

(a) Draw and label the x null clines in the quadrant of the xy-plane where both x and y are both nonnegative, and do the same for the y null clines.

(b) On the drawing for part (a), label the equilibrium point(s).

(c) Decide whether any equilibrium points are stable. Justify your answers.

(d) On the drawing for part (a), label where $\frac{dx}{dt}$ is positive and where $\frac{dx}{dt}$ is negative. Do the same for $\frac{dy}{dt}$. (Only consider the region of the xy-plane where both x and y are nonnegative.)

(e) What happens to $\begin{pmatrix} x(t) \\ y(t) \end{pmatrix}$ as t gets very large if $x(0) = 1$ and $y(0) = 0$?

9. Compute the partial derivative in the x-direction and in the y-direction for the following functions:
   (a) $x^2 + xy^3$
   (b) $x + \sin(xy^2)$
   (c) $xy + \sin(x)$
   (d) $x/(y^2 + 2)$
   (e) $x + y + 2$

10. Compute $\mathbf{MM}'$ and $\mathbf{M}'\mathbf{M}$ when
   (a) $\mathbf{M} = \begin{pmatrix} 1 & 2 \\ 3 & 4 \end{pmatrix}$ and $\mathbf{M}' = \begin{pmatrix} 4 & 3 \\ 2 & 1 \end{pmatrix}$
   (b) $\mathbf{M} = \begin{pmatrix} 3 & 1 \\ 4 & 2 \end{pmatrix}$ and $\mathbf{M}' = \begin{pmatrix} 2 & 4 \\ 1 & 3 \end{pmatrix}$

11. Compute $\mathbf{Mv}$ when the matrix $\mathbf{M}$ and the vector $\mathbf{v}$ are given by
   (a) $\mathbf{M} = \begin{pmatrix} 1 & 2 \\ 3 & 4 \end{pmatrix}$ and $\mathbf{v} = \begin{pmatrix} 1 \\ 2 \end{pmatrix}$
   (b) $\mathbf{M} = \begin{pmatrix} 3 & 1 \\ 4 & 2 \end{pmatrix}$ and $\mathbf{v} = \begin{pmatrix} 1 \\ -1 \end{pmatrix}$

12. Integrate the indicated function over the indicated rectangle. Do the x-integration first and the y-integration second. Then repeat the calculation by doing the y-integration first and the x-integration second.
   (a) $h(x, y) = y\cos(x)$ over the rectangle where $0 \le x \le \pi/2$ and $0 \le y \le 1$
   (b) $h(x, y) = x^2 + xy^2$ over the rectangle where $0 \le x \le 1$ and $0 \le y \le 1$
   (c) $h(x, y) = \cos(x + 2y)$ over the rectangle where $0 \le x \le \pi/2$ and $0 \le y \le \pi/2$

Answers to Chapter 1–12 Extra Exercises

1. (a) Equilibrium points are 0, 1 and -1.
    (b) 0 is stable, the others aren't.
    (c) x goes to ∞ as t goes to ∞.
    (d) x goes to $-\infty$ as t goes to $-\infty$.

2. (a) Equilibrium occur where $2x - x^2 - c = 0$, or $x = 1 \pm (1 - c)^{1/2}$. This is reasonable as long as $1 - c \ge 0$ or $1 \ge c$.
    (b) If $c > 0$, then x approaches $-\infty$ as $t \to \infty$.

3. Since the graph $y = -x^3$ is above the x-axis when $x < 0$ and below when $x > 0$, the equilibrium point 0 is stable.

4. **(a)** The x null clines are $y = x + 1$. The y null clines have $y = -2x$.

   (b) The equilibrium point has $x = -1/3$ and $y = 2/3$.

   (c) The stability matrix here is

$$\mathbf{D} = \begin{pmatrix} 1 & -1 \\ 2 & 1 \end{pmatrix}.$$

   The trace of $\mathbf{D}$ is 2, which is positive, so the equilibrium point is unstable.

   (d) Figure A.1 shows the phase plane diagram.

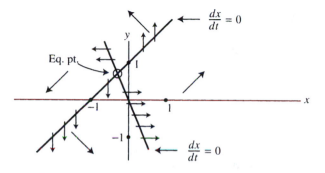

Figure A.1

5. **(a)** The x null clines are $y = x/3 + 2/3$. The y null clines have $x = 2$.

   (b) The equilibrium point has $x = 2$ and $y = 4/3$.

   (c) The stability matrix here is

$$\mathbf{D} = \begin{pmatrix} 1 & -3 \\ -1 & 0 \end{pmatrix}.$$

   The trace of $\mathbf{D}$ is 1, which is positive, so the equilibrium point is unstable.

   (d) Figure A.2 shows the phase plane diagram.

   (e) If $x(0) = 4$ and $y(0) = -1$, then $x(t) \to \infty$ and $y(t) \to -\infty$ as $t \to \infty$.

6. **(a)** The x null clines are $y = \pm(x^2 + 1)^{1/2}$. The y null clines have $x = 0$.

   (b) The equilibrium points are $x = 0$, $y = 1$ and $x = 0$, $y = -1$.

   (c) The stability matrix is

$$\mathbf{D} = \begin{pmatrix} 2x & -2y \\ -1 & 0 \end{pmatrix}.$$

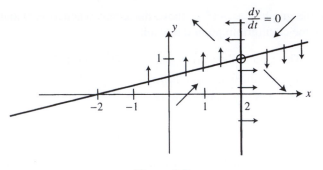

Figure A.2

At $x = 0$, $y = 1$, the trace of **D** is 0, which is nonnegative, so the equilibrium point is unstable. At the other equilibrium point, the trace of **D** is also 0, so the point $x = 0$, $y = -1$ is also unstable.

(d) Figure A.3 shows the phase plane diagram.

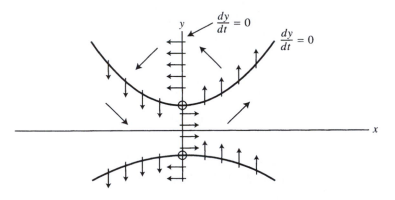

Figure A.3

(e) If $x(0) = 1$ and $y(0) = \sqrt{2}$, then $y(t)$ decreases and $x(t)$ slowly increases for small, positive time t.

7. (a) The x null clines are $x = 0$ and $y = 3x - 1$. The y null clines have $y = 0$ and $x = 1$.

(b) The equilibrium points are $x = 0$, $y = 0$ and $x = 1$, $y = 2$ and $x = 1/3$, $y = 0$.

(c) The stability matrix at the generic point (x, y) is

$$\mathbf{D} = \begin{pmatrix} 1 - 6x + y & x \\ -y & 1 - x \end{pmatrix}.$$

At $x = 0$, $y = 0$, this matrix has trace equals 2, so $x = 0$, $y = 0$ is unstable.

At $x = 1$, $y = 2$, this matrix has trace equals -3, det $= 2$, so $x = 1$, $y = 2$ is stable.

At $x = 1/3$, $y = 0$, this matrix has trace equals $-1/3$, det $= -2/3$, so $x = 1/3$, $y = 0$ is unstable.

(d) Figure A.4 shows the phase plane diagram.

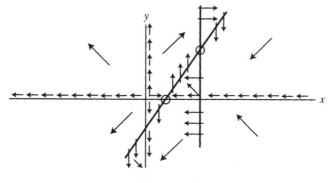

Figure A.4

(e) The predator is x; the prey is y; x apparently has an alternate food supply.

(f) Since the x-derivative is negative at $x = 2$, $y = 1$, $x(t) < 2$ for $t > 0$, but small. Also, the y-derivative is negative so $y(t) < 1$ for $t > 0$, but small.

(g) The x-derivative is negative at $x = 1$, $y = 1$, so $x(t) < 1$ for $t > 0$, but small. Also, the y-derivative is initially zero, but as $x(t) < 1$ for $t > 0$, the y-derivative turns positive, so $y(t) > 1$ for $t > 0$, but small.

8. **(a)** The x null cline has $y = 1/x$. The y null cline has $y = 0$ or $x = 1/3$.

(b) The equilibrium point has $y = 3$, $x = 1/3$.

(c) The stability matrix at general x and y is

$$\mathbf{D} = \begin{pmatrix} -y & -x \\ -3y & 1 - 3x \end{pmatrix}.$$

At the point $x = 1/3$, $y = 3$, this matrix has trace $= -3$ and det $= -3$ so this point is unstable.

(d) Figure A.5 shows the phase diagram where $x \geq 0$ and $y \geq 0$.

(e) If $y(t) = 0$ for any time t, then the y-derivative is zero (no matter what x is), so the trajectory cannot move off the $y = 0$ line (the x-axis). On this line, the x-derivative is equal to 1, so $x(t) = t + 1$ and x increases toward ∞ as t does.

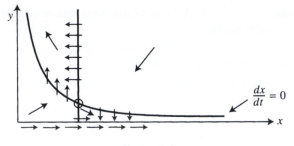

Figure A.5

9. **(a)** The x partial derivative is $2x + y^3$. The y partial derivative is $3xy^2$.

    (b) The x partial derivative is $1 + y^2 \cos(xy^2)$. The y partial derivative is $2xy \cos(xy^2)$.

    (c) The x partial derivative is $y + \cos(x)$. The y partial derivative is x.

    (d) The x partial derivative is $1/(y^2 + 2)$. The y partial derivative is $-2xy/(y^2 + 2)^2$.

    (e) The x partial derivative is 1. The y partial derivative is also 1.

10. **(a)** $\mathbf{MM'} = \begin{pmatrix} 8 & 5 \\ 20 & 13 \end{pmatrix}$ and $\mathbf{M'M} = \begin{pmatrix} 13 & 20 \\ 5 & 8 \end{pmatrix}$

    (b) $\mathbf{MM'} = \begin{pmatrix} 7 & 15 \\ 10 & 22 \end{pmatrix}$ and $\mathbf{M'M} = \begin{pmatrix} 22 & 10 \\ 15 & 7 \end{pmatrix}$

11. **(a)** $\mathbf{Mv} = \begin{pmatrix} 5 \\ 11 \end{pmatrix}$

    (b) $\mathbf{Mv} = \begin{pmatrix} 2 \\ 2 \end{pmatrix}$

12. **(a)** Doing the x-integral first gives y, which then integrates to $1/2$. Doing the y-integral first gives $\cos(x)/2$, which then integrates to $1/2$.

    (b) Doing the x-integral first gives $1/3 + y^2/2$. Then the y-integral gives $1/3 + 1/6 = 1/2$. Doing the y-integral first gives $x^2 + x/3$. Then the x-integral gives $1/3 + 1/6 = 1/2$.

    (c) Doing the x-intgral first gives $\cos(2y) - \sin(2y)$. Then the y-integral gives -1. Doing the y-integral first gives $-\sin(x)$. Then the x-integral gives -1.

Extra Exercises for Chapters 13–22*

1. Find the function $u(t, x)$ that solves the equation

$$\frac{\partial}{\partial t} u = -5 \frac{\partial}{\partial x} u$$

and obeys $u(0, x) = \dfrac{1}{(1 + x^2)}$.

2. Find the function $u(t, x)$ that solves the equation

$$\frac{\partial}{\partial t} u = -2 \frac{\partial}{\partial x} u$$

and obeys $u(t, 0) = \dfrac{1}{(1 + e^t)}$.

3. Find the function $u(t, x)$ that solves the equation

$$\frac{\partial}{\partial t} u = -2 \frac{\partial}{\partial x} u$$

and obeys $u(t, 1) = \dfrac{1}{(1 + e^t)}$. (*Note:* This constraint takes place at $x = 1$, not at $x = 0$.)

4. Find the function $u(t, x)$ that solves the equation

$$\frac{\partial}{\partial t} u = -3 \frac{\partial}{\partial x} u + 4u$$

and obeys $u(0, x) = \dfrac{1}{(1 + x^2)}$.

5. Show that the function $u(t, x) = \frac{1}{(1 + e^{-(x+t)/2})}$ solves the equation

$$\frac{\partial}{\partial t} u = -\frac{\partial}{\partial x} u + u(1 - u).$$

6. Find a nonnegative function $u(t, x)$ that is not everywhere zero and that solves the equation

$$\frac{\partial}{\partial t} u = \frac{\partial^2}{\partial x^2} u$$

for all t and for $0 \le x \le 1$, and that obeys $u(t, 0) = 0$ and $u(t, 1) = 0$. [*Hint:* Try $u(t, x)$ of the form $A(t)B(x)$. Also, see the remarks that follow Exercise 14.] You will have to consider which of your solutions is nonnegative.

---

*Answers are provided on page 491.

7. Find a function $u(t, x)$ that is not everywhere zero and that solves the equation

$$\frac{\partial}{\partial t}u = \frac{\partial^2}{\partial x^2}u$$

for all t and for $0 \le x \le 2$, and that obeys $u(t, 0) = 0$ and $(\frac{\partial}{\partial x}u)(t, 2) = 0$. [*Hint:* Try $u(t, x)$ of the form $A(t)B(x)$. Also, note that the boundary conditions are at $x = 0$ and 2.]

8. Is there a function $u(t, x)$ that solves the equation

$$\frac{\partial}{\partial t}u = \frac{\partial^2}{\partial x^2}u - 3u$$

for all t and for $-1 \le x \le 1$ and that obeys $u(t, -1) = 0$ and $u(t, 1) = 0$ and grows in size as t gets large? [Just check functions of the form $u(t, x) = A(t)B(x)$.] Justify your answer.

9. Is there a function $u(t, x)$ that solves the equation

$$\frac{\partial}{\partial t}u = \frac{\partial^2}{\partial x^2}u + 3u$$

for all t and for $0 \le x \le 1$ and that obeys $(\frac{\partial}{\partial x}u)(t, 0) = 0 = (\frac{\partial}{\partial x}u)(t, 1)$, and grows in size as t gets large? [Just check for functions of the form $u(t, x) = A(t)B(x)$.] Justify your answer.

10. Find the minimum positive number L with the following property: There is a solution

$$\frac{\partial}{\partial t}u = \frac{\partial^2}{\partial x^2}u + 3u$$

for all t and for $0 \le x \le L$ that obeys $u(t, 0) = 0$ and $u(t, L) = 0$ and that grows in size as t gets large. [Just look for solutions of the form $u(t, x) = A(t)B(x)$.]

11. Consider solutions to the equation

$$\frac{\partial}{\partial t}u = \frac{\partial^2}{\partial x^2}u + 5u(2 - u)$$

for all t and for $0 \le x \le 1$ that obey $(\frac{\partial}{\partial x}u)(t, 0) = 0 = (\frac{\partial}{\partial x}u)(t, 1)$.

  (a) Find the solutions that are independent of t and x.

  (b) Which of your solutions are stable?

  Justify your answers.

12. Consider solutions to the equation

$$\frac{\partial}{\partial t}u = \frac{\partial^2}{\partial x^2}u + 5u(2-u)$$

for all t and for $0 \le x \le 1$ that obey $u(t, 0) = 0 = u(t, 1)$. Is the constant solution $u = 0$ stable? Justify your answer.

13. Compute $(\frac{d}{dt}h)(g(x))$ when h and g are given by

(a) $h(t) = \cos(t)$ and $g(x) = \frac{1}{(1+x^2)}$

(b) $h(t) = \frac{1}{(1+t^2)}$ and $g(x) = \cos(x)$

14. Show that $u(t, x) = \frac{4}{(e^x + e^{-x})^2}$ is a time-independent solution to the equation

$$\frac{\partial}{\partial t}u = \frac{\partial^2}{\partial x^2}u - 4u + 6u^2.$$

Remarks: Remember that the general solution to the equation $\frac{d^2}{dx^2}h = ch$ has the following form:

1. If $c > 0$: $h(x) = \alpha e^{\sqrt{c}x} + \beta e^{-\sqrt{c}x}$.

2. If $c = 0$: $h(x) = \alpha + \beta x$.

3. If $c < 0$: $h(x) = \alpha \cos(\sqrt{-c}x) + \beta \sin(\sqrt{-c}x)$.

Here, α and β are constants that you are free to specify. (They are typically constrained by the boundary conditions.)

Answers to Chapter 13–22 Extra Exercises

1. Since $u(t, x) = u(0, x - 5t)$, we have $u(t, x) = (1 + (x - 5t)^2)^{-1}$.

2. Since $u(t, x) = u(t - x/2, 0)$, we have $u(t, x) = (1 + e^{(2t-x)/2})^{-1}$.

3. In this case, $u(t, x) = f(x - 2t)$, so we are told that $u(t, 1) = f(1 - 2t) = (1 + e^t)^{-1}$. Set $s = 1 - 2t$. Then $t = (1 - s)/2$ and $f(s) = (1 + e^{(1-s)/2})^{-1}$. Thus, $u(t, x) = (1 + e^{(1-x+2t)/2})^{-1}$.

4. Since $u(t, x) = e^{4t}u(0, x - 3t)$, the answer is $u(t, x) = e^{4t}(1 + (x - 3t)^2)^{-1}$.

5. This is an exercise in differentiating:

$$\frac{\partial}{\partial t}u = 2^{-1}(1 + e^{-(x+t)/2})^{-2}e^{-(x+t)/2} = \frac{\partial}{\partial x}u.$$

Then $u(1-u) = (1 + e^{-(x+t)/2})^2 e^{-(x+t)/2}$. Note that this is $2\frac{\partial}{\partial t}u$. Thus, since $\frac{\partial}{\partial x}u = \frac{\partial}{\partial t}u$, the right-hand side of the equation is $-\frac{\partial}{\partial t}u + 2\frac{\partial}{\partial t}u = \frac{\partial}{\partial t}u$, which is the left-hand side of the equation.

6. Try $u(t, x) = A(t)B(x)$. Then $A = A_0 e^{\lambda t}$ for some real number λ. Also, B obeys $\frac{d^2}{dx^2} B = \lambda B$ and B is to vanish at $x = 0$ and $x = 1$. A solution to these constraints has $\lambda = -\pi^2$, whence $B = \beta \sin(\pi x)$ for $\beta > 0$. Note that B obeys the boundary conditions and is nonnegative. Thus, $u(t, x) = e^{-\pi^2 t} \sin(\pi x)$ is a solution.

7. Try $u(t, x) = A(t)B(x)$. Then $A = A_0 e^{\lambda t}$ for some real number λ. Also, if $\lambda < 0$, then $B(x)$ has the form $B = \alpha \cos((-\lambda)^{1/2} x) + \beta \sin((-\lambda)^{1/2} x)$. Then AB obeys the differential equation. The boundary condition at $x = 0$ forces $\alpha = 0$. The boundary condition at $x = 2$ forces $\cos((-\lambda)^{1/2} 2) = 0$. Thus, $(-\lambda)^{1/2} 2 = (n+1/2)\pi$ for any whole number n. Thus, $\lambda = -(n+1/2)^2 \pi^2/4$. With this choice, $u(t, x) = e^{-(n+1/2)^2 \pi^2 t/4} \sin((n + 1/2)\pi x/2)$.

8. If a function $A(t)B(x)$ solves the equation plus boundary conditions, then $A = e^{\lambda t}$ for some real number $\lambda > 0$, and $B(x)$ obeys $\lambda B = \frac{d^2}{dx^2} B - 3B$ with boundary conditions $B(-1) = B(1) = 0$. Thus, $\frac{d^2}{dx^2} B = cB$, where $c = (\lambda + 3)$. This equation and the boundary conditions are definitely not satisfied if $c > 0$ unless $B = 0$ everywhere. If $c = 0$, the equations and boundary conditions are still not satisfied unless $B = 0$ everywhere. If $c < 0$, then the boundary conditions can be obeyed for particular choices of c. However, $c < 0$ requires $\lambda < 0$ and the resulting $u = e^{\lambda t} B(x)$ will shrink with time.

9. If a function $A(t)B(x)$ solves the equation plus boundary conditions, then $A = e^{\lambda t}$ for some real number $\lambda > 0$, and $B(x)$ obeys $\lambda B = \frac{d^2}{dx^2} B + 3B$. Thus, $\frac{d^2}{dx^2} B = cB$, where $c = (\lambda - 3)$. This equation and the boundary conditions are definitely not satisfied if $c > 0$. If $c = 0$, the equations and boundary conditions are satisfied by $B = 1$. Thus, there is a growing solution with time, $u(t, x) = e^{3t}$.

10. If a function $A(t)B(x)$ solves the equation plus boundary conditions $u = 0$ and $x = 0$ and $x = L$, then $A = e^{\lambda t}$, where λ is a real number, $B(x)$ obeys $\lambda B = \frac{d^2}{dx^2} B + 3B$, and $B = 0$ at $x = 0$ and $x = L$. Thus, B obeys $\frac{d^2}{dx^2} B = cB$, where $c = \lambda - 3$. The boundary conditions cannot be satisfied unless $c < 0$, and in this case, B must have the form $B = \sin(\pi x/L)$ and $c = -\pi^2/L^2$. Since $c = \lambda - 3$, we see that $\lambda = 3 - \pi^2/L^2$. Since we want u to grow with time, we want $\lambda > 0$, which demands $L^2 > \pi^2/3$ or $L > \pi/\sqrt{3}$.

11. The solutions that are independent of t and x are $u = 0$ and $u = 2$. To check stability, we ask whether there is a pair (g, λ), where $\lambda \geq 0$ and where $g(x)$ is not zero everywhere and solves the equation $\lambda g = \frac{d^2}{dx^2} g + f'(u_e)g$. Here, $f(u) = 5u(2 - u)$ and $u_e = 0$ or $u_e = 2$, depending on the equilibrium solution in question. Also, g must obey $\frac{d}{dx} g = 0$ at $x = 0$ and $x = 1$. If such a (g, λ) exists, then the equilibrium solution u_e is unstable.

In the case where $u_e = 0$, $f'(u_e) = 10$. Thus, we look for (g, λ) with $\lambda \geq 0$ and with g obeying $\lambda g = \frac{d^2}{dx^2}g + 10g$ with the boundary conditions. This has the usual form $\frac{d^2}{dx^2}g = cg$, where $c = \lambda - 10$. The only possible case where the boundary conditions can be obeyed has $c < 0$, where $g = \alpha \cos((-c)^{1/2}x) + \beta \sin((-c)^{1/2}x)$. To satisfy the boundary condition at $x = 0$, we need $\beta = 0$. To satisfy the boundary condition at $x = 1$, we need $(-c)^{1/2} = n\pi$ with n a whole number. The smallest possibility is $c = -\pi^2$ and thus $\lambda = 10 - \pi^2$. This is positive since $\pi^2 < (3.15)^2 = 9.9225 < 10$. Thus, case $u_e = 0$ is unstable.

For the case $u_e = 2$, the relevant $f'(u_e) = -10$. Then the equation in quesion for the pair (g, λ) is $\lambda \geq 0$ and $\lambda g = \frac{d^2}{dx^2}g - 10g$ with the same boundary conditions as before. This last has the form $\frac{d^2}{dx^2}g = cg$, where $c = \lambda + 10$. If $\lambda > 0$, then $c > 0$ and we are forced to take $g = \alpha e^{\sqrt{c}x} + \beta e^{-\sqrt{c}x}$, and there are no solutions to the boundary conditions except $\alpha = \beta = 0$, which doesn't count. Thus, $u_e = 2$ is stable.

12. The stability criterion is as follows: If there exists a pair (g, λ), where $\lambda \geq 0$ and $g(x)$ not everywhere zero, that satisfies $\lambda g = \frac{d^2}{dx^2}g + f'(u_e)g$, with $g(0) = g(1) = 0$, then the solution u_e is not stable. If no such (g, λ) exists, then the solution u_e is stable. Here, $u_e = 0$ and $f'(u_e) = 10$. Thus, we consider (λ, g) so that $\lambda g = \frac{d^2}{dx^2}g + 10g$. This has the form $\frac{d^2}{dx^2}g = cg$, where $c = \lambda - 10$. There are no solutions of this equation that also obey the boundary conditions except possibly in the case $c < 0$. In this case, the solution g must have the form $g = \alpha \cos((-c)^{1/2}x) + \beta \sin((-c)^{1/2}x)$. The boundary condition at $x = 0$ forces $\alpha = 0$. The boundary condition at $x = 1$ forces $c = -\pi^2$. Since $10 - \pi^2 > 0$, the solution is unstable.

13.   **(a)** $-\sin((1 + x^2)^{-1})$

     (b) $-2\cos(x)(1 + \cos(x)^2)^{-2}$

14. $\frac{d}{dx}u = -8(e^x + e^{-x})^{-3}(e^x - e^{-x})$. Then $\frac{d^2}{dx^2}u = 24(e^x + e^{-x})^{-4}$ $(e^x - e^{-x})^2 - 8(e^x + e^{-x})^{-2}$. This last expression is equal to $-96(e^x + e^{-x})^{-4} + 16(e^x + e^{-x})^{-2}$, as can be seen by writing the expression $(e^x - e^{-x})^2(e^{2x} - 2 + e^{-2x}) = (e^{2x} + 2 + e^{-2x}) - 4 = (e^x + e^{-x})^2 - 4$. Meanwhile, $-4u + 6u^2 = -16(e^x + e^{-x})^{-2} + 96(e^x + e^{-x})^{-4}$.

Extra Exercises for Chapters 23–28*

1. Which of the following choices of f and g has the square

$$\{(p, q) : 0 \leq p \leq 1, 0 \leq q \leq 1\}$$

---

*Answers are provided on page 495.

for a basin of attraction for the differential equation $\frac{d}{ds}\begin{pmatrix} p \\ q \end{pmatrix} = \begin{pmatrix} f(p,q) \\ g(p,q) \end{pmatrix}$?

    (a) $f(p, q) = q^2 - p^2$ and $g(p, q) = p + q$

    (b) $f(p, q) = q\sin(\pi p)$ and $g(p, q) = -q\sin(\pi p)$

    (c) $f(p, q) = q^2$ and $g(p, q) = -p$

    (d) $f(p, q) = -p^3 e^{p+q}$ and $g(p, q) = p^2 - q^3$

    (e) $f(p, q) = qp(q - p)$ and $g(p, q) = -q/(p^2 + 1)$

2. The following are some hypothetical stability matrices for an equilibrium point of a differential equation for two unknown functions of time. Classify the equilbrium point as being stable, repelling, or neither.

    (a) $\begin{pmatrix} -1 & -2 \\ 5 & 3 \end{pmatrix}$ (b) $\begin{pmatrix} -1 & -2 \\ -5 & 2 \end{pmatrix}$

    (c) $\begin{pmatrix} 5 & 6 \\ -7 & -6 \end{pmatrix}$ (d) $\begin{pmatrix} -2 & -4 \\ 3 & 1 \end{pmatrix}$

    (e) $\begin{pmatrix} -1 & 6 \\ 2 & -3 \end{pmatrix}$ (f) $\begin{pmatrix} -2 & -4 \\ 3 & 4 \end{pmatrix}$

3. Consider an equation for a function of time, $x(t)$, of the form $\frac{dx}{dt} = f(x) + c$, where c is a constant and where $f(\cdot)$ is a function. For the functions f listed, find all values of c where the number of equilibrium points changes.

    (a) $f(x) = (1 - x^2)/(1 + x^2)$

    (b) $f(x) = x^3 - 6x^2$

    (c) $f(x) = 3x^5 - 25x^3 + 60x$

    (d) $f(x) = e^x \sin(x)$

    (e) $f(x) = -3(x - 2)^2$

    (f) $f(x) = x^3 - 12x$

    (g) $f(x) = x^4 - 8x^2$

    (h) $f(x) = 3x^3 + x$

4. Consider the equation $\frac{dx}{dt} = f(x)$ for the functions f listed. In each case, suppose that $x(0) = 0$ and then find upper and lower bounds for the time t where $x(t) = 1$.

    (a) $f(x) = 2 + \sin(\pi x)$

    (b) $f(x) = e^{2x} + e^{2-2x}$

    (c) $f(x) = (2 - x^2)/(1 + x^2)$

    (d) $f(x) = 2x^4 - x + 2$

    (e) $f(x) = 5 - e^x$

    (f) $f(x) = \cos^2(\pi x) + 1$

5. In the following cases, decide which (if any) of the functions of time, $x(t)$ or $y(t)$, represent "slow subsytem" unknowns and which represent "fast subsystem" unknowns. Assume that your interest in these equations involves only those values for x and y that lie between -10 and 10.

(a) $x = 0.01x - 0.003xy$,

   $y = y - 0.014xy$

(b) $x = 3x - 2xy$,

   $y = 25y + 30xy$

(c) $x = 0.001x - 0.0007xy$,

   $y = y + 0.0007xy$

6. Compute the Fourier coefficients and the power spectrum $f(v)$ for the following cases of $m(t)$. Make a rough graph of $f(v)$ (if you are stumped here, obtain access to a graphing calculator).

(a) $m(t) = \begin{cases} \cos(t) & \text{when } -100\pi \le t \le 100\pi \\ 0 & \text{when } |t| >\le 100\pi \end{cases}$

(b) $m(t) = \begin{cases} \cos(2t) & \text{when } -100\pi \le t \le 100\pi \\ 0 & \text{when } |t| >\le 100\pi \end{cases}$

(c) $m(t) = \begin{cases} \cos(10t) & \text{when } -100\pi \le t \le 100\pi \\ 0 & \text{when } |t| >\le 100\pi \end{cases}$

(d) $m(t) = \begin{cases} \sin(100t) + \sin(10t) & \text{when } -100\pi \le t \le 100\pi \\ 0 & \text{when } |t| >\le 100\pi \end{cases}$

(e) $m(t) = \begin{cases} \sin(100t) + \cos(10t) & \text{when } -100\pi \le t \le 100\pi \\ 0 & \text{when } |t| >\le 100\pi \end{cases}$

Answers to Chapter 23–28 Extra Exercises

1. The cases in (b), (d), and (e) have the indicated square as a basin of attraction.

2.   (a) repelling
   (b) neither
   (c) stable
   (d) stable
   (e) neither
   (f) repelling

3.   (a) $c = -1$
      (b) $c = 0$ and $c = 32$
      (c) $c = -38$, $c = -32$, $c = 32$, and $c = 38$
      (d) $c = -e^{\pi(n+3/4)} \sin(\pi(n+3/4))$ for each $n = \{\ldots, -2, -1, 0, 1, 2, \ldots\}$
      (e) $c = 0$
      (f) $c = -8$ and $c = 8$
      (g) $c = 0$ and $c = 32$
      (h) No values of c

4.   (a) $1/3 \le t \le 1/2$. [Replace $\sin(\pi x)$ by 1 to get the lower bound and by 0 for the upper.]
      (b) $(1 + e^2)^{-1} \le t \le (2e)^{-1}$. [The function $e^{2x} + e^{2-2x}$ takes on its maximum where $0 \le x \le 1$ with value $1 + e^2$, and its minimum is at $x = 1/2$ with value $2e$.]
      (c) $1/2 \le t \le 2$. [The function $(2 - x^2)/(1 + x^2)$ takes on its maximum at $x = 0$ with value 2, and its minimum where $0 \le x \le 1$ is at $x = 1$ with value $1/2$.]
      (d) $1/3 \le t \le 8/13$. (The function $2x^4 - x + 2$ takes its maximum where $0 \le x \le 1$ at $x = 1$ with value 3, and its minimum where $0 \le x \le 1$ is at $x = 1/2$ with value $13/8$.)
      (e) $1/4 \le t \le (5 - e)^{-1}$. (The function $5 - e^x$ takes its maximum where $0 \le x \le 1$ at $x = 0$ with value 4, and its minimum on this interval is at $x = 1$ with value $5 - e$.)
      (f) $1/2 \le t \le 1$. [The function $\cos^2(\pi x) + 1$ takes its maximum where $0 \le x \le 1$ at $x = 0, 1$ with value 2, and its minimum on this interval is at $x = 1/2$ with value 1.]

5.   (a) The x-variable is slow and the y-variable is fast since $|\frac{d}{dt}x|$ is always less than 0.4 where both $|x| \le 10$ and $|y| \le 10$. Meanwhile, $|\frac{d}{dt}y|$ can be as large as 11 for these x- and y-values.
      (b) The decomposition into fast and slow subsystems makes no sense here.
      (c) The x-variable is slow and the y-variable is fast since $|\frac{d}{dt}x|$ is always less than 0.02 where both $|x| \le 10$ and $|y| \le 10$. Meanwhile, $|\frac{d}{dt}y|$ can be as large as 10 for these x- and y-values.

6.   (a) The Fourier coefficients are $a = 100^{-1}(4\pi^2 v^2 - 1)^{-1} v \sin(200\pi^2 v)$; $b = 0$. The number $\sigma = 1/2$. The power spectrum is $f(v) = (5000)^{-1}(4\pi^2 v^2 - 1)^{-2} v^2 \sin^2(200\pi^2 v)$.
      (b) The Fourier coefficients are $a = 100^{-1}(4\pi^2 v^2 - 4)^{-1} v \sin(200\pi^2 v)$; $b = 0$. The number $\sigma = 1/2$. The power spectrum is $f(v) = (5000)^{-1}(4\pi^2 v^2 - 4)^2 v^2 \sin^2(200\pi^2 v)$.

(c) The Fourier coefficients are $a = 100^{-1}(4\pi^2 v^2 - 100)^{-1} v \sin(200\pi^2 v)$, $b = 0$. The number $\sigma = 1/2$. The power spectrum is $f(v) = (5000)^{-1}(4\pi^2 v^2 - 100)^{-2} v^2 \sin^2(200\pi^2 v)$.

(d) The Fourier coefficients are

$$a = 0$$

$$b = (20\pi)^{-1} \sin(200\pi^2 v)\{10/(4\pi^2 v^2 - 10,000) + 1/(4\pi^2 v^2 - 100)\}.$$

The constant $\sigma = 1$. The power spectrum function is $f(v) = b^2 = (20\pi)^{-2} \sin^2(200\pi^2 v)\{10/(4\pi^2 v^2 - 10,000) + 1/(4\pi^2 v^2 - 100)\}^2$.

(e) The Fourier coefficients are

$$a = 100^{-1}(4\pi^2 v^2 - 100)^{-1} v \sin(200\pi^2 v);$$

and

$$b = (2\pi)^{-1}(4\pi^2 v^2 - 10,000)^{-1} \sin(200\pi^2 v).$$

The number $\sigma = 1$. The power spectrum is

$$f(v) =$$
$$[(100)^{-2}(4\pi^2 - 100)^{-2} v^2 + (2\pi)^{-2}(4\pi^2 v^2 - 10,000)^{-2}] \sin^2(200\pi^2 v).$$

Index

The letters "ff" following an index entry mean "and pages following".